Advanced Polymer Composite Materials: Processing, Modeling, Properties and Applications II

Advanced Polymer Composite Materials: Processing, Modeling, Properties and Applications II

Guest Editors

Giorgio Luciano
Maurizio Vignolo

Basel • Beijing • Wuhan • Barcelona • Belgrade • Novi Sad • Cluj • Manchester

Guest Editors

Giorgio Luciano
National Research Council
of Italy
SCITEC–CNR
Genova
Italy

Maurizio Vignolo
National Research Council
of Italy
SCITEC–CNR
Genova
Italy

Editorial Office
MDPI AG
Grosspeteranlage 5
4052 Basel, Switzerland

This is a reprint of the Special Issue, published open access by the journal *Polymers* (ISSN 2073-4360), freely accessible at: www.mdpi.com/journal/polymers/special_issues/D8B397QW9T.

For citation purposes, cite each article independently as indicated on the article page online and using the guide below:

Lastname, A.A.; Lastname, B.B. Article Title. *Journal Name* **Year**, *Volume Number*, Page Range.

ISBN 978-3-7258-3442-6 (Hbk)
ISBN 978-3-7258-3441-9 (PDF)
https://doi.org/10.3390/books978-3-7258-3441-9

© 2025 by the authors. Articles in this book are Open Access and distributed under the Creative Commons Attribution (CC BY) license. The book as a whole is distributed by MDPI under the terms and conditions of the Creative Commons Attribution-NonCommercial-NoDerivs (CC BY-NC-ND) license (https://creativecommons.org/licenses/by-nc-nd/4.0/).

Contents

About the Editors . . . . **vii**

Giorgio Luciano and Maurizio Vignolo
Editorial: Advanced Polymer Composite Materials: Processing, Modeling, Properties and Applications II
Reprinted from: *Polymers* **2024**, *16*, 2650, https://doi.org/10.3390/polym16182650 . . . . **1**

Yesica Vanesa Rojas-Muñoz, Patricio Román Santagapita and María Ximena Quintanilla-Carvajal
Probiotic Encapsulation: Bead Design Improves Bacterial Performance during In Vitro Digestion
Reprinted from: *Polymers* **2023**, *15*, 4296, https://doi.org/10.3390/polym15214296 . . . . **3**

Haji Akbar Sultani, Aleksandr Sokolov, Arvydas Rimkus and Viktor Gribniak
Quantifying the Residual Stiffness of Concrete Beams with Polymeric Reinforcement under Repeated Loads
Reprinted from: *Polymers* **2023**, *15*, 3393, https://doi.org/10.3390/polym15163393 . . . . **17**

Ahmed M. H. Ibrahim, Mohanad Idrees, Emine Tekerek, Antonios Kontsos, Giuseppe R. Palmese and Nicolas J. Alvarez
Engineered Interleaved Random Glass Fiber Composites Using Additive Manufacturing: Effect of Mat Properties, Resin Chemistry, and Resin-Rich Layer Thickness
Reprinted from: *Polymers* **2023**, *15*, 3189, https://doi.org/10.3390/polym15153189 . . . . **37**

Madara Žiganova, Remo Merijs-Meri, Jānis Zicāns, Ivan Bochkov, Tatjana Ivanova and Armands Vīgants et al.
Visco-Elastic and Thermal Properties of Microbiologically Synthesized Polyhydroxyalkanoate Plasticized with Triethyl Citrate
Reprinted from: *Polymers* **2023**, *15*, 2896, https://doi.org/10.3390/polym15132896 . . . . **52**

Baomin Zhao, Meng Tian, Xingsheng Chu, Peng Xu, Jie Yao and Pingping Hou et al.
Dopant-Free Hole-Transporting Material Based on Poly(2,7-(9,9-bis(N,N-di-p-methoxylphenylamine)-4-phenyl))-fluorene for High-Performance Air-Processed Inverted Perovskite Solar Cells
Reprinted from: *Polymers* **2023**, *15*, 2750, https://doi.org/10.3390/polym15122750 . . . . **71**

Hyun-Do Yun, Sun-Hee Kim and Wonchang Choi
Determination of Mechanical Properties of Sand-Coated Carbon Fiber Reinforced Polymer (CFRP) Rebar
Reprinted from: *Polymers* **2023**, *15*, 2186, https://doi.org/10.3390/polym15092186 . . . . **86**

Xiaoyu Jin, Yuning Hao, Zhuo Su, Ming Li, Guofu Zhou and Xiaowen Hu
Dual-Function Smart Windows Using Polymer Stabilized Cholesteric Liquid Crystal Driven with Interdigitated Electrodes
Reprinted from: *Polymers* **2023**, *15*, 1734, https://doi.org/10.3390/polym15071734 . . . . **99**

Qing Yin, Fangong Kong, Shoujuan Wang, Jinbao Du, Ling Pan and Yubo Tao et al.
3D Printing of Solar Crystallizer with Polylactic Acid/Carbon Composites for Zero Liquid Discharge of High-Salinity Brine
Reprinted from: *Polymers* **2023**, *15*, 1656, https://doi.org/10.3390/polym15071656 . . . . **109**

Yaoli Huang, Cong Zheng, Jinhua Jiang, Huiqi Shao and Nanliang Chen
A Flexible Bi-Stable Composite Antenna with Reconfigurable Performance and Light-Responsive Behavior
Reprinted from: *Polymers* **2023**, *15*, 1585, https://doi.org/10.3390/polym15061585 121

Vitaliy Solodilov, Valentin Kochervinskii, Alexey Osipkov, Mstislav Makeev, Aleksandr Maltsev and Gleb Yurkov et al.
Structure and Thermomechanical Properties of Polyvinylidene Fluoride Film with Transparent Indium Tin Oxide Electrodes
Reprinted from: *Polymers* **2023**, *15*, 1483, https://doi.org/10.3390/polym15061483 133

Bolin Tang, Miao Cao, Yaru Yang, Jipeng Guan, Yongbo Yao and Jie Yi et al.
Synthesis of KH550-Modified Hexagonal Boron Nitride Nanofillers for Improving Thermal Conductivity of Epoxy Nanocomposites
Reprinted from: *Polymers* **2023**, *15*, 1415, https://doi.org/10.3390/polym15061415 147

Qingping Wang, Longtao Zhu, Chunyang Lu, Yuxin Liu, Qingbo Yu and Shuai Chen
Investigation on the Effect of Calcium on the Properties of Geopolymer Prepared from Uncalcined Coal Gangue
Reprinted from: *Polymers* **2023**, *15*, 1241, https://doi.org/10.3390/polym15051241 160

Ahmed A. Al-Ghamdi, Ahmed A. Galhoum, Ahmed Alshahrie, Yusuf A. Al-Turki, Amal M. Al-Amri and S. Wageh
Superparamagnetic Multifunctionalized Chitosan Nanohybrids for Efficient Copper Adsorption: Comparative Performance, Stability, and Mechanism Insights
Reprinted from: *Polymers* **2023**, *15*, 1157, https://doi.org/10.3390/polym15051157 174

Tilun Shan, Huiguang Bian, Donglin Zhu, Kongshuo Wang, Chuansheng Wang and Xiaolong Tian
Study on the Mechanism and Experiment of Styrene Butadiene Rubber Reinforcement by Spent Fluid Catalytic Cracking Catalyst
Reprinted from: *Polymers* **2023**, *15*, 1000, https://doi.org/10.3390/polym15041000 203

Yan Tan, Ben Zhao, Jiangtao Yu, Henglin Xiao, Xiong Long and Jian Meng
Effect of Cementitious Capillary Crystalline Waterproofing Materials on the Mechanical and Impermeability Properties of Engineered Cementitious Composites with Microscopic Analysis
Reprinted from: *Polymers* **2023**, *15*, 1013, https://doi.org/10.3390/polym15041013 218

Octavian Danila and Barry M. Gross
Towards Highly Efficient Nitrogen Dioxide Gas in Humid and Wet Environments Using Triggerable-Polymer Metasurfaces
Reprinted from: *Polymers* **2023**, *15*, 545, https://doi.org/10.3390/polym15030545 238

Xing Xie and Dan Yang
Achieving High Thermal Conductivity and Satisfactory Insulating Properties of Elastomer Composites by Self-Assembling BN@GO Hybrids
Reprinted from: *Polymers* **2023**, *15*, 523, https://doi.org/10.3390/polym15030523 251

Yuna Oh, Kwak Jin Bae, Yonjig Kim and Jaesang Yu
Analysis of the Structure and the Thermal Conductivity of Semi-Crystalline Polyetherketone/Boron Nitride Sheet Composites Using All-Atom Molecular Dynamics Simulation
Reprinted from: *Polymers* **2023**, *15*, 450, https://doi.org/10.3390/polym15020450 266

About the Editors

Giorgio Luciano

Dr. Giorgio Luciano is a researcher at the National Research Council of Italy University. He completed his PhD studies in Chemical Sciences at the University of Genoa (Italy) in 2005. He has published 60 papers in various journals as well as took part in several research projects mainly related to material science, and his interests include material science, polymers, and thermal analysis.

Maurizio Vignolo

Maurizio Vignolo earned his degree in Chemistry from the University of Genoa in 2001 and his doctorate in Materials Science and Technology in 2006. He is an affirmed researcher in the field of Physics and Chemistry, currently employed as a senior technologist at the National Research Council of Italy (CNR). From 2001 to 2020, he studied superconducting materials and their applications. In particular, he focused on nano-boron production and superconducting tape fabrication and characterization, working on nanofabrication, ball milling, X-ray diffraction, and cold mechanical deformation. Currently (2020–present), Maurizio works at the CNR's "Giulio Natta" SCITEC Institute, where he focuses on composite polymeric materials, particularly their characterization and processability. He has published numerous articles in high-level scientific journals. Some of his most cited works involve the large hadron collider and the lepton collider. He has also conducted research on high-resolution X-ray spectroscopy and the preparation of self-assembled films of L-cysteine, thiols, and calixarene. His most recent publications involve microwave-assisted sintering for the preparation of composite materials.

Editorial

Editorial: Advanced Polymer Composite Materials: Processing, Modeling, Properties and Applications II

Giorgio Luciano * and Maurizio Vignolo *

National Research Council of Italy—Institute of Chemical Sciences and Technologies "Giulio Natta" CNR SCITEC, Via De Marini 6, 16149 Genova, Italy
* Correspondence: giorgio.luciano@scitec.cnr.it (G.L.); maurizio.vignolo@scitec.cnr.it (M.V.)

Building on the success of our first Special Issue, we are pleased to present this second collection dedicated to the multifaceted world of composite materials. The application of these materials continues to span numerous domains of human endeavor, from agriculture and industry to environmental protection, biomedicine, and transportation [1–18]. In the face of ongoing energy and climate challenges, the importance of research into polymer-based composite materials has become increasingly evident. Such studies have the potential to profoundly impact our current society and, more critically, shape a sustainable future for generations to come.

This second Special Issue further showcases the evolving landscape of composite materials research, presenting a diverse array of high-quality scientific contributions. The selected articles cover a wide spectrum of topics, from fundamental materials science to practical applications, highlighting the versatility and indispensability of composite materials in our daily lives.

A notable feature of this collection, building upon themes explored in our first issue, is the increased focus on biopolymers, such as chitosan and polylactic acid [6,10]. This reflects the growing trend towards sustainable and environmentally friendly materials. These biopolymer-based composites offer promising solutions for various applications, including water treatment, drug delivery, and biodegradable packaging.

The variety of applications explored in this second issue is truly remarkable and expands upon the foundations laid in our previous collection. We see innovative composite materials being developed for electronic applications, such as flexible antennas and smart windows [8,13], showcasing the potential of these materials in the realm of next-generation technologies. In the field of construction and civil engineering, novel composite reinforcements for concrete structures are presented, offering improved durability and performance [5,14].

Environmental applications continue to be well-represented, with studies on composite materials [2,3,5,10,17]. In the energy sector, we find exciting developments in materials for solar cells and thermal management, addressing crucial challenges in renewable energy and energy efficiency [4].

The breadth of characterization techniques and methodologies employed in these studies, from advanced microscopy to molecular dynamics simulations, further emphasizes the multidisciplinary approach required in modern materials science [2,5].

This second Special Issue not only demonstrates the current state-of-the-art in composite materials research but also highlights the rapid progress being made in the field since our first issue. It points towards future directions and challenges, showing how the landscape of composite materials is continuously evolving. It is our hope that this collection will inspire further interdisciplinary collaborations and innovations, driving the development of next-generation composite materials that can address the complex challenges of our time [14].

Citation: Luciano, G.; Vignolo, M. Editorial: Advanced Polymer Composite Materials: Processing, Modeling, Properties and Applications II. *Polymers* **2024**, *16*, 2650. https://doi.org/10.3390/polym16182650

Received: 31 July 2024
Accepted: 13 September 2024
Published: 20 September 2024

Copyright: © 2024 by the authors. Licensee MDPI, Basel, Switzerland. This article is an open access article distributed under the terms and conditions of the Creative Commons Attribution (CC BY) license (https://creativecommons.org/licenses/by/4.0/).

Funding: This research received no external funding.

Conflicts of Interest: The authors declare no conflict of interest.

References

1. Shan, T.; Bian, H.; Zhu, D.; Wang, K.; Wang, C.; Tian, X. Study on the Mechanism and Experiment of Styrene Butadiene Rubber Reinforcement by Spent Fluid Catalytic Cracking Catalyst. *Polymers* **2023**, *15*, 1000. [CrossRef] [PubMed]
2. Oh, Y.; Bae, K.J.; Kim, Y.; Yu, J. Analysis of the Structure and the Thermal Conductivity of Semi-Crystalline Polyetheretherketone/Boron Nitride Sheet Composites Using All-Atom Molecular Dynamics Simulation. *Polymers* **2023**, *15*, 450. [CrossRef] [PubMed]
3. Danila, O.; Gross, B.M. Towards Highly Efficient Nitrogen Dioxide Gas Sensors in Humid and Wet Environments Using Triggerable-Polymer Metasurfaces. *Polymers* **2023**, *15*, 545. [CrossRef] [PubMed]
4. Xie, X.; Yang, D. Achieving High Thermal Conductivity and Satisfactory Insulating Properties of Elastomer Composites by Self-Assembling BN@GO Hybrids. *Polymers* **2023**, *15*, 523. [CrossRef] [PubMed]
5. Tan, Y.; Zhao, B.; Yu, J.; Xiao, H.; Long, X.; Meng, J. Effect of Cementitious Capillary Crystalline Waterproofing Materials on the Mechanical and Impermeability Properties of Engineered Cementitious Composites with Microscopic Analysis. *Polymers* **2023**, *15*, 1013. [CrossRef] [PubMed]
6. Al-Ghamdi, A.A.; Galhoum, A.A.; Alshahrie, A.; Al-Turki, Y.A.; Al-Amri, A.M.; Wageh, S. Superparamagnetic Multifunctionalized Chitosan Nanohybrids for Efficient Copper Adsorption: Comparative Performance, Stability, and Mechanism Insights. *Polymers* **2023**, *15*, 1157. [CrossRef] [PubMed]
7. Wang, Q.; Zhu, L.; Lu, C.; Liu, Y.; Yu, Q.; Chen, S. Investigation on the Effect of Calcium on the Properties of Geopolymer Prepared from Uncalcined Coal Gangue. *Polymers* **2023**, *15*, 1241. [CrossRef] [PubMed]
8. Huang, Y.; Zheng, C.; Jiang, J.; Shao, H.; Chen, N. A Flexible Bi-Stable Composite Antenna with Reconfigurable Performance and Light-Responsive Behavior. *Polymers* **2023**, *15*, 1585. [CrossRef] [PubMed]
9. Solodilov, V.; Kochervinskii, V.; Osipkov, A.; Makeev, M.; Maltsev, A.; Yurkov, G.; Lokshin, B.; Bedin, S.; Shapetina, M.; Tretyakov, I.; et al. Structure and Thermomechanical Properties of Polyvinylidene Fluoride Film with Transparent Indium Tin Oxide Electrodes. *Polymers* **2023**, *15*, 1483. [CrossRef] [PubMed]
10. Yin, Q.; Kong, F.; Wang, S.; Du, J.; Pan, L.; Tao, Y.; Li, P. 3D Printing of Solar Crystallizer with Polylactic Acid/Carbon Composites for Zero Liquid Discharge of High-Salinity Brine. *Polymers* **2023**, *15*, 1656. [CrossRef] [PubMed]
11. Tang, B.; Cao, M.; Yang, Y.; Guan, J.; Yao, Y.; Yi, J.; Dong, J.; Wang, T.; Wang, L. Synthesis of KH550-Modified Hexagonal Boron Nitride Nanofillers for Improving Thermal Conductivity of Epoxy Nanocomposites. *Polymers* **2023**, *15*, 1415. [CrossRef] [PubMed]
12. Zhao, B.; Tian, M.; Chu, X.; Xu, P.; Yao, J.; Hou, P.; Li, Z.; Huang, H. Dopant-Free Hole-Transporting Material Based on Poly(2,7-(9,9-Bis(N,N-Di-p-Methoxylphenylamine)-4-Phenyl))-Fluorene for High-Performance Air-Processed Inverted Perovskite Solar Cells. *Polymers* **2023**, *15*, 2750. [CrossRef] [PubMed]
13. Jin, X.; Hao, Y.; Su, Z.; Li, M.; Zhou, G.; Hu, X. Dual-Function Smart Windows Using Polymer Stabilized Cholesteric Liquid Crystal Driven with Interdigitated Electrodes. *Polymers* **2023**, *15*, 1734. [CrossRef] [PubMed]
14. Sultani, H.A.; Sokolov, A.; Rimkus, A.; Gribniak, V. Quantifying the Residual Stiffness of Concrete Beams with Polymeric Reinforcement under Repeated Loads. *Polymers* **2023**, *15*, 3393. [CrossRef] [PubMed]
15. Do Yun, H.; Kim, S.H.; Choi, W. Determination of Mechanical Properties of Sand-Coated Carbon Fiber Reinforced Polymer (CFRP) Rebar. *Polymers* **2023**, *15*, 2186. [CrossRef] [PubMed]
16. Žiganova, M.; Merijs-Meri, R.; Zicāns, J.; Bochkov, I.; Ivanova, T.; Vīgants, A.; Ence, E.; Štrausa, E. Visco-Elastic and Thermal Properties of Microbiologically Synthesized Polyhydroxyalkanoate Plasticized with Triethyl Citrate. *Polymers* **2023**, *15*, 2896. [CrossRef] [PubMed]
17. Ibrahim, A.M.H.; Idrees, M.; Tekerek, E.; Kontsos, A.; Palmese, G.R.; Alvarez, N.J. Engineered Interleaved Random Glass Fiber Composites Using Additive Manufacturing: Effect of Mat Properties, Resin Chemistry, and Resin-Rich Layer Thickness. *Polymers* **2023**, *15*, 3189. [CrossRef] [PubMed]
18. Rojas-Muñoz, Y.V.; Santagapita, P.R.; Quintanilla-Carvajal, M.X. Probiotic Encapsulation: Bead Design Improves Bacterial Performance during In Vitro Digestion. *Polymers* **2023**, *15*, 4296. [CrossRef] [PubMed]

Disclaimer/Publisher's Note: The statements, opinions and data contained in all publications are solely those of the individual author(s) and contributor(s) and not of MDPI and/or the editor(s). MDPI and/or the editor(s) disclaim responsibility for any injury to people or property resulting from any ideas, methods, instructions or products referred to in the content.

Article

Probiotic Encapsulation: Bead Design Improves Bacterial Performance during In Vitro Digestion

Yesica Vanesa Rojas-Muñoz [1], Patricio Román Santagapita [2] and María Ximena Quintanilla-Carvajal [1,3,*]

[1] Maestría en Diseño y Gestión de Procesos, Facultad de Ingeniería, Campus Universitario del Puente del Común, Universidad de La Sabana, Chía 250001, Colombia; yesicaromu@unisabana.edu.co

[2] Departamento de Química Orgánica, Facultad de Ciencias Exactas y Naturales, Universidad de Buenos Aires & Centro de Investigación en Hidratos de Carbono (CIHIDECAR, UBA-CONICET), Buenos Aires 1428, Argentina; patricio.santagapita@qo.fcen.uba.ar

[3] Grupo de Investigación de Procesos Agroindustriales (GIPA), Facultad de Ingeniería, Campus Universitario del Puente del Común, Universidad de La Sabana, Chía 250001, Colombia

* Correspondence: maria.quintanilla1@unisabana.edu.co

Citation: Rojas-Muñoz, Y.V.; Santagapita, P.R.; Quintanilla-Carvajal, M.X. Probiotic Encapsulation: Bead Design Improves Bacterial Performance during In Vitro Digestion. *Polymers* 2023, 15, 4296. https://doi.org/10.3390/polym15214296

Academic Editors: Giorgio Luciano and Maurizio Vignolo

Received: 20 September 2023
Revised: 14 October 2023
Accepted: 17 October 2023
Published: 1 November 2023

Copyright: © 2023 by the authors. Licensee MDPI, Basel, Switzerland. This article is an open access article distributed under the terms and conditions of the Creative Commons Attribution (CC BY) license (https://creativecommons.org/licenses/by/4.0/).

Abstract: The stability and release properties of all bioactive capsules are strongly related to the composition of the wall material. This study aimed to evaluate the effect of the wall materials during the encapsulation process by ionotropic gelation on the viability of *Lactobacillus fermentum* K73, a lactic acid bacterium that has hypocholesterolemia probiotic potential. A response surface methodology experimental design was performed to improve bacterial survival during the synthesis process and under simulated gastrointestinal conditions by tuning the wall material composition (gelatin 25% w/v, sweet whey 8% v/v, and sodium alginate 1.5% w/v). An optimal mixture formulation determined that the optimal mixture must contain a volume ratio of 0.39/0.61 v/v sweet whey and sodium alginate, respectively, without gelatin, with a final bacterial concentration of 9.20 $\log_{10}$ CFU/mL. The mean particle diameter was 1.6 ± 0.2 mm, and the experimental encapsulation yield was 95 ± 3%. The INFOGEST model was used to evaluate the survival of probiotic beads in gastrointestinal tract conditions. Upon exposure to in the vitro conditions of oral, gastric, and intestinal phases, the encapsulated cells of *L. fermentum* decreased only by 0.32, 0.48, and 1.53 $\log_{10}$ CFU/mL, respectively, by employing the optimized formulation, thereby improving the survival of probiotic bacteria during both the encapsulation process and under gastrointestinal conditions compared to free cells. Beads were characterized using SEM and ATR-FTIR techniques.

Keywords: probiotic; encapsulation; ionotropic gelation; functional food; INFOGEST

1. Introduction

Currently, consumer awareness is increasingly interested in the health impact of food consumption [1]. Consequently, some foods have added bioactives in their matrix, potentially reducing the risk of suffering specific diseases and providing better health benefits. This is the definition of functional foods. Some examples include products enriched with vitamins, minerals, or fiber, or supplemented with probiotics [2,3]. The most widely accepted definition of a probiotic is "live microorganisms that, when administered in adequate amounts, confer a benefit for the health of the host" [4,5]. The global probiotics market size was estimated at USD 40 billion in 2017, and it is projected to reach USD 65.87 billion by 2024 [6].

The most used probiotics are lactic acid bacteria (LAB) of the *Lactobacillus* and *Bifidobacterium* genera [7,8]. *Lactobacillus fermentum* K73 is a LAB isolated from fermented food consumed on the Colombian Atlantic coast. According to studies performed in vitro, *L. fermentum* has probiotic and hypocholesterolemic potential [9]. Probiotic foods are required to have a minimum of 1×10^6 CFU/g of viable bacteria until the end of their shelf life [10]. However, producing probiotic foods on a large scale presents a significant challenge as probiotic strains require special handling and production methods to maintain their viability and functionality. The production process can potentially damage the probiotic cells, affecting the final concentration of viable bacteria in the food product [11]. In addition, for probiotics to be effective, they must survive the harsh physiological conditions of the gastrointestinal tract [12,13]. Therefore, ensuring the viability and functionality of probiotic cells in functional food is critical for the food industry. To overcome this challenge, one strategy is the encapsulation of cells by using a carrier system that protects and releases them at the site of action [14,15].

Currently, the main encapsulation technologies for probiotics in the food industry are extrusion, emulsion, and spray-drying. The extrusion technique has been well received by the industry due to the benefits it brings at low cost, such as the avoidance of elevated temperatures or organic solvents, and most of the materials that can be used being GRAS (generally recognized as safe) [16]. In its simplest form, the encapsulation process is based on a gelling solution of a biopolymer mixed directly with the bioactive system of interest, which is then added into a gelling solution to form beads via a crosslinking reaction between the ion and the polyelectrolyte [17,18]. Wall materials are often polymers that aim to separate the internal phase from the surrounding matrix. There is a wide spectrum for the selection of food-grade polymers suitable for application in microencapsulation by extrusion, among which are sodium alginate, chitosan, milk proteins, and gums [15,19]. Sodium alginate is a linear anionic and hydrophilic biopolymer of natural origin. Its compatibility, biodegradability, and low-cost characteristics have made it the most widely used material in the extrusion technique. It is composed of blocks of β-D-mannuronic acid (M) and α-L glucuronic acid (G). In solution, sodium alginate forms a cross-linked structure in which its anionic acid groups can react with divalent (Zn^{2+}, Ca^{2+}, Ba^{2+}, and others) or polyvalent cations, forming alginate hydrogels with a structure commonly known as "egg-box" [20,21].

The development of Ca-alginate beads for encapsulating bioactive has been a major research focus. However, these beads have several drawbacks, including high porosity, susceptibility to acidic environments, and scaling problems. To address these issues and optimize the encapsulation process, researchers have developed beads by incorporating other polymers to achieve a synergistic effect [22]. In previous studies in which probiotics have been encapsulated using this technique, the unitary operation of centrifugation has been used to concentrate bacterial biomass, which is later mixed with encapsulating materials [22–24]. However, this operation increases production costs, making it challenging for the food industry to scale up production of functional foods based on probiotics. This study presents a novel approach where the whey culture medium in which *L. fermentum* K73 grows is used as the bead wall material, eliminating the need for centrifugation [25].

The objective of this work was to evaluate the effect of the formulation of wall materials in the extrusion encapsulation process on the viability of *Lactobacillus fermentum* K73, improving the performance of the bacterial counts in their passage through the simulated model of gastrointestinal conditions (INFOGEST). The results obtained from the optimal mixture response surface methodology (RSM) allowed for the selection of the volume ratio of each wall material evaluated (type A gelatin, sweet whey, sodium alginate) in the optimal formulation, and the beads were subsequently characterized. This approach represents a significant advance in the development of encapsulation techniques for probiotics, with potential applications in functional foods.

2. Materials and Methods
2.1. Materials

The materials used in the present study were de Man Rogosa and Sharpe (MRS) agar, MRS broth, peptone water, and yeast extract obtained from Scharlau Microbiology (Barcelona, Spain). Glycerol and di-hydrated calcium chloride ($CaCl_2 \cdot 2H_2O$) were purchased from PanReac AppliChem (Barcelona, Spain). Sodium alginate (M/G 0.9; MW 1.40×10^4 g/mol) and the enzymes used in the in vitro digestion assays (pancreatin P7545, lipase L3126, and ox-bile extract 70168) were acquired from Sigma-Aldrich (Saint Louis, MI, USA). Sweet whey (crude protein 11%, crude fat 1.5%, and lactose 61%) was obtained from Saputo Ingredients (Lincolnshire, IL, USA), and gelatin (type A, Bloom 270 g, and MW~100 kDa) was obtained from Cimpa S.A.S. (Bogotá, Colombia).

2.2. Methods

2.2.1. Strain and Bacterial Conservation

The strain *L. fermentum* K73 used in this work was obtained from the collection Usab-Bio of the Department of Engineering, Universidad de La Sabana (Colombia). To preserve the stock cultures, 2 mL vials containing MRS broth and glycerol as a cryoprotective agent, at 40% (v/v) in a volume ratio of 1:1, were stored at $-80\ °C$ (Ultra-freezer, Precisa, Hangzhou, China). Prior to use, the bacterial culture was incubated in MRS broth for 12 h at 37 °C under static aerobic conditions [9].

2.2.2. Biomass Production

Biomass production was carried out in a 1.3 L bioreactor (BioFlo 110; New Brunswick Scientific Co., Enfield, CT, USA). Operational conditions were the following: temperature at 37 °C and agitation speed of 100 rpm for 10 h. The culture medium was prepared with 8% (w/v) sweet whey and 0.22% (w/v) yeast extract adjusted to a final pH of 5.5 with 1 M HCl. After sterilizing the culture medium at 121 °C for 15 min, *L. fermentum* K73 was inoculated at 10% (v/v) [25].

2.3. Formulation of Wall Materials

2.3.1. Preparing Solutions

Sodium alginate 1.5% (w/w) and gelatin type A at 25% (w/w) solutions were prepared with deionized water at 90 °C for 30 min under magnetic stirring (250 rpm) [24,26]. Calcium chloride hardener solution (100 mM) was prepared with deionized water, and it was adjusted to a final pH of 4.5 according to preliminary assays. All solutions were sterilized.

2.3.2. Mixture Design

The experimental design for the formulation of the bead wall materials was performed using an Optimal Mixture response surface methodology (RSM) through Design-Expert software version 11.0.0 (Stat-Ease Inc., Minneapolis, MN, USA). The design included 16 runs (as shown in Table 1) with 5 replicas. The response variable was viability ($\log_{10}$ CFU/mL). The following numerical factors were wall materials volume ratios: sodium alginate (0.40–0.90), gelatin type A (0.00–0.50), and sweet whey (0.00–0.50). Each mixture was inoculated at 20% (v/v). The ranges of the proportion of each material were selected according to preliminary evaluations.

The RSM was used to determine the optimal mix of bead wall materials while considering the maximization of probiotic viability (results are included in Table 1). The suggested optimal mixture was replicated in the laboratory, and the error between the predicted and observed response variable was calculated using Equation (1).

$$\%Error = \frac{Predicted\ variable - Observed\ variable}{Predicted\ variable} \times 100 \qquad (1)$$

Table 1. Optimal bead wall material selection using a mixture experimental design.

Run	Factors			Response Variable
	A: Alginate	B: Gelatin	C: Sweet Whey	Viability ($\log_{10}$ CFU/mL)
1	0.80	0.02	0.18	9.12
2	0.50	0.26	0.24	9.05
3	0.55	0.00	0.45	9.11

Table 1. Cont.

Run	Factors			Response Variable
	A: Alginate	B: Gelatin	C: Sweet Whey	Viability ($\log_{10}$ CFU/mL)
4	0.40	0.10	0.50	8.91
5	0.66	0.34	0.00	9.14
6	0.90	0.06	0.04	8.47
7	0.50	0.26	0.24	8.95
8	0.52	0.37	0.11	9.18
9	0.66	0.34	0.00	9.04
10	0.40	0.50	0.10	8.38
11	0.65	0.15	0.20	9.05
12	0.40	0.10	0.50	8.91
13	0.76	0.18	0.06	8.87
14	0.50	0.26	0.24	9.04
15	0.68	0.00	0.32	9.26
16	0.40	0.50	0.10	8.65

2.4. Probiotic Encapsulation

The encapsulation process consisted of loading the mixture into a syringe with a 0.6 mm diameter needle. The resulting string of drops was dropped into the hardener solution at a height of 3 cm. The instantly formed beads were left stirring in a $CaCl_2$ solution with gentle magnetic agitation for 30 min to promote efficient gelling of the particles and the resulting beads were rinsed with distilled water [14].

2.4.1. Cell Release

The release of the probiotic cells from the bead matrix was achieved using the technique documented by (Bevilacqua et al., 2020 [27]). Briefly, 5.0 g of beads was mixed with a 5.0% (w/v) sodium citrate solution at a dilution ratio of 1:10 (v/v) and homogenized using a vortex VG 3 (IKA, Werke, Germany) at 800 rpm for 60 s until the beads were completely dissolved. Subsequently, cell viability was determined using the plate count method.

2.4.2. Cell Viability

The cell count of *L. fermentum* K73 for all the proposed experiments was carried out by making 1:10 (v/v) serial dilutions in peptone water 0.1% (w/v). Plating was performed on MRS agar from dilution 1 to 7 in triplicate. Incubation conditions were 37 °C for 24 h under aerobic conditions. Plates containing up to 250 colonies were numbered. The result was reported as the logarithm of the final colony-forming units per milliliter (CFU/mL).

2.4.3. Encapsulation Efficiency (EE)

Cell entrapment efficiency was calculated as the difference between viable cell count before and after the microencapsulation process (Graff et al., 2008 [28]).

$$EE\% = \left(log(CFU/mL)_{N_2} / log(CFU/mL)_{N_1} \right) \times 100 \qquad (2)$$

where N_1 and N_2 indicate the number of viable bacteria in the mixture and released from the beads, respectively.

2.5. Characterization of Beads

2.5.1. Attenuated Total Reflectance–Fourier Transform-Infrared Spectroscopy (ATR-FTIR)

Fourier Transform Infrared (FTIR) spectra were acquired using a spectrometer (Varian model 630-IR, Agilent-Tech Inc., Santa Clara, CA, USA), with 16 scans collected from 675 to 4000 cm^{-1} at 4 cm^{-1}. All samples were studied using a single reflection ATR system (Cary 630 ATR-TIR Instrument, Agilent-Tech Inc., Santa Clara, CA, USA) of Ge crystal and with an incident angle of 45°. Control reagents (sodium alginate and prepared culture medium) and beads were freeze-dried prior to measurement.

Data were processed by using the free-license Spectragryph v1.2.15 software (developed by Dr. Friedrich Menges, Oberstdorf, Germany). Spectra were baseline-corrected (adaptative correction; coarseness: 50; offset 10), and were normalized between 0 and 1 and smoothed (Savitzky-Golay, 15 points, 2) for figure presentation. The FWHM (full width at half maximum) of the peak was calculated by using the software tool.

2.5.2. Scanning Electron Microscopy (SEM)

Some morphological parameters of the beads were observed by scanning electron microscopy (FE-MEB LYRA3 Tescan, Bollebergen, Belgium). Samples were subjected to freeze-drying and air-drying to compare bead morphological changes between drying methods. For the freeze-drying process, beads were washed and frozen at −80 °C for 24 h, then were placed in the freeze-dryer (Labconco, FreeZone 4.5L, Fullerton, CA, USA) at −40 °C and 0.05 mbar pressure for 24 h. For the air-drying process, beads were ventilated in an air chamber at 40 °C for 4 h. Dried beads were stored in desiccators over phosphorus pentoxide (P_2O_5) for two days and were coated with gold before observation. SEM was operated at 10 kV and 60–5000× magnification. In addition, wet beads were observed using a high-performance scanning electron microscope (JSM-6490LV, JEOL, Zaventem, Belgium) in low vacuum mode.

2.5.3. Particle Size

One hundred fresh beads were randomly selected and placed on dark paper, and a digital camera (Canon EOS 70D) was used to photograph them. The diameter of the beads was determined by processing the images in ImageJ software (V 1.50i, National Institutes of Health, Bethesda, MD, USA) [21].

2.6. Probiotic Cell Viability under INFOGEST Simulated Gastrointestinal Model

The INFOGEST protocol was used to simulate the conditions of the gastrointestinal tract proposed by [29,30]. Briefly, the oral phase was prepared by mixing 5 g of the sample (beads) and 5 mL of simulated salivary fluid (SSF). The final oral phase pH was brought to 7.0 and was shaken (100 rpm) at 37 °C for 2 min. Then, the gastric phase consisted of adding 10 mL of simulated gastric fluid (SGF) to the oral phase and adjusting it to a final pH of 3.0 with the required volume of HCl (1 M). The sample was under agitation (100 rpm) for 2 h at 37 °C. Finally, the intestinal phase was performed by mixing the gastric phase, 20 mL simulated intestinal fluid (SIF), and enzymes bile extract (10 mM), pancreatin (100 U/mL), and pancreatic lipase (2000 U/mL). The final pH of the intestinal phase was adjusted to 7.0, and it was incubated under the same conditions as the gastric phase. After each phase, a viability assay was carried out according to the 2.4.1 numeral. Bacteria survival was determined according to Equation (3).

$$Survival\% = \left(log(CFU/mL)_{final} / log(CFU/mL)_{initial} \right) \times 100 \qquad (3)$$

2.7. Statistical Analysis

The significant test of the designs was performed by analysis of variance (ANOVA) with a confidence level of 95%. The coefficient of determination (R^2) was used to evaluate the fit of the measurements to the regression models. Three replications were performed in all assays and the data were presented as the mean ± standard deviation. The RSM Optimal Mixture design was performed using the Design-Expert software (version 8.1.0, Stat-Ease Inc., Minneapolis, MN, USA). For the optimization of the response variable, the desirability criterion of the specialized software was used.

3. Results and Discussion

3.1. Formulation of Wall Materials: Mixture Design

The selection of suitable wall materials for the probiotic cell encapsulation process is closely related to the viability, stability, and release of the bioactive, since these materials protect the compound of interest from external factors, such as processing, storage, and conditions of the GI tract [15,31,32]. In the present study, the mixture's optimal ratio between gelatin, sodium alginate, and sweet whey as bead wall materials was evaluated. The results of the experimental design of the mixture (Table 1) were adjusted to a statistically significant quadratic model ($p \leq 0.05$). The coefficient of determination (R^2) was 0.80, and the lack of fit of the model was not significant ($p \geq 0.05$), as shown in Table 2. The analysis of variance (ANOVA) of the regression for the response variable "viability" showed two statistically significant double interactions between the sodium alginate and the other two materials.

Table 2. ANOVA of the mixture design for the response variable: viability ($\log_{10}$ CFU/mL).

	Sum of Squares	Degrees of Freedom	p-Value
Model	0.751	5	0.003
Linear	0.126	2	0.078
AB	0.491	1	0.005
AC	0.359	1	0.001
BC	0.001	1	0.834
Residuals	0.190	10	
Lack of fit	0.149	6	0.208
Pure Error	0.041	4	
R^2	0.80		

The second order polynomial equation in terms of the coded factors is:

$$Y = 7.96A + 8.46B + 8.91C - 3.31AB - 2.98AC + 0.18BC \qquad (4)$$

where Y is the predicted response variable (viability), and A, B, and C are the coded values of the proportion of alginate, gelatin, and whey in the mixture, respectively.

Figure 1 shows the graphic optimization of the effect of the different wall materials on bacterial viability. The region between sodium alginate and sweet whey (A–C) shows a strong interaction with maximum viability. In the numerical optimization, alginate is present in a high proportion at the optimal point, and the alginate–whey mixture presents the maximum value of the response variable. This may be because increasing the proportion of alginate would allow for more Ca^{2+} ion binding sites and, consequently, for a greater number of alginate strands to be held together in the microcapsule structure, effectively protecting bacterial cells [33]. Moreover, an increasing sodium alginate concentration leads to higher retention of bioactive assuring the microstructural stability of the beads [34]. On the other hand, milk proteins are a wall material capable of forming gels. They can also have interactions with other polymers to form complexes, in this case with alginate [35]. The model obtained from the experimental mixture design allowed for selection of the optimal ratio for the formulation of the wall materials based on the desirability criterion. The optimal point was a mixture composed of sweet milk whey (0.604) and sodium alginate (0.396) with a predicted viability of 9.261 $\log_{10}$ CFU/mL. The optimal point was replicated experimentally as a validation run to evaluate the predictive capacity of the model, obtaining a low error rate (0.63%) considering the experimental viability of 9.20 $\log_{10}$ CFU/mL. This observation is consistent with the study of [36], in which the wall materials were studied individually (whey protein and alginate) and in a mixture, with the finding that the optimal proportion that maximized the viability of the microorganism was 0.62/0.38, respectively. Dehkordi et al., (2020) [37] investigated the increase in the viability of *L. acidophilus* encapsulated in alginate–sweet whey capsules compared to alginate–whey protein isolate (WPI) capsules. In their study, the cell suspension was centrifuged at $3000\times g$ for 15 min, washed with saline solution, and then resuspended in deionized water. However, in the present article, this step is not required due to the novel composition of the *L. fermentum* culture medium used. Moreover, WPI contains 93% protein content, which can increase the final cost of the capsule. In contrast, sweet milk whey is an agro-industrial waste and could potentially be a more economical option.

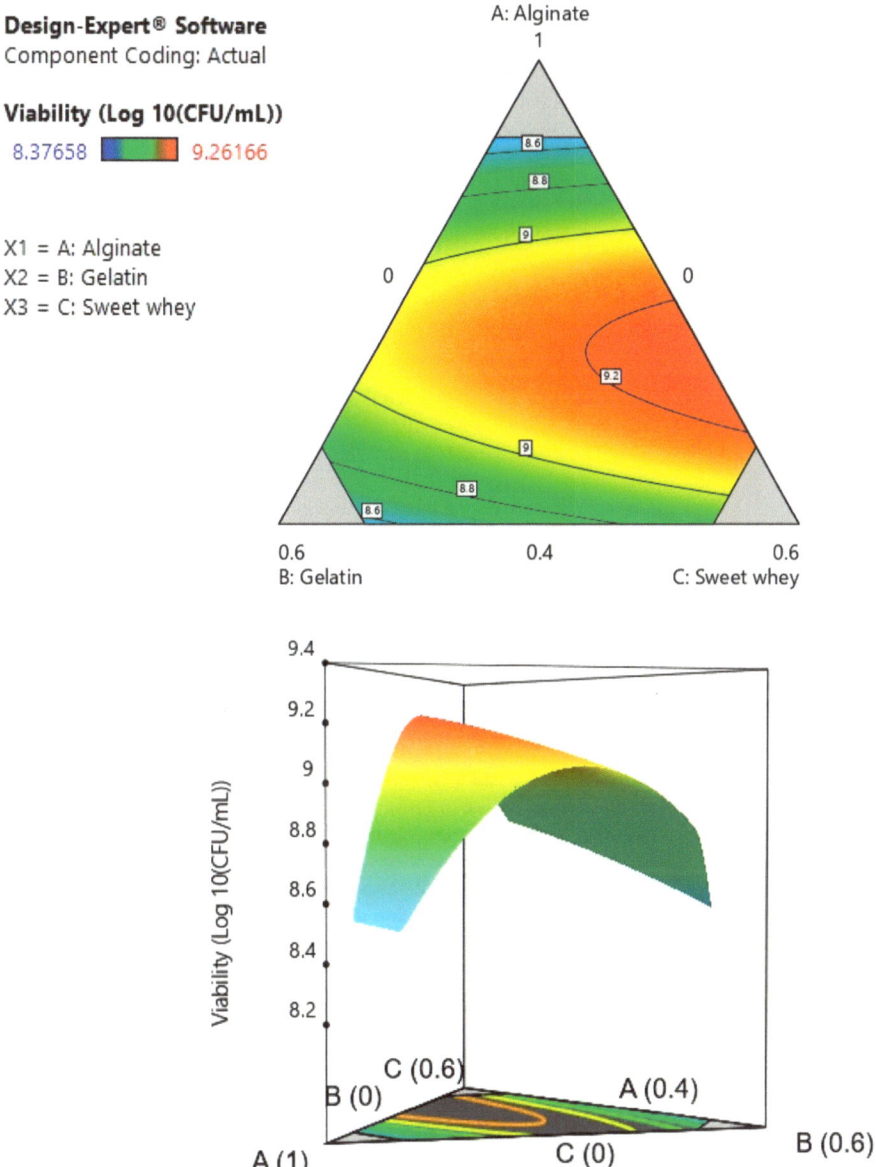

Figure 1. Contour plot and response surface for the variable viability of *Lactobacillus fermentum* K73.

In addition, the encapsulation efficiency of the extruded beads with the optimal mixture in this study was 94.8%, a value consistent with the existing literature. Donthidi et al. (2010) [38] reported that encapsulation with alginate and whey protein enhanced the survival of probiotics with a 90.9% encapsulation efficiency. In another study [39], microencapsulated *L. bulgaricus* cells in alginate–milk microspheres using vibratory technology reported a high encapsulation yield (99%).

It is interesting to highlight that gelatin was excluded from the optimal point by the model since its inclusion reduced the encapsulation efficiency at any of the assayed concentrations, even though the encapsulation efficiency values were rather high (as an absolute value). Gelatin has been used in the food industry as a vehicle for probiotic encapsulation in alginate–gelatin beads [40,41] and was an excellent excipient for the encapsulation of *Lactobacillus fermentum* K73 by electrospinning. However, gelatin-containing beads were more challenging to produce since the viscosity of gelatin changes with temperature, and since the probiotic bacteria is mesophilic, it was encapsulated with gelatin at 37 °C. The addition of secondary components and their interaction is critical to tune the fine structure of Ca(II)-alginate beads [42,43] since the interactions among them could affect the network, its pore size, and the interactions established within the wall materials the cells. The interaction among the materials and the characteristics of the beads are included in the next section.

3.2. Characterization of Beads

3.2.1. Attenuated Total Reflectance–Fourier Transform-Infrared Spectroscopy (ATR-FTIR)

Figure 2A,B show the ATR-FTIR spectra of the beads and the wall materials. Considering the components of the beads, the culture medium (CM) shows the typical signal corresponding to sweet whey, which is composed of lactose and oligosaccharides, proteins, and fat [44]. Moreover, the symmetric and asymmetric stretching of carboxylic acids from sodium alginate at 1593 and 1405 cm^{-1} are also observed, respectively (Figure 2A). Prior to analyzing the spectrum of the beads, is important to keep in mind the type of interactions that can be established among the wall constituents. Considering the pH of the optimal mix solution prior to gelation (4.82), electrostatic interactions between sodium alginate (SA) and sweet whey (SW) are expected. The isoelectric point of SW main protein (β-lactoalbumin) is 5.1–5.2, given an overall positive charge in the protein, as also reported [45]. On the other hand, sodium alginate is negatively charged at a pH higher than 3.65 [46]. Several peaks shifted in the beads with respect to the plain components; the symmetric and asymmetric stretching of carboxylic acids changed from 1593 to 1598 and 1405 to 1422 cm^{-1}, respectively, as shown in Figure 2B. There are also changes at 3000–3600 (OH stretching) in both position and relative intensities, revealing rearrangements of the hydrogen bonds (which change their average length) to 3235 from 3216 and 3267 cm^{-1} for SA or CM, respectively (Figure 2A). Moreover, the FWHM (full width at half maximum) of the peak of SW-ALG beads is thinner than those of SA or CM (380 vs. ~420 cm^{-1} for individual components), which is linked to a higher degree of homogeneities of the intermolecular interactions, which reduce the dispersion of the vibrational levels and higher conformational selectivity [47]. By comparing to sodium alginate (the main component), even though the increase in wavenumber indicates a reduction in hydrogen-bond density and strength, this change has also been linked to a decreased molecular packing, hence the greater protection in the dried state [47].

Moreover, some changes were also observed in the main peaks of the sugars, particularly for C-O-C and C-O-H of the sugar ring, showing several displacements between 1200 and 800 cm^{-1}: the maximum (COC stretching) was shifted from 1025 and 1017 (of SA or CM, respectively) to 1011 cm^{-1} in beads; and the C-OH stretching from 1083 and 1068 to 1077 cm^{-1} (Figure 2B). Furthermore, a new peak at 939 from the bands at 947 and 933 cm^{-1} of CM and SA, respectively, and a shift from 989 (of CM) to 993 cm^{-1} in beads were observed (Figure 2B), accounting for more interactions (mainly hydrogen bonding and van der Waals forces) between components in the formulated beads.

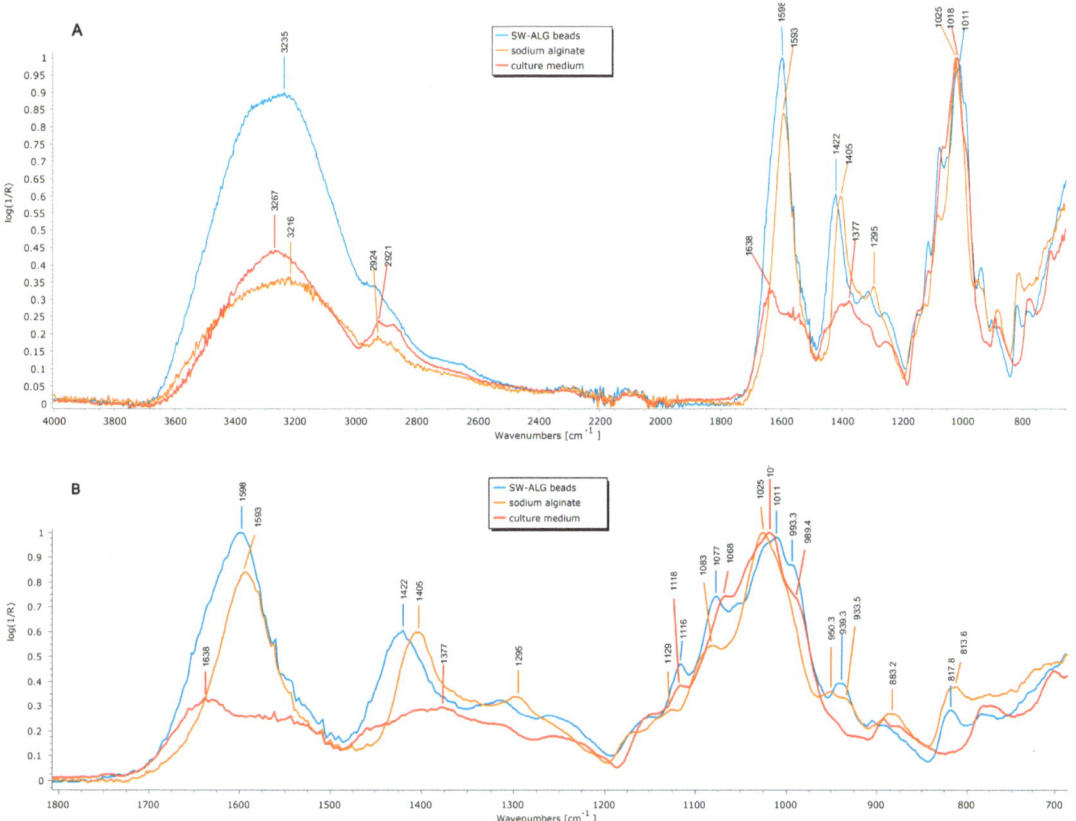

Figure 2. FTIR analysis of sodium alginate (ALG), culture medium—mainly sweet whey (SW)—and SW-ALG freeze-dried microcapsules: (**A**) full-range spectrum; (**B**) detailed region between 1700 and 700 cm^{-1}.

3.2.2. Particle Size and Morphological Characterization

The morphological characterization of the beads was performed immediately after encapsulation. Fresh beads are shown in Figure 3A. The encapsulated probiotics were characterized by performing image analysis. SW-SA beads showed spherical morphologies with an average diameter (±standard deviation) of 1.6 ± 0.2 mm and high circularity (0.8 ± 0.1). The obtained beads exhibited adequate mechanical stability under handling. Different authors have reported that beads loaded with probiotics have diameter values between 1.5 and 1.9 mm. However, these sizes vary widely depending on the encapsulation materials and extrusion diameter [16]. *Bifidobacterium longum* encapsulated by [48] in alginate–dairy matrices depicted varied sizes of microcapsules (2.3–3.1 mm) due to the type of encapsulation material. Lopes et al. (2017) [49] extruded alginate–gelatin beads to encapsulate *L. rhamnosus*, obtaining regular- and spherical-shape beads with a size ranging between 1.53 and 1.90 mm.

The morphology of the wet and dried beads was observed by SEM (Figure 3). Wet microcapsules showed a spherical shape without agglomerations and with a continuous surface without hollow areas or deep-crack morphology (Figure 3B), confirming particle size analysis. However, frozen and air-dried microcapsules slightly lose this spherical structure (Figure 3C,D, respectively), showing that the drying method affects the morphology of the beads. These results are in line with another study [43], which found that the loss of water determines the morphology of the final structure in each drying, as revealed by the contractions and small eruptions on the surface as a result of the massive loss of water (from 0.98 to 0.04 g H$_2$O/g dry weight). The encapsulated *L. fermentum* K73 cells on the surface of the bead was directly observed to be covered by the matrix (Figure 3E).

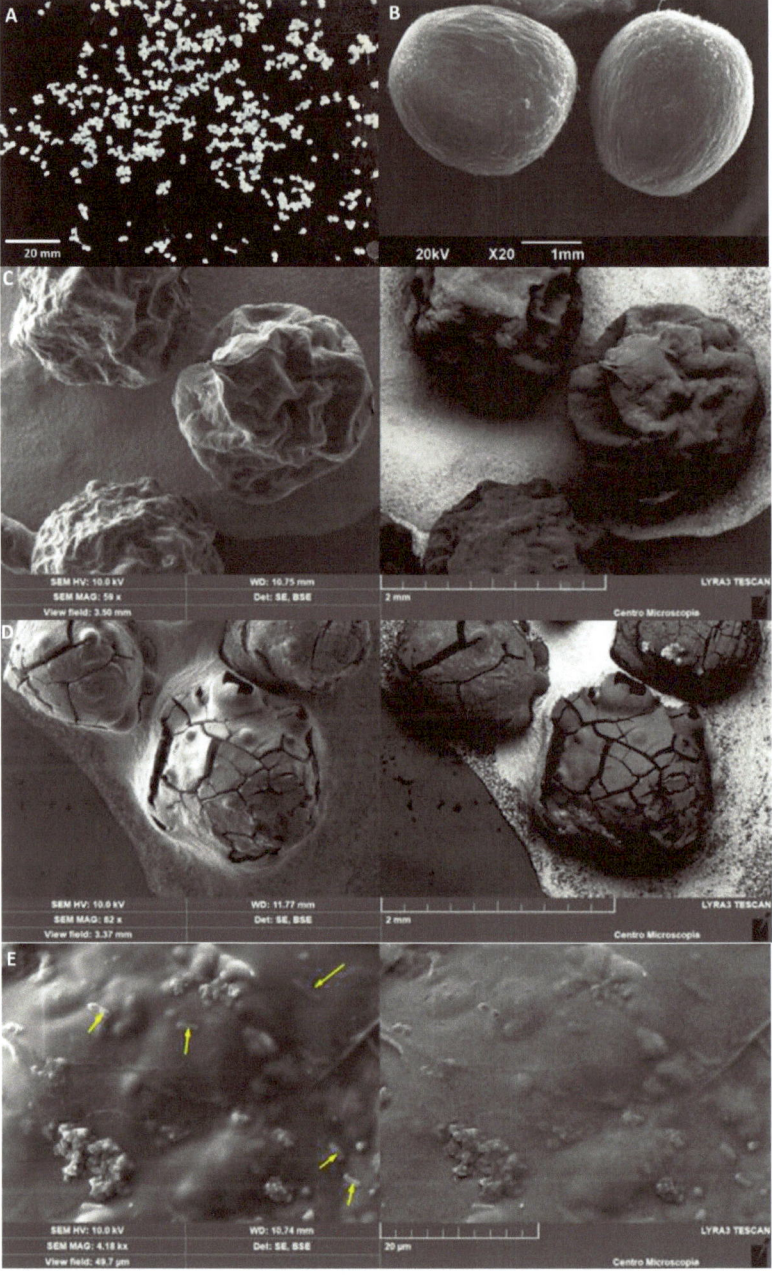

Figure 3. *L. fermentum* K73 encapsulated in (**A**) fresh beads, (**B**) wet beads, (**C**) freeze-dried beads, and (**D**) air-dried beads. (**E**) Bacteria on the surface of freeze-dried bead.

3.3. Probiotic Cell Viability under INFOGEST-Simulated Gastrointestinal Model

Probiotics play a significant role in human health by providing a protective effect on the microbiota in the gastrointestinal tract. Thus, ensuring the minimum dose at which the microorganism is capable of colonizing and exerting its beneficial activity is essential [50]. For this reason, the viability of free and encapsulated *L. fermentum* K73 cells was evaluated under the standardized static in vitro digestion protocol developed by the INFOGEST international network, as shown in Figure 4, for each GI tract phase. The viability of free probiotic cells significantly decreased ($p \leq 0.05$) compared to encapsulated cells in SW-SA beads.

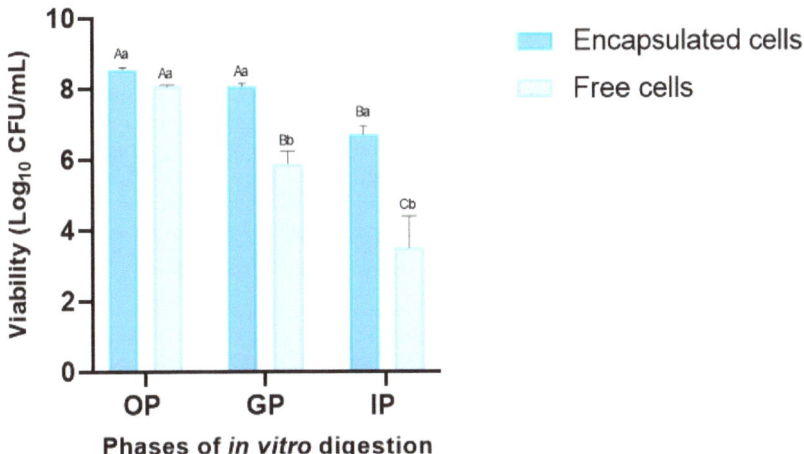

Figure 4. Viability of free and encapsulated *L. fermentum* K73 cells during OP (oral phase: 2 min, pH 7.0), GP (gastric phase: 120 min, pH 3.0), and IP (intestinal phase: 120 min, pH 7.0) according to the INFOGEST in vitro model. The mean value ± standard deviation of at least three independent measurements is included. (A–C) Different letters within the same treatment (encapsulated or free cells) indicate statistical significance between in vitro digestion phases ($p < 0.05$). (a,b) Different letters with the same in vitro digestion phase indicate statistical significance between treatments (encapsulated or free cells) ($p < 0.05$).

Survival rates showed no significant decrease in the oral phase of the two treatments evaluated, the viability of the encapsulated cells was 97.6% (0.21 log CFU/mL), and for free cells, it was 92.1%, (0.69 log CFU/mL) compared with the initial bacteria viability. After 2 h of incubation in the gastric phase (pH 3), free cells showed a significant decrease (67.2% or 2.87 cycles) compared with SW-ALG encapsulated cells (92.1% or 0.69 cycles). Encapsulation with alginate and sweet whey was effective in protecting probiotic cells since the reduction in the viability of the encapsulated bacteria after digestion was significantly lower (76.5%—1.91 log CFU/mL) than that observed in free cells (40.1%, 5.11 log CFU/mL). These results agree with previously reported studies, where alginate encapsulation in combination with other materials helped increase the viability of probiotic microorganisms after subjecting them to adverse conditions [51–54]. Eckert et al. (2018) [55] found that the non-encapsulated *Lactobacillus* spp. cells were sensitive to the conditions of the simulated gastrointestinal tract, and some even died after the gastric phase. However, the reduction in viability of whey–pectin–alginate probiotic microparticles during digestion was 2–3 log cycles, showing that encapsulation effectively protected cells by keeping viability above the requirements for probiotic foods. Lee et al. (2019) [17] encapsulated *L. acidophilus* KBL409 with alginate and chitosan, and observed a dramatic decrease in the survival rate (46.5%—4.95 log reduction) of free cells, while more than 80% of chitosan–alginate encapsulated cells were viable after digestion.

In addition, it is important to point out that the count of viable encapsulated cells in the intestinal phase was 6.7 CFU/mL, which is higher than the limit required to classify a product as a probiotic (>6 log CFU/mL). This result highlights the protective effect of the optimal mixture of the beads under the digestion conditions. Considering the wall components and the established interactions among them (Figure 2), it seems that a favorable barrier against rapid degradation and/or diffusion at the periphery of the microspheres between whey–alginate may occur, protecting the microorganisms from the GI conditions and producing probiotic particles of high encapsulation efficiency.

4. Conclusions

In conclusion, the encapsulation of *Lactobacillus fermentum* K73 by ionotropic gelation using sodium alginate and sweet whey as wall materials was successful. The optimization process leads to a combination of two widely and well-known materials with the advantage of using the same culture medium as one of the wall materials. Bead design provided an additional barrier that led to obtaining a high viability and encapsulation yield of the probiotic (9.261 $\log_{10}$ CFU/mL, with a 94.8% of encapsulation efficiency). The particles produced with the optimal mixture exhibited improved structure and enhancements to probiotic toleration to simulated gastrointestinal conditions, conforming to the requirements for probiotic foods.

Author Contributions: Conceptualization, Y.V.R.-M. and M.X.Q.-C.; Methodology, Y.V.R.-M. and P.R.S.; Software, P.R.S.; Validation, Y.V.R.-M.; Formal analysis, Y.V.R.-M., P.R.S. and M.X.Q.-C.; Investigation, Y.V.R.-M.; Resources, M.X.Q.-C.; Data curation, P.R.S.; Writing—original draft, Y.V.R.-M.; Writing—review & editing, P.R.S. and M.X.Q.-C.; Supervision, M.X.Q.-C.; Funding acquisition, M.X.Q.-C. All authors have read and agreed to the published version of the manuscript.

Funding: This research was funded by grant 1112, ING 136-2013, from the Research Department of the University of La Sabana.

Institutional Review Board Statement: Not applicable.

Data Availability Statement: The data presented in this study are available on request from the corresponding author.

Acknowledgments: This research did not receive any specific grant from funding agencies in the public, commercial, or not-for-profit sectors. This project was supported by grant 1112, ING 136-2013, from the Research Department of the University of La Sabana. The authors appreciate the support of the Graduate Assistant scholarship.

Conflicts of Interest: The authors declare no conflict of interest.

References

1. AlaAlam, M.S.; Aslam, R. Extrusion for the Production of Functional Foods and Ingredients. In *Innovative Food Processing Technologies*; Elsevier: Amsterdam, The Netherlands, 2020. [CrossRef]
2. Helkar, P.B.; Sahoo, A.; Patil, N. Review: Food Industry By-Products used as a Functional Food Ingredients. *Int. J. Waste Resour.* **2016**, *6*, 1–6. [CrossRef]
3. Reque, P.M.; Brandelli, A. Encapsulation of probiotics and nutraceuticals: Applications in functional food industry. *Trends Food Sci. Technol.* **2021**, *114*, 1–10. [CrossRef]
4. Haffner, F.B.; Diab, R.; Pasc, A. Encapsulation of probiotics: Insights into academic and industrial approaches. *AIMS Mater. Sci.* **2016**, *3*, 114–136. [CrossRef]
5. Hill, C.; Guarner, F.; Reid, G.; Gibson, G.R.; Merenstein, D.J.; Pot, B.; Morelli, L.; Canani, R.B.; Flint, H.J.; Salminen, S.; et al. Expert consensus document: The international scientific association for probiotics and prebiotics consensus statement on the scope and appropriate use of the term probiotic. *Nat. Rev. Gastroenterol. Hepatol.* **2014**, *11*, 506–514. [CrossRef]
6. Abid, M.B.; Koh, C.J. Probiotics in health and disease: Fooling Mother Nature? *Infection* **2019**, *47*, 911–917. [CrossRef] [PubMed]
7. Bultosa, G. Functional Foods: Dietary Fibers, Prebiotics, Probiotics, and Synbiotics. In *Encyclopedia of Food Grains*, 2nd ed.; Elsevier: Amsterdam, The Netherlands, 2015; Volume 2–4. [CrossRef]
8. Rodríguez-Sánchez, S.; Fernández-Pacheco, P.; Seseña, S.; Pintado, C.; Palop, M.L. Selection of probiotic Lactobacillus strains with antimicrobial activity to be used as biocontrol agents in food industry. *LWT* **2021**, *143*, 111142. [CrossRef]

9. Cueto, C.; Aragón, S. Evaluation of probiotic potential of lactic acid bacteria to reduce in vitro cholesterol. *Sci. Agropecu.* **2012**, *1*, 45–50. [CrossRef]
10. Qi, X.; Simsek, S.; Chen, B.; Rao, J. International Journal of Biological Macromolecules Alginate-based double-network hydrogel improves the viability of encapsulated probiotics during simulated sequential gastrointestinal digestion: Effect of biopolymer type and concentrations. *Int. J. Biol. Macromol.* **2020**, *165*, 1675–1685. [CrossRef]
11. Bauer-Estrada, K.; Sandoval-Cuellar, C.; Rojas-Muñoz, Y.; Quintanilla-Carvajal, M.X. The modulatory effect of encapsulated bioactives and probiotics on gut microbiota: Improving health status through functional food. *Food Funct.* **2022**, *14*, 32–55. [CrossRef]
12. Kim, J.; Muhammad, N.; Jhun, B.H.; Yoo, J.-W. Probiotic delivery systems: A brief overview. *J. Pharm. Investig.* **2016**, *46*, 377–386. [CrossRef]
13. Misra, S.; Pandey, P.; Mishra, H.N. Novel approaches for co-encapsulation of probiotic bacteria with bioactive compounds, their health benefits and functional food product development: A review. *Trends Food Sci. Technol.* **2021**, *109*, 340–351. [CrossRef]
14. Martín, M.J.; Lara-Villoslada, F.; Ruiz, M.A.; Morales, M.E. Microencapsulation of bacteria: A review of different technologies and their impact on the probiotic effects. *Innov. Food Sci. Emerg. Technol.* **2015**, *27*, 15–25. [CrossRef]
15. Rodrigues, F.J.; Cedran, M.F.; Bicas, J.L.; Sato, H.H. Encapsulated probiotic cells: Relevant techniques, natural sources as encapsulating materials and food applications—A narrative review. *Food Res. Int.* **2020**, *137*, 109682. [CrossRef] [PubMed]
16. Krasaekoopt, W.; Watcharapoka, S. Effect of addition of inulin and galactooligosaccharide on the survival of microencapsulated probiotics in alginate beads coated with chitosan in simulated digestive system, yogurt and fruit juice. *LWT-Food Sci. Technol.* **2014**, *57*, 761–766. [CrossRef]
17. Lee, Y.; Ji, Y.R.; Lee, S.; Choi, M.J.; Cho, Y. Microencapsulation of probiotic lactobacillus acidophilus kbl409 by extrusion technology to enhance survival under simulated intestinal and freeze-drying conditions. *J. Microbiol. Biotechnol.* **2019**, *29*, 721–730. [CrossRef]
18. Posbeyikian, A.; Tubert, E.; Bacigalupe, A.; Escobar, M.M.; Santagapita, P.R.; Amodeo, G.; Pellurini, M. Evaluation of calcium alginate bead formation kinetics: An integrated analysis through light microscopy, rheology and microstructural SAXS. *Carbohydr. Polym. J.* **2021**, *269*, 118293. [CrossRef] [PubMed]
19. Brahma, S.; Sadiq, M.B.; Ahmad, I. *Probiotics in Functional Foods*; Elsevier: Amsterdam, The Netherlands, 2019; ISBN 9780081005965.
20. Pedroso-Santana, S.; Fleitas-Salazar, N. Ionotropic gelation method in the synthesis of nanoparticles/microparticles for biomedical purposes. *Polym. Int.* **2020**, *69*, 443–447. [CrossRef]
21. Zazzali, I.; Rocio, T.; Calvo, A.; Manuel, V.; Ruíz-henestrosa, P.; Santagapita, P.R.; Perullini, M. Effects of pH, extrusion tip size and storage protocol on the structural properties of Ca(II)-alginate beads. *Carbohydr. Polym.* **2019**, *206*, 749–756. [CrossRef]
22. Silva, M.P.; Tulini, F.L.; Martins, E.; Penning, M.; Fávaro-Trindade, C.S.; Poncelet, D. Comparison of extrusion and co-extrusion encapsulation techniques to protect Lactobacillus acidophilus LA3 in simulated gastrointestinal fluids. *LWT-Food Sci. Technol.* **2018**, *89*, 392–399. [CrossRef]
23. Olivares, A.; Silva, P.; Altamirano, C. Microencapsulation of probiotics by efficient vibration technology. *J. Microencapsul.* **2017**, *34*, 667–674. [CrossRef]
24. Benucci, I.; Cerreti, M.; Maresca, D.; Mauriello, G.; Esti, M. Yeast cells in double layer calcium alginate–chitosan microcapsules for sparkling wine production. *Food Chem.* **2019**, *300*, 125174. [CrossRef]
25. Aragón-Rojas, S.; Quintanilla-Carvajal, M.X.; Hernández-Sánchez, H. Multifunctional Role of the Whey Culture Medium in the Spray-Drying Microencapsulation of Lactic Acid Bacteria. *Food Technol. Biotechnol.* **2018**, *56*, 381–397. [CrossRef] [PubMed]
26. Ricaurte, L.; Santagapita, P.R.; Díaz, L.E.; Quintanilla-Carvajal, M.X. Edible gelatin-based nanofibres loaded with oil encapsulating high-oleic palm oil emulsions. *Colloids Surfaces A Physicochem. Eng. Asp.* **2020**, *595*, 124673. [CrossRef]
27. Bevilacqua, A.; Campaniello, D.; Speranza, B.; Racioppo, A.; Altieri, C.; Sinigaglia, M.; Corbo, M.R. Microencapsulation of *Saccharomyces cerevisiae* into Alginate Beads: A Focus on Functional Properties of Released Cells. *Foods* **2020**, *9*, 1051. [CrossRef] [PubMed]
28. Graff, S.; Hussain, S.; Chaumeil, J.C.; Charrueau, C. Increased intestinal delivery of viable *Saccharomyces boulardii* by encapsulation in microspheres. *Pharm. Res.* **2008**, *25*, 1290–1296. [CrossRef]
29. Brodkorb, A.; Egger, L.; Alminger, M.; Alvito, P.; Assunção, R.; Ballance, S.; Bohn, T.; Bourlieu-Lacanal, C.; Boutrou, R.; Carrière, F.; et al. INFOGEST static in vitro simulation of gastrointestinal food digestion. *Nat. Protoc.* **2019**, *14*, 991–1014. [CrossRef]
30. Minekus, M.; Alminger, M.; Alvito, P.; Ballance, S.; Bohn, T.; Bourlieu, C.; Carriere, F.; Boutrou, R.; Corredig, M.; Dupont, D.; et al. A standardised static in vitro digestion method suitable for food—An international consensus. *Food Funct.* **2014**, *5*, 1113–1124. [CrossRef]
31. Labuschagne, P. Impact of wall material physicochemical characteristics on the stability of encapsulated phytochemicals: A review. *Food Res. Int.* **2018**, *107*, 227–247. [CrossRef]
32. Yoha, K.S.; Nida, S.; Dutta, S.; Moses, J.A.; Anandharamakrishnan, C. Targeted Delivery of Probiotics: Perspectives on Research and Commercialization. *Probiotics Antimicrob. Proteins* **2021**, *14*, 15–48. [CrossRef]
33. Chew, S.; Nyam, K. Microencapsulation of kenaf seed oil by co-extrusion technology. *J. Food Eng.* **2016**, *175*, 43–50. [CrossRef]
34. Aguirre-Calvo, T.R.; Aguirre-Calvo, D.; Perullini, M.; Santagapita, P.R. A detailed microstructural and multiple responses analysis through blocking design to produce Ca(II)-alginate beads loaded with bioactive compounds extracted from by-products. *Food Hydrocoll. Health* **2021**, *1*, 100030. [CrossRef]

35. Głąb, T.K.; Boratyński, J. Potential of Casein as a Carrier for Biologically Active Agents. *Top. Curr. Chem.* **2017**, *375*, 71. [CrossRef] [PubMed]
36. Hébrard, G.; Hoffart, V.; Beyssac, E.; Cardot, J.M.; Alric, M.; Subirade, M. Coated whey protein/alginate microparticles as oral controlled delivery systems for probiotic yeast. *J. Microencapsul.* **2010**, *27*, 292–302. [CrossRef] [PubMed]
37. Dehkordi, S.S.; Alemzadeh, I.; Vaziri, A.S.; Vossoughi, A. Optimization of Alginate-Whey Protein Isolate Microcapsules for Survivability and Release Behavior of Probiotic Bacteria. *Appl. Biochem. Biotechnol.* **2020**, *190*, 182–196. [CrossRef]
38. Donthidi, A.R.; Tester, R.F.; Aidoo, K.E. Effect of lecithin and starch on alginate-encapsulated probiotic bacteria. *J. Microencapsul.* **2010**, *27*, 67–77. [CrossRef]
39. Shi, L.; Li, Z.; Li, D.; Xu, M.; Chen, H.; Zhang, Z.; Tang, Z. Encapsulation of probiotic Lactobacillus bulgaricus in alginate—Milk microspheres and evaluation of the survival in simulated gastrointestinal conditions. *J. Food Eng.* **2013**, *117*, 99–104. [CrossRef]
40. Li, X.Y.; Chen, X.G.; Cha, D.S.; Park, H.J.; Liu, C.S. Microencapsulation of a probiotic bacteria with alginategelatin and its properties. *J. Microencapsul.* **2009**, *26*, 315–324. [CrossRef]
41. Khalil, K.A.; Mustafa, S.; Mohammad, R.; Bin Ariff, A.; Ahmad, S.A.; Dahalan, F.A.; Manap, M.Y.A. Encapsulation of *Bifidobacterium pseudocatenulatum* Strain G4 within Bovine Gelatin-Genipin-Sodium Alginate Combinations: Optimisation Approach Using Face Central Composition Design-Response Surface Methodology (FCCD-RSM). *Int. J. Microbiol.* **2019**, *2019*, 4208986. [CrossRef]
42. Aguirre-Calvo, T.R.; Molino, S.; Perullini, M.; Rufián-Henares, J.Á.; Santagapita, P.R. Effect of in vitro digestion-fermentation of Ca(II)-alginate beads containing sugar and biopolymers over global antioxidant response and short chain fatty acids production. *Food Chem.* **2020**, *333*, 127483. [CrossRef]
43. Santagapita, P.R.; Mazzobre, M.F.; Buera, P. Invertase stability in alginate beads Effect of trehalose and chitosan inclusion and of drying methods. *FRIN* **2012**, *47*, 321–330. [CrossRef]
44. Hernández-Ledesma, B.; Ramos, M.; Gómez-Ruiz, J.Á. Bioactive components of ovine and caprine cheese whey. *Small Rumin. Res.* **2011**, *101*, 196–204. [CrossRef]
45. Corfield, R.; Martínez, K.D.; Allievi, M.C.; Santagapita, P.; Mazzobre, F.; Schebor, C.; Pérez, O.E. Whey proteins-folic acid complexes: Formation, isolation and bioavailability in a *Lactobacillus casei* model. *Food Struct.* **2020**, *26*, 100162. [CrossRef]
46. Smidsrød, O.; Larsen, B.; Painter, T.; Haug, A.R. The role of intramolecular autocatalysis in the acid hydrolysis of polysaccharides containing 1,4-linked hexuronic acid. *Acta Chem. Scand.* **1969**, *23*, 1573–1580. [CrossRef]
47. Wolkers, W.F.; Oliver, A.E.; Tablin, F.; Crowe, J.H. A Fourier-transform infrared spectroscopy study of sugar glasses. *Carbohydr. Res.* **2004**, *339*, 1077–1085. [CrossRef] [PubMed]
48. Prasanna, P.H.P.; Charalampopoulos, D. Food Bioscience Encapsulation of Bi fi dobacterium longum in alginate-dairy matrices and survival in simulated gastrointestinal conditions, refrigeration, cow milk and goat milk. *Food Biosci.* **2018**, *21*, 72–79. [CrossRef]
49. Lopes, S.; Bueno, L.; Júnior, F.D.A.; Finkler, C. Preparation and characterization of alginate and gelatin microcapsules containing *Lactobacillus rhamnosus*. *An. Acad. Bras. Cienc.* **2017**, *89*, 1601–1613. [CrossRef]
50. Parker, E.A.; Roy, T.; D'Adamo, C.R.; Wieland, L.S. Probiotics and gastrointestinal conditions: An overview of evidence from the Cochrane Collaboration. *Nutrition* **2018**, *45*, 125–134.e11. [CrossRef]
51. Beldarrain Iznaga, T.; Villalobos Carvajal, R.; Leiva Vega, J.; Sevillano Armesto, E. Food and Bioproducts Processing Influence of multilayer microencapsulation on the viability of Lactobacillus casei using a combined double emulsion and ionic gelation approach. *Food Bioprod. Process.* **2020**, *124*, 57–71. [CrossRef]
52. De Prisco, A.; van Valenberg, H.J.F.; Fogliano, V.; Mauriello, G. Microencapsulated Starter Culture During Yoghurt Manufacturing, Effect on Technological Features. *Food Bioprocess Technol.* **2017**, *10*, 1767–1777. [CrossRef]
53. Ramos, P.E.; Abrunhosa, L.; Pinheiro, A.; Cerqueira, M.A.; Motta, C.; Castanheira, I.; Chandra-Hioe, M.V.; Arcot, J.; Teixeira, J.A.; Vicente, A.A. Probiotic-loaded microcapsule system for human in situ folate production: Encapsulation and system validation. *Food Res. Int.* **2016**, *90*, 25–32. [CrossRef]
54. Ta, L.P.; Bujna, E.; Antal, O.; Ladányi, M.; Juhász, R.; Szécsi, A.; Kun, S.; Sudheer, S.; Gupta, V.K.; Nguyen, Q.D. Effects of various polysaccharides (alginate, carrageenan, gums, chitosan) and their combination with prebiotic saccharides (resistant starch, lactosucrose, lactulose) on the encapsulation of probiotic bacteria *Lactobacillus casei* 01 strain. *Int. J. Biol. Macromol.* **2021**, *183*, 1136–1144. [CrossRef] [PubMed]
55. Eckert, C.; Agnol, W.D.; Dallé, D.; Serpa, V.G.; Maciel, M.J.; Lehn, D.N.; Volken de Souza, C.F. Development of alginate-pectin microparticles with dairy whey using vibration technology: Effects of matrix composition on the protection of *Lactobacillus* spp. from adverse conditions. *Food Res. Int.* **2018**, *113*, 65–73. [CrossRef] [PubMed]

Disclaimer/Publisher's Note: The statements, opinions and data contained in all publications are solely those of the individual author(s) and contributor(s) and not of MDPI and/or the editor(s). MDPI and/or the editor(s) disclaim responsibility for any injury to people or property resulting from any ideas, methods, instructions or products referred to in the content.

Article

Quantifying the Residual Stiffness of Concrete Beams with Polymeric Reinforcement under Repeated Loads

Haji Akbar Sultani, Aleksandr Sokolov, Arvydas Rimkus and Viktor Gribniak *

Laboratory of Innovative Building Structures, Vilnius Gediminas Technical University (VILNIUS TECH), Sauletekio av. 11, LT-10223 Vilnius, Lithuania; haji-akbar.sultani@vilniustech.lt (H.A.S.); aleksandr.sokolov@vilniustech.lt (A.S.); arvydas.rimkus@vilniustech.lt (A.R.)
* Correspondence: viktor.gribniak@vilniustech.lt; Tel.: +370-6-134-6759

Abstract: Current technology development ensures a variety of advanced materials and options for reinforcing concrete structures. However, the absence of a uniform testing methodology complicates the quantification and comparative analysis of the mechanical performance of the composite systems. The repeated mechanical loads further complicate the issue. This research extends the recently developed residual stiffness assessment concept to the repeated loading case. It provides an engineer with a simplified testing layout and analytical model to quantify the residual flexural stiffness of standardized laboratory specimens subjected to repeated cycling loads. This model explicitly relates the particular moment and curvature values, requiring neither iterative calculations nor the load history. Thus, this feature allows residual stiffness quantification under repeated loading conditions, including complete reloading of the beam samples imitating the structural strengthening procedure; the proposed technique is equally efficient in quantifying the residual stiffness of the beam samples with any combinations of fiber-reinforced polymer (FRP) reinforcements, i.e., embedded bars, near-surface-mounted strips, and externally bonded sheets. This study employs 12 flexural elements with various reinforcement and loading layouts to illustrate the proposed methodology's efficiency in quantifying the residual strength of the tension concrete, which estimates the efficiency of the reinforcement system. The explicit quantifying of the residual resistance of the FRP reinforcement systems under repeated load cycles describes the essential novelty of this work.

Keywords: reinforced concrete; fiber-reinforced polymer (FRP); bending test; analytical model; residual strength; repeated loads

Citation: Sultani, H.A.; Sokolov, A.; Rimkus, A.; Gribniak, V. Quantifying the Residual Stiffness of Concrete Beams with Polymeric Reinforcement under Repeated Loads. *Polymers* **2023**, *15*, 3393. https://doi.org/10.3390/polym15163393

Academic Editors: Giorgio Luciano and Maurizio Vignolo

Received: 14 July 2023
Revised: 5 August 2023
Accepted: 11 August 2023
Published: 13 August 2023

Copyright: © 2023 by the authors. Licensee MDPI, Basel, Switzerland. This article is an open access article distributed under the terms and conditions of the Creative Commons Attribution (CC BY) license (https://creativecommons.org/licenses/by/4.0/).

1. Introduction

1.1. Literature Review

Technology development provides various materials for reinforcing and strengthening concrete structures [1,2]. Fiber-reinforced polymers (FRPs) are a promising steel alternative because they are high-strength, lightweight, immune to corrosion, and electromagnetically transparent [3–6]. Still, the relatively low resistance to ultraviolet radiation, elevated temperatures, and humidity reduce the mechanical performance of FRP materials with time [7–9]; the cycling loads complicate the issue, reducing the mechanical resistance of the reinforcement systems even more [10,11]. Regarding FRP bars, most experimental works consider the pull-out behavior under cyclic loads, e.g., Mohamed et al. [12], Kim and Lee [13], Liu X. et al. [14], Shen et al. [15], Xiao et al. [16], Shi et al. [17], and Pan et al. [18], which could transform into the structural analysis. However, this "transformation" describes a non-trivial problem mainly related to the local bond treatment. For instance, in the tests [12–18], the bond length varied from 60 mm to 130 mm for the bar diameters 10 mm, 12 mm, 13 mm, and 14 mm. Notwithstanding the apparent efficiency in estimating the bond performance, these test results are unrepresentative of structural elements, which do not face such severe bond deformations in the service conditions [19]. Still, these tests isolate a single bar behavior, which is unrealistic in most structural applications.

Various FRP materials, i.e., embedded bars, near-surface mounted (NSM) strips, and externally bonded reinforcement (EBR) sheets, exhibit diverse mechanical performances when reinforcing composite systems. The literature has extensively documented these performances [20]. Studies by Rimkus et al. [21] and Gribniak et al. [22] have shown that combining FRPs with steel reinforcement effectively addresses engineering challenges. Anas et al. [23] investigated the structural performance of square slab specimens reinforced with different types of FRPs, and they found that samples with CFRP bars exhibited exceptional impact resistance. Using a four-point bending test, Yuan et al. [24] examined the flexural behavior of reinforced concrete (RC) beams strengthened with GFRP tubes. The results demonstrated that the strengthening solution improved the composite element's flexural strength and bending stiffness under investigation. Farahi et al. [25] conducted three-point bending tests on concrete beams reinforced with composite materials, demonstrating such beams' ductility and energy dissipation potential under monotonic and repeated loading. However, the variety of the specimen shapes and loading conditions does not allow comparing the test outcomes to optimize the reinforcement parameters. For instance, Godat et al. [26] concluded that the effective axial strains in composite reinforcement are higher in smaller specimens, while larger samples exhibit lower deformation in the FRP component. Therefore, a unified testing procedure is necessary to compare different reinforcement systems adequately.

Concrete structures often experience cyclic rather than static loading in practical engineering applications. Li et al. [27] investigated the influence of concrete strength and reinforcement parameters on the flexural behavior and ductility of concrete beams reinforced with basalt FRP bars under repeated loading. The experimental results showed a reduction in the peak load of the beams with an increase in the number of loading cycles at a constant deflection. Previous research by Kargaran and Kheyroddin [28] and Sultani et al. [29] supports the notion that repeated mechanical loading introduces complexities in the behavior of structural composites. Therefore, it is crucial to examine the structural performance of concrete beams reinforced with FRP bars under cyclic loading to advance the development of composite reinforcement in concrete structures.

An alternative approach to studying the flexural behavior of RC specimens under various loading conditions is using the acoustic emission method. Mat Saliah and Md Nor [30] extensively investigated this method in the context of assessing the structural integrity of concrete beams reinforced with composite materials. This technique evaluates microscopic damage within concrete elements caused by different external loading conditions. However, its predictive capability for the structural performance of the investigated specimens is limited. Another approach is to predict flexural behavior using numerical simulations. For example, Sun et al. [31] numerically studied the static and dynamic performances of a concrete beam reinforced with FRP bars using a simplified spectral model. This numerical approach shows promise for dynamic analysis by considering changes in dynamic characteristics. However, it is not suitable for static analysis.

The absence of a standardized methodology for quantifying reinforcement efficiency in composite systems subjected to repeated loads motivated this study. Several studies have examined the behavior of RC beams with composite reinforcement under repeated loading. For instance, Fathuldeen and Qissab [32] investigated the mechanical performance of RC beams strengthened with a CFRP NSM system under low cycle repeated loads. This study proved the efficiency of the hybrid reinforcement systems in resisting repeated load, highlighting the optimization importance of the steel and CFRP reinforcement proportions. These outcomes align with the monotonic loading test results and conclusions [21]. Zhu et al. [33] investigated the fiber effect in improving the mechanical resistance of RC beams with steel and FRP bars under repeated loads. The tests revealed the fiber efficiency in improving the concrete's tensile resistance and compression ductility. These results also agree with the monotonic test outcomes [34]. The research in [33] also observed a substantial decrease in the flexural strength (up to 90%) of the beams with FRP reinforcement after increasing the load intensity. Song et al. [35] identified the adverse effect of cycling loads on the ultimate resistance of CFRP bars used in concrete frames in combination with steel reinforcement. However, the repeated loading effect on the residual stiffness of RC elements, which is predominant for normal (service) structural conditions, lacks adequate attention in the literature.

1.2. The Proposed Standardized Analysis Concept

Unlike the structural analysis design, the proposed procedure focuses on the mechanical performance of composite materials, particularly reinforcement systems. As described in the previous section, recent achievements in engineering, developing advanced fibers and reinforcement materials (steel and non-metallic), and other reinforcement solutions caused problems in setting an efficient solution for a particular situation. Furthermore, the combination of fibers and continuous reinforcement (either embedded bars, near-surface-mounted strips, or externally bonded laminates) in hybrid systems (e.g., [10,11]) complicates the reinforcement efficiency analysis.

As mentioned in Section 1.1 and reported in references [4,36], the absence of a uniform testing methodology complicates the comparative analysis of the efficiency of alternative materials and reinforcement layouts reported in the literature. Furthermore, the analysis of full-scale objects is possible but relevant only for the verification of particular solutions. However, it is unpractically expensive for studies involving a variety of potentially feasible solutions.

Therefore, Gribniak et al. [20,22] developed a "standardized" testing procedure to ensure the comparative analysis and simulation of various possible loading situations to select several feasible solutions for further investigation, e.g., full-scale tests. The monotonic loading tests verified the adequacy and reliability of the developed testing layout, chosen geometry of the sample, and analytical model, which explicitly quantifies the residual stiffness of the reinforced specimen expressed in terms of equivalent stresses in the tension concrete. Reaching zero, the latter value corresponds to the total loss of the bonding performance of the reinforcement, also known as the composite action or tension-stiffening effect [37]. Thus, the term "standardized" describes the peculiar geometry of the laboratory samples and testing methodology developed to satisfy the simplified modeling assumption (i.e., the rectangular distribution of stresses in the concrete in tension). In other words, the "exact" average stress–average strain tension-stiffening diagram [38] has a rectangular shape close to the rectangular approximation assumed in this study. Thus, the equivalent stress–strain relationships do not represent analytical material models but a quantitative estimate of the tension-stiffening effect.

At the same time, the comparative analysis of the equivalent stress–strain relationships of alternative reinforcement systems can estimate the improvement of the reference solution. For instance, Gribniak and Sokolov [36] evaluated the fiber-bridging effect in the presence of bar reinforcement by comparing the plain concrete and fiber-reinforced concrete "standardized" samples. The obtained difference between equivalent stress–strain relationships determined the material model suitable for the numerical simulation of the fiber-bridging effect in full-scale beams. The considered small-size samples can also verify finite element models representing a peculiar geometry case [39]. However, notwithstanding the apparent benefits of laboratory testing, reducing the length of the "standardized" samples can cause reinforcement anchorage problems [17], which require particular care in designing the test programs.

Jakubovskis et al. [40] transformed this approach to quantify the bacterial healing effect in RC samples. It also can be extended to determining the residual strength of the fiber-reinforced concrete with bar reinforcement [36]. Furthermore, this testing procedure ensures the residual stiffness analysis of elements subjected to temperature (including an open environment), ultraviolet radiation, and long-term (creep) effects [41]. In other words, the proposed tool helps quantify the reinforced material performance in composite systems. The term "composite" determines concrete with various combinations of reinforcements (e.g., steel and non-metallic fibers, embedded bars, near-surface-mounted strips, and externally bonded reinforcement in different combinations).

This study adapts the proposed testing layout and the simplified analytical model to analyze the residual flexural stiffness of the laboratory specimens subjected to repeated mechanical loads. It employs the concept of the equivalent residual stresses acting on concrete in tension to measure the structural performance of alternative composite reinforcement systems and their variation with the load cycles of various intensities. This analytical model explicitly relates the bending moment and curvature values, quantifying the equivalent stresses acting in the concrete under the assumption of the rectangular stress distribution. This solution requires neither iterative calculations nor the load history definition.

At the same time, the simplified analytical model [22] does not define concrete's constitutive law in its traditional sense. The simplified nature of the model only ensures approximating the equivalent stresses with a sufficient degree of accuracy—from the analysis [22], the average approximation error regarding the exact solution [38] does not exceed 7%. However, the latter inverse analysis procedure is inapplicable to the specimens under cycled loads, making the proposed simplified analysis concept irreplaceable for analyzing the repeated load effects. This experimental study employs the flexural test results of 12 "standardized" beam samples with different arrangements and combinations of FRP reinforcement. This manuscript provides only an illustrative example of the proposed analysis procedure and tends to cover only some possible structural situations and loading conditions. The explicit quantifying of the residual resistance of the composite reinforcement systems under repeated load cycles describes this study's essential novelty.

2. Testing Method and Analytical Model

The explicit nature of the analytical expressions, which do not require any additional modifications regarding the previous publications [20,22,29], ensures the residual stiffness analysis of elements facing the load repetitions. Therefore, Section 2.1 briefly describes the simplified analytical model (Figure 1); reference [22] provides further explanations and the model verification results. Still, the unloading repetitions generate residual (permanent) deformations, which are mandatory for the proposed stiffness analysis procedure, as Section 2.2 discusses in detail. Section 2.3 describes the experimental campaign, illustrating the analysis technique.

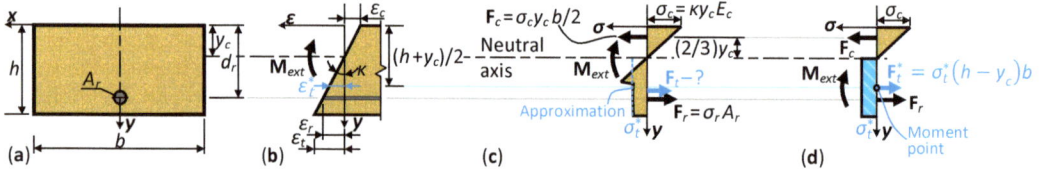

Figure 1. Analytical model of the stress distribution in the pure-bending zone [22]: (**a**) a transformed cross-section; (**b**) the bending-induced strain distribution along the cross-section height; (**c**) the corresponding stress distribution; (**d**) the equivalent stress approximation in the tension concrete zone. Note: the blue color shows the approximation procedure.

2.1. Analytical Model

The model (Figure 1) considers the transformed reinforcement approach, which combines different materials in one component (Figure 1a). The following equations define the transformed reinforcement characteristics:

$$d_r = \sum_{i=1}^{n} E_i A_i d_i / \sum_{i=1}^{n} E_i A_i, \quad A_r = \frac{1}{E_r}\sum_{i=1}^{n} E_i A_i, \quad E_r = E_1, \qquad (1)$$

where d_r is the effective depth; A_r is the cross-section area; E_r is the modulus of elasticity; n is the number of the different reinforcement parts; E_i, A_i, and d_i are the modulus of elasticity, cross-section area, and effective depth of the i-th reinforcement component.

The proposed analytical model uses the following assumptions:

- The strain distribution follows the Euler–Bernoulli hypothesis (Figure 1b). Numerous literature sources proved the adequacy of this assumption for RC members (e.g., [42,43]).
- The smeared crack model describes the stress–strain behavior of the tension concrete (Figure 1b,c). Various literature examples (e.g., [37,38,43]) proved the correctness of this modeling concept.
- Idealized elastic material laws define the mechanical behavior of the reinforcement and the compressed concrete. This modeling approach substantially simplifies the mathematical expressions and ensures a straightforward solution.
- The rectangular distribution of the tensile stresses in concrete defines the equivalent stress σ_t^* (Figure 1d). This center simplification ensures formulating the exact relationship between the bending moment and curvature and avoiding iterative solutions.

The latter two assumptions allow quantifying the stiffness of the beams subjected to repeated loads with the solution expressed in terms of the equivalent tensile stresses. The following equilibrium equations of internal forces and bending moments for the centroid of the equivalent stress diagram (Figure 1d) define the analytical model:

$$\mathbf{F}_t^* + \mathbf{F}_r - \mathbf{F}_c = 0; \quad \mathbf{F}_r\left(d_r - \frac{h+y_c}{2}\right) + \mathbf{F}_c\left(\frac{h+y_c}{2} - \frac{y_c}{3}\right) - \mathbf{M}_{ext} = 0, \quad (2)$$

where $\mathbf{F}_t^*$ is the equivalent resultant force in the tensile concrete; $\mathbf{F}_r$ and $\mathbf{F}_c$ are the internal forces acting on the tensile reinforcement and the compressed concrete; d_r is the efficient depth of the transformed reinforcement (Equation 1); h and y_c are the height and gravity center coordinate of the cross-section in Figure 1. The model also allows the efficient depth to exceed the cross-section height (i.e., the $d_r > h$ condition is acceptable).

The above equation system relates the internal forces and stresses acting on the cross-section, employing the strain compatibility condition (Figure 1b). Reference [22] defines the intermediate solution steps; the final equation describes the third-order polynomial of the neutral axis coordinate y_c for the particular curvature κ and bending moment $\mathbf{M}_{ext}$:

$$\sum_{i=0}^{3} \mathcal{K}_i y_c^i = 0, \quad (3a)$$

with coefficients

$$\mathcal{K}_3 = \frac{E_c b}{6 E_r A_r}, \quad \mathcal{K}_2 = 1 + \frac{E_c b h}{2 E_r A_r}, \quad \mathcal{K}_1 = h - 3d_r, \quad \mathcal{K}_0 = 2d_r^2 - hd_r - \frac{2\mathbf{M}_{ext}}{\kappa E_r A_r}, \quad (3b)$$

where E_c is the modulus of elasticity of concrete; h and b are the height and width of the cross-section in Figure 1; κ is the average curvature of the pure-bending zone subjected to the external bending moment $\mathbf{M}_{ext}$; Equation 1 describes the remaining parameters.

The above polynomial has three roots, and the $[0 < y_c \leq h]$ condition describes the neutral axis position:

$$y_c = \frac{2\mathcal{K}}{3\mathcal{K}_3}\cos\left(\frac{1}{3}\cos^{-1}\left[-\frac{13.5\mathcal{K}_3^2 \mathcal{K}_0 - 4.5\mathcal{K}_3 \mathcal{K}_2 \mathcal{K}_1 + \mathcal{K}_2^3}{\mathcal{K}^3}\right]\right) - \frac{\mathcal{K}_2}{3\mathcal{K}_3}, \quad \mathcal{K} = \sqrt{\mathcal{K}_2^2 - 3\mathcal{K}_3 \mathcal{K}_1}. \quad (4)$$

The following explicit formulas define the equivalent stress (Figure 1d) and average strain of the concrete in tension (Figure 1b):

$$\sigma_t^* = \kappa \frac{y_c^2 b E_c - 2(d_r - y_c) E_r A_r}{2(h - y_c)b}, \quad \varepsilon_t^* = 0.5\kappa(h - y_c). \quad (5)$$

The above expressions solve the residual strength problem independently of the loading conditions, explicitly relating the equivalent stress σ_t^* with the bending moment M_{ext}, average curvature in the pure-bending zone κ, cross-section geometry (Figure 1a), and material parameters discussed in this section above. Thus, the proposed analytical model becomes applicable for quantifying the residual mechanical performance of the composite reinforcement system (expressed in terms of the equivalent stress σ_t^* and equivalent strain ε_t^* relationship) under repeated loading conditions considered in this work.

2.2. Sample Geometry and Testing Layout

Gribniak et al. [20,22] and Sultani et al. [29] established the geometry and the testing layout of the bending samples, following the formation of multiple cracks in a relatively small laboratory sample and thus reducing the discrete cracking effect on the curvature estimation result. Therefore, this study considers the 1000 mm slab-shaped beams loaded in the four-point scheme, as Figure 2a shows. The 200 × 100 mm cross-section was reinforced with near-surface mounted (NSM) strips, embedded bars, and externally bonded reinforcement (EBR) sheets in various combinations. This investigation includes 12 standardized beam samples to illustrate applying the proposed analysis procedure.

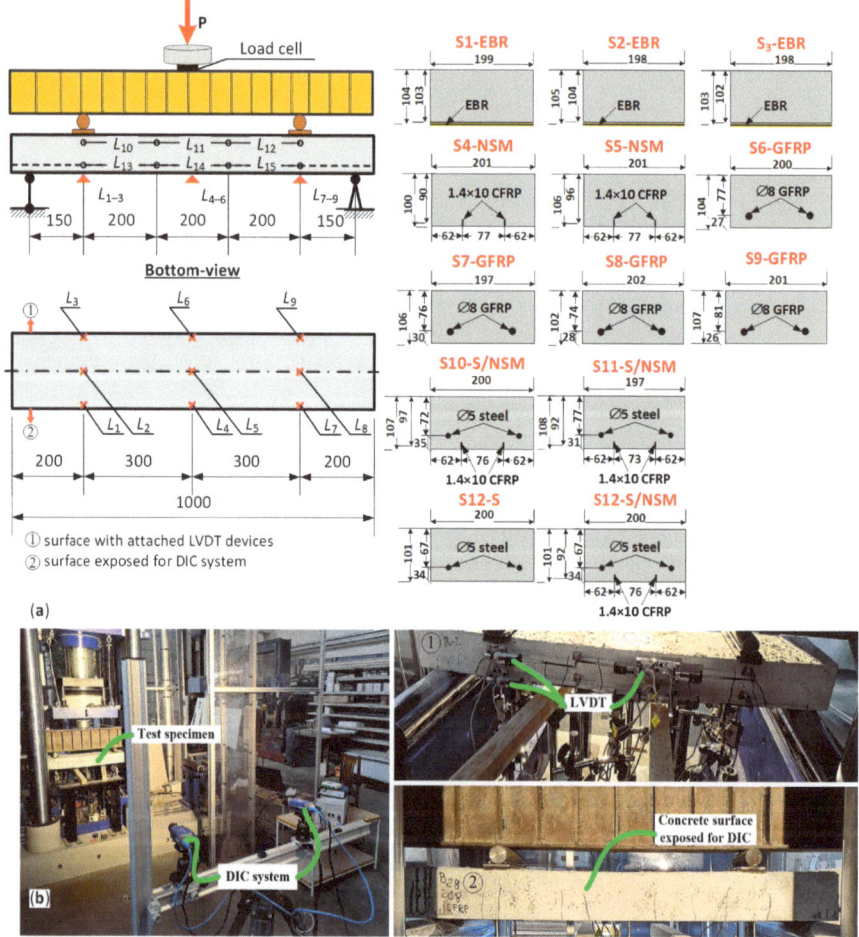

Figure 2. Test specimens (dimensions are in mm): (**a**) loading scheme; (**b**) experimental view.

The verification capability of the deformation monitoring results defined the second condition for developing the testing setup shown in Figure 2a. This study employs three independent groups of monitoring devices of the pure-bending zone: two sets of linear variable displacement transducers (LVDTs) capture the vertical and longitudinal displacements, and a digital image correlation (DIC) system monitors the cracking process and surface deformations; the LVDTs and DIC system monitor deformations of the opposite sides denoted as "①" and "②" in Figure 2b. The LVDT devices L_{10}–L_{15} estimate deformations of the side surface ①, and L_1–L_3, L_4–L_6, and L_7–L_9 indicators monitor the vertical displacements. The DIC captures the deformations and crack patterns on the side surface, designated as "②". Figure 2b illustrates the DIC setup, employing two LaVision VC-Image E-Lite 5M cameras placed on a tripod 3.0 m from the exposition samples and 0.5 m apart. The cameras with charge-coupled device (CCD) detectors have a 2456 × 2085 pixel resolution and operate at 12.2 frames per second. The lighting equipment Arri ensures the quality and accuracy of the digital images.

Table 1 describes the beam specimens' geometry and material properties. In this table, h and b represent the height and width of the cross-section in Figure 1a; d, A, and E are the reinforcement parameters (effective depth, cross-section area, and modulus of elasticity); f_t describes the tensile strength of the reinforcement, and the subscripts "1" and "2" correspond to the transformed reinforcement components in Equation 1. Table 1 also shows the compressive strength of the standard ⌀150 × 300 mm concrete cylinder (f'_c) on the testing day; the column "Age" specifies the testing age. The column "f'_c" specifies the average value and standard deviation determined for four identical cylinders; the remaining parameters of the concrete necessary for the analysis (i.e., the modulus of elasticity and tensile strength) were determined using the average compressive strength values (Table 1) and the Eurocode 2 formulas [44]. Gribniak et al. [45,46] proved the adequacy of such an approach for the numerical analysis of RC elements. The manufacturers reported mechanical properties and cross-section parameters of the polymeric reinforcement materials for EBR, NSM, and GFRP systems; the tensile tests of three identical samples [47] determined the characteristics of steel reinforcement listed in Table 1.

Table 1. Geometry and material parameters of the beam samples.

Beam	h	b	d_1	d_2	A_1	A_2	E_1	E_2	$f_{t,1}$	$f_{t,2}$	$n\rho$	f'_c	Age
	(mm)				(mm²)		(GPa)		(MPa)		(%)	(MPa)	(Days)
S1-EBR	103	199	104	–	32.8	–	230	–	4830	–	1.03	50.69 ± 1.53	47
S2-EBR	104	198	105	–	32.8	–	230	–	4830	–	1.13	35.12 ± 2.63	22
S3-EBR	102	198	103	–	32.8	–	230	–	4830	–	1.18	32.98 ± 3.15	21
S4-NSM	100	201	90	–	28.0	–	170	–	2800	–	0.83	34.12 ± 2.48	21
S5-NSM	106	201	96	–	28.0	–	170	–	2800	–	0.78	34.62 ± 2.92	21
S6-GFRP	104	200	77	–	100.6	–	60	–	1490	–	1.22	35.12 ± 2.61	22
S7-GFRP	106	197	76	–	100.6	–	60	–	1490	–	1.22	34.62 ± 2.48	21
S8-GFRP	102	202	74	–	100.6	–	60	–	1490	–	1.11	40.80 ± 1.63	13
S9-GFRP	107	201	81	–	100.6	–	60	–	1490	–	1.26	40.80 ± 1.63	13
S10-S/NSM	107	200	72	97	38.11	28	206	170	503.9	2800	2.45	32.98 ± 3.15	22
S11-S/NSM	108	197	77	92	38.11	28	206	170	503.9	2800	2.30	40.80 ± 1.63	14
S12-S	101	200	67	–	38.11	–	206	–	503.9	–	1.83	34.12 ± 2.92	19
S12-S/NSM	101	200	67	92	38.11	28	206	170	503.9	2800	2.45	35.23 ± 3.05	26

For comparison purposes of the composite reinforcement systems, Table 1 also includes the transformed modular ratio and reinforcement ratio product [34]:

$$n\rho = \frac{E_r}{E_c} \cdot \frac{A_r}{bd_r}. \qquad (6)$$

Figure 2b shows the loading setup. A servo-hydraulic testing machine LVF5000 with a 5 MN capacity (WALTER + BAI AG., Löhningen, Switzerland) loads the test samples with a 0.4 mm/min velocity; DION 7 software ensures the loading system control and data acquisition process. The load cycles are formed in a semi-automatic manner with manual load reduction to the minimum cycle boundary. The data logger ALMEMO 5690-2 collects the load cell and LVDT outputs every second.

This study focuses on the stiffness decrease under repeated mechanical loads. Still, the analytical model (Section 2.1) does not require load history specification—it explicitly relates the bending moment and curvature values in the pure-bending zone. Thus, in any combination, it remains equally efficient for high cycle and repeated loads, including temperature, creep, aggressive chemicals, and mechanical loads. The considered load cycles are essential for analyzing repeated factors when the mechanical load is necessary to estimate the residual stiffness of the beam sample, e.g., after the harsh environmental impacts. Figure 3 schematically depicts the loading application cycles—each loading stage consists of five load cycles with a 15% fluctuation about the target load referred to as the service moment, M_{ser}. These load cycling numbers were set arbitrarily in this study to avoid measurement errors and provide several data points for averaging.

Table 2. Loading parameters of the beam samples (kNm).

Beam	Load Type	M_{min}	M_{max}	M_{ser}	M_{ult}
S1-EBR	A	-	-	3.710	7.035
S2-EBR	B	3.075	4.125	3.710	≡S1(*)
S3-EBR	B	3.075	4.125	3.710	≡S1
S4-NSM	A	-	-	2.220	4.005
S5-NSM	B	1.875	2.475	2.220	≡S4
S6-GFRP	B	2.850	3.750	3.375	6.527(†)
S7-GFRP	B	2.850	3.750	3.375	≡S6
S8-GFRP	C	2.175	2.975	2.625	≡S6
		2.850	3.750	3.375	
S9-GFRP	C	2.175	2.975	2.625	≡S6
		2.850	3.750	3.375	
S10-S/NSM	B	1.875	2.475	2.220	≡S4
S11-S/NSM	C	0.975	1.350	1.215	1.42(‡)
		1.875	2.475	2.220	≡S4
		3.075	4.125	3.710	≡S1
S12-S	B	0.975	1.350	1.215	1.42(‡)
S12-S/NSM	C	0.975	1.350	1.215	1.42(‡)
		1.875	2.475	2.212	≡S4
		3.075	4.125	3.710	≡S1

Note: (*) the sign "≡" means "equivalent to"; (†) Gribniak et al. [20] tested the identical element until failure under monotonic load; (‡) the theoretical moment corresponding to the steel yielding limits the upper cycle boundary.

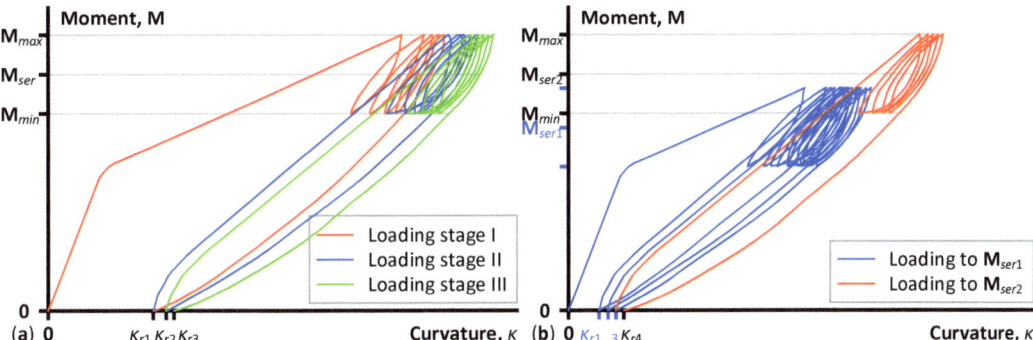

Figure 3. Loading protocols: (**a**) cycling loads over the service moment (load type "B", Table 2); (**b**) different intensity loads (type "C", Table 2).

During the physical tests, the loading process starts automatically at 0.4 mm/min until the maximum load of the cycle M_{max} (exceeding M_{ser} by 15%). Subsequently, the operator interrupts the loading process when it reaches the upper boundary M_{max}, manually reducing the load to the minimum cycle boundary M_{min} (corresponding to approximately 85% of M_{ser}). After that, the bending moment rises again under computer control to the maximum cycle load M_{max}, continuing the loading cycles. The complete unloading finalizes the first loading stage, resulting in the residual deformation expressed as curvature κ_{r1}. Typically, this loading process is repeated three times (with five load cycles in each repetition) with the complete unloading of the beam samples, investigating the residual deformation trends. Each subsequent loading stage starts at the same speed as the previous loading round. However, the curvature analysis accounts for the residual deformations from all past loading stages (κ_{r1}, κ_{r2}, κ_{r3} ...), shifting the curvature diagrams as shown in Figure 3. To determine the loading conditions in terms of the ultimate load M_{ult}, Table 2 uses the following notation:

- The sign "≡" relates the loading conditions to the particular beam sample. For instance, the symbol "≡S1" refers to the loading condition of the S1-EBR beam, determining the exact service moment (M_{ser}) and the cycle boundaries (M_{min} and M_{max}).
- Gribniak et al. [20] tested the identical element to the S6-GFRP sample until failure under monotonic load, which determines the loading conditions in this study.
- In the element with steel reinforcement (S12-S), the theoretical moment, corresponding to the steel yielding, limits the ultimate cycle load M_{max}; the service load M_{ser} was set to exceed the cracking moment calculated by Eurocode 2 formulas [44].
- The target loadings of the elements, combining steel and NSM reinforcements, were set to represent the loading conditions of alternative test samples (S1-EBR, S4-NSM, and S12-S) for comparison purposes.

Figure 3 schematically depicts the loading layouts "B" and "C" from Table 2. The monotonic tests (type "A") or previous experimental program [20] determined the service load, representing approximately 55% of the load-bearing capacity of nominally identical samples, tested under monotonic load until failure (except for the S12-S sample, whose loading parameters are discussed above). This limitation came from the design principles of steel reinforcement when the efficient design of the bending sample ensures 50–60% stresses in the tension reinforcement regarding the yielding strength of the steel (e.g., [37]). Gribniak et al. [45] conducted a detailed statistical analysis of the service load conditions. However, the definition of FRP reinforcement systems' service load is a more complex problem. The necessity to compare the deformation of the elements with steel and GFRP bars defines the typical analysis issue.

Gribniak et al. [46] proposed the analysis methodology, which compares the statistical data corresponding to the identical stresses in the tensile reinforcement. However, this investigation cannot employ the latter approach mainly because of these simplified laboratory samples' limited dimensions and reinforcement detailing restrictions. Therefore, this study makes no difference between the test samples' failure mechanisms determining the ultimate load. Independently of the reinforcement parameters, the service moment represents 55% of the load-bearing capacity of the identical beam sample tested under monotonic load. The limitations of this assumption are evident since a 30% magnitude also appears in the literature (e.g., [18,48]). However, any possible approaches still result in comparison subjectivity, and the efficiency analysis of the reinforcement should account for specific limitations in real projects.

2.3. Experimental Program

The illustrative experimental program employs laboratory-mixed concrete, using the same mix design as the previous studies [20,22,29]. The proportions for a cubic meter are the following: cement CEM I 42.5 R = 356 kg, water = 163 L, limestone powder = 177 kg, 0/4 mm sand = 890 kg, 4/16 mm crushed dolomite aggregates = 801 kg, superplasticizer Mapei Dynamon XTend = 1.97% (by the cement weight), and admixture SCP 1000 Optimiser = 3.5 kg. In addition, two types of synthetic fibers (0.9 kg of CRACKSTOP M ULTRA ⌀0.022 × 13 mm and 4.2 kg of DURUS EASYFINISH ⌀0.7 × 40 mm) were used in the concrete to avoid a sudden failure of the shear zone. All the beams were produced in steel forms and were unmolded 2–3 days after the casting. After that, all samples were stored in laboratory conditions at an average of 73% humidity and 20 °C before the testing day or forming external FRP reinforcement systems.

As Table 1 shows, the test program includes four specimen types. In particular, the abbreviation EBR corresponds to externally bonded reinforcement using carbon fiber (CF) sheets. Thus, two unidirectional MAPEWRAP C UNI-AX sheets with the dry fabric's 100 × 0.164 mm equivalent thickness were attached to the most tensioned surface of the beam samples. These sheets were placed along the beam and attached to the concrete surface using a two-component MAPEWRAP 31 epoxy resin (Figure 4a–d). Before the EBR bonding, the cleaned concrete surface was leveled with epoxy putty and primer. The adhesive was allowed to dry for seven days before conducting the tests.

Figure 4. Installing reinforcement systems: (**a**) CF sheets; (**b**) epoxy adhesive; (**c**) prepared surface of specimens; (**d**) specimen with adhesively bonded CF sheets; (**e**) S&P C-LAMINATE CFRP strip; (**f**) S&P RESIN 220 epoxy adhesive; (**g**) arrangement of the NSM grooves; (**h**) specimen with embedded CFRP strips. Note: the beam samples are shown in an inverted position to clarify the installation details.

The NSM abbreviation in Table 1 describes a near-surface-mounted reinforcement system. The notation S/NSM corresponds to the NSM system formed on the beam sample with two 5 mm embedded steel bars, creating a hybrid reinforcement system. The NSM component consisted of two pultruded 10 × 1.4 mm carbon fiber-reinforced polymer (CFRP) strips (S&P C-Laminate) installed in the 12 × 4 mm grooves milled at the bottom surface of the specimens (Figure 4e–h). Before placing the strips, the grooves were filled with a two-component S&P RESIN 220 epoxy adhesive. Excess epoxy was removed with a spatula to ensure the test samples' surface was even. The adhesive was allowed to cure for seven days before conducting the mechanical tests. Remarkably, the NSM system of the S12-S/NSM sample was formed after testing the S12-S beam with two 5 mm steel bars, simulating an RC structure strengthening consequence.

This study also includes beams with two embedded 8 mm glass fiber-reinforced polymer (GFRP) bars. The GFRP reinforcement system employs COMBAR bars from SHÖCK. Figure 5 shows the surface treatment of all the reinforcement materials in this work.

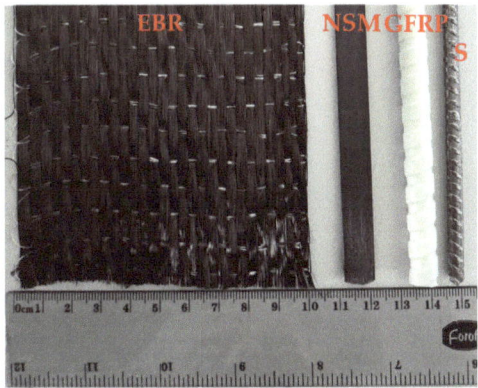

Figure 5. The surface treatment of the reinforcement materials.

3. Results

Figure 6 demonstrates the crack propagation of several selected beam samples during the five load cycles corresponding to succeeding load repetitions (*R1*, *R2*, *R3*, etc.) captured with the DIC system. However, these cracking results could only qualitatively illustrate the composite performance of the reinforcement system in involving the tension concrete resisting the load. Thus, the number of cracks (in the post-cracking stage) could reveal the tension-stiffening effect [49]. Remarkably, the proposed setup produces multiple cracks in the 600 mm long pure-bending zone, which makes it suitable for the tension-stiffening analysis using the smeared crack model, e.g., as described in reference [38].

Gribniak et al. [22] also demonstrated the feasibility of the simplified analytical model (Section 2.1) for the residual strength analysis. Thus, the stiffness analysis employs monitoring results of the pure-bending zone. The indicator distribution scheme (Figure 2) ensures the curvature estimation using three independent measurement sets, i.e., the vertical displacements registered by the L_1–L_9 LVDT devices and the surface deformations captured by DIC and LVDTs. This instrumentation allows cross-verifying the test measurements, which is vital during the cyclic tests because of a certain inertness of the LVDT devices under reversed load and DIC sensitivity to the cameras' movements [20,50]. Gribniak et al. [22] described the curvature analysis procedure in detail; the present study indicated that vertical displacements produce reliable curvature values, defining the analysis object. Following this approach, the averaged curvature over the pure-bending zone can be determined as follows:

$$\kappa = 8\Delta/\left(l^2 + 4\Delta^2\right), \quad \Delta = (u_4 + u_5 + u_6)/3 - (u_1 + u_2 + u_3 + u_7 + u_8 + u_9)/6, \tag{7}$$

where Δ is the deflection over the pure-bending zone having length l equal to 600 mm, as Figure 2a shows; u_i is the displacement obtained by the L_i device ($i = 1\ldots9$).

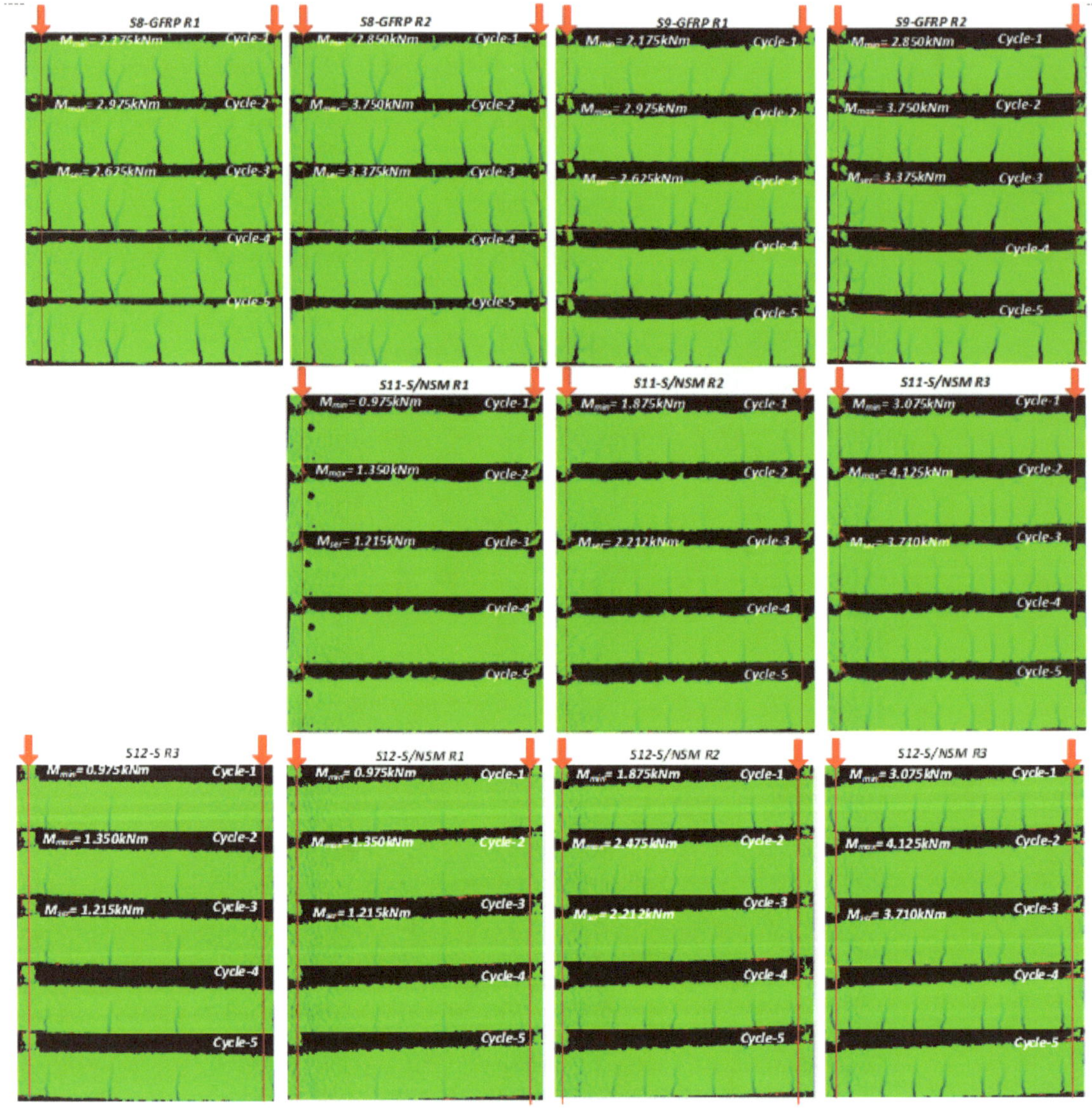

Figure 6. Cracking patterns of the selected samples identified by the DIC system. Note: the red arrows indicate the load application points.

Figure 7 shows the corresponding moment–curvature diagrams of all beam samples from Tables 1 and 2. The sub-charts in this figure have identical ordinate scales, but the abscissa scales in each row differ. This figure also indicates the service moments for the residual stiffness comparison. However, this analysis is barely possible straightforwardly because of the differences in the modulus of elasticity of reinforcement and geometry characteristics (Table 1). Therefore, this study uses the equivalent stress (σ_t^*) to measure and compare the stiffness decay with the load.

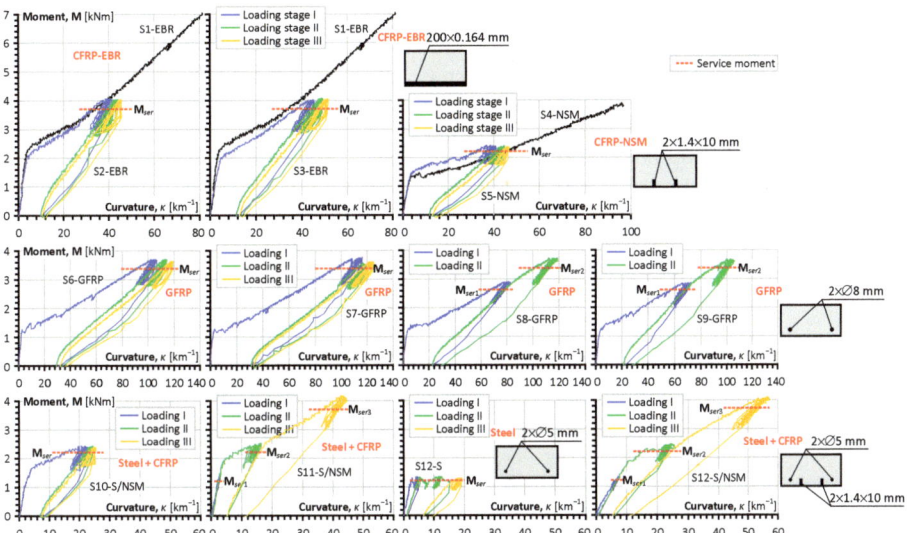

Figure 7. Moment–curvature results of the test samples from Table 1.

Remarkably, complete reloading followed all load cycling sets to measure the residual deformation of the beam samples. The latter values are vital for the residual stiffness analysis—the origin of each consequent diagram (Figures 3 and 7) coincides with the reloading deformation (residual curvature) from the preceding loading stage, producing the adequate resultant curvature suitable for Equation (5).

Table 3 shows the residual deformations (curvatures) determined for all samples subjected to unloading repetitions. These values are mandatory for adequate analysis when the mechanical load, necessary to estimate the residual stiffness of the beam sample, describes the consequential shift of the moment–curvature diagrams after reloading cycles (Figure 3). Thus, Table 3 determines the essential outcome valuable for further analysis and the modeling of the composite reinforcement systems similar to those considered in this work.

Table 3. Residual curvatures (κ_{res}) of the beams subjected to repeated loadings.

Beam	Load Stage I		Load Stage II		Load Stage III		Total Result	
	M_{ser} (kNm)	κ_{res} (km^{-1})	M_{ser} (kNm)	κ_{res} (km^{-1})	M_{ser} (kNm)	κ_{res} (km^{-1})	M_{max} (kNm)	$\Sigma\kappa_{res}$ (km^{-1})
S2-EBR	3.710	9.893	3.710	0.719	3.710	0.500	4.125	11.11
S3-EBR	3.710	11.23	3.710	1.448	3.710	0.863	4.125	13.54
S5-NSM	2.220	11.75	2.220	1.285	2.220	1.059	2.475	14.10
S6-GFRP	3.375	29.08	3.375	3.052	3.375	3.989	3.750	36.12
S7-GFRP	3.375	30.88	3.375	2.159	3.375	4.274	3.750	37.32
S8-GFRP	2.625	22.64	3.375	6.693	–	–	3.750	29.33
S9-GFRP	2.625	21.32	3.375	6.819	–	–	3.750	28.14
S10-S/NSM	2.212	7.084	2.212	0.522	2.212	1.937	2.475	9.544
S11-S/NSM	1.215	0.548	2.212	4.585	3.710	5.859	4.125	10.99
S12-S	1.215	1.100	1.215	5.570	1.215	2.767	1.350	9.437
S12-S/NSM	1.215	0.604	2.212	4.856	3.710	5.744	4.125	11.20

4. Discussion

4.1. The Cyclic Load Effect

The residual curvature analysis of the results from Table 3 reveals similar resultant deformations of all beam samples (except for elements with GFRP reinforcement), which were almost independent of the $n\rho$ ratio (Table 1) and loading layout and intensity (Figure 7). In addition, the steel-reinforced sample (S12-S) under the 1.215 kNm service moment demonstrates the 9.4 km^{-1} residual curvature comparable to the 11.2 km^{-1} result of the same element after strengthening (S12-S/NSM) subjected to 3.710 kNm load. This outcome demonstrates the efficiency of the strengthening solution.

Increasing the sample numbers will help identify more meaningful tendencies. However, Table 3 already shows the exceptional vulnerability of GFRP reinforcement systems to repeated load. For example, the S6 to S9 beams demonstrate tripled residual curvatures regarding the elements with EBR sheets, and hybrid reinforcement systems reached higher bending moments comparing the GFRP counterparts. Still, the S11-S/NSM and S12-S/NSM samples had a doubled $n\rho$ value, i.e., 2.45% vs. 1.22% (Table 1), which could explain the residual resistance increase. However, the EBR elements had a similar $n\rho$ ratio (1.18%). This allows the authors to hypothesize that the embedded GFRP bars cannot ensure sufficient resistance to repeated loads because of high deformability, corrupting the bond with concrete. The following equivalent stiffness analysis checks this hypothesis.

The moment–curvature diagrams in Figure 7 produce the initial data for the analysis, and Equation (5) determines the equivalent stresses (σ_t^*) corresponding to the service load. Thus, Figure 8 shows the residual stress diagrams corresponding to the load cycles, grouped by the reinforcement type and loading layout. Therefore, analyzing the charts in Figure 8 requires referring to Table 2, which indicates the loading conditions for every load stage.

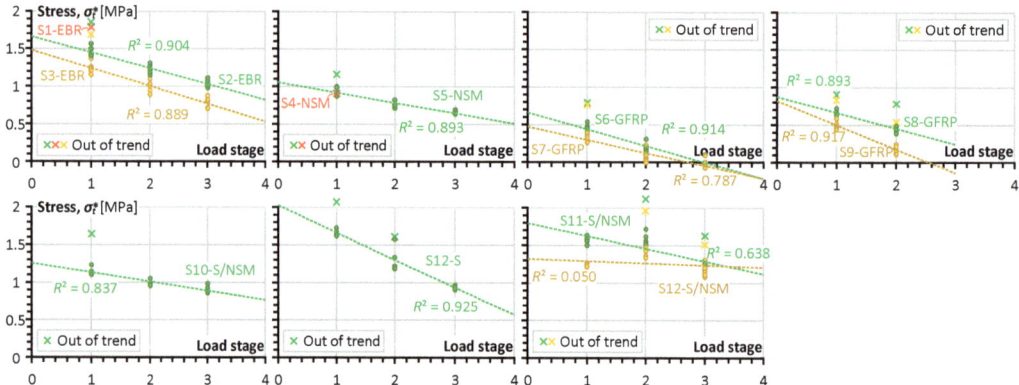

Figure 8. The decrease in the equivalent tensile stress (σ_t^*) with loading. Note: the red color corresponds to the samples subjected to monotonic load; the yellow and green colors show the test results of alternative counterparts.

Figure 9 schematically depicts the curvature estimation procedure, where only ascending loading diagram branches produce the analysis points. This sketch represents a smoothed view of the experimental diagram and, thus, is a reliable illustrative example. The analysis of the regression trends in Figure 8 reveals the essential differences between the refused points, and the remaining points belong to the first loading cycle. The further load repetitions do not include such outliers (except for the S12 sample after strengthening).

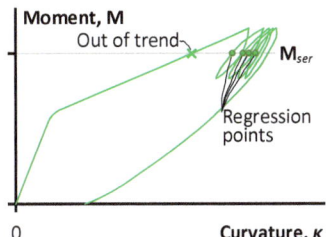

Figure 9. Schematic identification of the moment–curvature points for regression analysis in Figure 8.

The schematic in Figure 9 illustrates the situation when the first branch of the moment–curvature diagram does not generate the data point for determining the trend line of the stress degradation because of the essential difference in the element stiffness before and after the load cycles. The inclusion of such exceptional results will corrupt the stiffness trends. Therefore, only four data points (except for the first loading point) are employed for the residual strength regression analysis.

4.2. The Reinforcement Effect

In addition to the tripled residual curvatures (Table 3), Figure 8 demonstrates the minimal efficiency of GFRP bars in terms of the equivalent tensile stresses, which describes the composite behavior of reinforced concrete [20,22,29]. In particular, the S6-GFRP and S7-GFRP elements, which were subjected to cycling load with a 3.375 kNm service moment, demonstrate the complete disappearance of the tensile stresses in the concrete. A less eager loading in the S8-GFRP and S9-GFRP beams (Table 2) explains an improvement of equivalent stress values estimated at the first loading cycle (Figure 8). However, the identified tendency indicates the concrete contribution disappearing corresponding to the third loading stage. The mechanical bond degrading in concrete because of a relatively low stiffness of the GFRP bars describes the possible explanation for this effect. This observation aligns with the literature results (e.g., [12,13]). Unfortunately, such experimental investigations are rare in the literature; most such experimental works consider the pull-out behavior of GFRP bars. At the same time, the bonding problem of the embedded GFRP bars becomes apparent only for repeated load situations—Gribniak et al. [22] did not identify the bond deterioration problems of GFRP bars under monotonic load for the same reinforcement configuration. On the other hand, this drawback is untypical for alternative reinforcement schemes, making them acceptable to replace embedded GFRP bars.

A sharp inclination of the trend line of the equivalent stresses of the S12-S sample (with only steel reinforcement) highlights another bond-degrading mechanism characteristic of embedded bars. This stiffness decrease results from the cover cracking and the corresponding reduction in the concrete efficiency—only the concrete between the bar reinforcement and compressive zone (Figure 1b) contributes to the mechanical resistance of the cross-section. This observation aligns with the previous test findings [20,22,29], but the load cycles made this degradation mechanism more transparent. Although the load intensity of the S12-S beam was relatively high, i.e., the service load represents 85% of the theoretical moment of the steel yielding (Table 2), this load is well below the service loads faced by the remaining test samples (Table 3). On the other hand, this element possesses the highest $n\rho$ ratio (1.83%) among the test samples (except for the hybrid reinforcement systems). This outcome proclaims the inefficiency of the typical steel reinforcement for resisting cycling load, requiring a further unpractical $n\rho$ ratio increase. Furthermore, it reveals the need to renew structural design principles and tailor the materials' mechanical properties for construction [51]. This analysis procedure opposes the current practice, associating standardized engineering solutions with existing materials, the physical characteristics of which are imperfectly suiting the structural requirements and leading to an inefficient increase in the material amount for safety's sake.

In this context, the CFRP reinforcement systems demonstrate outstanding performance under the cycling load. The equivalent stresses in the S1 to S3 EBR samples reach the 1.86 MPa value and do not decrease below 0.70 MPa under the 3.71 kNm service load cycling; the NSM systems demonstrate the 1.16 MPa stress and preserve the minimum 0.64 MPa stresses, though these values correspond to the 2.22 kNm bending moment. The S10-S/NSM hybrid system improves the latter values correspondingly to 1.65 MPa and 0.87 MPa with the same tendency of stress decay as the S5-NSM beam. Under the increased load cycles, the S11-S/NSM proves a further increase in mechanical performance: the maximum and minimum equivalent stresses are equal to 2.11 MPa and 1.15 MPa.

The S12-S/NSM beam represents the strengthening situation of the S12-S element. At the load stages "2" and "3", the S12-S/NSM sample demonstrates very similar equivalent stresses to the S11-S/NSM beam (Figure 8). However, the identified trend line reveals a remarkable tendency—the loading cycles do not affect the residual resistance of the strengthened sample, preserving the averaged equivalent stresses at an approximately 1.3 MPa level. This finding supports the efficiency of the NSM strengthening systems for the mechanical load cycles. Further studies should reveal the hybrid reinforcement system's efficient layout and steel-to-CFRP proportions.

Figure 10 illustrates the results by relating the equivalent stress and strain values, i.e., Equation (5). This figure, including only "regression points" from Figure 8, demonstrates the stress decrease tendency with strain. This tendency is apparent for the GFRP beam samples (green-filled markers). Thus, seemingly, the relatively low modulus of elasticity of GFRP bars (Table 1) increases the concrete strains, causing a loss of the bond performance. The remaining reinforcement systems limit the deformations, which do not exceed a third of GFRP values. However, only the hybrid systems (S11-S/NSM and S12-S/NSM) prevent the reduction in the equivalent stresses under the load repetitions, ensuring reinforcement efficiency.

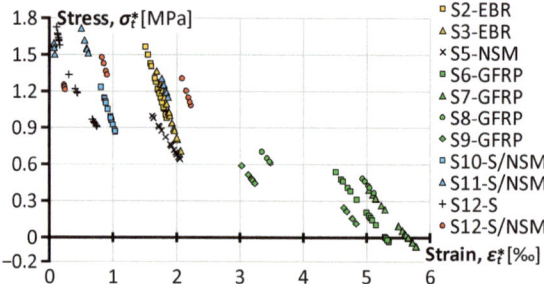

Figure 10. The equivalent stress–strain values correspond to the results of Figure 8.

4.3. The Load Intensity Effect

The results of Figures 8 and 10 demonstrate the essential importance of the load intensity on the residual stiffness decay. In particular, this effect is apparent in the S12-S beam sample subjected to 85% of the maximum theoretical load. The sharp decrease in the equivalent stresses results from this almost ultimate behavior.

Table 4 summarizes the results of Figure 10 in average terms. In addition, this table estimates the alteration of the equivalent stresses as a function of the equivalent strains. The following expression determines the stiffness alteration ratio:

$$\frac{\Delta \sigma_t^*}{\Delta \varepsilon_t^*} = \frac{\sigma_{t,i}^* - \sigma_{t,i-1}^*}{\varepsilon_{t,i}^* - \varepsilon_{t,i-1}^*}, \quad i = 2, 3. \tag{8}$$

Table 4. The average equivalent stresses and strains and the stiffness alteration ratio in Figure 10.

Beam	Load Stage I		Load Stage II			Load Stage III		
	σ_t^* (MPa)	ε_t^* (‰)	σ_t^* (MPa)	ε_t^* (‰)	$\Delta\sigma_t^*/\Delta\varepsilon_t^*$ (GPa)	σ_t^* (MPa)	ε_t^* (‰)	$\Delta\sigma_t^*/\Delta\varepsilon_t^*$ (GPa)
S2-EBR	1.474	1.570	1.224	1.710	−1.787	1.040	1.815	−1.757
S3-EBR	1.268	1.749	1.000	1.899	−1.791	0.771	2.029	−1.763
S5-NSM	0.949	1.679	0.761	1.903	−0.838	0.660	2.026	−0.825
S6-GFRP	0.472	4.610	0.196	5.014	−0.683	0.009	5.290	−0.680
S7-GFRP	0.348	5.107	0.075	5.546	−0.621	−0.021	5.700	−0.619
S8-GFRP	0.655	3.418	0.431	5.018	−0.140	-	-	-
S9-GFRP	0.514	3.137	0.180	4.743	−0.208	-	-	-
S10-S/NSM	1.160	0.857	0.990	0.963	−1.613	0.902	1.018	−1.581
S11-S/NSM	1.573	0.073	1.590	0.572	0.033	1.222	1.824	−0.294
S12-S	1.674	0.131	1.654	0.318	−1.712	0.937	0.715	−1.050
S12-S/NSM	1.240	0.239	1.394	0.873	0.243	1.173	2.166	−0.171

Here, the subscript i describes the values corresponding to successive loading stages; the negative ratio corresponds to the average stress reduction because of the load repetitions.

In addition to the beam S12-S discussed above, Table 4 reveals the most substantial decrease in the residual stiffness of the EBR (S2 and S3) samples expressed in the ratio $\Delta\sigma_t^*/\Delta\varepsilon_t^*$. A relatively high service moment (Table 2) in combination with a relatively low resistance of CF sheets to transverse (shear) load [20] could explain the intensive stress decrease. At the same time, these specimens still demonstrate substantial residual stiffness expressed in the equivalent stresses' terms because of the significant bonding area of EBR (Figure 4h).

Comparing the stiffness decay tendency (Table 4) of the S10-S/NSM and S11-S/NSM samples reveals a surprising outcome related to the positive correlation between the equivalent stresses and the service load ($\Delta\sigma_t^*/\Delta\varepsilon_t^* > 0$) of the S11 specimen. Analysis of Figure 7 can explain this issue—the service load of the S10 beam sample belongs to the crack formation stage, which predominantly controls the stiffness decrease. Analyzing the mechanical response of the S6 to S9 GFRP-reinforced beams provides the opposite case when the deformations exceed the concrete bonding limit, making the reinforcement inefficient ($\sigma_t^* \approx 0$). However, this work only exemplifies the proposed residual stiffness analysis procedure. Further studies should consider the load intensity effect and form the corresponding testing protocols.

5. Conclusions

This manuscript proposes composite reinforcement systems' residual stiffness analysis procedure under repeated mechanical loads. The experimental program demonstrates the effectiveness of the proposed methodology, analyzing the bending test results of 12 beam samples with various reinforcement types. The following essential conclusions result from this work:

- The proposed testing procedure is suitable for quantifying the residual stiffness decrease under repeated mechanical load, including the complete load removal between the loading cycles. The quantification employs the equivalent stresses acting in the concrete in tension under the assumption of the rectangular stress distribution. This simplified model approximates the tensile stresses with sufficient accuracy—the average approximation error (regarding the "exact" solution) does not exceed 7%. On the other hand, the "exact" inverse analysis reported in the literature is inapplicable to the

- specimens under cycling loads, which makes the proposed methodology irreplaceable for this study's purpose.
- This study reveals a limited ability of glass fiber-reinforced polymer (GFRP) bars to ensure the bonding performance under repeated loads. The concrete deformations exceed the bonding limit, making the reinforcement inefficient because of a relatively low modulus of elasticity (60 GPa) of the GFRP bars. Moreover, the bonding problem becomes apparent only for repeated loads—the previous tests did not identify the bond deterioration problems of GFRP bars under monotonic load for the same reinforcement configurations. On the other hand, this drawback is untypical for alternative reinforcement schemes considered in this study, proving the viability of the proposed analysis methodology.
- The carbon fiber (CF) reinforced materials demonstrate outstanding mechanical performance under repeated loads. The externally bonded reinforcement (EBR) system ensures the equivalent stresses, which do not decrease below 0.70 MPa; the near-surface mounted (NSM) system preserves the minimum 0.64 MPa stresses. The hybrid reinforcement system, combining steel bars and NSM CFRP strips, improves this value to 0.87 MPa. Under the increased load cycles, the hybrid reinforcement demonstrates a further increase in mechanical performance—the equivalent stresses exceed 1.15 MPa, exceeding 50% of the tensile resistance of the concrete.
- The NSM reinforcement system efficiently strengthened the beam sample with steel reinforcement bars tested until 85% of the theoretical load-bearing capacity. The load repetitions did not affect the residual resistance of the strengthened specimen, preserving the averaged equivalent stresses at an approximately 1.3 MPa level. Further studies should reveal the hybrid reinforcement system's efficient layout and steel-to-CFRP ratio.

Author Contributions: Conceptualization, methodology, software, and validation, V.G. and A.R.; formal analysis, H.A.S.; investigation, A.S.; resources, V.G.; data curation and writing—original draft preparation, H.A.S. and A.R.; writing—review and editing, V.G.; visualization, H.A.S. and V.G.; project administration, A.S.; supervision and funding acquisition, V.G. All authors have read and agreed to the published version of the manuscript.

Funding: This project has received funding from the European Regional Development Fund (Project No. 01.2.2-LMT-K-718-03-0010) under a grant agreement with the Research Council of Lithuania (LMTLT). VILNIUS TECH funded the APC.

Data Availability Statement: The authors will provide the raw data of this work upon request.

Conflicts of Interest: The authors declare no conflict of interest.

References

1. Obaidat, A.T. Flexural behavior of reinforced concrete beam using CFRP hybrid system. *Eur. J. Environ. Civ. Eng.* **2022**, *26*, 6165–6187. [CrossRef]
2. Dang, H.V.; Phan, D.N. Experimental investigation and analysis of pure bending plastic hinge zone in hybrid beams rein-forced with high reinforcement ratio under static loads. *Eur. J. Environ. Civ. Eng.* **2022**, *26*, 6188–6210. [CrossRef]
3. Garnevičius, M.; Gribniak, V. Developing a hybrid FRP-concrete composite beam. *Sci. Rep.* **2022**, *12*, 16237. [CrossRef]
4. Gribniak, V. Material-oriented engineering for eco-optimized structures—A new design approach. *Adv. Mater. Lett.* **2023**, *14*, 23011713. [CrossRef]
5. Yun, H.-D.; Kim, S.-H.; Choi, W. Determination of mechanical properties of sand-coated carbon fiber reinforced polymer (CFRP) rebar. *Polymers* **2023**, *15*, 2186. [CrossRef] [PubMed]
6. Wdowiak-Postulak, A.; Wieruszewski, M.; Bahleda, F.; Prokop, J.; Brol, J. Fibre-reinforced polymers and steel for the reinforcement of wooden elements—Experimental and numerical analysis. *Polymers* **2023**, *15*, 2062. [CrossRef] [PubMed]
7. Liu, T.; Liu, X.; Feng, P. A comprehensive review on mechanical properties of pultruded FRP composites subjected to long-term environmental effects. *Compos. Part B Eng.* **2020**, *191*, 107958. [CrossRef]
8. Bazli, M.; Jafari, A.; Ashrafi, H.; Zhao, X.-L.; Bai, Y.; Singh Raman, R.K. Effects of UV radiation, moisture and elevated temperature on mechanical properties of GFRP pultruded profiles. *Constr. Build. Mater.* **2020**, *231*, 117137. [CrossRef]
9. Xu, J.; Gu, Y.; Fu, T.; Zhang, X.; Zhang, H. Research on the heating process of CFRP circular tubes based on electromagnetic induction heating method. *Polymers* **2023**, *15*, 3039. [CrossRef]

10. Nehdi, M.; Said, A. Performance of RC frames with hybrid reinforcement under reversed cyclic loading. *Mater. Struct.* **2005**, *38*, 627–637. [CrossRef]
11. Muciaccia, G.; Khorasani, M.; Mostofinejad, D. Effect of different parameters on the performance of FRP anchors in combination with EBR-FRP strengthening systems: A review. *Constr. Build. Mater.* **2022**, *354*, 129181. [CrossRef]
12. Mohamed, N.; Farghaly, A.S.; Benmokrane, B. Beam-testing method for assessment of bond performance of FRP bars in concrete under tension-compression reversed cyclic loading. *ASCE J. Compos. Constr.* **2016**, *21*, 06016001. [CrossRef]
13. Kim, B.; Lee, J. Resistance of interfacial debonding failure of GFRP bars embedded in concrete reinforced with structural fibers under cycling loads. *Compos. Part B Eng.* **2019**, *156*, 201–211. [CrossRef]
14. Liu, X.; Wang, X.; Xie, K.; Wu, Z.; Li, F. Bond behavior of basalt fiber-reinforced polymer bars embedded in concrete under mono-tensile and cyclic loads. *Int. J. Concr. Struct. Mater.* **2020**, *14*, 19. [CrossRef]
15. Shen, D.; Wen, C.; Zhu, P.; Li, M.; Ojha, B.; Li, C. Bond behavior between basalt fiber-reinforced polymer bars and concrete under cyclic loading. *Constr. Build. Mater.* **2020**, *258*, 119518. [CrossRef]
16. Xiao, L.; Dai, S.; Jin, Q.; Peng, S. Bond performance of GFRP bars embedded in steel-PVA hybrid fiber concrete subjected to repeated loading. *Struct. Concr.* **2023**, *24*, 1597–1611. [CrossRef]
17. Shi, J.; Sun, S.; Cao, X.; Wang, H. Pullout behaviors of basalt fiber-reinforced polymer bars with mechanical anchorages for concrete structures exposed to seawater. *Constr. Build. Mater.* **2023**, *373*, 130866. [CrossRef]
18. Pan, Y.; Yu, Y.; Yu, J.; Lu, Z.; Chen, Y. Effects of simulated seawater on static and fatigue performance of GFRP bar–concrete bond. *J. Build. Eng.* **2023**, *68*, 105985. [CrossRef]
19. Jakubovskis, R.; Kaklauskas, G.; Gribniak, V.; Weber, A.; Juknys, M. Serviceability analysis of concrete beams with different arrangement of GFRP bars in the tensile zone. *ASCE J. Compos. Constr.* **2014**, *18*, 04014005. [CrossRef]
20. Gribniak, V.; Sokolov, A.; Rimkus, A.; Sultani, H.A.; Tuncay, M.C.; Torres, L. A novel approach to residual stiffness of flexural concrete elements with composite reinforcement. In Proceedings of the IABSE Symposium—Towards a Resilient Built Environment Risk and Asset Management, Guimarães, Portugal, 27–29 March 2019; IABSE: Zurich, Switzerland, 2019; pp. 46–51.
21. Rimkus, A.; Barros, J.A.O.; Gribniak, V.; Rezazadeh, M. Mechanical behavior of concrete prisms reinforced with steel and GFRP bar systems. *Compos. Struct.* **2019**, *220*, 273–288. [CrossRef]
22. Gribniak, V.; Sultani, H.A.; Rimkus, A.; Sokolov, A.; Torres, L. Standardised quantification of structural efficiency of hybrid reinforcement systems for developing concrete composites. *Compos. Struct.* **2021**, *274*, 114357. [CrossRef]
23. Anas, S.M.; Alam, M.; Isleem, H.F.; Najm, H.M.; Sabri, M.M.S. Ultra high performance concrete and C-FRP tension rebars: A unique combinations of materials for slabs subjected to low-velocity drop impact loading. *Front. Mater.* **2022**, *9*, 1061297. [CrossRef]
24. Yuan, J.S.; Gao, D.; Zhu, H.; Chen, G.; Zhao, L. Flexural behavior of reinforced concrete beams reinforced with glass fiber reinforced polymer rectangular tubes. *Front. Mater.* **2020**, *7*, 577299. [CrossRef]
25. Farahi, B.; Esfahani, M.R.; Sabzi, J. Experimental investigation on the behavior of reinforced concrete beams retrofitted with NSM-SMA/FRP. *Amirkabir J. Civ. Eng.* **2019**, *51*, 209–212. [CrossRef]
26. Godat, A.; Qu, Z.; Lu, X.Z.; Labossiere, P.; Ye, L.P.; Neale, K.W. Size effects for reinforced concrete beams strengthened in shear with CFRP strips. *ASCE J. Compos. Constr.* **2010**, *14*, 260–271. [CrossRef]
27. Li, Z.; Zhu, H.; Zhen, X.; Wen, C.; Chen, G. Effects of steel fiber on the flexural behavior and ductility of concrete beams reinforced with BFRP rebars under repeated loading. *Compos. Struct.* **2021**, *270*, 114072. [CrossRef]
28. Kargaran, A.; Kheyroddin, A. Experimental and numerical investigation of seismic retrofitting of RC square short columns using FRP composites. *Eur. J. Environ. Civ. Eng.* **2022**, *26*, 4619–4642. [CrossRef]
29. Sultani, H.A.; Rimkus, A.; Sokolov, A.; Gribniak, V. A new testing procedure to quantify unfavourable environmental effect on mechanical performance of composite reinforcement system. In Proceedings of the 14th fib International Ph.D. Symposium in Civil Engineering, Rome, Italy, 5–7 September 2022; fib: Lausanne, Switzerland, 2022; pp. 377–384.
30. Mat Saliah, S.N.; Md Nor, N. Assessment of the integrity of reinforced concrete beams strengthened with carbon fibre reinforced polymer using the acoustic emission technique. *Front. Mech. Eng.* **2022**, *8*, 885645. [CrossRef]
31. Sun, R.; Perera, R.; Gu, J.; Wang, Y. A simplified approach for evaluating the flexural response of concrete beams reinforced with FRP bars. *Front. Mater.* **2021**, *8*, 765058. [CrossRef]
32. Fathuldeen, S.W.; Qissab, M.A. Behavior of RC beams strengthened with NSM CFRP strips under flexural repeated loading. *Struct. Eng. Mech.* **2019**, *70*, 67–80. [CrossRef]
33. Zhu, H.; Li, Z.; Wen, C.; Cheng, S.; Wei, Y. Prediction model for the flexural strength of steel fiber reinforced concrete beams with fiber-reinforced polymer bars under repeated loading. *Compos. Struct.* **2020**, *250*, 112609. [CrossRef]
34. Gribniak, V.; Kaklauskas, G.; Torres, L.; Daniunas, A.; Timinskas, E.; Gudonis, E. Comparative analysis of deformations and tension-stiffening in concrete beams reinforced with GFRP or steel bars and fibers. *Compos. Part B Eng.* **2013**, *50*, 158–170. [CrossRef]
35. Song, S.; Wang, G.; Min, X.; Duan, N.; Tu, Y. Experimental study on cyclic response of concrete frames reinforced by Steel-CFRP hybrid reinforcement. *J. Build. Eng.* **2021**, *34*, 101937. [CrossRef]
36. Gribniak, V.; Sokolov, A. Standardized RC beam tests for modeling the fiber bridging effect in SFRC. *Constr. Build. Mater.* **2023**, *370*, 130652. [CrossRef]

37. Kaklauskas, G.; Gribniak, V.; Bacinskas, D. Discussion of "Tension stiffening in lightly reinforced concrete slabs" by R. I. Gilbert. *ASCE J. Struct. Eng.* **2008**, *134*, 1261–1262. [CrossRef]
38. Gribniak, V.; Kaklauskas, G.; Juozapaitis, A.; Kliukas, R.; Meskenas, A. Efficient technique for constitutive analysis of reinforced concrete flexural members. *Inverse Probl. Sci. Eng.* **2017**, *25*, 27–40. [CrossRef]
39. Rimkus, A.; Cervenka, V.; Gribniak, V.; Cervenka, J. Uncertainty of the smeared crack model applied to RC beams. *Eng. Fract. Mech.* **2020**, *233*, 107088. [CrossRef]
40. Jakubovskis, R.; Jankutė, A.; Urbonavičius, J.; Gribniak, V. Analysis of mechanical performance and durability of self-healing biological concrete. *Constr. Build. Mater.* **2020**, *260*, 119822. [CrossRef]
41. Sultani, H.A.; Rimkus, A.; Sokolov, A.; Gribniak, V. Analyzing environmental effects on the mechanical performance of composite reinforcement systems. *Lect. Notes Civ. Eng.* **2023**, *349*, 618–627. [CrossRef]
42. Dulinskas, E.; Gribniak, V.; Kaklauskas, G. Influence of steam curing on high-cyclic behaviour of prestressed concrete bridge elements. *Balt. J. Road Bridge Eng.* **2008**, *3*, 115–120. [CrossRef]
43. Gribniak, V.; Kaklauskas, G.; Bacinskas, D. Experimental investigation of shrinkage influence on tension stiffening of RC beams. In Proceedings of the Eighth International Conference: Creep, Shrinkage and Durability of Concrete and Concrete Structures (ConCreep 8), Ise-Shima, Japan, 30 September–2 October 2008; CRC Press: London, UK, 2009; Volume 1, pp. 571–577. [CrossRef]
44. EN 1992-1-1:2004; Eurocode 2: Design of Concrete Structures—Part 1: General Rules and Rules for Buildings. CEN: Brussels, Belgium, 2004.
45. Gribniak, V.; Cervenka, V.; Kaklauskas, G. Deflection prediction of reinforced concrete beams by design codes and computer simulation. *Eng. Struct.* **2013**, *56*, 2175–2186. [CrossRef]
46. Gribniak, V.; Rimkus, A.; Torres, L.; Hui, D. An experimental study on cracking and deformations of concrete in tension elements reinforced with multiple GFRP bars. *Compos. Struct.* **2018**, *201*, 477–485. [CrossRef]
47. Rimkus, A.; Gribniak, V. Experimental data of deformation and cracking behaviour of concrete ties reinforced with multiple bars. *Data Brief* **2017**, *13*, 223–229. [CrossRef] [PubMed]
48. Zhu, H.; Li, C.; Cheng, S.; Yuan, J. Flexural performance of concrete beams reinforced with continuous FRP bars and discrete steel fibers under cyclic loads. *Polymers* **2022**, *14*, 1399. [CrossRef] [PubMed]
49. Gribniak, V.; Rimkus, A.; Torres, L.; Jakstaite, R. Deformation analysis of RC ties: Representative geometry. *Struct. Concr.* **2017**, *18*, 634–647. [CrossRef]
50. Gribniak, V.; Rimkus, A.; Misiunaite, I.; Zakaras, T. Improving local stability of aluminium profile with low-modulus stiffeners: Experimental and numerical web buckling analysis. *Thin-Walled Struct.* **2022**, *172*, 108858. [CrossRef]
51. Gribniak, V. Special Issue "Advanced Composites: From Materials Characterization to Structural Application". *Materials* **2020**, *13*, 5820. [CrossRef]

Disclaimer/Publisher's Note: The statements, opinions and data contained in all publications are solely those of the individual author(s) and contributor(s) and not of MDPI and/or the editor(s). MDPI and/or the editor(s) disclaim responsibility for any injury to people or property resulting from any ideas, methods, instructions or products referred to in the content.

Article

Engineered Interleaved Random Glass Fiber Composites Using Additive Manufacturing: Effect of Mat Properties, Resin Chemistry, and Resin-Rich Layer Thickness

Ahmed M. H. Ibrahim [1], Mohanad Idrees [1], Emine Tekerek [2], Antonios Kontsos [2], Giuseppe R. Palmese [1] and Nicolas J. Alvarez [1,*]

1. Department of Chemical and Biological Engineering, Drexel University College of Engineering, Philadelphia, PA 19104, USA
2. Department of Mechanical Engineering and Mechanics, Drexel University College of Engineering, Philadelphia, PA 19104, USA
* Correspondence: alvarez@drexel.edu

Abstract: Standard lay-up fabrication of fiber-reinforced composites (FRCs) suffer from poor out-of-plane properties and delamination resistance. While advanced manufacturing techniques (e.g., interleaving, braiding, and z-pinning) increase delamination resistance in FRCs, they typically result in significant fabrication complexity and limitations, increased manufacturing costs, and/or overall stiffness reduction. In this work, we demonstrate the use of facile digital light processing (DLP) technique to additively manufacture (AM) random glass FRCs with engineered interleaves. This work demonstrates how vat photo-polymerization techniques can be used to build composites layer-by-layer with controlled interleaf material, thickness, and placement. Note that this engineering control is almost impossible to achieve with traditional manufacturing techniques. A range of specimens were printed to measure the effect of interleaf thickness and material on tensile/flexural properties as well as fracture toughness. One important observation was the ≈60% increase in interlaminar fracture toughness achieved by using a tough resin material in the interleaf. The comparison between AM and traditionally manufactured specimens via vacuum-assisted resin transfer molding (VARTM) highlighted the limitation of AM techniques in achieving high mat consolidation. In other words, the volume fraction of AM parts is limited by the wet fiber mat process, and engineering solutions are discussed. Overall, this technique offers engineering control of FRC design and fabrication that is not available with traditional methods.

Keywords: additive manufacturing; interleaving; random chopped fibers; thermosetting polymers; interlaminar properties

Citation: Ibrahim, A.M.H.; Idrees, M.; Tekerek, E.; Kontsos, A.; Palmese, G.R.; Alvar, N.J. Engineered Interleaved Random Glass Fiber Composites Using Additive Manufacturing: Effect of Mat Properties, Resin Chemistry, and Resin-Rich Layer Thickness. *Polymers* 2023, 15, 3189. https://doi.org/10.3390/polym15153189

Academic Editors: Giorgio Luciano and Maurizio Vignolo

Received: 5 July 2023
Revised: 21 July 2023
Accepted: 22 July 2023
Published: 27 July 2023

Copyright: © 2023 by the authors. Licensee MDPI, Basel, Switzerland. This article is an open access article distributed under the terms and conditions of the Creative Commons Attribution (CC BY) license (https://creativecommons.org/licenses/by/4.0/).

1. Introduction

Fiber-reinforced composite (FRC) materials consist of high-strength, high-modulus fibers embedded in a polymeric matrix. In an FRC, both fiber and matrix retain their physical and chemical identities, yet produce a combination of properties not obtainable by either component separately. Generally speaking, fibers act as load carriers and the matrix transfers the loads from fiber to fiber [1]. While FRCs offer significant advantages over metals, they have limitations. For example, glass fiber-reinforced composites (GFRCs) suffer from poor out-of-plane properties and tend to delaminate. The latter stems from a mismatch in properties between resin and fiber, i.e., significant differences in Poisson ratio, and mechanical/thermal properties, that lead to interlaminar normal/shear stresses and fiber delamination [2].

Significant effort has gone into enhancing the delamination resistance of FRCs, such as the use of toughened polymeric matrices [3], z-pinning [4], stitching [5], braiding [6], and optimizing stacking sequences [7]. Unfortunately, these ideas led to significant increases in fabrication costs and composite weight and/or loss of in-plane properties [2]. The most promising technique to reduce delamination is interleaving, which is the process of adding a discrete resin layer between consecutive fiber plies [8]. The interleaves, also known as resin-rich layers (RRLs), are either composed of a brittle or a ductile polymeric matrix. Interleaving has been experimentally proven to enhance modes I and II delamination resistance [9–18], as it leads to the creation of a plastic yield zone capable of absorbing a large amount of energy through plastic deformation [19,20].

One important drawback of RRL delamination resistance is the reduction in composite fiber volume fraction (FVF) with increasing RRL thickness, which inadvertently decreases mechanical properties. Furthermore, there is an optimum thickness of RRL whereby additional thickness does not offer any further delamination resistance. However, this thickness is not well known, but is argued to depend on the plastic zone length of the toughening resin [21]. The trade-off between delamination resistance and reduction in fiber volume fraction highlights the necessity of optimizing δ_{RRL} thickness.

Traditional fabrication techniques of interleaved composites are very limited to certain types of resins and typically are only achieved through pre-preg lay-up techniques. Furthermore, interleaving using traditional manufacturing techniques is labor intensive, limits the types of resins usable, and is of prohibitively high cost. The state-of-the-art would greatly benefit from an automated interleaving process, whereby a broad range of resins can be used, and additive manufacturing (AM) presents such an opportunity. AM techniques offer a novel approach to traditional manufacturing methods, allowing for multi-material manufacturing and geometric control and are attracting attention in a wide array of industries [22,23]. Recently, we have developed a method of producing GRFCs using digital light processing (DLP) stereolithography, a vat photo-polymerization AM technique utilizing liquid resins that cure selectively upon exposure to ultra violet (UV) light [21]. In a previous work, we showed that this technique is capable of producing GRFCs using woven (PW) glass fiber mats with comparable volume fractions of traditional lay-up techniques, but significant higher delamination resistance. Fiber plies were manually introduced into the printing system and printed as typical print layers to form a composite part. The DLP method offers several advantages over traditional hand lay-up techniques, such as selective spatial reinforcement of a part, control over interlayer spacing, and the use of multiple resins in a single part [21].

In this work, we utilize the abovementioned AM technique to determine the effect of RRL thickness and resin properties on composites made with chopped strand glass fiber (CSGF) mats. The method is used to control the interleaf spacing, i.e., RRL thickness, and the resin material in flat composite structures that could not be achieved with traditional manufacturing methods. This offers a unique opportunity to study design parameters of interleaved composites. Unique composite test specimens were prepared and tested in short beam shear (SBS), flexure, tensile, and mode II delamination. Interestingly, we show that there are significant differences in the trends for CSGF mats and PW fabrics for varying RRL thickness. Furthermore, this work examines the important printer design criteria for achieving high fiber volume fractions of CSGF. Although CSGF is utilized in many applications for its low cost and isotropic properties, the random orientation of the fiber mats lead to significant processing issues that must be overcome to achieve successful consolidation. This work quantitatively demonstrates the issues with wet fiber mat consolidation using AM, and discusses engineering solutions.

2. Materials

This work primarily utilized a benchmark additive resin called DA-2, previously discussed in [24], which is composed of: bisphenol A glycidyl methacrylate 'Bis-GMA' (37.5 wt%), ethoxylated bisphenol A dimethacrylate 'Bis-EMA' (37.5 wt%), 1,6-hexanediol dimethacrylate 'HDDMA' (25 wt%), and a photoinitiator, Phenylbis (2,4,6-trimethylbenzoyl) phosphine oxide 'PPO' (0.7 wt%). Tenacious resin was purchased from Siraya Tech and used as received. The reinforcing material was randomly oriented glass fiber, purchased from Orca Composites. Properties of the resins and fiber are provided in Tables 1 and 2, respectively.

Table 1. Properties of DA-2 and Tenacious.

Property	DA2	Tenacious
Density (25 °C)	1.20	-
Glass Transition Temperature [T_g] (°C)	99	41.97
Tensile Modulus (GPa)	2.80 ± 0.10	1.9 ± 0.09
Tensile Strength (MPa)	61.9 ± 6.30	37.3 ± 2.80
Tensile Strain at Failure (%)	2.5 ± 0.60	44.06 ± 1.90
Flexural Modulus (GPa)	3 ± 0.10	1.6 ± 0.20
Flexural Strength (MPa)	110 ± 10	51 ± 5.20
Fracture Toughness [G_{Ic}] (J/m^2)	58.80 ± 0.30	1580

Table 2. Properties of glass fiber.

CSGF Properties	
Areal Weight (g/m^2 or gsm)	900
Density (g/cm^3 or g/cc)	2.68 [25]
Tensile Strength (MPa)	3100–3800 [25]
Tensile Modulus (GPa)	80–81 [25]
Elongation at Break (%)	4.5–4.9 [25]

3. Experimental Methods

Figure 1 below demonstrates the printing process and shows images of RRLs in the final printed parts.

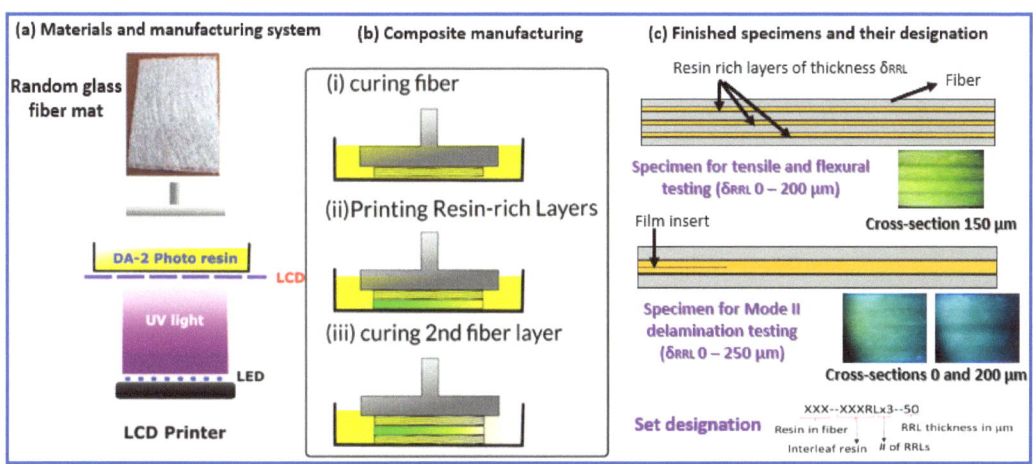

Figure 1. Demonstration of RRL incorporation in 3D-printed CSGF composite bars.

3.1. Composite Bars Printing

An ELEGOO MARS LCD printer (405 nm light) was used to produce all composite specimens. In all cases, a neat layer of resin was printed initially to ensure good quality surface finish, good adhesion to the build platform, and ease of sample removal. Subsequently, the platform was sent to a raised position such that the first fiber ply was added by hand to the resin vat. Note that the fiber ply is pre-wetted with resin to minimize void fraction in the final 3D printed composite. The building platform was then sent to full down position to consolidate the fiber mat. Once consolidated, the LCD screen shined blue light through the bottom of the vat for a determined exposure time to selectively cure the resin in the fiber mat. Once the resin is cured, the building platform was raised and the printing paused to ensure that the cured fiber ply adhered well to the building platform, and any excess fiber along the print edges were removed. The latter step is important to ensure that excess fiber does not prevent the build platform from pressing down on subsequent mats with a uniform pressure. All fiber layers are composed of four mats. Thus, the above process is repeated four times until a fiber-reinforced section was completed. Once completed, either another fiber layer was added or an RRL was printed. An RRL with a desired thickness was added by zeroing the printer height with the printed specimen and then printing resin-only layers until the desired RRL thickness was achieved. This process was utilized to print tensile and flexural specimens/bars with three RRLs and four fiber-reinforced sections. Selected δ_{RRL} values are 50, 100, 150, and 200 μm. Two more sets for tensile and flexural testing were manufactured via vacuum-assisted resin transfer molding (VARTM) to compare their properties with the 3D-printed specimens. Mode II delamination test specimens were printed with only one central RRL ranging in thickness from 0 to 250 μm (50 μm increments), and a film insert was introduced in the mid-plane of the RRL for the Mode II delamination studies. Additionally, 0-μm RRL specimens were printed for tensile, flexural, and Mode II delamination testing to determine the effect of δ_{RRL} on mechanical properties, as well as a short beam shear (SBS) testing set. All sets tested with their respective dimensional measurements are provided in Table 3. Microscopy imaging of the individual interleaved specimens show that resin layers were printed with good dimensional accuracy; their details are provided in Table 4.

Table 3. Specimen dimensions for mechanical testing, see Section 3.3 for respective ASTM standards.

Manufacturing Method	Test	Length (mm)	Width (mm)	Height (mm)
3D Printing	Short beam shear (SBS)	40	11.50	6.40–6.70
	Tensile	100	12	3.90–6.20
	Flexure	110	12.35	4.00–6.60
	Mode II delamination	110	19	4.30–6.00
VARTM	Tensile	100	12	2.80–2.90
	Flexure	110	14	2.85–3.10

3.2. Composite Post-Processing

All printed parts underwent a post-curing procedure inside a Formlabs 405 nm UV light oven (Somerville, MA) at a temperature of 75 °C for 2 h. Afterwards, the bars were polished to remove excess fiber on the edges and also polish the faces of the specimens to minimize defects that could impact the mechanical testing results.

3.3. Testing Conditions

3.3.1. Short Beam Shear (SBS) Testing

SBS testing was conducted on rectangular specimens following standard ASTM D2344/D2344M-16 [26], where the cross-head speed was set to 1 mm/min, the span length-to-thickness ratio was kept constant at 2:1, and the SBS stress was determined from:

$$\sigma_{SBS} = \frac{3P}{4bh} \tag{1}$$

where P is the load measured and b and h are the specimen's width and thickness, respectively.

Table 4. RRL measurements for interleaved sets.

Test	Set	Average Measured δ_{RRL} (μm)
Tensile	DA2-DA2RLx3-50	67.28 ± 2.09
	DA2-DA2RLx3-100	114.67 ± 4.22
	DA2-DA2RLx3-150	140.27 ± 4.99
	DA2-DA2RLx3-200	195.08 ± 8.41
Flexure	DA2-DA2RLx3-50	61.33 ± 1.28
	DA2-DA2RLx3-100	97.29 ± 2.34
	DA2-DA2RLx3-150	128.43 ± 1.57
	DA2-DA2RLx3-200	205.43 ± 5.46
	DA2-TENRLx3-100	109 ± 2.5
Mode II Delamination	DA2-DA2RLx1-50	70 ± 2.89
	DA2-DA2RLx1-100	108 ± 6.90
	DA2-DA2RLx1-150	146 ± 8.72
	DA2-DA2RLx1-200	200.4 ± 6.45
	DA2-DA2RLx1-250	240 ± 9.76
	DA2-TENRLx1-100	111 ± 5.00

3.3.2. Flexure Testing

Flexural testing was conducted on rectangular bars according to standard ASTM D790-17 [27], where the cross-head speed was set to 1 mm/min, and the span length-to-thickness ratio was kept constant at 16:1. Note that all test specimens were composed of two fiber ply regions (top and bottom) separated by one RRL. Flexural strength and modulus were calculated using the equations below:

$$\sigma_f = \frac{3PL}{2bd^2} \quad (2)$$

$$E_B = \frac{mL^3}{4bd^3} \quad (3)$$

where P is the measured load, L is span length, m is the slope of the linear portion of the load-displacement curve, and b and d are the specimen's width and thickness, respectively.

3.3.3. Tensile Testing

Tensile testing was conducted on rectangular bars following the structure ASTM D3039/3039M-17 [28]. Cross-head speed was set to 0.5 mm/min, gauge length was 50 mm, and the applied gripping pressure was 350 psi. Mechanical testing was performed using an MTS 370.10 servo hydraulic frame equipped with a 100 kN load cell. Specimen displacement and strain were measured by using a 3D Digital Image Correlation (DIC) system comprising of two 5 megapixel Baumer TXG50 monochrome cameras with 2/3″ Charged Coupled Device (CCD) sensors (manufactured by GOM-3D Metrology). A field of view of 55 × 44 mm² was chosen, while image capturing was performed at a rate of 0.2 Hz. The deformation measurements were obtained using the subset method with 40 × 40 pixels facet size and 20 pixels step size. Tensile strength, modulus, and strain were determined using:

$$\sigma_T = \frac{P_{max}}{A} \quad (4)$$

$$\epsilon = \frac{\delta}{L_g} \quad (5)$$

$$E_T = \frac{\Delta\sigma}{\Delta\epsilon} \quad (6)$$

where P_{max} is the measured load before failure, A is the specimen's cross-sectional area, δ is the measured displacement via an extensometer, L_g is the gauge length, and $\Delta\sigma/\Delta\epsilon$ is the initial slope of the stress strain curve.

3.3.4. Mode II Delamination

Delamination testing in mode II was performed using a three point bending set up. The samples were printed with only one resin layer in the mid-plane with a given thickness in the range described in the previous section. A polyimide film of 40 mm in length was printed in the center of the RRL layer to initiate the crack. Specimen geometry and testing conditions were selected following ASTM D7905/D7905M standard [29]. The mode II interlaminar fracture toughness (G_{IIc}) was calculated using:

$$G_{IIc} = \frac{3mP_{max}^2 a_0^2}{2B} \quad (7)$$

where m is a parameter obtained from compliance testing, P_{max} is the maximum load reached during delamination testing, a_0 is the position of the delamination test marking, which was chosen to be 30 mm from the insert's end, and B is the specimen's width.

3.4. Fiber Volume Fraction (FVF) and Void Fraction Measurement

The true volume fraction of the printed specimens was experimentally determined via ignition loss experiments following ASTM standard D2548-18 [30]. The specimens were weighed, W_b, and then heated to 600 °C and cooked for one hour. Specimens were weighed post heating, W_a, and the actual FVF, $\phi_{f,a}$, was calculated via:

$$\phi_{f,a} = \frac{\frac{W_a}{\rho_g}}{\frac{W_a}{\rho_g} + \frac{W_b - W_a}{\rho_r}} \quad (8)$$

The void fraction of the specimens, ϕ_v, was calculated using equation:

$$\phi_v = 1 - \frac{\frac{W_a}{\rho_g}}{\frac{W_a}{\rho_g} + \frac{W_b - W_a}{\rho_r}} - \frac{\frac{W_b - W_a}{\rho_r}}{V_c} \quad (9)$$

where V_c is the sample's total volume.

4. Results and Discussion

4.1. FVF Results

The volume fraction of reinforcement material is directly related to the mechanical properties of the composite. In traditional manufacturing methods, there is a fiber consolidation step to increase the fiber volume fraction before curing. In the AM method here, the consolidation step is facilitated by the downward pressure of the build platform on the fiber mat prior to photo-cure. Figure 2 shows fiber volume and void fraction measured for flexural and tensile specimens as a function of RRL thickness. In Figure 2a,c, there are three important observations: (i) the AM specimens have much lower volume fractions than VARTM samples, (ii) the flexural and tensile specimens have FVF that does not decrease with RRL thickness for $\delta_{RRL} < 125$ µm, and (iii) that the volume fractions for flexural and tensile specimens are very similar. Point (iii) is a testament to the reproducibility of this manufacturing method, while points (i) and (ii) were somewhat surprising and will be discussed in detail below.

Another important aspect of composite manufacturing is the introduction of voids during manufacturing [31,32]. The voids act as defects that can cause premature failure of the specimen. Vacuum-assisted methods are advantageous, as they minimize the amount of trapped air. The AM method is done in ambient conditions and thus void fraction is a reasonable concern. Figure 2b,d show the void fraction measured in the different test specimens printed using AM. While all void fractions are non-zero, the void fractions are relatively constant with increasing RRL thickness for tensile and flexural specimens. This suggests that all voids are coming from the process of introducing the fiber mat. The large error bars indicate large variations in void distribution for different printed specimens. Note that the introduction of a dry fiber mat into the resin vat introduced significantly higher void fractions (data not shown). Thus, pains were taken to pre-wet the fiber mat before introduction into the resin vat to avoid the trapping of air bubbles during mat placement. Regardless of these efforts, the void fraction for the specimens could not be reduced below 4–5%. To minimize the void fraction further, one could imagine introducing a fiber pre-wetting step that is carried out under vacuum conditions, which is currently in the works.

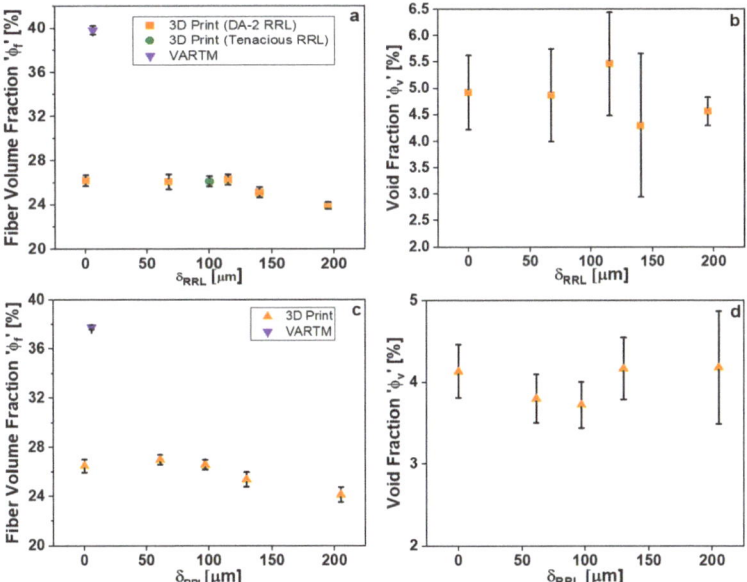

Figure 2. FVF measurements (**a**) and void fraction (**b**) for flexure specimens; FVF measurements (**c**) and void fraction (**d**) for tensile specimens

4.1.1. Physics of Mat Consolidation

As fiber volume fraction is arguably the most important parameter in determining mechanical properties, we investigated the reasons behind the lower volume fraction between AM and VARTM, and the lack of decrease in FVF below $\delta_{RRL} < 125$ μm. There are two reasons for the lower FVF in AM specimens compared to VARTM: (1) lower applied pressure during consolidation and (2) the larger pressure required to consolidate a wet mat versus a dry mat. In the case of (1), the z-motor in a DLP printer has a finite amount of torque, which limits the applied pressure. The maximum downward force measured on the printer was 150 N, which gives a maximum consolidation pressure of 0.03 MPa considering a 110 mm by 45 mm fiber mat. In VARTM, the pressure is uniform and equal to 0.1 MPa everywhere.

A simple experiment was conducted to demonstrate this point more clearly. Figure 3a shows a compression strain as a function of pressure for a single dry and wet fiber mat measured using a parallel plate geometry on an Instron machine via a moving top-plate and stationary bottom plate. The compression strain for a given downward force was measured via a change in height of the top plate accounting for compliance of the machine. The theoretical fiber volume fraction can be calculated via a modified Composites Research Advisory Group (CRAG) Equation [33] given by:

$$\phi_{f,theoretical} = \frac{\rho_{areal}}{\rho_g} \frac{n_{plies}}{h_{dry}(1 - \epsilon_{compression})} \quad (10)$$

where ρ_{areal} is the areal weight of the fiber mat and $\epsilon_{compression}$ is the measured compression strain of the fiber mat. Figure 3b shows the calculated FVF using Equation (10) as a function of consolidation pressures. This graph clearly shows that a dry fiber mat is capable of achieving a little less than twice the FVF of a consolidated wet mat at the same pressure. This important result shows the difficulty in achieving high FVF of random chopped fiber mats with AM methods.

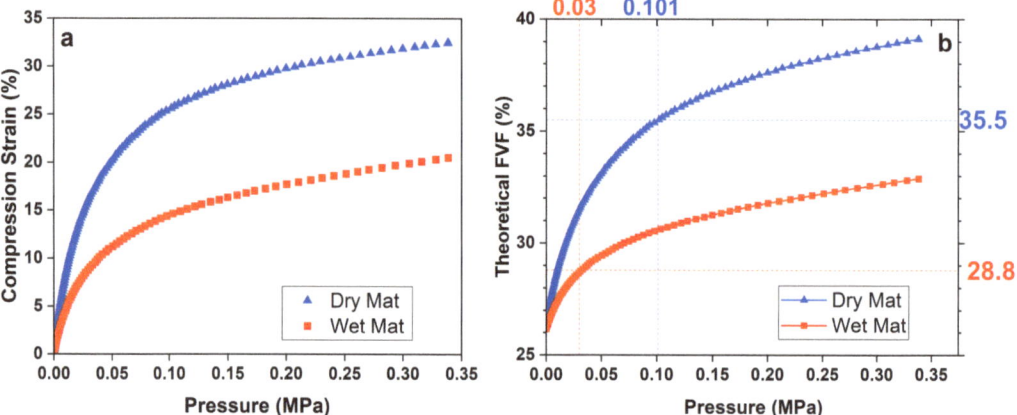

Figure 3. Fiber compression results and fiber volume fraction prediction.

Taking into account the different pressures applied by the VARTM and AM, the consolidation experiment predicts a FVF in AM of 28–29% and 36% in VARTM; see the dotted lines on Figure 3b. Recall from Figure 2a that VARTM and AM achieved a FVF of 40% and 26%, respectively. Overall, the theoretical predictions via compliance are very close to the experimentally determined values in both cases, which supports the argument that the consolidation of wet random chopped glass fiber mats in AM requires significantly higher pressures/forces than VARTM. Figure 3 clearly shows that FVF of AM CSGF composites can be increased by using higher torque motors in the z-axis to increase pressure. However, one must be cognisant that additional pressure could lead to fiber breakage, limiting the amount of pressure allowed. This work is currently under way.

The fact that VARTM resulted in higher mat consolidation than the theoretical prediction in Figure 3 can be explained by the fact that VARTM performs its consolidation with uniform pressure compared to the Instron and AM methods, which apply pressure via a solid platform and are, thus, subject to stress variations. Unfortunately, this issue cannot be overcome in AM methods, and thus the theoretical wet mat consolidation curve is expected to hold. A more accurate dry mat consolidation theoretical curve could be achieved using VARTM at various vacuum pressures. However, the dry mat results in Figure 3 offer a lower limit and suffice to point out the differences between the VARTM method and AM method. Interestingly, there was very little discrepancy in VARTM and AM FVF with PW mats due to the fact that they require much less consolidation [21].

The last result that needs explaining is the unchanged FVF with increasing RRL thickness for $\delta_{RRL} < 125$ µm. Ideally, one would have expected to see that the inclusion of resin-rich domains would decrease the FVF of the specimen proportionately. However, the fact that the FVF stays constant for relatively small δ_{RRL} means that the consolidation of the fiber mat is dependent on RRL. In other words, the only explanation for a constant FVF with larger resin-rich domains is that the fiber mat layers have increasing FVF with increasing RRL thickness. This can only mean that the presence of an RRL increases the consolidation of the subsequently printed fiber mat. One reason for the better mat consolidation is that the RRL provides a more compliant surface by which to apply pressure to the subsequent mat layer. In other words, the mechanical properties of the layer just before mat consolidation is important in ensuring a uniform distribution of stress when the build platform compresses the mat. This dependence should be taken into account when designing an automated AM method of mat placement and consolidation. We now look at how the mechanical properties of AM composite specimens depend on RRL thickness.

4.2. Short Beam Shear/Interlaminar Shear Strength

SBS test was conducted to evaluate the contact strength between CSGF and DA-2 resin (i.e., interlaminar shear strength). During this test, the loading roller applied a compressive force on the beam, leading to formation and propagation of cracks at the center of the specimen from the bottom to the top, illustrated by Figure 4a–c. Note that the test was stopped upon the recording of first load drop (Figure 5), corresponding to diagonal crack formation as in Figure 4d, as per ASTM standard D2344/D2344M-16, and for facilitation of understanding the composite's deformation behavior. Figure 5 shows SBS versus displacement curves for five specimens

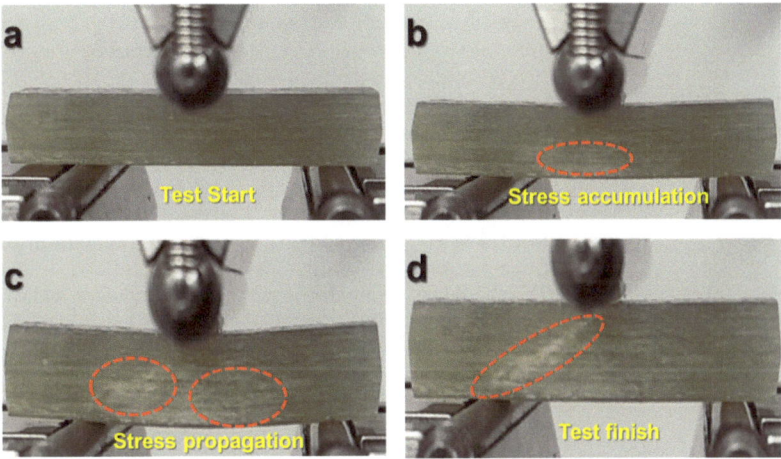

Figure 4. (**a**) SBS test start. (**b**) Cracking signs appearing due to accumulated stress. (**c**) Crack and stress propagation. (**d**) SBS test finish.

The tested specimens did not show any delamination failure. This was discussed in the work by [34,35], where they attributed the observed specimen damages to mixed shearing and compressive buckling caused by the loading roller. These results indicate that stress distribution through the thickness of the SBS specimen deviates from classical beam theory, where the stress is expected to be highest at beam's mid-plane. The reason for this discrepancy is unknown and is still being investigated. From the curves in Figure 5, the measured SBS for DA-2/CSGF composites was 20 MPa, which is remarkably higher than the literature values (i.e., 10–15 MPa) of VARTM epoxy composites using the same fiber mat [36]. This finding confirms the good adhesion and contact between DA-2 and CSGF.

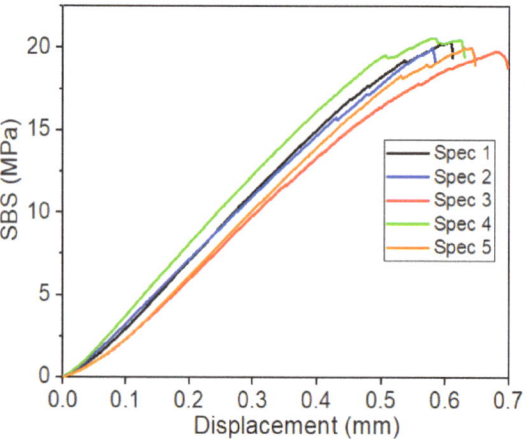

Figure 5. SBS-displacement curves for the tested specimens.

4.3. Static Mechanical Properties Testing

Flexural and Tensile Testing

Figure 6 shows the flexural properties measured for different δ_{RRL} and two different resins, DA-2 and Tenacious (see Table 1 for material properties), compared to a set made using a traditional composite manufacturing technique, VARTM. Unlike the case of PW fiber fabric reported in Idrees et al., the VARTM modulus and strength results are considerably higher than the 0-µm RRL samples printed with CSGF mats [21]. This difference can be explained by the very different FVF that is achieved using VARTM versus AM, see Figure 2. This is discussed in depth in Section 4.1.1.

Figure 6a,b show a relatively constant modulus and strength with increasing δ_{RRL}, despite the drop in FVF at > 100-µm δ_{RRL}. As explained above, a thicker RRL is more efficient in applying uniform pressure on the imperfect mat landscape than a hard reinforcing fiber layer. Thus, the laminate layers have an increased modulus in the fiber-reinforced zones, which contributes to the overall increase in composite modulus due to the presence of RRLs. This is further validated by specimens printed with a Tenacious RRL. The Tenacious RRL specimens have almost identical modulus and strength to the pure DA2 specimens. This should not be the case given the lower modulus and strength of Tenacious; see Table 1. Again, this can only be explained by an increased consolidation of the mat layer by the inclusion of a softer more compliant RRL. The data strongly suggest that the additive manufacturing of GFRC using DLP should consider RRL for improved laminate consolidation Note that this is not the case for oriented fiber mats, as described in Idrees et al. [21], where consolidation is not so important. Figure 6c,d show the modulus and strength normalized by the FVF. We can see from the normalized properties that the non-monotonic behavior is exaggerated and is different to the constant trend observed for oriented fiber mats [21].

As in flexural testing, two batches were made for testing tensile properties of CSGF DA-2 composites, one 3D printed and one VARTM. Figure 7a,b show that the tensile modulus remained almost unchanged for all δ_{RRL}, whereas the strength slightly declined for $\delta_{RRL} = 100$ µm, but remained constant for larger thicknesses. Overall, the trends between flexural and tensile specimens are very similar. The normalized properties shown in Figure 7c,d clearly highlight the non-monotonic behavior of the random fiber mat composites with increase in δ_{RRL}, reflecting the effect of laminate consolidation on the observed trends. Interestingly, Figure 7c,d show that the normalized properties of all AM specimens are higher than the VARTM processed samples.

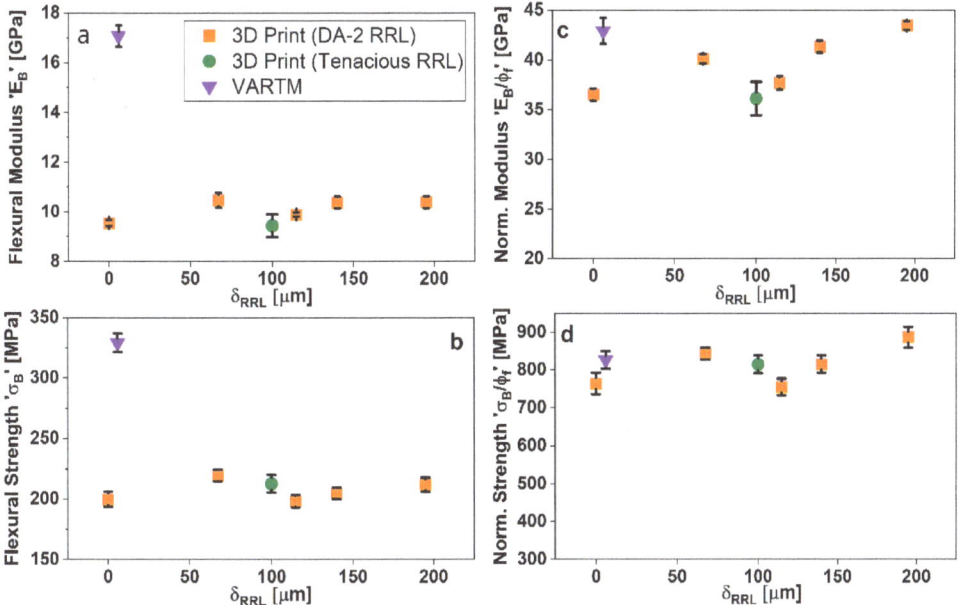

Figure 6. Flexural properties comparison with respect to RRL thickness.

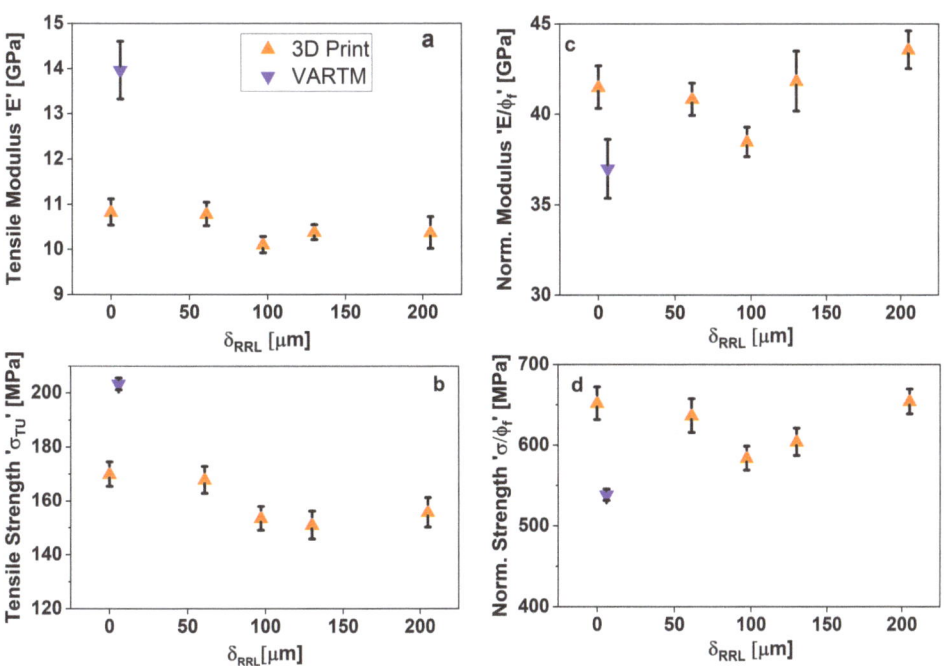

Figure 7. Tensile properties comparison with respect to RRL thickness.

4.4. Mode II Delamination

Mode II delamination was used to quantify the effect of RRL thickness on the delamination resistance, i.e., fracture toughness. Note that mode II measures delamination resistance under predominately shear loading. For mode II delamination testing, six sets were made with δ_{RRL} ranging from 0 to 250 µm using DA-2 as both fiber matrix and RRL. An additional 100 µm RRL thickness set was printed using DA-2 as the fiber matrix and the tough resin, i.e., Tenacious, for the RRL. The propagation of the crack was monitored during the test and the crack for all samples tested was observed to initiate at a displacement between 2.5 and 3 mm. Furthermore, Figure 8 shows that the final crack propagation length was relatively independent of the RRL thickness and resin chemistry. The only major difference between samples was the load required to initiate and propagate the crack.

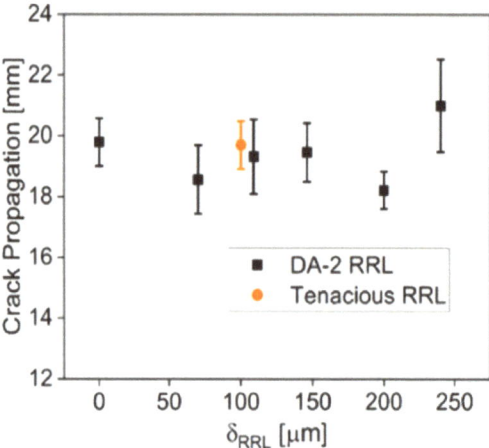

Figure 8. Crack propagation versus δ_{RRL}.

Figure 9 shows that the G_{IIc} is a weak function of RRL thickness using DA-2. The magnitude of G_{IIc} is independent of RRL thickness up to $\delta_{RRL} = 150$ µm, and only slightly increases at 200 and 250 µm. This very small increase in delamination resistance with δ_{RRL} is below that of other published works, where δ_{RRL} showed G_{IIc} improvement upwards of 60% [11,13]. Thus, these results underline the importance of the RRL chemistry. This is further exemplified in the use of Tenacious resin in the RRL, which has a profound impact on the G_{IIc}, i.e., almost a two-fold increase in G_{IIc}. Furthermore, these results are for the most part consistent with a previous study using DA2/PW GF woven fiber [21], except for some key differences. For example, unlike DA2/PW GF composites, CSGF mats show no clear correlation between the peak load and overall crack propagation. Another important difference was the higher G_{IIc} values for CSGF composites compared to PW composites. The average G_{IIc} value was 0.94 kJ/m^2 for 0 µm RRL using PW fabric, whereas the average G_{IIc} value was 1.317 kJ/m^2 using CSGF mats. Although in both fiber materials, the measured G_{IIc} decreases for $\delta_{RRL} = 50$ µm, the decrease using CSGF mats was less sharp, i.e., 6% from 1.317 to 1.245 kJ/m^2 compared to 46% from 0.94 to 0.5 kJ/m^2. Moreover, CSGF mats showed an increasing G_{IIc} with δ_{RRL} above the G_{IIc} of $\delta_{RRL} = 0$ µm, while PW fiber composites showed the highest G_{IIc} measured at $\delta_{RRL} = 0$ µm.

One explanation for the differences between woven fabric and random mats is the different FVF. More specifically, the PW composites were printed with FVF around 40%, compared to the approximately 27% presented here. Davies et al. showed that the delamination resistance decreases with increasing FVF [37]. However, more work is needed to better understand the importance of mat architecture and FVF on the overall delamination resistance of interleaved composites.

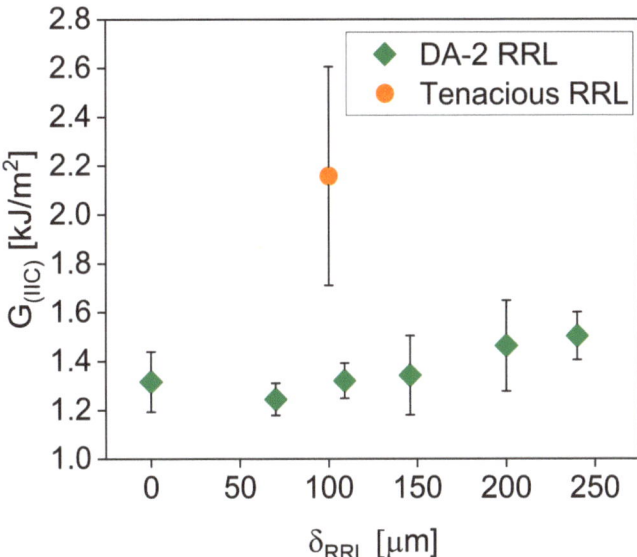

Figure 9. G_{IIc} versus δ_{RRL}.

5. Conclusions

This paper demonstrated successful fabrication of interleaved random glass fiber-reinforced thermosetting polymer composites via DLP, a highly promising vat photo-polymerization AM technique. The key advantage of this technique is efficient and accurate control over interleaf location, chemistry, and dimension. Note that the specimens presented here could not be easily achieved with traditional manufacturing methods. While there are clearly many advantages to the presented AM composite manufacturing process, there exist several challenges that must be considered. Namely, random chopped glass fiber is not easily consolidated using the AM build platform due to the higher pressures required compared to VARTM. This work clearly demonstrated the quantitative differences in the consolidation of wet and dry fiber mats, which is an important consideration when determining FVF of printed parts. Furthermore, it was not possible to eliminate void defects in this process. New strategies would need to be developed to avoid the inclusion of air pockets during the AM composite manufacturing process, such as a vacuum chamber.

The AM composite manufacturing process was used to to study the effect of RRL thickness and chemistry on the tensile, flexure, and interlaminar toughness of printed composite specimens. From the obtained results, we concluded the following:

1. DA-2/ CSGF composites exhibit remarkably higher intelaminar shear strength than other same fiber composites reported in the literature.
2. Additively manufactured random glass fiber composites are about 50% lower in FVF than VARTM composites due to printer motor limitations and the much higher pressures needed to consolidate pre-wetted fiber mats.
3. The presence of RRLs increases fiber mat layer consolidation by distributing the applied consolidation stress more evenly across the mat.
4. Interleaving using brittle resins does not significantly increase mode II delamination resistance. However, significant increases are observed when a ductile resin was used for the RRL. Thus, the resin used for interleaving strongly determines the overall interlaminar fracture toughness of the part.
5. DA-2/CSGF composites have higher Mode II delamination resistance than woven glass fiber composite parts. However, this difference could be due to the very different FVF.

Overall, additive manufacturing is a reliable method for incorporating RRLs in composite parts using a layup technique. The layup process could be automated and incorporated into stereolithographic methods for the facile production of stiff, tough parts using multiple resins and selective incorporation of interleaved domains. More work is required to understand the optimum resin properties for the interleaf and the limitations on toughness and failure mechanisms. These questions are the subject of ongoing investigations.

Author Contributions: A.M.H.I.: draft writing, sample manufacturing and characterization, mechanical testing experiments (short beam shear, flexure and delamination), and respective analysis. M.I.: methodology and data accuracy validation. E.T.: tensile testing conduction and data analysis. A.K.: resources and facilities provision for tensile testing. N.J.A. and G.R.P.: funding provision and research supervision. All authors reviewed the manuscript. All authors have read and agreed to the published version of the manuscript.

Funding: Research was sponsored by the Army Research Laboratory and was accomplished under Cooperative Agreement Number W911NF-17-2-0227. The views and conclusions contained in this document are those of the authors and should not be interpreted as representing the official policies, either expressed or implied, of the Army Research Laboratory or the U.S. Government. The U.S. Government is authorized to reproduce and distribute reprints for Government purposes not withstanding any copyright notation herein.

Data Availability Statement: Data are available upon request.

Acknowledgments: Research was sponsored by the Army Research Laboratory and was accomplished under Cooperative Agreement Number W911NF-17-2-0227. The views and conclusions contained in this document are those of the authors and should not be interpreted as representing the official policies, either expressed or implied, of the Army Research Laboratory or the U.S. Government. The U.S. Government is authorized to reproduce and distribute reprints for Government purposes notwithstanding any copyright notation herein.

Conflicts of Interest: The authors declare that there is no conflict of interest.

References

1. Mallick, P.K. *Fiber- Reinforced Composites Materials, Manufacturing, and Design*, 3rd ed.; CRC Press: Boca Raton, FL, USA, 2007.
2. Shivakumar, K.; Panduranga, R. Interleaved polymer matrix composites—A review. In Proceedings of the 54th AIAA/ASME/ASCE/AHS/ASC Structures, Structural Dynamics, and Materials Conference, Boston, MA, USA, 8–11 April 2013; pp. 1–13. [CrossRef]
3. Odagiri, N.; Kishi, H.; Yamashita, M. Development of torayca prepreg p2302 carbon fiber reinforced plastic for aircraft primary structural materials. *Adv. Compos. Mater.* **1996**, *5*, 249–254. [CrossRef]
4. Wang, S.; Zhang, Y.; Wu, G. Interlaminar shear properties of z-pinned carbon fiber reinforced aluminum matrix composites by short-beam shear test. *Materials* **2018**, *11*, 1–14. [CrossRef] [PubMed]
5. Mignery, L.A.; Tan, T.M.; Sun, C.T. *Use of Stitching To Suppress Delamination in Laminated Composites*; ASTM Special Technical Publication: West Conshohocken, PA, USA, 1985; pp. 371–385. [CrossRef]
6. Pagano, N.J.; Pipes, R.B. The Influence of Stacking Sequence on Laminate Strength. *J. Compos. Mater.* **1971**, *5*, 50–57. [CrossRef]
7. Dow, M.B.; Dexter, H.B. Development of Stitched, Braided and Woven Composite Structures in the ACT Program and at Langley Research Center. 1997; pp. 1–73. Available online: https://dl.acm.org/doi/pdf/10.5555/888217 (accessed on 4 July 2023).
8. Masters, J.E. Improved impact and delamination resistance through interleafing. *Key Eng. Mater.* **1989**, *37*, 317–348. [CrossRef]
9. Sela, N.; Ishai, O.; Banks-Sills, L. The effect of adhesive thickness on interlaminar fracture toughness of interleaved cfrp specimens. *Composites* **1989**, *20*, 257–264. [CrossRef]
10. Grande, D.H.; Ilcewicz, L.B.; Avery, W.B.; Bascom, W.D. Effects of intra- and inter-laminar resin content on the mechanical properties of toughened composite materials. In Proceedings of the NASA Advanced Composites Technology Conference, Seattle, WA, USA, 29 October–1 November 1990; Volume 53, pp. 455–475.
11. Hojo, M.; Ando, T.; Tanaka, M.; Adachi, T.; Ochiai, S.; Endo, Y. Modes I and II interlaminar fracture toughness and fatigue delamination of CF/epoxy laminates with self-same epoxy interleaf. *Int. J. Fatigue* **2006**, *28*, 1154–1165. [CrossRef]
12. Garcia, E.J.; Wardle, B.L.; John Hart, A. Joining prepreg composite interfaces with aligned carbon nanotubes. *Compos. Part A Appl. Sci. Manuf.* **2008**, *39*, 1065–1070. [CrossRef]
13. Yasaee, M.; Bond, I.P.; Trask, R.S.; Greenhalgh, E.S. Mode II interfacial toughening through discontinuous interleaves for damage suppression and control. *Compos. Part A Appl. Sci. Manuf.* **2012**, *43*, 121–128. [CrossRef]
14. Yasaee, M.; Bond, I.P.; Trask, R.S.; Greenhalgh, E.S. Mode I interfacial toughening through discontinuous interleaves for damage suppression and control. *Compos. Part A Appl. Sci. Manuf.* **2012**, *43*, 198–207. [CrossRef]

15. Kim, J.W.; Lee, J.S. Influence of interleaved films on the mechanical properties of carbon fiber fabric/polypropylene thermoplastic composites. *Materials* **2016**, *9*, 1–12. [CrossRef]
16. Lewin, G. The influence of resin rich volumes on the mechanical properties of glass fibre reinforced polymer composites. *Plymouth Stud. Sci.* **2016**, *9*, 123–159.
17. Sacchetti, F.; Grouve, W.J.; Warnet, L.L.; Villegas, I.F. Effect of resin-rich bond line thickness and fibre migration on the toughness of unidirectional Carbon/PEEK joints. *Compos. Part A Appl. Sci. Manuf.* **2018**, *109*, 197–206. [CrossRef]
18. Damodaran, V.; Castellanos, A.G.; Milostan, M.; Prabhakar, P. Improving the Mode-II interlaminar fracture toughness of polymeric matrix composites through additive manufacturing. *Mater. Des.* **2018**, *157*, 60–73. [CrossRef]
19. Singh, S.; Partridge, I.K. Mixed-mode fracture in an interleaved carbon-fibre/epoxy composite. *Compos. Sci. Technol.* **1995**, *55*, 319–327. [CrossRef]
20. Dowling, N. *Mechanical Behavior of Materials: Engineering Methods for Deformation, Fracture, and Fatigue*, 4th ed.; Pearson: Reston, VA, USA, 2013.
21. Idrees, M.; Ibrahim, A.M.H.; Tekerek, E.; Kontsos, A.; Palmese, G.R.; Alvarez, N.J. The effect of resin-rich layers on mechanical properties of 3D printed woven fiber-reinforced composites. *Compos. Part A Appl. Sci. Manuf.* **2021**, *144*, 106339. [CrossRef]
22. Singh, N.; Siddiqui, H.; Koyalada, B.S.R.; Mandal, A.; Chauhan, V.; Natarajan, S.; Kumar, S.; Goswami, M.; Kumar, S. Recent Advancements in Additive Manufacturing (AM) Techniques: A Forward-Looking Review. *Met. Mater. Int.* **2023**, *29*, 2119–2136. [CrossRef]
23. Muhindo, D.; Elkanayati, R.; Srinivasan, P.; Repka, M.A.; Ashour, E.A. Recent Advances in the Applications of Additive Manufacturing (3D Printing) in Drug Delivery: A Comprehensive Review. *AAPS PharmSciTech* **2023**, *24*, 1–22. [CrossRef]
24. Tu, J.; Makarian, K.; Alvarez, N.J.; Palmese, G.R. Formulation of a model resin system for benchmarking processing-property relationships in high-performance photo 3D printing applications. *Materials* **2020**, *13*, 4109. [CrossRef]
25. Wallenberger, F.T.; Watson, J.C.; Li, H. *Glass Fibers*; ASM International: Materials Park, OH, USA, 2001; Volume 21. [CrossRef]
26. ASTM International. Standard Test Method for Short-Beam Strength of Polymer Matrix Composite Materials. *Annu. Book Astm Stand.* **2011**, 1–8. [CrossRef]
27. ASTM International. Standard Test Methods for Flexural Properties of Unreinforced and Reinforced Plastics and Electrical Insulating Materials. D790. *Annu. Book Astm Stand.* **2002**, 1–12. [CrossRef]
28. ASTM International. Standard Test Method for Tensile Properties of Polymer Matrix Composite Materials. *Annu. Book Astm Stand.* **2014**, 1–13. [CrossRef]
29. ASTM International. Standard test method for determination of the mode II interlaminar fracture toughness of unidirectional fiber-reinforced polymer matrix composites. *Annu. Book Astm Stand.* **2014**, 1–18. [CrossRef]
30. ASTM International. Standard test method for ignition loss of cured reinforced resins. *Annu. Book Astm Stand.* **2018**, 1–3. [CrossRef]
31. Kashfi, M.; Tehrani, M. Effects of void content on flexural properties of additively manufactured polymer composites. *Compos. Part C Open Access* **2021**, *6*, 1–8. [CrossRef]
32. Hetrick, D.R.; Sanei, S.H.R.; Ashour, O. Void Content Reduction in 3D Printed Glass Fiber-Reinforced Polymer Composites through Temperature and Pressure Consolidation. *J. Compos. Sci.* **2022**, *6*, 128. [CrossRef]
33. Curtis, P.T. *Crag Test Methods for the Measurement of the Engineering Properties of Fibre Reinforced Plastics*; Technical Report; Royal Aerospace Establishment: Farnborough, UK, 1988.
34. Figliolini, A.M.; Carlsson, L.A. Mechanical Properties of Carbon Fiber/Vinylester Composites Exposed to Marine Environments. *Polym. Compos.* **2014**, *35*, 1559–1569.
35. Whitney, J.M.; Browning, C.E. On short-beam shear tests for composite materials. *Exp. Mech.* **1985**, *25*, 294–300. [CrossRef]
36. Heckadka, S.S.; Nayak, S.Y.; Narang, K.; Vardhan Pant, K. Chopped Strand/Plain Weave E-Glass as Reinforcement in Vacuum Bagged Epoxy Composites. *J. Mater.* **2015**, *2015*, 957043. [CrossRef]
37. Davies, P.; Casari, P.; Carlsson, L.A. Influence of fibre volume fraction on mode II interlaminar fracture toughness of glass/epoxy using the 4ENF specimen. *Compos. Sci. Technol.* **2005**, *2*, 295–300. [CrossRef]

Disclaimer/Publisher's Note: The statements, opinions and data contained in all publications are solely those of the individual author(s) and contributor(s) and not of MDPI and/or the editor(s). MDPI and/or the editor(s) disclaim responsibility for any injury to people or property resulting from any ideas, methods, instructions or products referred to in the content.

Article

Visco-Elastic and Thermal Properties of Microbiologically Synthesized Polyhydroxyalkanoate Plasticized with Triethyl Citrate

Madara Žiganova [1,*], Remo Merijs-Meri [1], Jānis Zicāns [1], Ivan Bochkov [1], Tatjana Ivanova [1], Armands Vīgants [2], Enno Ence [3] and Evita Štrausa [3]

1. Institute of Polymer Materials, Faculty of Materials Science and Applied Chemistry, Riga Technical University, 3 Paula Valdena Street, LV-1048 Riga, Latvia; remo.merijs-meri@rtu.lv (R.M.-M.); janis.zicans@rtu.lv (J.Z.); ivans.bockovs@rtu.lv (I.B.); tatjana.ivanova@rtu.lv (T.I.)
2. Laboratory of Bioconversion of Carbohydrates, University of Latvia, 1 Jelgavas Street, LV-1050 Riga, Latvia; armands.vigants@lu.lv
3. SIA MILZU!, LV-3040 Riga, Latvia; enno.ence@rtu.lv (E.E.); evita.strausa@rtu.lv (E.Š.)
* Correspondence: madara.ziganova@rtu.lv; Tel.: +371-67-089-2525

Abstract: The current research is devoted to the investigation of the plasticization of polyhydroxybutyrate (PHB) and polyhydroxybutyrate-co-hydroxyvalerate (PHBV) with triethyl citrate (TEC). Three different PHB or PHBV-based systems with 10, 20, and 30 wt.% of TEC were prepared by two-roll milling. The effect of TEC on the rheological, thermal, mechanical, and calorimetric properties of the developed compression-molded PHB and PHBV-based systems was determined. It was revealed that the addition of TEC significantly influenced the melting behavior of both polyhydroxyalkanoates (PHA), reducing their melting temperatures and decreasing viscosities. It was also revealed that all the investigated systems demonstrated less than 2% weight loss until 200 °C and rapid degradation did not occur until 240–260 °C in an oxidative environment. Apart from this, a remarkable increase (ca 2.5 times) in ultimate tensile deformation ε_B was observed by increasing the amount of TEC in either PHB or PHBV. A concomitant, considerable drop in ultimate strength σ_B and modulus of elasticity E was observed. Comparatively, the plasticization efficiency of TEC was greater in the case of PHBV.

Keywords: biopolymers; polyhydroxyalkanoates; plasticization; triethyl citrate; polyhydroxybutyrate

Citation: Žiganova, M.; Merijs-Meri, R.; Zicāns, J.; Bochkov, I.; Ivanova, T.; Vīgants, A.; Ence, E.; Štrausa, E. Visco-Elastic and Thermal Properties of Microbiologically Synthesized Polyhydroxyalkanoate Plasticized with Triethyl Citrate. *Polymers* **2023**, *15*, 2896. https://doi.org/10.3390/polym15132896

Academic Editors: Giorgio Luciano and Maurizio Vignolo

Received: 20 April 2023
Revised: 26 June 2023
Accepted: 27 June 2023
Published: 29 June 2023

Copyright: © 2023 by the authors. Licensee MDPI, Basel, Switzerland. This article is an open access article distributed under the terms and conditions of the Creative Commons Attribution (CC BY) license (https://creativecommons.org/licenses/by/4.0/).

1. Introduction

Huge amounts of annually generated synthetic plastic waste critically affect the environment. Since 2009 the waste quantity has increased by 24%, whereas in 2019 34.4 kg of plastic waste per person on average was generated in the EU [1]. The environmental issues predominantly are caused by the daily consumption of synthetic polymer products with short life cycles (packaging and disposables). Many products with short life cycles are often mixed in a waste stream making their separation and recycling complicated [2]. Consequently, it is important to develop environmentally sustainable alternatives, primarily for products of short life cycles.

Microbially synthesized polyhydroxyalkanoates (PHAs) are polyesters produced by microorganisms as intracellular granules under nutrient stress. In 1925, Lemognie discovered the simplest form of PHAs, polyhydroxybutyrate (PHB), as a source of energy and carbon storage in microorganisms. Under optimal conditions, above 80% of the dry weight of *Alcaligenis eutrophus* is of PHB [3]. Other most studied strains for PHB production are *Ralstonia eutropha* (also known as *Cupriavidus necator*) [4], *Alcaligenes* spp., *Azotobacter* spp. [5], *Bacillus* spp., *Nocardia* spp., *Pseudomonas* spp., and *Rhizobium* spp. [6]. These strains are suitable for the production of not only PHB, but also other members of the PHA family such as poly(3-hydroxyvalerate) (PHV), poly(3-hydroxybutyrate-co-3-hydroxyvalerate) copolymer (PHBV), poly(3-hydroxyoctanoate) (PHO), and poly(3-hydroxynonanoate) (PHN) [4,5,7,8]. Copolymers, such as PHBV, have better stress-strain characteristics than PHB and, therefore, they are more attractive for practical use. A fully biodegradable PHBV has been commercially available since 1990 by the company ICI "Biopol". However, the share of PHAs is still negligible (1.7% [9]) from the biopolymers global market, which in 2019 was estimated to be 3.8 million tons, in sharp contrast to the fossil-based polymer market of 372 million tons [10].

Neat PHB is brittle without modification due to high stereoregularity degree and formation of very large and overlapped spherulites with a high tendency to crack. To resolve this issue PHB is plasticized, which undoubtedly affects its rheological, thermal, and mechanical properties [4,11–13]. The most common plasticizers for PHB are (1) esters, such as citrates [14], adipates [13], phthalates [15], diols and (2) polyols, such as poly(ethylene glycol) (PEG) [11] and Laprol [16], respectively, (3) vegetable oils, such as epoxidized soybean oil [16], and (4) terpenes [17]. The influence of some plasticizers on the thermal and mechanical properties of PHB and PHBV is reported in Table 1. Unemura et al. [14] found that TEC is an efficient plasticizer for PHB, gradually changing its mechanical and thermal properties. Requena et al. [11] evaluated the efficiency of adding PEG200, PEG1000, and PEG4000 to PHB by achieving minor improvement in ultimate elongation at break though at decreased stiffness and tensile strength. Scalioni et al. [18] reported that the addition of TEC at a mass fraction of 0.3 resulted in a decrease in the elastic modulus of PHB from 230 MPa to 120 MPa. Rapa et al. [19] obtained samples of PHB/TEC blends displaying maximum elongation at a break of 3.1% by loading of 30% TEC.

Table 1. Selected properties of some plasticized PHB systems.

PHAs	TEC(wt. Parts)	Plasticizer	T_m (°C)	X (%)	ε_B (%)	Reference
PHB	0	Epoxidized linseed oil (ELO)	175	52	9.7	[20]
	0.05		173	47	12.7	
	0.1		171	46	13.6	
	0.05	Epoxidized soybean oil (ESBO)	173	47	9.2	
	0.1		172	48	8.9	
PHB	0	Triethyl citrate (TEC)	180	81	5.8 ± 0.6	[14]
	0.1		173	71	5.6 ± 0.4	
	0.2		171	62	7.4 ± 0.9	
	0.3		162	53	6.9 ± 1.6	
PHB	0	Dioctyl (o-)phthalate (DOS)	169	56 *	2.5 ± 0.5	[15]
	0.25		164	54 *	3.9 ± 0.3	
	0.3		163	57 *	4.3 ± 0.6	
	0.35		165	50 *	5.4 ± 0.9	
	0.4		165	57 *	5.2 ± 0.6	
	0.5		165	51 *	-	
	0	Acetyl tributyl citrate (ATBC)	169	56 *	2.5 ± 0.5	
	0.1		163	61 *	6.1 ± 0.8	
	0.2		160	58 *	8.5 ± 0.9	
	0.25		158	60 *	8.8 ± 0.9	
	0.3		157	59 *	9.7 ± 0.7	

Table 1. Cont.

PHAs	TEC(wt. Parts)	Plasticizer	T_m (°C)	X (%)	ε_B (%)	Reference
PHBV	0	Biodegradable oligomeric polyester based on lactic acid, adipic acid, and 1,2-propanediol at a molar ratio of 20:40:40 (PLAP)	174	53	8 ± 0.4	[21]
	0.1		173	55	8.2 ± 0.4	
	0.2		170	56	8.1 ± 0.4	
	0.3		174	61	6.6 ± 0.5	

* Crystallinity calculated from DSC data by assuming the enthalpy of 100% crystalline PHB = 146 J/g as reported by [12].

Although citrate-plasticized PHB and PHBV systems have been widely investigated, not all the aspects have been completely resolved, for example, thermooxidative behavior at elevated temperatures in the air environment. Apart from this, high price and complicated synthesis of technologically competitive PHA copolymers with high stress-strain characteristics still are one of the main limiting factors for increasing production amounts of the polymer.

Consequently, in the current research, we have performed a synthesis of PHB using a simple low-cost approach. To reduce brittleness, we have performed melt plasticization of PHB using TEC as a cheap and environmentally friendly plasticizer. To evaluate plasticization efficiency at different TEC contents we have investigated structural, thermogravimetric, rheological, and mechanical properties over a broad temperature range. For comparison, the effect of TEC on the above-mentioned properties of commercially available PHB copolymer (PHBV) with small (1%) hydroxyvalerate content has been investigated.

2. Materials and Methods

2.1. Materials

PHB homopolymer was obtained from bacteria *Cupriavidus necator NCIMB 11,599* by fermentation on glucose in a fed-batch process with phosphate limitation according to Haas et al. [22]. The PHB was recovered by a modified method of Yang et al. [23]. In short, cells of *Cupriavidus necator* were separated by centrifugation at 4500 rpm for 25 min, and obtained biomass was freeze-dried. PHB was extracted from dry biomass by resuspending in 7% SDS solution and incubating for 20 h at 70 °C. After centrifugation at 8000 rpm for 10 min, the PHB sediment was washed with water four times and freeze-dried.

Poly(3-hydroxybutyrate-co-3-hydroxyvalerate was a commercial product (PHBV, China, Ningbo City, TianAn Biopolymer: ENMAT Y1000) with 1 mol% HV 3-(hydroxyvalerate) content [24].

Triethyl citrate (TEC, Burlington, MA, USA, Sigma Aldrich, M_w = 276.28 g·mol^{-1}, ρ = 1.137 g L^{-1}) was used as a plasticizer.

2.2. Preparation of Plasticized Systems

Both biopolymers, before plasticization, were dried at 60 °C in a vacuum oven for 24 h. As shown in Table 2, plasticized systems with TEC weight concentrations of 10%, 20%, and 30% were mixed using a two-roll mill LRM-S-110/3E from Lab Tech Engineering Company Ltd. Mixing time was 3 min and the roll temperatures were 165 °C and 175 °C. Furthermore, the plasticized systems were milled at room temperature and 700 rpm using a Retsch cutting mill SM300 with a 6 mm sieve. The obtained flakes (see Figure 1) with average dimensions of 3 mm × 2 mm were used for manufacturing test specimens using compression molding. Test specimens for mechanical property tests were cut from ~0.5 mm thick plates with dimensions of 60 mm × 100 mm obtained by hot pressing at 190 °C. Samples for oscillatory shear rheology tests were cut from 1 mm thick plates with dimensions of 60 mm × 100 mm, similarly obtained by hot pressing.

Table 2. Codes and the composition of the plasticized PHB and PHBV systems.

Code	PHBV (wt.%)	PHB (wt.%)	TEC (wt.%)
PHBV	100	—	0
PHBV10	90	—	10
PHBV20	80	—	20
PHBV30	70	—	30
PHB	—	100	0
PHB10	—	90	10
PHB20	—	80	20
PHB30	—	70	30

Figure 1. Image of PHBV flakes after milling.

2.3. Characterization Methods

2.3.1. Molecular Weight (M_w)

The viscosity average molecular weight was determined using Ubbelohde viscometer type 1C with diameter of the capillary 0.56 mm (Schott—Instruments GmbH, Mainz, Germany) at 30 °C temperature following the guidelines of ISO 1628. All the samples were dissolved in chloroform to obtain solutions with five different concentrations in the range between 50 mg and 250 mg of PHAs per 100 mL of the solvent. The viscosity average molecular weight for each sample was obtained using the Mark–Houwink equation with K and values of 1.18×10^{-2} and 0.780, respectively, as reported elsewhere [4,5,20]:

$$[\eta] = k\, M_w^{\,a} \qquad (1)$$

where [η] is the intrinsic viscosity of PHAs solutions in chloroform and M_w is viscosity average molecular weight.

2.3.2. Fourier Transform Infrared Spectroscopy (FT−IR)

FT-IR spectra were obtained by Thermo Fisher Scientific Nicolet 6700 spectrometer (Thermo Fisher Scientific Inc., Waltham, MA, USA) by Attenuated Total Reflectance (ATR) technique. All the spectra were recorded in the range 650–4000 cm^{-1} with a resolution of 4 cm^{-1}.

2.3.3. Thermogravimetric Analysis (TGA)

Therogravimetric properties were analyzed using a Mettler Toledo thermogravimetric analyzer TGA1/SF (Mettler Toledo, Greifensee, Switzerland). Samples of approximately 10 mg were heated from ambient temperature to 600 °C at a heating rate of 10 °C/min under an air atmosphere. The material weight loss was calculated using the original software following the ASTM D3850.

2.3.4. Differential Scanning Calorimetry (DSC)

Melting/crystallization behavior was evaluated using a Mettler Toledo differential scanning calorimeter DSC 1/200W. The specimen of approximately 10 mg was sealed in an aluminum pan and subjected to the following temperature cycles: (1) heating from—50 °C to 200 °C at a rate of 10 °C/min and holding at the corresponding target temperature for 5 min, (2) cooling to 25 °C at a rate of 10 °C/min and holding at the corresponding target temperature for 5 min, followed by (3) second heating from 25 °C to 200 °C at a rate of 10 °C/min. The DSC measurements were performed underneath a nitrogen atmosphere. The degree of crystallinity (χ) was calculated using the following equation:

$$\chi = \frac{\Delta H_c}{\Delta H_m^o (1-W)} \times 100 \qquad (2)$$

where ΔH_c is the measured specific melt enthalpy of the compound and ΔH_m^o is the melting enthalpy of the 100% crystalline PHB = 146 J/g [12].

2.3.5. Oscillatory Shear Rheology

Discs (ca 1.0 mm (h) × 25 mm ∅) were cut from compression-molded plates of both PHAs and plasticized systems using a die-cutting press and circular die with appropriate dimensions. Complex viscosity η^* was measured as a function of angular frequency ω in the oscillatory mode at 190 °C at 1% strain, and within the frequency range of 0.01 Hz to 100 Hz (ω = 0.0628 to 628 rad/s) using a Modular Compact Rheometer SmartPave 102 (Anton Paar GmbH, Graza, Austria—Europe) equipped with 25 mm diameter parallel plate configuration.

2.3.6. Tensile Properties

Tensile stress—strain characteristics were determined at a temperature of 20 °C in accordance with EN ISO 527 using Zwick Roell material testing equipment BDO—FB020TN (Zwick Roell Group, Ulm, Germany) equipped with pneumatic grips. Type 5A test specimens were stretched at a constant deformation speed of 50 mm/min. Demonstrated values represent the averaged results of the measurements performed on 10 test specimens for each type of plasticized system.

2.3.7. Dynamic Mechanical Thermal Analysis (DMTA)

Dynamic mechanical thermal analysis was carried out using METTLER TOLEDO DMA/SDTA861^e (METTLER TOLEDO GmbH, Analytical, Schwerzenbach, Switzerland) operating in a tensile mode at 10 N of maximum stress, 10 µm of maximum strain, and a frequency of 1 Hz. Tests were run within a temperature range from −50 °C to +105 °C at a heating rate of 2 °C/min.

3. Results

3.1. Molecular Weight

Viscosity average molecular weights of PHBV copolymer and PHB homopolymer were calculated from intrinsic viscosity values which were determined using a Ubbelohde type viscometer type 1C with diameter of the capillary 0.56 mm (Schott—Instruments GmbH, Mainz, Germany). The intrinsic viscosity of a polymer is related to its molecular weight, side chain length, and degree of branching. In general, polymers with higher molecular weights and longer side chain lengths have higher intrinsic viscosities, as there is a greater degree of chain entanglement. Intrinsic viscosity also provides information about the conformations of a polymer by reflecting the degree of chain entanglement and intermolecular interactions that occur in a solution.

Linear extrapolation trendlines of PHA viscosity as a function of its solution in chloroform concentration are reported in Supplementary Materials Figure S1. As seen from Figure S1 and Table 3, PHBV shows considerably higher viscosity than PHB, consequently, M_w of PHBV is approximately eight times higher than that of PHB. Quagliano et al. [13] have reported that molecular weight, yield, composition, and purity of PHB largely depend on the carbon source and its concentration. For example, it has been observed that by increasing glucose or molasses concentration from 10–50 g/L in the isolated rhizospheric soil samples from the Agronomy Faculty Campus (Buenos Aires, Argentina), the molecular weight of PHB after 24 h of fermentation gradually increased from 55–80 kDa to 300–400 kDa, and 500–700 kDa in the case of glucose and molasses carbon source, respectively [13]. In turn, Luigi-Jules Vandi et al. [24] have reported that the Mw of commercial PHBV with 1 mol% HV 3-(hydroxyvalerate) content, purchased in a powder form from TianAn Biopolymer, China, under the trade name of ENMAT Y1000, usually ranges from 550–650 kDa as analyzed by gel permeation chromatography. The molecular weight of PHB is typically lower than that of PHBV due to differences in their chemical composition and polymerization mechanisms [25,26].

Table 3. Molecular weight and instinct viscosity of PHAs.

Sample Code	M_w (kDa)	η (Pa·s)
PHBV	540	0.69
PHB	66	3.86

3.2. Fourier-Transform Infrared Spectroscopy (FT-IR)

FTIR-ATR spectroscopy was used to assess the structural changes in the plasticized systems after the introduction of TEC. The collected FTIR spectra are shown in Figure 2. In Table 4, representative absorption bands of PHB, PHBV, and TEC are summarized. There is no great difference between the FTIR spectra of PHB and PHBV due to a small amount of HV units in PHBV (1 mol%). The addition of TEC also did not change FTIR spectra dramatically due to structural similarities between the plasticizer and the polymer. The appearance of no new peaks in the FTIR spectra of the plasticized systems also confirms no chemical interaction between TEC and PHB or PHBV. The greatest changes after the addition of TEC have been observed in the carbonyl absorption region; while for TEC, this peak is shifted to the direction of longer wavelengths in comparison to those of PHB and PHBV. However, several bands attributed to C–O–C groups' asymmetric stretching (1180 cm^{-1} and 1181 cm^{-1} for PHB and PHBV, respectively), C–O–C groups' symmetric stretching (1130 cm^{-1} and 1129 cm^{-1} for PHB and PHBV, respectively), C-O groups' vibrations (1226 cm^{-1} and 1274 cm^{-1}), and also –C=O groups' vibrations have been previously related to the ratio of amorphous and crystalline parts of PHAs [27]. Therefore, the main attention was devoted to the assessment of the changes in the crystalline structure of PHB and PHBV after plasticization with TEC. In the current case, respective bands are not significantly shifted due to the addition of TEC. However, the intensity of the bands at 1181 cm^{-1}, 1129 cm^{-1}, 1226 cm^{-1}, and 1274 cm^{-1} of PHBV decreased to a greater extent in comparison to that of the TEC plasticized PHB (see Figure 3). This may be indicative of a larger influence of TEC in the crystalline structure of PHBV in comparison to that of PHB, resulting in more effective plasticization of the copolymer along with addition of TEC.

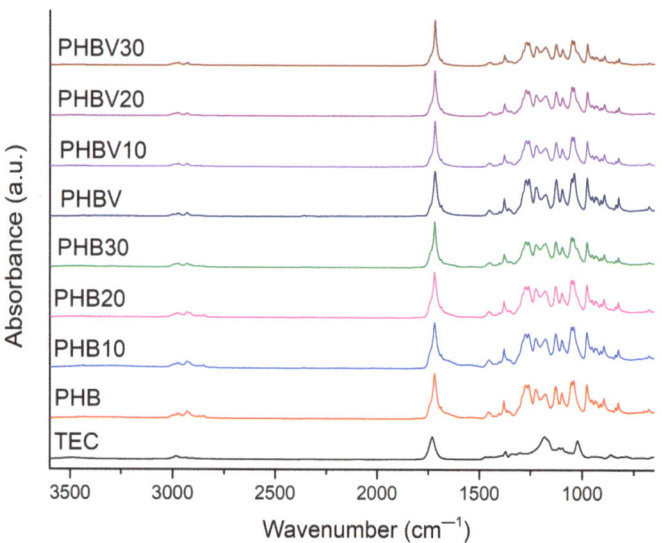

Figure 2. FT-IR spectra of PHAs and their plasticized composites basic functional groups.

Table 4. Representative FT-IR absorption intensities of PHAs [28–31].

Mode of Molecular Vibration	PHB		PHBV		TEC	
	Current Research	Reference	Current Research	Reference	Current Research	Reference
C–C backbone stretching	978	—	977	977 [28]	—	—
O–C–C stretching	1043/1054	—	1044/1054	1054 [28]	1023	—
O–C–C asymmetric stretching	1100		1099	1099 [28]	1096 1113	1097 [31] 1114 [31]
C–O–C symmetric stretching	1130		1129	1129 [28]	—	
C–O–C asymmetric stretching	1180	1000–1300 [29]	1181	1179 [28]	1182	
C–O symmetric stretching	1226		1226	1226 [28]	—	1050-1300 [31]
C–O symmetric stretching of aliphatic esters	1260		1261	1261 [28]	—	
C–O symmetric stretching	1274		1274	1275 [28]	—	
C–H symmetric bending of methyl (-CH$_3$) groups	1379	1377 [29]	1379	1379 [28]	1370	1373 [31]
C–H asymmetric stretching and bending vibrations of methyl (-CH$_3$) and methylene (-CH$_2$-) groups	1453	1452 [29]	1452	1452 [28]	—	—
C=O stretching of ester groups	1718	1727 [29]	1718	1720 [28] 1722 [30]	1730	1735 [31]
—CH$_3$ symmetric stretching	2851/2873	—	2851/2873	2881 [30]	—	—
—CH$_2$ symmetric stretching	2390	—	2932	2933 [28] 2925/2945 [30]	—	—
C–H asymmetric vibration of methyl (-CH$_3$) groups	2975	2927/2969 [29]	2976	2975 [28]	2982	2983 [31]
Terminal –OH group	3434	3434 [29]	3435	3434 [28]	3484	3502 [31]

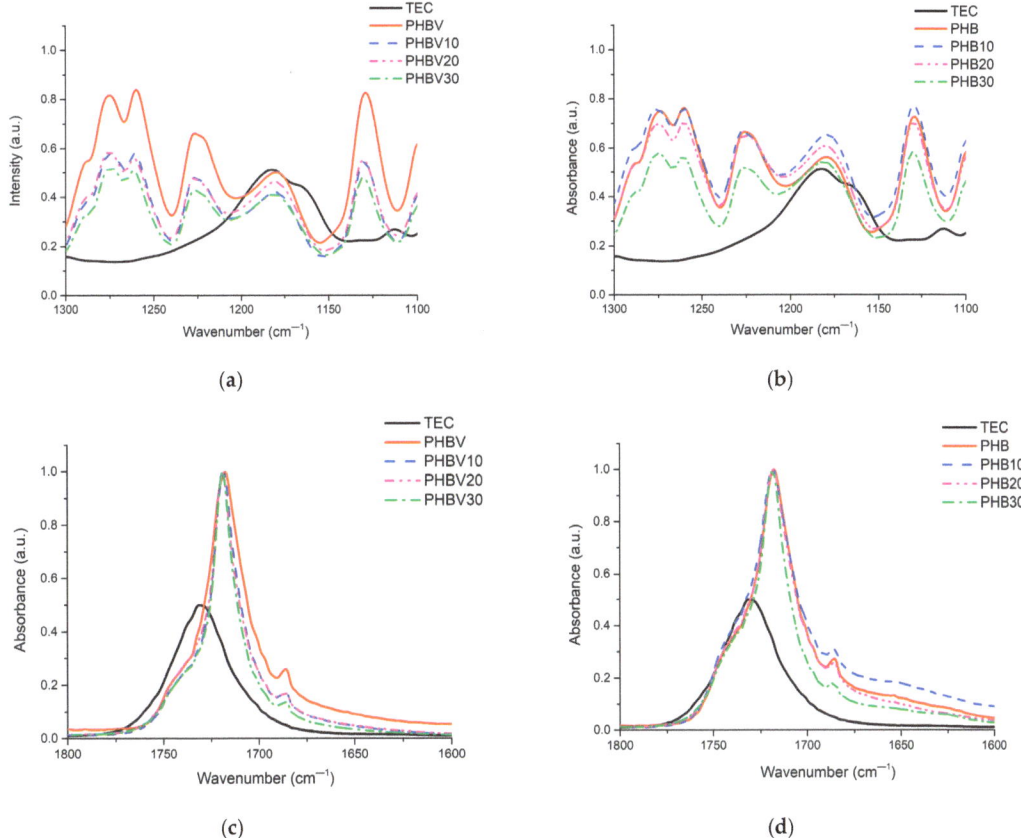

Figure 3. FTIR-ATR spectra of PHBV, PHB, and their plasticized systems within wavenumber range 1725-1740 cm^{-1} (**a**,**b**), and 1150-1240 cm^{-1} (**c**,**d**).

3.3. Thermal Gravimetric Analysis (TGA)

Although many research groups have investigated the thermal behavior of PHAs by TGA, only a few of these investigations have been performed in an oxidative environment, disregarding the fact that even in a closed system, such as an extruder barrel, there is a certain amount of dissolved oxygen [32]. Consequently, TGA tests in the current research have been performed in an oxidative environment. The TGA thermograms of the investigated PHB- and PHBV-based systems are shown in Figure 4. The thermal stability of neat PHB is higher than that of PHBV, which could be explained by the lower deactivation energy of the latter (177 kJ/mol and 136 kJ/mol, respectively) as reported by Yun Chen et al. [27]. As expected, the addition of TEC, which has lower thermal stability, decreased the thermal resistance of the investigated plasticized systems. By increasing the content of TEC, the onset thermal degradation temperature T_{on} decreases. A relatively larger decrease in T_{on} is the case for PHB-based systems resulting in the fact that both plasticized systems with 30 wt.% of TEC show almost identical T_{on}. However, the slope of TGA curves within the main mass loss region for the plasticized systems decreased, testifying that TEC contributes to the formation of the gas-impermeable char layer, reducing the diffusion of oxygen to the zone of burning and decreasing the combustion rate. It should, however, be mentioned that there is negligible mass loss (less than 1%) of the investigated systems if heated up to 190 °C, which was the processing temperature of the investigated systems. In spite of this slight mass loss, the decrease in the molecular weight of PHAs during 30 min of isothermal heating at 180 °C is more than 20% [33], which repeatedly testifies that the processing of PHAs base systems should be performed at possibly low temperatures and short cycle times in Table 5.

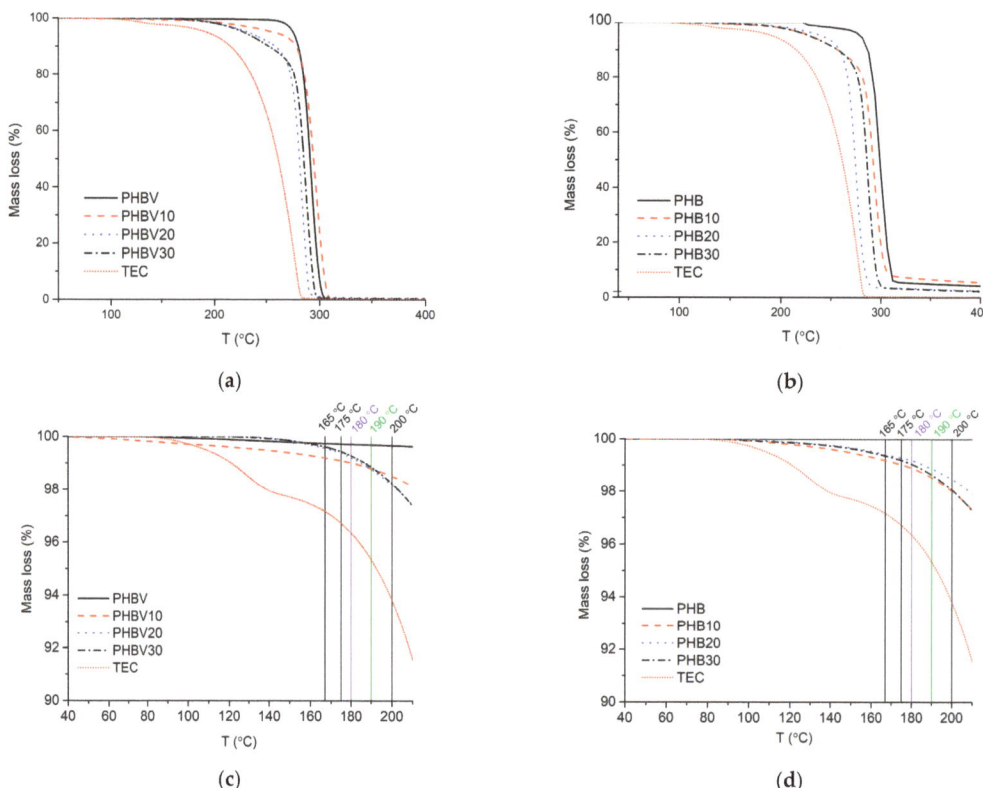

Figure 4. TGA thermograms of TEC plasticized PHBV and PHB systems within the full temperature range (**a**,**b**, respectively) and within the temperature range of 40 °C and 200 °C (**c**,**d**, respectively).

Table 5. Percent mass loss temperatures of PHBV, PHB, and its plasticized systems.

Sample Code	Residual Mass at Fixed Temperature, wt.%			T_{on}, °C	Percent Mass Loss Temperatures, °C		
	180 °C	190 °C	200 °C		$T_{1\%}$	$T_{5\%}$	T_{deg}
PHB	100	100	100	288	227	279	298
PHB10	100	99	99	280	202	246	258
PHB20	100	99	98	264	185	241	274
PHB30	100	99	98	275	181	229	286
PHBV	100	100	100	283	259	276	295
PHBV10	99	99	99	285	208	257	299
PHBV20	99	99	98	279	184	234	283
PHBV30	99	99	98	275	187	228	287
TEC	96	95	94	276	120	190	276

3.4. Differential Scanning Calorimetry (DSC)

DSC thermograms of the first heating run of all the investigated systems are summarized in Figure 5, whereas the main calorimetric data of the thermograms are given in Table 6. DSC thermograms of the subsequent cooling and second heating runs of PHB,

PHBV, and their plasticized systems are reported in the Supplementary Materials Figures S2–S5, whereas the main calorimetric data are summarized in Table 7 for the cooling run and Table 8 for the second heating run.

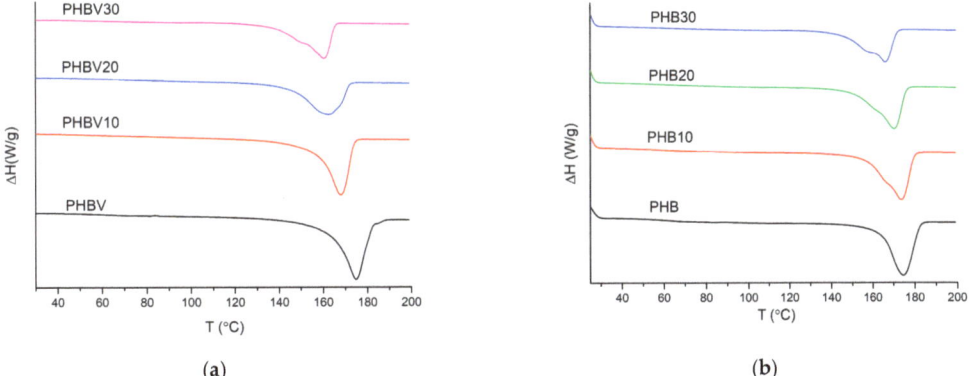

Figure 5. First heating DSC thermograms of the plasticized systems based on PHBV (**a**) and PHB (**b**).

Table 6. Results of the first heating run of neat PHBV, PHB, and TEC plasticized systems.

Sample Code	wt. %	1st Heating Run						
		χ (%)	ΔH$_m$ (J/g)	T$_m^1$ (°C)	T$_m^2$ (°C)	T$_{onset}$ (°C)	T$_{offset}$ (°C)	ΔT (°C)
PHB	0	53	77	-	174	136	183	47
PHB10TEC	10	58	76	165 *	172	133	181	48
PHB20TEC	20	59	69	160 *	169	127	177	50
PHB30TEC	30	60	61	157 *	166	122	173	51
PHBV	0	64	94	-	175/185 *	135	192	57
PHBV10%TEC	10	59	78	-	168	118	177	59
PHBV20%TEC	20	57	66	-	163	114	174	60
PHBV30%TEC	30	59	60	151*	161	106	168	62

* Relates to inflection point.

Table 7. Results of the cooling run of neat PHBV, PHB, and TEC plasticized systems.

Sample Code	wt.%	Cooling Run					
		χ (%)	ΔH$_m$ (J/g)	T$_m^1$ (°C)	T$_{onset}$ (°C)	T$_{offset}$ (°C)	ΔT (°C)
PHB	0	45	66	91	67	112	45
PHB10TEC	10	44	58	71	51	100	49
PHB20TEC	20	47	55	75	49	100	51
PHB30TEC	30	45	46	69	43	96	53
PHBV	0	46	68	83	55	108	53
PHBV10%TEC	10	47	61	81	52	106	54
PHBV20%TEC	20	44	51	75	44	99	55
PHBV30%TEC	30	45	46	71	31	98	67

Table 8. Results of the second heating run of neat PHBV, PHB, and TEC plasticized systems.

Sample Code	wt.%	2nd Heating Run									
		χ_{cc} (%)	ΔH_{cc} (J/g)	T_{cc} (°C)	χ * (%)	ΔH_m (J/g)	T_m^1 (°C)	T_m^2 (°C)	T_{onset} (°C)	T_{offset} (°C)	ΔT (°C)
PHB	0	3	4	99	54 (51)	78	169	173	139	183	44
PHB10TEC	10	2	2	89	59 (57)	77	156	167	131	176	45
PHB20TEC	20	3	4	89	59 (56)	69	154	166	129	175	46
PHB30TEC	30	4	4	83	62 (58)	63	157	165	127	174	47
PHBV	0	5	8	95	61 (56)	90	167	172	128	185	57
PHBV10%TEC	10	3	4	94	62 (58)	81	162	170	118	176	58
PHBV20%TEC	20	5	6	91	63 (58)	74	154	165	117	175	58
PHBV30%TEC	30	4	5	90	62 (58)	64	148	162	112	172	60

* Value in the brackets reflects the initial crystallinity of the PHB or PHBV by considering the cold crystallization effect.

It is known that the crystallization of PHAs is affected by nucleation acts and spherulite growth dynamics, which often results in the formation of multimodal exothermic peaks due to the irregular release of heat [34,35]. Consequently, multimodal melting behavior is observed for the investigated PHBV in the first heating run demonstrating one expressed major melting peak at 175 °C, which overlaps with a minor melting peak at 185 °C. The presence of double peaks in melting endotherms is generally explained by two mechanisms: (1) double lamellar thickness population model [36] and (2) melting and recrystallization model [37]. Most probably, the major melting peak of PHBV is attributed to the melting of initially present crystalline structures, which tend to recrystallize into thicker more perfect lamellas. Due to the low co-monomer content, it is believed that the melting peak of HV moieties is overlapped with the melting of dominating HB moieties. This results in a broader melting interval of PHBV in comparison to PHB. During the cooling run, single melting peak of PHBV is observed around 83 °C. In the case of the second run, the bimodal melting behavior of PHBV is observed, whereas the melting peaks are shifted to lower temperatures, which may be because of less crystallization time for the PHA sample, as previously observed by Yun Chen et al. [27]. Similar trends may also be observed from PHB scans, whereas the observed melting/crystallization peak temperatures are somewhat higher in comparison to those of PHBV.

If TEC is added, the melting endotherm of PHBV systems shifts to the left side resulting in lower peak temperatures of melting of the polymers' crystalline fractions. It is also worth noting that the addition of TEC, even at its lowest amount (10 wt.%), contributes to the development of multimodal melting behavior, i.e., by increasing TEC concentration, the lower temperature melting peak becomes more separate. This testifies that TEC influences the nucleation process of PHBV. In a similar way, the crystallization peak temperature of the plasticized PHA compositions decreases by TEC addition, and in the case of PHBV-based systems crystallization occurs in a broader range in comparison to PHB/TEC.

As it is demonstrated in Table 8, controlled crystallization of the investigated systems at the rate of 10 °C/min in the DSC cell, initiated the development of frozen structures, resulting in the appearance of a cold crystallization peak during the second heating run. It may be observed that cold crystallization peak temperature T_{cc} is lowered by the addition of TEC, whereas cold crystallization enthalpy is only slightly affected. Concomitant, cold crystallization affects the initial crystallinity of PHB or PHBV crystalline fraction not more than by a couple of percent. In general, the crystallinity of PHBV is somewhat higher than that of PHB. This is not common behavior; however, it can be explained by a greater amount of crystallizable fractions due to the higher molecular weight of the copolymer. The addition of plasticizer is known to reduce the crystallinity of polymers due to the penetration of plasticizer between polymer macromolecules by reducing intermolecular interaction strength. The observed increase in crystallinity most probably is related to the plasticizer acting as a nucleating agent, promoting the growth of new crystallites or facilitating the aggregation of existing crystallites. Similar behavior has been observed by Jost et al. for a number of different plasticizers including TEC already at small concentrations (5%) [38].

3.5. Oscillatory Shear Rheology

By considering that offset melting temperatures of the investigated PHA systems were within the interval between 168 and 192 °C, oscillatory shear rheology tests were made at 190 °C, close to the highest T_{offset} value. This temperature was also used during the compression molding of the investigated plasticized systems. It has been determined that by increasing shear rates, the complex viscosity η^* values of neat PHBV, and PHB as well as the TEC plasticized systems decrease demonstrating shear thinning behavior typical for non-Newtonian fluids (see Figure 6). As already expected, η^* of neat PHBV at low angular velocity values ω is higher in comparison to PHB, which is determined evidently by its higher molecular mass. However, at high ω values η^* of PHBV becomes smaller than that of PHB, which is explained by the lower thermal stability of PHBV and easier disruption of molecular entanglements due to greater mobility of macromolecular chain of the copolymer caused by valerate moieties. Thus, one may conclude that PHBV is more sensitive to shear stresses than PHB. The addition of TEC decreases conformational rigidity, lowers viscosity, and, hence, eases the processability of the plasticized systems by reducing the intermolecular interactions and disrupting the crystalline structure of the polymers. A relatively smaller decrease in η^* for the systems with 10 wt.% of TEC is because the plasticizer molecules may not be efficiently adsorbed between PHBV or PHB chains, leading to a less pronounced reduction in viscosity compared to the systems with higher TEC concentrations.

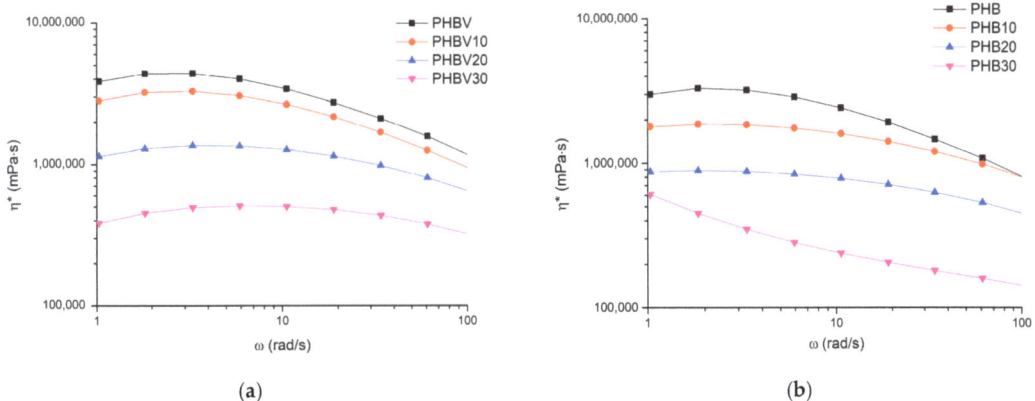

Figure 6. Complex viscosity (η^*) versus angular velocity (ω) plots for PHBV (**a**), PHB (**b**), and their plasticized systems (190 °C).

Besides it has been observed that at the beginning of the oscillatory test (the highest angular velocity value) storage modulus G' exceeds the loss modulus G''. For example, at an angular frequency of 628 rad/s G' and G'' values are 158 kPa and 87 kPa for neat PHBV and 143 kPa and 59 kPa for neat PHB, respectively. The modulus cross-over point is reached at 100 rad/s for PHBV (60 kPa) and 30 rad/s for PHB (30 kPa), after which G'' starts to dominate over G'. At the lowest angular frequency (0.1 rad/s) respective G' and G'' values are 0.11 Pa and 40 Pa for PHBV, and 35 Pa and 108 Pa for PHB, testifying that the copolymer has higher shear stress sensitivity. As demonstrated in Figure 7, G' and G'' of the investigated systems decrease by increasing the TEC content in the polymer composition, especially in the case of plasticized PHBV, following the same trend as matrix polymers.

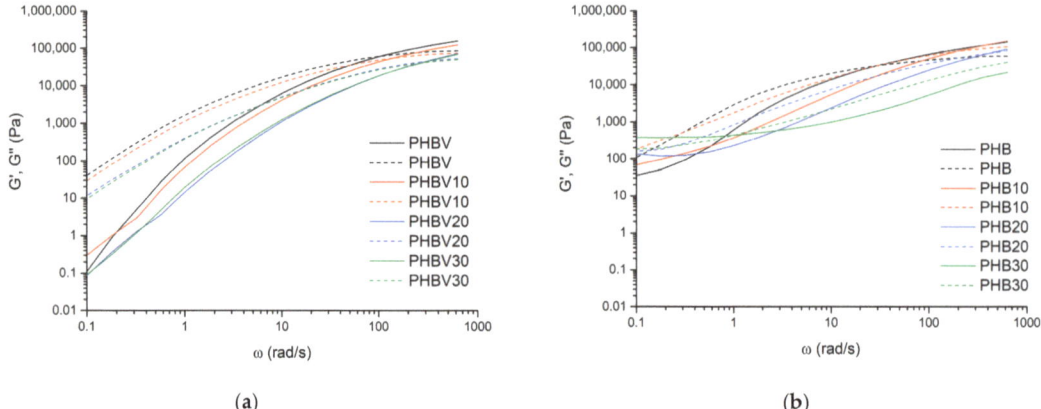

Figure 7. Storage (G″) and loss (G′) modules versus angular velocity (ω) plots for the plasticized systems based on PHBV (**a**), and PHB (**b**) (190 °C).

3.6. Tensile Properties

As seen in Figure 8a,b, by increasing the TEC content up to 30 wt.%, the modulus of elasticity E of the plasticized systems experiences a nearly twofold decrease from 2546 MPa to 1236 MPa and from 3559 MPa to 712 MPa for PHB- and PHBV-based systems, respectively. This indicates that the addition of structurally bulky TEC considerably affects the rigidity of the polymer matrix, especially in the case of PHBV. Consequently, although E of neat PHBV, mainly due to its higher molecular weight, is ca 1.3 times higher than that of neat PHB, after plasticization with 30 wt.% of TEC E of PHBV-based system becomes 2.2 times lower than that of its PHB-based counterpart. Similarly, the addition of TEC has also led to a considerable decrement of stress at break σ_B of all the plasticized systems, especially in the case PHBV-based systems. Thus, due to plasticization with 30 wt.% of TEC σ_B of PHBV- and PHB-based systems decrease by 60% and 40%, respectively. This suggests that PHBV is more efficiently plasticized by TEC in comparison to PHB. Hence, plasticized PHBV-based systems demonstrate 2.5 times larger ultimate elongation values in comparison to PHB-based systems. Disregarding this, ultimate deformation ε_B of PHB and PHBV due to plasticization increases to a similar extent, i.e., approximately 2.5 times at the maximum TEC concentration.

3.7. Dynamic Mechanical Analysis (DMA)

Loss factor tan δ and storage modulus E′ versus temperature T relationships of the investigated plasticized systems are shown in Figures 9 and 10. The tanδ(T) relationships demonstrate well-expressed relaxation region within temperature intervals −10 °C–+55 °C with maxima at 22 °C for PHBV and between −10 °C and +40 °C with maxima at 19 °C for PHB. This relaxation is associated with glass transition in the amorphous phases of PHB or PHBV. The breadth of this relaxation region is associated with the presence of crystalline fraction in both polymers as previously stated by Scandola et al. [39]. As shown in Table 9, the addition of TEC causes a considerable negative shift in glass transition maxima by 26 °C for both PHBV plasticized with 30 wt.% of TEC and its PHB-based counterpart. This is because the plasticizer causes the weakening of intermolecular forces that contribute to the stiffness of the material.

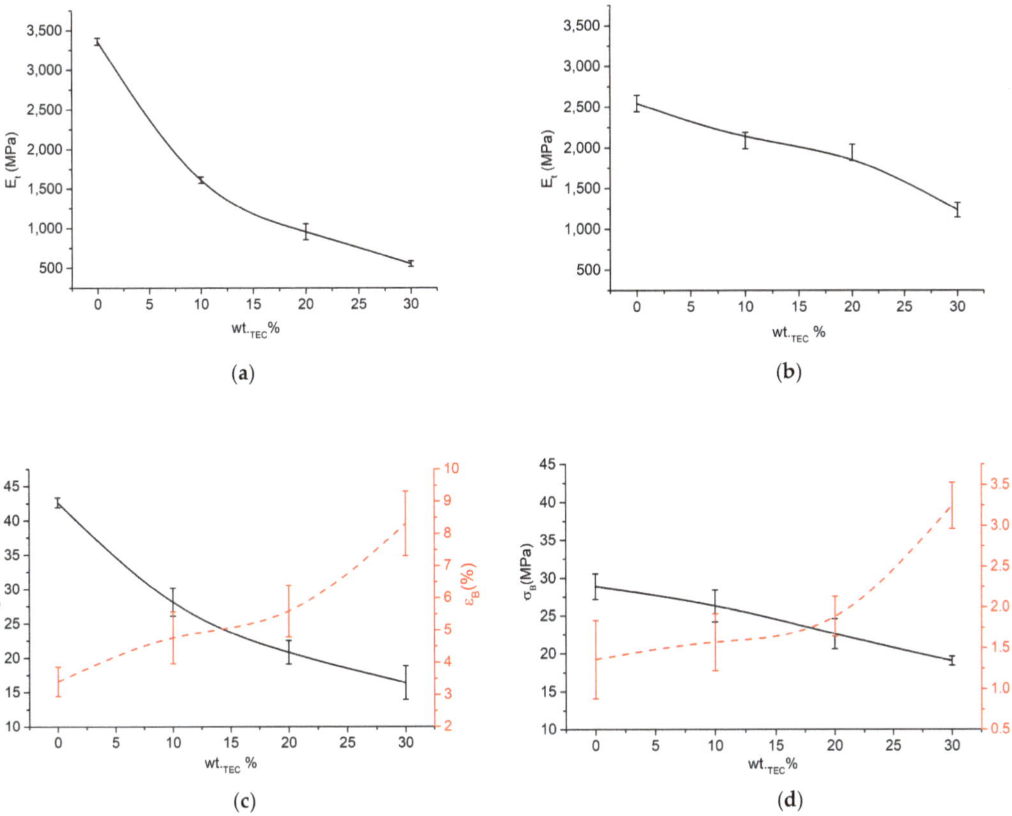

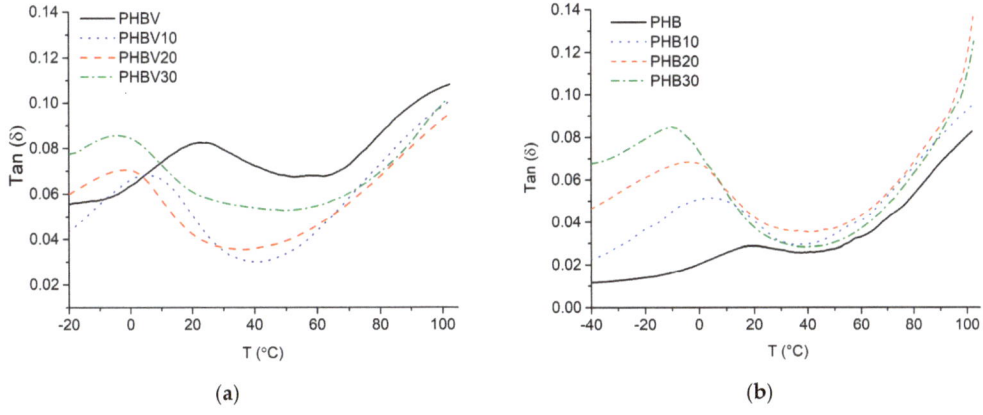

Figure 8. Young's modulus E (**a**,**b**), stress at break σ_B (**c**,**d**), and ultimate deformation ε_B (**c**,**d**), of the PHBV (**a**,**c**), and PHB (**b**,**d**) based systems.

Figure 9. tan δ-(T) relationships of the plasticized systems based on PHBV (**a**) and PHB (**b**).

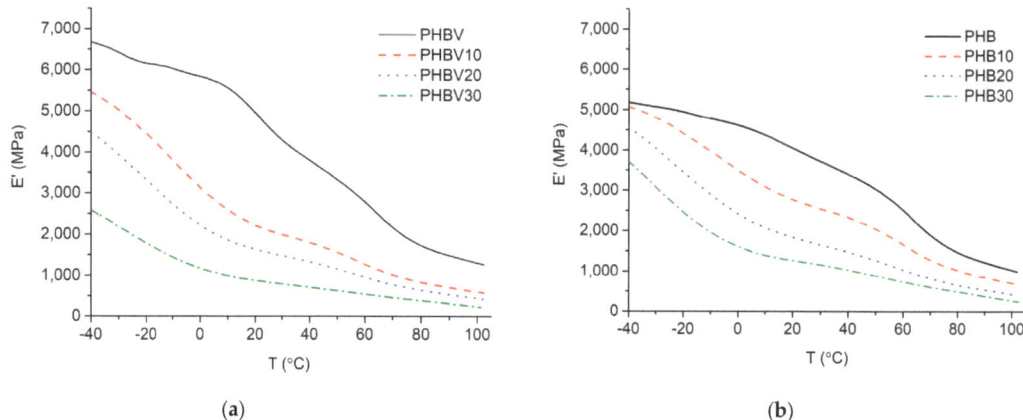

Figure 10. E′(T) relationships of the plasticized systems based on PHBV (**a**), and PHB (**b**).

Table 9. Tg from tan δ peak maximum.

Sample	T$_g$, °C	Sample	T$_g$, °C
PHBV	22	PHB	19
PHBV10	5	PHB10	12
PHBV20	−1	PHB20	−5
PHBV30	−4	PHB30	−7

Apart from the T$_g$ peak, another well-expressed intensity is observed in the tanδ(T) relationships with onset at ca 63 °C and ca 45 °C for PHBV and PHB, respectively. This intensity may be related to the beginning of the crystal–crystal slippage occurring in semicrystalline polymers just below melting as stated by Madbouly et al. [40] and McDonald et al. [6] who observed high-temperature relaxation of PHB at about 110 °C. This transition may also be related to the α′ relaxation of the amorphous–crystalline interphase [27]. The addition of TEC promoted this relaxation process to occur at somewhat lower temperatures, especially in the case of PHBV-based systems, confirming that TEC affects the structure of PHBV to a greater extent in comparison to PHB.

In correspondence with tan δ data and trends in tensile properties of the investigated PHA-based systems, storage modulus temperature relationships E′(T) are shifted to the direction of lower temperatures and lower modulus values by increasing TEC content in the plasticized system. As already expected, larger E′ changes have been observed for PHBV-based systems. Figure 11 depicts the E′ change of PHBV and PHB plasticized systems below and above the glass transition region (−45 °C and +45 °C respectively). As expected, below T$_g$ the change of E′ of PHBV as a result of TEC addition up to 30 wt.%, is around 45%, whereas the change of the counterpart PHB-based system is only 26%. However, if the temperature is raised above T$_g$, the decrement of E′ is much greater, i.e., by ca 70% and 80% for PHB and PHBV-based systems, respectively. Consequently, the TEC addition drop of E′ of PHBV-based systems are more intensive, similarly as it was observed in the case of tensile tests.

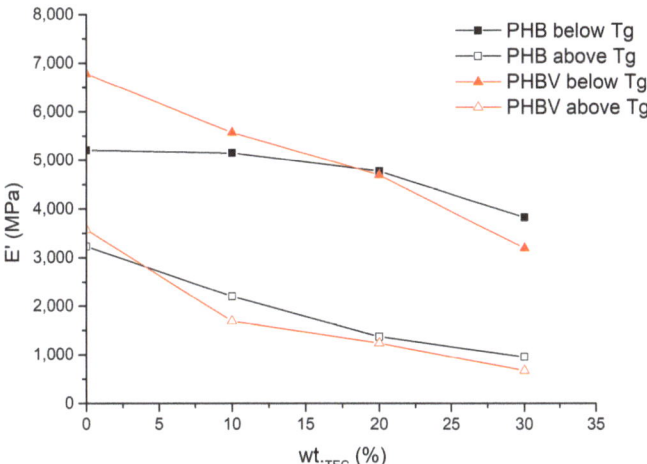

Figure 11. E′ of plasticized systems based on PHBV and PHB before and after T_g.

4. Conclusions

In this research, the efficiency of the plasticization of PHB and PHBV with TEC (10, 20, and 30 wt.%) as an environmentally friendly plasticizer is demonstrated. The following results have been obtained due to plasticization with TEC:

(1) Considerable thermooxidative degradation in the air of the investigated plasticized systems does not occur until 240–260 °C, while the minimum onset thermal degradation temperature is 264 °C;
(2) The rate of thermooxidative degradation of the plasticized systems is decreased to a certain extent due to the contribution of TEC in the building of the gas-impermeable char layer;
(3) Increased shear forces cause decrement of melt viscosity as well as storage and loss modules of both PHB and especially PHB- based systems due to lower activation energy of the latter and weakened interaction between the polymer chains because of plasticization;
(4) The melting range of the plasticized systems is considerably decreased (by ca 10 °C at the maximum peak value), thus relieving the processability of the investigated systems;
(5) Ultimate elongation ε_B values of the investigated plasticized systems increase on average 2.5 times by increasing TEC content, reaching values as high as 9% (for PHBV-based systems);
(6) Modulus of elasticity E as well as tensile strength σ_B values experience certain decrements, especially for PHBV-based systems above glass transition temperature T_g.

Consequently, plasticized low molecular PHB has improved use potential due to reduced brittleness, making it similar to the commercial PHBV in respect to ultimate elongation value. Apart from this, the possibility to process plasticized PHB and plasticized PHBV at somewhat lower temperatures potentially reduces the potential of thermooxidative decomposition of the polymers during melt processing; thus, making them more suitable for the manufacturing of environmentally sound packaging, which is the expected target market of the investigated PHA compositions. To achieve this aim, it is expected to investigate the long-term stability of the developed composites under the influence of different factors of the external environment. It is also expected to assess further modification potential of the developed plasticized systems by using agricultural residues.

Supplementary Materials: The following supporting information can be downloaded at: https://www.mdpi.com/article/10.3390/polym15132896/s1, Figure S1: Viscosity of PHB and PHBV versus solution concentration; Figure S2: Thermograms of DSC cooling run of the PHBV-based systems; Figure S3: Thermograms of DSC cooling run of the PHB based systems; Figure S4: Thermograms of DSC 2nd heating run of the PHBV-based systems; Figure S5: Thermograms of DSC 2nd heating run of the PHB-based systems.

Author Contributions: Conceptualization, R.M.-M. and J.Z.; methodology, M.Ž., R.M.-M., I.B. and A.V.; investigation, M.Ž., I.B. and A.V.; formal analysis, J.Z.; resources, J.Z., E.E. and E.Š.; data curation, M.Ž., I.B. and A.V.; writing—original draft preparation, M.Ž.; writing—review and editing, R.M.-M. and A.V.; visualization, M.Ž. and T.I.; supervision, R.M.-M.; project administration, T.I.; funding acquisition, J.Z., E.E. and E.Š. All authors have read and agreed to the published version of the manuscript.

Funding: This research/publication was supported by Riga Technical University's 2021 Doctoral Grant programme. This work has been supported by the European Social Fund within the Project No 8.2.2.0/20/I/008 «Strengthening of PhD students and academic personnel of Riga Technical University and BA School of Business and Finance in the strategic fields of specialization» of the Specific Objective 8.2.2 «To Strengthen Academic Staff of Higher Education Institutions in Strategic Specialization Areas» of the Operational Programme «Growth and Employment» as well as project No.1.2.1.1/18/A/002 (Project 41), research of MILZU! Ltd.

Data Availability Statement: The data presented in this study are available on request from the corresponding author.

Conflicts of Interest: The authors declare no conflict of interest.

References

1. Eurostat. EU Recycled 41% of Plastic Packaging Waste in 2019. 2021. Available online: https://ec.europa.eu/eurostat/web/products-eurostat-news/-/ddn-20211027-2 (accessed on 15 May 2023).
2. van der Harst, E.; Potting, J. A critical comparison of ten disposable cup LCAs. *Environ. Impact Assess. Rev.* **2013**, *43*, 86–96. [CrossRef]
3. Utriainen, M.; Application, F.; Data, P.; Oy, E. (12) Patent Application Publication (10) Pub. No.: US 2009/0312954 A1. *D Pat. Appl. Publ.* **2009**, *1*, 1–6. Available online: https://patentimages.storage.googleapis.com/9d/30/f7/7b725f3f41be3c/US20090082491A1.pdf (accessed on 28 March 2023).
4. Aramvash, A.; Moazzeni Zavareh, F.; Gholami Banadkuki, N. Comparison of different solvents for extraction of polyhydroxybutyrate from Cupriavidus necator. *Eng. Life Sci.* **2018**, *18*, 20–28. [CrossRef]
5. Myshkina, V.L.; Nikolaeva, D.A.; Makhina, T.K.; Bonartsev, A.P.; Bonartseva, G.A. Effect of growth conditions on the molecular weight of poly-3- hydroxybutyrate produced by Azotobacter chroococcum 7B. *Appl. Biochem. Microbiol.* **2008**, *44*, 482–486. [CrossRef]
6. McAdam, B.; Fournet, B.M.; McDonald, P.; Mojicevic, M. Production of polyhydroxybutyrate (PHB) and factors impacting its chemical and mechanical characteristics. *Polymers* **2020**, *12*, 2908. [CrossRef] [PubMed]
7. Sirohi, R.; Prakash Pandey, J.; Kumar Gaur, V.; Gnansounou, E.; Sindhu, R. Critical overview of biomass feedstocks as sustainable substrates for the production of polyhydroxybutyrate (PHB). *Bioresour. Technol.* **2020**, *311*, 123536. [CrossRef]
8. Aragosa, A.; Specchia, V.; Frigione, M. Isolation of two bacterial species from argan soil in morocco associated with polyhydroxybutyrate (Phb) accumulation: Current potential and future prospects for the bio-based polymer production. *Polymers* **2021**, *13*, 1870. [CrossRef]
9. Nova and Institute. Global Production Capacities. 2021. Available online: https://primebiopol.com/aumenta-produccion-bioplasticos/?lang=en (accessed on 16 May 2023).
10. Nova-Institute. For the First Time: Growth Rate for Bio-based Polymers with 8% CAGR Far above Overall Polymer Market Growth. 2021. Available online: https://renewable-carbon.eu/news/for-the-first-time-growth-rate-for-bio-based-polymers-with-8-cagr-far-above-overall-polymer-market-growth/ (accessed on 16 May 2023).
11. Requena, R.; Jiménez, A.; Vargas, M.; Chiralt, A. Effect of plasticizers on thermal and physical properties of compression-moulded poly[(3-hydroxybutyrate)-co-(3-hydroxyvalerate)] films. *Polym. Test.* **2016**, *56*, 45–53. [CrossRef]
12. Stanley, A.; Murthy, P.S.K.; Vijayendra, S.V.N. Characterization of Polyhydroxyalkanoate Produced by Halomonas venusta KT832796. *J. Polym. Environ.* **2020**, *28*, 973–983. [CrossRef]
13. Quagliano, J.C.; Amarilla, F.; Fernandes, E.G.; Mata, D.; Miyazaki, S.S. "Effect of simple and complex carbon sources, low temperature culture and complex carbon feeding policies on poly-3-hydroxybutyric acid (PHB) content and molecular weight (Mw) from Azotobacter chroococcum 6B. *World J. Microbiol. Biotechnol.* **2001**, *17*, 9–14. [CrossRef]

14. Umemura, R.T.; Felisberti, M.I. Plasticization of poly(3-hydroxybutyrate) with triethyl citrate: Thermal and mechanical properties, morphology, and kinetics of crystallization. *J. Appl. Polym. Sci.* **2021**, *138*, 49990. [CrossRef]
15. Wang, L.; Zhu, W.; Wang, X.; Chen, X.; Chen, G.-Q. Processability Modifications of Poly(3-hydroxybutyrate) by Plasticizing, Blending, and Stabilizing. *J. Appl. Polym. Sci.* **2008**, *107*, 166–173. [CrossRef]
16. Nosal, H.; Moser, K.; Warzała, M.; Holzer, A.; Stańczyk, D.; Sabura, E. Selected Fatty Acids Esters as Potential PHB-V Bioplasticizers: Effect on Mechanical Properties of the Polymer. *J. Polym. Environ.* **2020**, *29*, 38–53. [CrossRef]
17. Mangeon, C.; Michely, L.; Rios De Anda, A.; Thevenieau, F.; Renard, E.; Langlois, V. Natural Terpenes Used as Plasticizers for Poly(3-hydroxybutyrate). *ACS Sustain. Chem. Eng.* **2018**, *6*, 16160–16168. [CrossRef]
18. Scalioni, L.V.; Gutiérrez, M.C.; Felisberti, M.I. Green composites of poly(3-hydroxybutyrate) and curaua fibers: Morphology and physical, thermal, and mechanical properties. *J. Appl. Polym. Sci.* **2017**, *134*, 1–13. [CrossRef]
19. Râpă, M.; Darie-Nita, R.; Grosu, E.; Popa, E.; Trifoi, A.; Pap, T.; Vasile, C. Effect of plasticizers on melt processability and properties of PHB. *J. Optoelectron. Adv. Mater.* **2015**, *17*, 1778–1784.
20. Garcia-Garcia, D.; Ferri, J.M.; Montanes, N.; Lopez-Martinez, J.; Balart, R. Plasticization effects of epoxidized vegetable oils on mechanical properties of poly(3-hydroxybutyrate). *Polym. Int.* **2016**, *65*, 1157–1164. Available online: http://hdl.handle.net/10251/82871 (accessed on 17 May 2023). [CrossRef]
21. Barbosa, J.L.; Perin, G.B.; Felisberti, M.I. Plasticization of Poly(3-hydroxybutyrate- co-3-hydroxyvalerate) with an Oligomeric Polyester: Miscibility and Effect of the Microstructure and Plasticizer Distribution on Thermal and Mechanical Properties. *ACS Omega* **2021**, *6*, 3278–3290. [CrossRef]
22. Haas, R.; Jin, B.; Zepf, F.T. Production of Poly(3-hydroxybutyrate) from Waste Potato Starch. *Biosci. Biotechnol. Biochem.* **2008**, *72*, 253–256. [CrossRef]
23. Yang, Y.H.; Brigham, C.; Willis, L.; Rha, C.K.; Sinskey, A. Improved detergent-based recovery of polyhydroxyalkanoates (PHAs). *Biotechnol. Lett.* **2011**, *33*, 937–942. [CrossRef]
24. Vandi, L.J.; Chan, C.M.; Werker, A.; Richardson, D.; Laycock, B.; Pratt, S. Extrusion of wood fibre reinforced poly(hydroxybutyrate-co-hydroxyvalerate) (PHBV) biocomposites: Statistical analysis of the effect of processing conditions on mechanical performance. *Polym. Degrad. Stab.* **2018**, *159*, 1–14. [CrossRef]
25. Turco, R.; Santagata, G.; Corrado, I.; Pezzella, C.; Di Serio, M. In vivo and Post-synthesis Strategies to Enhance the Properties of PHB-Based Materials: A Review. *Front. Bioeng. Biotechnol.* **2021**, *8*, 619266. [CrossRef] [PubMed]
26. Ino, K.; Sato, S.; Ushimaru, K.; Saika, A.; Fukuoka, T.; Ohshiman, K.; Morita, T. Mechanical properties of cold-drawn films of ultrahigh-molecular-weight poly(3-hydroxybutyrate-co-3-hydroxyvalerate) produced by Haloferax mediterranei. *Polym. J.* **2020**, *52*, 1299–1306. [CrossRef]
27. Chen, Y.; Chou, I.N.; Tsai, Y.H.; Wu, H.S. Thermal degradation of poly(3-hydroxybutyrate) and poly(3-hydroxybutyrate- co-3-hydroxyvalerate) in drying treatment. *J. Appl. Polym. Sci.* **2013**, *130*, 3659–3667. [CrossRef]
28. Abbasi, M.; Pokhrel, D.; Coats, E.R.; Guho, N.M.; McDonald, A.G. Effect of 3-Hydroxyvalerate Content on Thermal, Mechanical, and Rheological Properties of Poly(3-hydroxybutyrate-co-3-hydroxyvalerate) Biopolymers Produced from Fermented Dairy Manure. *Polymers* **2022**, *14*, 4140. [CrossRef]
29. Ramezani, M.; Amoozegar, M.A.; Ventosa, A. Screening and comparative assay of poly-hydroxyalkanoates produced by bacteria isolated from the Gavkhooni Wetland in Iran and evaluation of poly-β-hydroxybutyrate production by halotolerant bacterium Oceanimonas sp. GK1. *Ann. Microbiol.* **2015**, *65*, 517–526. [CrossRef]
30. Chotchindakun, K.; Pathom-Aree, W.; Dumri, K.; Ruangsuriya, J.; Pumas, C.; Pekkoh, J. Low Crystallinity of Poly(3-Hydroxybutyrate-co-3-Hydroxyvalerate) Bioproduction by Hot Spring Cyanobacterium Cyanosarcina sp. AARL T020. *Plants* **2021**, *10*, 503. [CrossRef]
31. Teixeira, S.C.; Silva, R.R.A.; de Oliveira, T.V.; Stringheta, P.C.; Pinto, M.R.M.R.; de Soares, F.F.N. Glycerol and triethyl citrate plasticizer effects on molecular, thermal, mechanical, and barrier properties of cellulose acetate films. *Food Biosci.* **2021**, *42*, 101202. [CrossRef]
32. Sánchez, K.D.T.; Allen, N.S.; Liauw, C.M.; Johnson, B. Effects of type of polymerization catalyst system on the degradation of polyethylenes in the melt state. Part 1: Unstabilized polyethylenes (including metallocene types). *J. Vinyl Addit. Technol.* **2011**, *17*, 8–39. [CrossRef]
33. Xiang, H.; Wen, X.; Miu, X.; Li, Y.; Zhou, Z.; Zhu, M. Thermal depolymerization mechanisms of poly(3-hydroxybutyrate-co-3-hydroxyvalerate). *Prog. Nat. Sci. Mater. Int.* **2016**, *26*, 58–64. [CrossRef]
34. Di Lorenzo, M.L.; Sajkiewicz, P.; La Pietra, P.; Gradys, A. Irregularly shaped DSC exotherms in the analysis of polymer crystallization. *Polym. Bull.* **2006**, *57*, 713–721. [CrossRef]
35. Briassoulis, D.; Tserotas, P.; Athanasoulia, I.G. Alternative optimization routes for improving the performance of poly(3-hydroxybutyrate) (PHB) based plastics. *J. Clean. Prod.* **2021**, *318*, 128555. [CrossRef]
36. Hsiao, C.S.B.S.; Zuo, F.; Mao, Y. *Handbook of Polymer Crystallization*; John Wiley & Sons: Hoboken, NJ, USA, 2013; Volume 15.
37. Sauer, B.B.; Kampert, W.G.; Blanchard, N.E.; Threefoot, S.A.; Hsiao, B.S. Temperature modulated DSC studies of melting and recrystallization in polymers exhibiting multiple endotherms. *Polymer* **2000**, *41*, 1099–1108. [CrossRef]
38. Jost, V.; Langowski, H.C. Effect of different plasticisers on the mechanical and barrier properties of extruded cast PHBV films. *Eur. Polym. J.* **2015**, *68*, 302–312. [CrossRef]

39. Scandola, M.; Ceccorulli, G.; Doi, Y. Viscoelastic relaxations and thermal properties of bacterial poly(3-hydroxybutyrate-co-3-hydroxyvalerate) and poly(3-hydroxybutyrate-co-4-hydroxybutyrate). *Int. J. Biol. Macromol.* **1990**, *12*, 112–117. [CrossRef]
40. Madbouly, S.A.; Mansour, A.A.; Abdou, N.Y. Molecular dynamics of amorphous/crystalline polymer blends studied by broadband dielectric spectroscopy. *Eur. Polym. J.* **2007**, *43*, 1892–1904. [CrossRef]

Disclaimer/Publisher's Note: The statements, opinions and data contained in all publications are solely those of the individual author(s) and contributor(s) and not of MDPI and/or the editor(s). MDPI and/or the editor(s) disclaim responsibility for any injury to people or property resulting from any ideas, methods, instructions or products referred to in the content.

Article

Dopant-Free Hole-Transporting Material Based on Poly(2,7-(9,9-bis(N,N-di-p-methoxylphenylamine)-4-phenyl))-fluorene for High-Performance Air-Processed Inverted Perovskite Solar Cells

Baomin Zhao [1,*], Meng Tian [1], Xingsheng Chu [1], Peng Xu [1], Jie Yao [1], Pingping Hou [2], Zhaoning Li [1] and Hongyan Huang [2,*]

[1] State Key Laboratory for Organic Electronics and Information Displays & Jiangsu Key Laboratory for Biosensors, Institute of Advanced Materials (IAM), Nanjing University of Posts and Telecommunications, 9 Wenyuan Road, Nanjing 210023, China

[2] School of Electronic Information, Nanjing Vocational College of Information Technology, 99 Wenyuan Road, Nanjing 210023, China

* Correspondence: iambmzhao@njupt.edu.cn (B.Z.); 20210031@njcit.cn (H.H.)

Citation: Zhao, B.; Tian, M.; Chu, X.; Xu, P.; Yao, J.; Hou, P.; Li, Z.; Huang, H. Dopant-Free Hole-Transporting Material Based on Poly(2,7-(9,9-bis(N,N-di-p-methoxylphenylamine)-4-phenyl))-fluorene for High-Performance Air-Processed Inverted Perovskite Solar Cells. *Polymers* **2023**, *15*, 2750. https://doi.org/10.3390/polym15122750

Academic Editor: Rong-Ho Lee

Received: 22 May 2023
Revised: 8 June 2023
Accepted: 16 June 2023
Published: 20 June 2023

Copyright: © 2023 by the authors. Licensee MDPI, Basel, Switzerland. This article is an open access article distributed under the terms and conditions of the Creative Commons Attribution (CC BY) license (https://creativecommons.org/licenses/by/4.0/).

Abstract: It is a great challenge to develop low-cost and dopant-free polymer hole-transporting materials (HTM) for PSCs, especially for efficient air-processed inverted (p-i-n) planar PSCs. A new homopolymer HTM, poly(2,7-(9,9-bis(N,N-di-p-methoxyphenyl amine)-4-phenyl))-fluorene (denoted as PFTPA), with appropriate photo-electrochemical, opto-electronic and thermal stability, was designed and synthesized in two steps to meet this challenge. By employing PFTPA as dopant-free hole-transport layer in air-processed inverted PSCs, a champion power conversion efficiency (PCE) of up to 16.82% (0.1 cm^2) was achieved, much superior to that of commercial HTM PEDOT:PSS (13.8%) under the same conditions. Such a superiority is attributed to the well-aligned energy levels, improved morphology, and efficient hole-transporting, as well as hole-extraction characteristics at the perovskite/HTM interface. In particular, these PFTPA-based PSCs fabricated in the air atmosphere maintain a long-term stability of 91% under ambient air conditions for 1000 h. Finally, PFTPA as the dopant-free HTM was also fabricated the slot-die coated perovskite device through the same fabrication condition, and a maximum PCE of 13.84% was obtained. Our study demonstrated that the low-cost and facile homopolymer PFTPA as the dopant-free HTM are potential candidates for large-scale production perovskite solar cell.

Keywords: fluorene-based hole transporting polymer; dopant-free; air-processed; inverted perovskite solar cells

1. Introduction

Over the past few years, encouraging progress has been made in the field of organic-inorganic hybrid perovskite solar cells [1–7]. The power conversion efficiency (PCE) of the organic-inorganic halide perovskite solar cells (PSCs) has rocketed to a certified record of 25.8% [8,9], indicating its great potential to compete with traditional silicon solar cells in near future. Substantial effort has been carried out to push the PSC performance to its theoretical limit, including fabrication techniques, device architectures, functional components based on new materials in perovskite layer and charge-transporting layers [10–14]. Two types of device architectures have been widely employed for PSCs: normal (n-i-p) and inverted (p-i-n) configurations, with each type featuring different advantages and challenges. The n-i-p PSC devices currently have superior PCE but typically use transparent electron transport layers (ETLs) of metal oxides that require high-temperature fabrication methods and doped hole transport materials (HTMs) that can introduce device degradation pathways. Moreover, these n-i-p architecture PSCs suffer from a large degree of $J - V$

hysteresis. As an alternative, the emerging inverted PSCs with a p-i-n architecture use p-type and n-type materials deposited at relatively low temperatures by solution processing as bottom and top charge transport layers, respectively [15–18]. The inverted PSCs have shown many advantages, such as high efficiencies (as high as >25% using self-assembly hole-extraction layer) [19–23], low-temperature processing on flexible substrates, and, furthermore, negligible $J - V$ hysteresis effects. Thus, p-i-n PSCs are typically employed in silicon/perovskite [24] and perovskite/perovskite [25] two-terminal tandem devices, which is essentially important for the future of commercial PSC technology.

For inverted PSCs, HTMs not only greatly improve hole extraction and transport from the perovskite layer to electrode, but also have a significant impact on the crystallinities and morphologies of perovskite film, which could boost the PCE and stability of devices [26–28]. Especially to achieve superior long-term stability of PSCs, it is necessary to develop efficient dopant-free HTMs. Fundamentally, highly efficient dopant-free HTMs for inverted PSCs have several basic requirements, such as suitable energy levels for perovskite materials, high hole mobility and conductivity, high chemical and thermal stability, excellent film processing ability and film stability [29–31]. Since in the inverted devices, HTM is deposited before the perovskite layer, the surface properties of the HTM layer significantly affect the polycrystalline film quality of the perovskite layer. Thus, excellent wettability with perovskite precursor solutions is an essential prerequisite for promoting the crystallization process of perovskite [31,32]. In addition, a weak absorption coefficient in the visible to near-infrared region and high photostability in the ultra-violet region are highly desirable for dopant-free HTMs in p-i-n devices because the light passes through the HTM layer before being absorbed by the perovskite layer [31,32].

Dopant-free polymer HTMs have attracted much attention due to their advantages such as high heat resistance, high hydrophobicity, excellent film-processing ability, and compatibility with the scale roll-to-roll printing technique [32,33]. More importantly, the amorphous nature and strong intrachain charge transfer along the conjugated backbone result in a good balance between high mobility and good film quality, rather than a trade-off for small-molecule HTMs. Although plenty of efficient dopant-free D-A copolymeric HTMs have been reported [34–36], most of them were utilized in n-i-p device stacks. Except for their high costs, D-A copolymers also absorb some sunlight in visible region when they are applied to the p-i-n PSCs. It should be noted that the majority of highly efficient PSCs based on D-A type copolymers are fabricated inside a glove box filled with costly inert gas to avoid moisture, which is incompatible with the low-cost and large-scale manufacturing of PSCs in ambient conditions.

The state-of-the-art p-i-n device architectures use fullerene or related derivatives as ETLs and dopant-free polymeric HTMs such as poly[bis(4-phenyl)(2,3,6-trimethylphenyl)amine] (PTAA). However, its low hole transporting ability and high hydrophobicity require additional dopants and wetting treatment for PTAA, which unavoidably reduce the device performance reproducibility [37]. In the case of PEDOT:PSS, a widely used classic water-soluble conducting polymer HTL in inverted PSCs, its acidic and hygroscopic nature can degrade the perovskite layer and corrode anodes, reducing the stability of the solar cells [38]. Furthermore, the mismatched energy level between the valence band of perovskite materials and the work function of PEDOT:PSS could lead to severe non-radiative recombination, limiting the V_{OC} as well as the photovoltaic performance. To date, the inverted PSCs using pristine PEDOT:PSS as the HTL achieved a PCE of 15.05% [39], which is far less than the other dopant-free polymeric HTL, such as PTAA, PII2T8T and poly[3-(4-carboxybutyl)-thiophene-2,5-diyl] (P3CT) [40].

On the other hand, most of the reported highly efficient PSCs were fabricated in a well-controlled glovebox free from moisture and oxygen. Large-scale manufacturing in ambient processing conditions remain challenging both in lab and in factory, because polycrystalline perovskite materials are extremely sensitive to moisture and oxygen in ambient air. Several recent advances in the development of air-processed PSCs mainly focus on the control of the processing of perovskite materials. But little attention has been paid to dopant-free polymeric HTMs in the inverted PSC devices fabricated in air. Our aim is to develop low-cost, highly efficient dopant-free polymeric HTM, which is suitable for air-processed PSCs.

Herein, we introduce a novel homopolymer, poly(2,7-(9,9-bis(N,N-di-p-methoxylphenyl amine)-4-phenyl))-fluorene (denoted as PFTPA, see in Scheme 1). PFTPA exhibits matched energy alignment with adjacent perovskite, superior hydrophobicity, and high hole mobility. As a preliminary result, the dopant-free PFTPA-based air-processed p-i-n PSCs exhibit a champion power conversion efficiency (PCE) of 16.82% (0.1 cm^2) under a 100 mW cm^2 AM1.5G solar illumination and maintain a long-term stability of 91% under ambient air conditions for 1000 h. These results suggest the great potential of homopolymer PFTPA HTMs for future low-cost large-scale and flexible PSCs application.

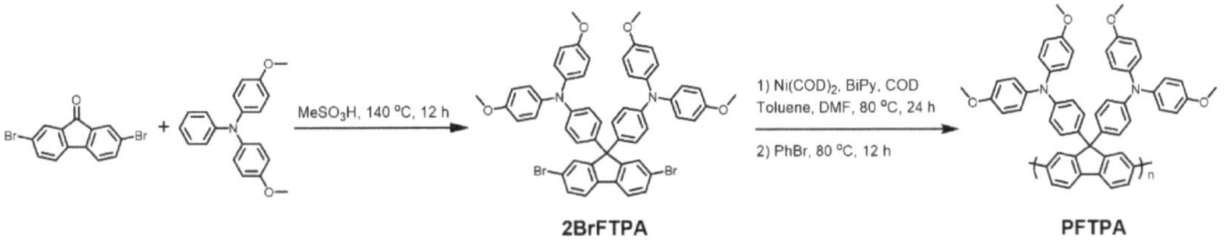

Scheme 1. The chemical structure and synthetic route of PFTPA.

2. Materials and Methods

2.1. Synthesis of PFTPA

2.1.1. Synthesis of 4,4'-(2,7-Dibromo-9H-fluorene-9,9-diyl)bis(N,N-bis(4-methoxyphenyl) aniline) (2BrFTPA)

In a two-necked round bottom flask (50 mL), 2,7-dibromofluorenone (1.50 g, 4.40 mmol), 4,4-dimethoxytriphenylamine (10.85 g, 35.5 mmol) and methylsulfonic acid (4 mL) were mixed. The mixture was heated to 140 °C and kept refluxing for 12 h. Then, the reaction system was cooled to room temperature and the solidified raw products were dissolved with dichloromethane. This mixture solution was washed by saturated saline in turn. The organic phase was dried with anhydrous sodium sulfate before dichloromethane was removed. The crude product was further purified by silica gel column chromatography and recrystallized from a mixed solvent of ethyl acetate/hexane to obtain 2BrFTPA as a white solid (4.10 g, yield 75%). ^{1}H NMR (400 MHz, CDCl$_3$) δ = 7.53 (d, J = 8.1 Hz, 2H), 7.49–7.41 (m, 4H), 7.03 (d, J = 8.8 Hz, 8H), 6.90 (d, J = 8.8 Hz, 4H), 6.79 (d, J = 9.0 Hz, 8H), 6.74 (d, J = 8.7 Hz, 4H), 3.76 (s, 12H) ppm. ^{13}C NMR (101 MHz, CDCl$_3$) δ = 156.06, 153.96, 147.54, 140.73, 138.01, 135.69, 130.72, 129.45, 128.60, 127.00, 121.78, 121.56, 119.67, 114.77, 64.55, 55.57 ppm. MALDI-TOF-MS (m/z): Calculated for C$_{53}$H$_{42}$Br$_2$N$_2$O$_4$: 928.15; Found: 928.11 [M]$^+$.

2.1.2. Polymerization of PFTPA

Bis-(1,5-cyclooctadiene) nickel (130 mg, 0.48 mmol), 2,2-Bipyridine (76 mg, 0.48 mmol), 1,5-cyclooctadiene (52 mg, 0.48 mmol) and anhydrous DMF (5 mL) were added into a 25 mL thick pressure-resistant Schlenk reaction tube, heated to 80 °C under the protection of nitrogen and stirred for 30 min. Monomer 2BrFTPA (370 mg, 0.40 mmol) was dissolved in 8 mL of anhydrous toluene, bubbled with nitrogen for 10 min to remove the air in the solution, and then added to the reaction tube. The reaction system was heated to 80 °C and stirred for 24 h. Bromobenzene (2 mL) was added for end capping and stirring continued at 80 °C for 12 h; then, it was dropped into the mixed solution of methanol/hydrochloric acid (v/v = 2/1). The catalyst was then filtered and removed. The filter cake was redissolved with chloroform (15 mL), dropped into the mixed solution of methanol/acetone (v/v = 4/1), and reprecipitated to obtain the solid crude product. The solid crude product was put into Soxhlet extractor, extracted successively with petroleum ether and dichloromethane, and yellow green PFTPA film (260 mg, yield 70%) was formed on the bottle wall after removing dichloromethane by rotary evaporation. ^{1}H NMR (400 MHz, CDCl$_3$) δ = 7.76 (m, 2H), 7.61 (m, 4H), 7.07 (m, 4H), 6.94 (m, 8H), 6.75 (m, 12H), 3.72 (s, 12H) ppm. GPC (THF, polystyrene standard, 35 °C): M_n = 12.2 kDa, M_w = 29.9 kDa, PDI = 2.46.

2.2. Device Fabrication and Measurement

Materials: A modified PEDOT:PSS solution was prepared by mixing 1 mL of PEDOT:PSS (HC Starck, Baytron P AI 4083), 60 mg of sodium polystyrene sulfonate (molecular weight ~70,000, Sigma-Aldrich, Shanghai, China), and 5 mL of deionized water and stirring for 10 min. A CH$_3$NH$_3$PbI$_3$ precursor solution was prepared by dissolving a 1.2 M PbI$_2$ (Sigma-Aldrich, Burlington, MA, USA) and a 1.2 M CH$_3$NH$_3$I (Greatcell Solar Materials, Beijing, China) in dimethylformamide (DMF) and stirring at 70 °C for 30 min. After cooling to room temperature, solid NH$_4$Cl was added to the solution with a concentration of 0–20 mg/mL, and the solution was then stirred for 30 min at room temperature. PFTPA was dissolved in toluene and a 0.5 mg/mL solution was prepared for standby.

Device Fabrication: Patterned ITO glass was cleaned in detergent (Deconex 12PA detergent solution), deionized water, acetone, and isopropanol sequentially by ultrasonication and then treated with UV-ozone for 15 min. For control devices, PEDOT:PSS was dropped onto the ITO glass substrate through a syringe filter (0.2 μm RC filter) and spin coated at 5000 rpm for 20 s. For target devices, PFTPA in toluene was deposited on ITO by spin coating at 5000 rpm for 30 s. The substrate was then heated on a hotplate at 150 °C for 10 min in air. After cooling to room temperature, the substrate was put on a piece of Halyard TERI Wiper. A total of 20 μL of CH$_3$NH$_3$PbI$_3$ solution was dropped on the substrate. Then, a N$_2$ gas flow was applied to the substrate from a plastic tube with an inner diameter of 4 mm. The tube was perpendicular to the substrate and the outlet of the tube was 1 cm above the substrate. The flow rate of the N$_2$ gas was adjusted using a flowmeter. The solution spread on the substrate and the superfluous solution flowed off the substrate while blowing. The color of the substrate changed from yellow to dark brown in 10 s. Then, the tube was moved around the substrate to dry the film at the edge. Next, the substrate was heated at 100 °C for 30 s. PC$_{61}$BM in chlorobenzene (20 mg mL^{-1}) and PEIE were spin coated onto the CH$_3$NH$_3$PbI$_3$ layer at 1000 rpm for 30 s. Finally, a 100 nm Ag was evaporated onto the BCP layer through a shadow mask to produce an active area of 0.1 cm^2 [33,38].

For the perovskite solar cells made using slot-die coating, ITO glass was cleaned and PFTPA was spin-coated on the substrate as described above. A 0.65 M CH3NH3PbI3 solution with a 10 mg/mL NH$_4$Cl additive was coated onto the modified PFTPA layer using a slot-die coater with a setting speed of 8 mm/s. The temperature of the printer bed was set to 60 °C. N$_2$ gas with a flow rate of 20 L/min was used to dry the film while coating. PC61BM, PEIE, and Ag layers were deposited as described above.

Device Characterizations: The optical absorption of the perovskite samples was measured using a UV-vis spectrophotometer (Shimadzu UV-1750, Kyoto, Japan). The steady-state photoluminescence (PL) spectra were obtained using a PL microscopic spectrometer (HITACHI, F4600, Tokyo, Japan). The time-resolved photoluminescence (TRPL) was measured at 780 nm using excitation with a 510 nm light pulse from Edinburgh FLS980. The photocurrent density-voltage curves of the perovskite solar cells were measured using a solar simulator (Oriel 94023A, 300 W) and a Keithley 2400 source meter. The intensity (100 mW/cm^2) was calibrated using a standard Si solar cell (Oriel, VLSI standards). All the devices were tested under AM 1.5G sun light (100 mW/cm^2) using a metal mask of 0.1 cm^2 with a scan rate of 10 mV/s.

3. Results and Discussion

3.1. Synthesis and Design Principle

The chemical structure and synthetic route toward PFTPA are shown in Scheme 1. The design of PFTPA was inspired by 2,2′,7,7′-tetrakis(N,N′-di-p-methoxyphenylamine)-

9,9′-spirobifluorene (Spiro-OMeTAD), the most explored HTM. Spiro-OMeTAD contains a spirobifluorene core and four diphenylamine end groups. Its twisted structure ensures high solubility and facilitates the process of the HTM film, but leads to large intermolecular distance and weak intermolecular interaction resulting in its low charge mobility. On the other hand, Spiro-OMeTAD possesses a wide bandgap and a deeper HOMO energy level than many other reported dopant-free polymeric HTMs. Therefore, PFTPA was successively designed by reducing the rigidity of the spirobifluorene core to 9,9-diphenylfluorene and extending the conjugated length from one spirobifluorene unit to polyfluorene. The key monomer of 2BrFTPA was obtained by refluxing 2,7-dibromofluorenone and 4,4-dimethoxytriphenylamine (OMeTPA) in methylsulfonic acid without any expensive catalyst with a high isolated yield. Eventually, the homopolymer, PFTPA, was obtained by Yamamoto polymerization according to monomer 2BrFTPA. Detailed synthetic procedures and structural characterizations of monomer and PFTPA were described in the Experimental Section. The chemical structure of 2BrFTPA was confirmed by ^{1}H NMR, ^{13}C NMR and MALDI-TOF mass spectrum measurements as shown in Figures S1–S3. The target polymeric HTM was readily soluble in common organic solvents such as chloroform, toluene, and chlorobenzene (CB). The average molecular weights (Mn and Mw) and polydispersity index (PDI) of PFTPA were measured via gel permeation chromatography (GPC). The weight-average molecular weight (M_W) was 29.9 kDa with a PDI of 2.46. Note that the lab synthesis and purification cost of 2BrFTPA and PFTPA are both very low. For example, the total cost for PFTPA is estimated at 16.74 USD/g (the lab synthesis and purification cost is summarized in Table S1), which is much lower than that of many other reported dopant-free polymeric HTMs and shows a promising scale-up strategy for commercial production.

3.2. Thermal, Photophysical and Electrochemical Properties

Thermal properties of polymeric HTMs have an important impact on the stability of PSC devices. Thermal properties of PFTPA were characterized via thermogravimetric analyzer (TGA) and differential scanning calorimeter (DSC). As shown in Figure 1a, PFTPA displayed an outstanding thermal stability with decomposition temperatures (T_d, 5% weight loss temperature) of 367 °C. The DSC curves display the temperature-time data of the second heating circle as shown in Figure S5; the glass phase transition temperature (T_g) of PFTPA is about 200 °C. Both TGA and DSC results indicate that PFTPA can be maintained in amorphous states during the thermal annealing process, which is essential for device fabrication and device operation stability.

Normalized ultraviolet visible (UV-Vis) absorption spectra and photoluminescence (PL) spectra of PFTPA in dilute dichloromethane solution (10^{-5} M) and spin-coated thin films are shown in Figure 1b, and the corresponding data are summarized in Table 1. The maximum absorption wavelengths of PFTPA in solution and film state are 307 nm and 310 nm, respectively. The long wavelength absorption edges are estimated to be 421 nm and 434 nm, respectively. The absorption spectra of PFTPA in a dichloromethane solution and thin film are very similar in shape, while the absorption spectrum in film is slightly broadened compared to the spectrum in the solution, indicating that no strong intermolecular π-π stacking exists. The fluorescence emission peaks of PFTPA in dichloromethane solution and thin film are 431 nm and 489 nm, respectively, and the corresponding Stokes shifts are 9397 cm^{-1} and 11,808 cm^{-1}, respectively. A larger Stokes shift will benefit the hole-injection of HTMs and the efficiency of PSC devices. According to the initial absorption wavelength of the absorption spectrum of PFTPA in dilute dichloromethane solution (λ_{Edge}), the optical band gap of PFTPA is calculated at 2.95 eV by equation $E_g^{opt} = 1240/\lambda_{edge}$. The absorbance of PFTPA as a wide bandgap homopolymer is mainly located in the near ultraviolet light region, and PFTPA displays strong blue emission. Thus, in p-i-n PSC devices, the competitive absorption of solar light by the hole transport layer and perovskite layer can be efficiently avoided. Meanwhile, by converting UV light to blue emission, PFTPA could reduce the damage of UV irradiation to perovskite layer and increase the solar light density in the visible region, leading to efficient absorption and conversion of solar light by PSC devices.

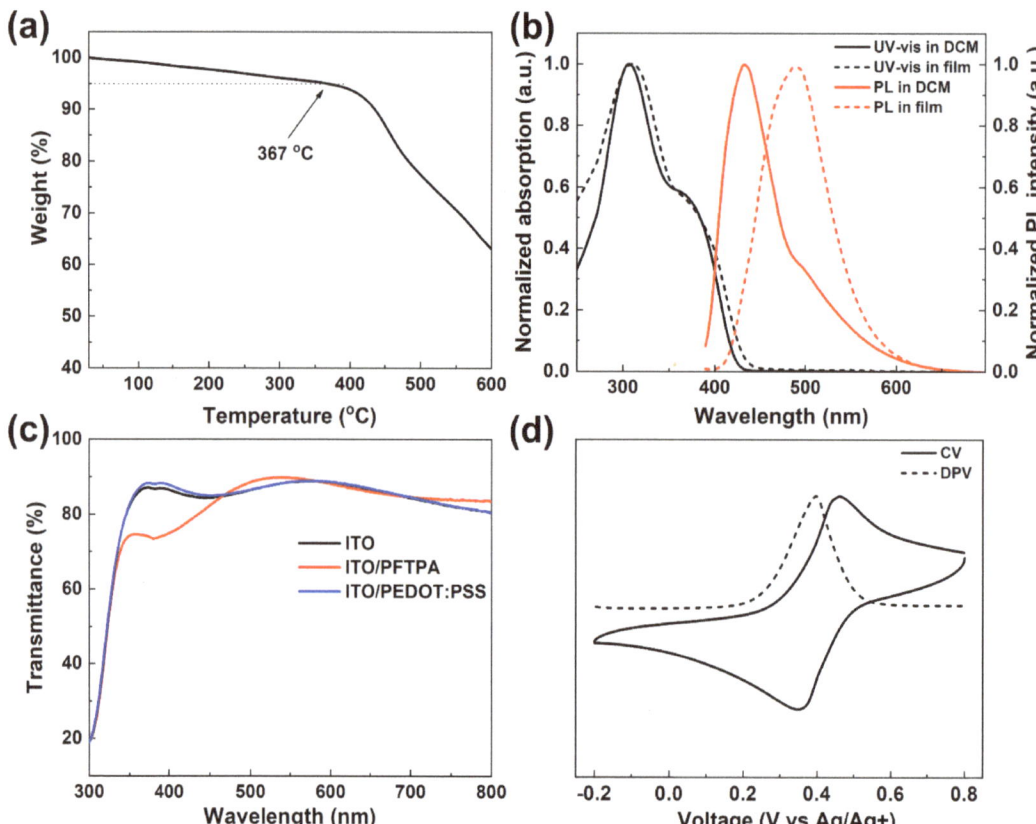

Figure 1. Thermal, photophysical and electrochemical properties of PFTPA. (**a**) TGA curves recorded at a heating rate of 10 °C/min; (**b**) Normalized UV-vis absorption spectra and photoluminescence (PL) spectra of PFTPA in dilute DCM solution (10^{-5} M) and in spin-coated films; (**c**) The transmittance spectra of ITO and ITO with different HTMs; (**d**) The cyclic voltammetry (CV) and differential pulse voltammetry (DPV) curves of PFTPA.

Table 1. The photovoltaic performance of perovskite solar cells.

	Scan Direction	V_{OC} (V)	J_{SC} (mA cm^{-2})	FF (%)	PCE (%) [a]
PFTPA	Reverse	1.13	19.66	75.7	16.82
	Forward	1.12	20.08	72.5	16.30
Slot-die	Reverse	1.08	19.50	65.7	13.84
PEDOT:PSS	Reverse	1.00	19.87	69.5	13.80

Note: [a], average value for PCE device according to 15 devices.

The transmittance of the hole transport layer in inverted PSC devices significantly affects the utilization efficiency of sunlight by the perovskite active layer. The light transmission spectra of bare ITO substrate, ITO/PFTPA and ITO/PEDOT:PSS films were measured (shown in Figure 1c). PFTPA has obvious light absorption in the range of 350–450 nm, resulting in low light transmittance in this range. However, at the wavelength ranging from 470 to 800 nm, its light transmittance is better than that of PEDOT:PSS, ensuring more sunlight can efficiently reach the perovskite active layer. Moreover, as discussed above, the absorbed sunlight in the range of 350–450 nm can be partially down-converted to the blue emission of PFTPA, which will also be absorbed by the perovskite layer.

Electrochemical cyclic voltammetry (CV) and differential pulse voltammetry (DPV) were performed to determine the frontier orbital energy levels of PFTPA experimentally. Figure 1d depicts the CV and DPV profile of PFTPA in a dilute DCM solution. As can be seen, PFTPA showed a reversible oxidation process in the positive range. The onset oxidation potential (E_{onset}^{OX}) of PFTPA is estimated to be 0.34 V (vs. Ag/Ag$^+$) with ferrocene as the reference and the oxidation potential value of ferrocene of 0.10 V (vs. Ag/Ag$^+$, see in Figure S6). Usually, the HOMO energy level is calculated according to the following Formula (1):

$$\text{HOMO} = -\left(E_{onset}^{OX} - E_{Fc/Fc+} + 5.1\right) \text{eV}. \quad (1)$$

In the end, the HOMO energy level of PFTPA is calculated to be −5.34 eV, which is deeper than that of spiro-OMeTAD (−5.01 eV) and PTAA (−5.1 eV). This value matches the VB energy level (5.40 eV) of MAPbI$_3$ well, which will benefit the efficient extraction of the hole at the interface between the hole transport layer and the perovskite active layer. In addition, the deeper HOMO levels of HTMs could benefit the higher open-circuit voltage (V_{OC}) of the PSC devices. The LUMO energy level of PFTPA is calculated to be −2.39 eV by adding its optical band-gap energy (E_g = 2.95 eV) to its HOMO energy. Obviously, PFTPA with much higher LUMO energy level than the conduction band (CB) energy level of MAPbI$_3$ (−3.9 eV) could block electron flow into perovskite more effectively.

3.3. Density Functional Theory Simulation of PFTPA

To further understand the molecular configuration and electron distribution of its frontier molecular orbitals of PFTPA, density functional theory (DFT) simulation was performed using Gaussian 09 at the B3LYP/6-31G(d,p) base set. In order to investigate the twist angle of monomers in conjugated backbone of PFTPA, a model polymer with three repeating units was adopted for theoretical simulation as shown in Figure 2. The optimized geometry shows that the dihedral angles between the middle fluorene unit and the left and right monomers are +33.74° and −33.72°, respectively. This implies that all the 4,4-dimethoxytriphenylamine groups are arranged in a helical manner around the polyfluorene backbone. Thus, these 4,4-dimethoxytriphenylamine groups suppress the π-π stacking between polyfluorene backbone, which explains the similar absorbance spectra of PFTPA in solution and film state. The LUMO orbital of PFTPA is distributed in the polyfluorene backbone, while the HOMO orbital is mainly located around the 4,4-dimethoxytriphenylamine groups in one repeat unit. By plotting the molecular orbits of HOMO-1, HOMO-2 and HOMO-3, all these HOMO orbitals are distributed in one 4,4-dimethoxytriphenylamine group due to the degenercy of energy levels among the monomers in PFTPA.

The electronic transition process and absorption spectrum of PFTPA in the dichloromethane solution was also theoretically simulated based on the time-dependent DFT (TD-DFT) method, as shown in Figure S7 and Table S2. The absorption peaks of PFTPA mainly originate from n-π* and π-π* transitions between 4,4-dimethoxytriphenylamine and the polyfluorene backbone. For example, the maximum absorption peak near 310 nm and the absorption shoulder account for the n-π* and π-π* transition absorption, respectively.

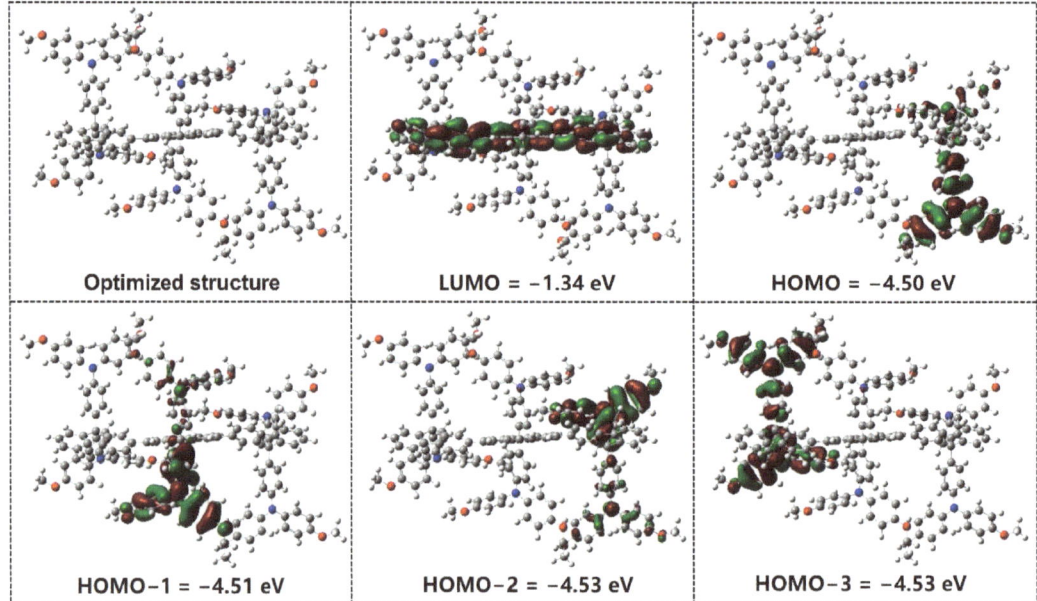

Figure 2. The optimized chemical structure and frontier orbitals distribution of PFTPA.

3.4. Photovoltaic Performance of the PFTPA-Based PSC Devices

To evaluate its behavior as dopant-free HTM, PFTPA was incorporated without any doping additives in inverted PSCs. PEDOT:PSS was used as HTM for control devices without additives as well. We fabricated an inverted PSC device with the planar configuration of ITO/HTLs/MAPbI$_3$/[6,6]-phenyl-C61-butyric acid methyl ester (PC$_{61}$BM)/PEIE/Ag as shown in Figure 3a. The perovskite layer was prepared in air environment by air blowing assisted drop coating (BADC) [41]. PC$_{61}$BM was used as an electron transporting layer, and ethoxylated polyethyleneimine (PEIE) was used as a cathode modification layer to improve charge extraction efficiency. As shown in Figure 3a, the HOMO energy level of PFTPA and the LUMO energy level of PC$_{61}$BM achieved good energy level matching with the valence band and conduction band of MAPbI$_3$, respectively, which effectively promotes the extraction and transmission of photogenerated carriers at interfaces.

The $J-V$ characteristic curves of the optimal perovskite device are shown in Figure 3b. The open-circuit voltage (V_{OC}) of the perovskite device based on the PFTPA is 1.13 V, the short-circuit current (J_{SC}) is 19.66 mA cm^{-2}, the filling factor (FF) is 75.7%, and the PCE reaches 16.82%. In comparison, the PCE of the control device based on PEDOT:PSS under the same preparation conditions is only 13.80% because of the much lower V_{OC} (1.00 V). According to the operating principle of the perovskite solar cells, it can be determined that the V_{OC} of the device is affected by the band gap of the photoactive layer, the HOMO energy level of the hole transporting layer and the LUMO energy level of the electron transporting layer. Therefore, compared with PFTPA, the higher HOMO energy level of PEDOT:PSS leads to the increased energy loss at the hole extraction interface and results in a much lower V_{OC} of the control device. Under a fixed bias voltage (0.96 V), the stable output efficiency of the champion device based on PFTPA is 16.65%, as shown in Figure 3c, which is consistent with the PCE obtained in the $J-V$ curves. The repeatability of device preparation is an important index to evaluate the preparation method and material performance. The PCEs of 15 perovskite solar cells based on PFTPA and PEDOT:PSS are counted, as shown in Figure 3d. Both of them show a narrow statistical distribution, indicating that the above devices have good repeatability. Additionally, PFTPA as the dopant-free HTM has been applied to the slot-die coated inverted PSCs. As a preliminary result, the slot-die coated inverted PSCs at a device area of 0.1 cm^2 showed a max PCE of 13.84% with a J_{SC} of 19.50 mA cm^{-2}, a V_{OC} of 1.08 V, and an FF of 65.7% (Figure S13 and Table 1). It should be mentioned that as a preliminary result, PCEs of the dopant-free PFTPA-based air-processed p-i-n PSCs maintain a long-term stability of 91% under ambient air conditions for 1000 h (see Figure S8). These results suggest the great potential of homopolymer PFTPA HTMs for future low-cost large-scale and flexible PSCs application.

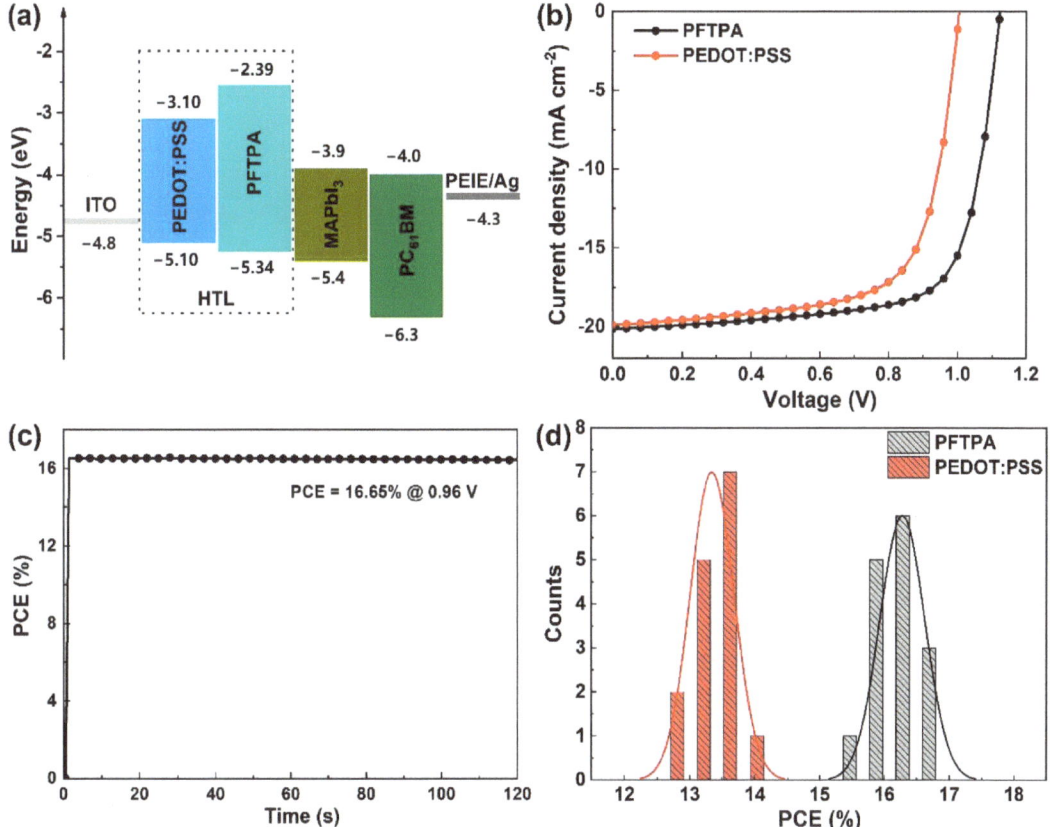

Figure 3. (**a**) The energy diagram of perovskite solar cells, (**b**) the $J-V$ curves of perovskite solar cells based on HTMs of PFTPA and PEDOT:PSS, (**c**) the stabilized PCE of PFTPA-based PSC device obtained near the maximum power point voltage at 0.96 V, (**d**) histogram of 15 devices based on PFTPA and PEDOT:PSS.

3.5. Morphology Analysis of Perovskite Films

The effects of PFTPA and PEDOT:PSS as HTM substrates on the growth quality of perovskite crystals were systematically studied by means of scanning electron microscopy (SEM) and thin film X-ray diffraction (XRD) as shown in Figure 4. Owing to the low solubility of PFTPA in polar solvents, it was possible to fabricate a high-quality perovskite crystalline film on its surface by using the solution-processing method. The contact angle between HTM and DMF was measured (Figure S9). The measured contact angles on ITO, PEDOT:PSS and PFTPA were 9.3°, 18.4° and 77.6°, respectively. This discrepancy in contact angle may result in different morphology of the solution-processed perovskite film. That is, the lesser wettability toward DMF of the PFTPA surface would suppress the heterogeneous nucleation and thus facilitate the grain boundary migration in grain growth, leading to large grain sizes of the resulting polycrystalline perovskite film. To further confirm their good crystal qualities, SEM images of the perovskite films deposited on the top of ITO, PEDOT:PSS or PFTPA substrate were given. The perovskite crystal sizes on PFTPA or PEDOT:PSS-modified ITO substrate were significantly larger (Figure 4a,b) than that on the bare ITO substrate (Figure 4c). The average grain size deduced from SEM images was 158.3 nm, 233. 3 nm and 271.9 nm for perovskite crystals on ITO, PEDOT:PSS and PFTPA, respectively (Figure S10). According to XRD patterns (Figure 4d), three films displayed similar diffraction peaks at 14.12°, 28.36° and 31.81°, assigned to the (110), (220) and (310) planes of perovskite, respectively. The appearance of the intense XRD peaks of the perovskite film deposited on PEDOT:PSS and PFTPA substrates further indicated their good crystallinity. The grain sizes of perovskite crystals on ITO, PEDOT:PSS and PFTPA substrates estimated according to XRD profiles were 163.7 nm, 231.9 nm and 276.5 nm, respectively. The grain size distribution trend based on XRD profiles is consistent with the SEM test results. It should be mentioned that the perovskite crystals grown on PFTPA substrates were more compact and uniform, while the perovskite crystals grown on PEDOT:PSS substrates had more obvious crystal boundary defects (dot-line circling in Figure 4a,b). These defects lead to carrier recombination and non-radiative recombination, which lead to energy loss, reduction in FF and V_{OC} of PSC devices, and ultimate effect on the photovoltaic performance of the devices. At the same time, boundary defects like pinholes provide channels for water vapor infiltration, which can often compromise the performance of PSC devices, especially their stability [42].

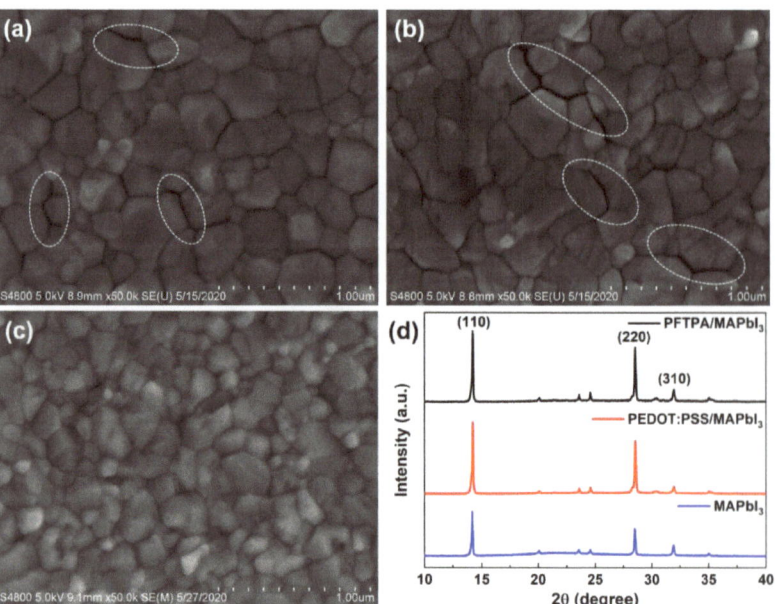

Figure 4. SEM images of perovskite on the top of (**a**) PFTPA, (**b**) PEDOT:PSS and (**c**) ITO substrates, (**d**) the X-ray diffraction patterns of perovskite on different substrates.

The morphological characteristics of perovskite films also include atomic force microscopy (AFM). As shown in Figure S11, the root mean square (RMS) roughness of perovskite films on PFTPA and PEDOT:PSS substrates are 16.22 nm and 17.22 nm, respectively. At the same time, more obvious holes can be observed in the latter. As crystal boundary defects, these holes also lead to the degradation of the performance of PSC devices. The above characterization results consistently show that PFTPA as the hole transport layer is more conducive to the growth of crystalline perovskite films on its surface. In the end, PSC devices based on PFTPA hole-transporting layer show higher FF and PCE.

3.6. SCLC Measurements of PFTPA for Its Hole-Transporting Morbility

In order to further verify the hole-transporting characteristics of polymeric HTM PFTPA, a hole-only device based on PFTPA was fabricated and tested. The device structure is ITO/PEDOT:PSS/PFTPA/MoO$_3$/Ag. Space charge limited current (SCLC) method was used for this measurement. The $J-V$ characteristic curve of the hole-only device is shown in Figure S12. The fitting calculation of hole mobility of PFTPA (μ_h) was carried out according to the simplified Mott-Gurney formula:

$$J = \frac{9}{8}\varepsilon_0\varepsilon_r\mu_h\frac{V^2}{L^3}, \qquad (2)$$

where J is the current density, ε_0 represents the vacuum permittivity, and ε_r is the relative dielectric constant (assuming $\varepsilon_r = 3$ for organic materials) [43], V represents the bias voltage applied to the device, L represents the thickness of the hole transport layer (about 80 nm), μ_h is the hole mobility. The calculated hole mobility of PFTPA is 1.12×10^{-5} cm^2 V^{-1} s^{-1}, which is higher than the reported hole mobility of PEDOT:PSS of 6.86×10^{-6} cm^2 V^{-1} s^{-1} [44]. Based on the above characterization results, PFTPA benefits from its unique 4,4-dimethoxytriphenylamine helical side group structure to enhance the carrier transport efficiency between molecules, which makes it show more efficient carrier transport capacity than PEDOT:PSS, and the corresponding perovskite cell devices achieve higher photovoltaic performance.

3.7. Steady-State PL and TRPL Measurements

Steady-state photoluminescence (PL) and time-resolved photoluminescence (TRPL) of the HTMs/MAPbI$_3$ films were conducted to further analyze and understand the charge extraction and transmission process at the hole transport layer/perovskite interface, as shown in Figure 5. For steady-state PL spectra, the quenching of luminescence intensity indicates the efficient separation of photogenerated carriers, so the carrier extraction efficiency at the interface can be qualitatively analyzed. Based on the above principle, the area integral of the steady-state PL spectrum emission peak was calculated. The photoluminescence intensity of PFTPA/MAPbI$_3$ and PEDOT:PSS/MAPbI$_3$ thin films decreased to 26.3% and 25.4% of that of pure MAPbI$_3$ thin films, respectively, indicating that both PFTPA and PEDOT:PSS showed high carrier extraction efficiency as hole transport materials. The photoluminescence lifetime of the hole transport material/perovskite films was characterized by TRPL spectra. The shorter the lifetime, the more efficient the hole transport process at the interface. As shown in Figure 5b, the TRPL spectrum is excited by a 510 nm wavelength and collected at a 770 nm wavelength. Both perovskite single-component films and HTM/perovskite bilayer films have a fast and a slow decay process, which are fitted by the following double exponential formula, and then the average luminescence lifetime (τ_{avg}) is calculated:

$$I(t) = A_1 exp\left(-\frac{t}{\tau_1}\right) + A_2 exp\left(-\frac{t}{\tau_2}\right) + A_0, \qquad (3)$$

$$\tau_{avg} = \frac{A_1\tau_1^2 + A_2\tau_2^2}{A_1\tau_1 + A_2\tau_2}. \qquad (4)$$

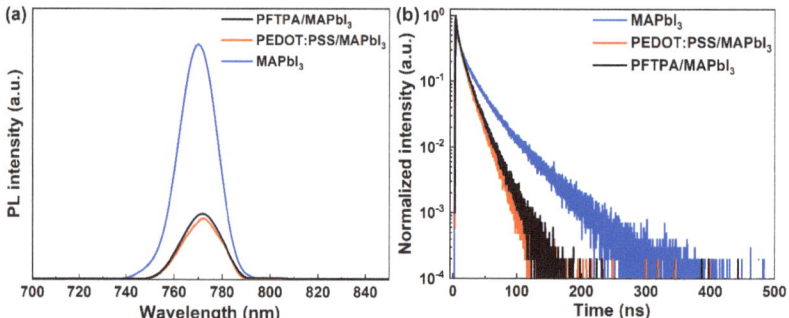

Figure 5. (a) The steady-state PL spectra and (b) time-resolved PL (TRPL) spectra of the HTMs/MAPbI$_3$ films with excitation at 510 nm.

Fast decay luminescence lifetime of perovskite single-component films (τ_1) and slow decay luminescence lifetime (τ_2) are 3.1 ns and 24.8 ns, respectively. The average luminescence lifetime (τ_{avg}) of a perovskite single-component film is 21.4 ns. For a PEDOT:PSS/MAPbI$_3$ bilayer film, τ_1 and τ_2 are 2.5 ns and 13.0 ns, respectively, with a τ_{avg} of 11.4 ns. The τ_1 and τ_2 of PFTPA/MAPbI$_3$ bilayer films are 2.8 ns and 14.5 ns, respectively, and PFTPA/MAPbI$_3$ bilayer films display a τ_{avg} of 12.5 ns. Compared with pure perovskite films, the luminescence lifetime of the hole transport layer decreases significantly, that is, the photogenerated hole transport process is more efficient. Meanwhile, PEDOT:PSS/MAPbI$_3$ bilayer films have shorter luminescence lifetime, which indicates that the charge transfer process at the interface is more efficient. The above steady-state PL and TRPL test results show that PEDOT:PSS, as a hole transport material, is slightly better than PFTPA in hole extraction and transport. However, the performance test results of perovskite solar cells prepared in this chapter show that the perovskite solar cells based on PFTPA have better photovoltaic performance, which may be related to the good film-forming property of PFTPA, the improvement of perovskite crystal growth quality and the more matching energy level structure.

4. Conclusions

In summary, a dopant-free polymeric HTM PFTPA was synthesized by a simple and efficient two-step method. The polymer takes polyfluorene as the main chain backbone and 4,4-dimethoxytriphenylamine as the side chain. Its hole mobility reaches 1.12×10^{-5} cm^2 V^{-1} s^{-1}. When PFTPA is applied to inverted planar perovskite solar cells, its good film-forming property promotes the growth of MAPbI$_3$ films with high crystallinity. At the same time, due to its matching HOMO energy level, the V_{OC} of the champion device reaches 1.13 V, and the PCE becomes 16.82%, which is much higher than the control device using PEDOT:PSS as HTM (PCE = 13.80%). This work provides a new idea for the design and synthesis of dopant-free polymeric HTMs. In terms of device preparation methods, the preparation of HTLs and a perovskite photoactive layer of devices are completed, performed in an air environment, with lower equipment requirements and manufacturing costs. This method is of great significance for the development of low-cost large-size perovskite solar cell devices.

Supplementary Materials: The following supporting information can be downloaded at: https://www.mdpi.com/article/10.3390/polym15122750/s1. Figure S1: ^{1}H NMR spectrum of 2BrFTPA in CDCl$_3$; Figure S2: ^{13}C NMR spectrum of 2BrFTPA in CDCl$_3$; Figure S3: MALDI-TOF MS spectrum of 2BrFTPA; Figure S4: ^{1}H NMR spectrum of PFTPA in CDCl$_3$; Figure S5: DSC curves of PFTPA recorded at a heating rate of 10 °C/min and a cooling rate of 20 °C/min; Figure S6: The cyclic voltammetry (CV) curve of Ferrocene measured with Ag/Ag$^+$ as the reference electrode in acetonitrile; Figure S7: The overlay of TD-DFT simulation and experimental UV-vis spectra of PFTPA; Figure S8: The longtime device stability of PFTPA-based PSC; Figure S9: The measured contact angles of DMF (a) on ITO substrate, (b) on PEDOT:PSS substrate and on PFTPA substrate; Figure S10: The perovskite crystal size distribution; Figure S11: AFM height images (size: 5 µm × 5 µm) of perovskite on (a) PFTPA and (b) PEDOT:PSS substrates; Figure S12: The $J - V$ curve of hole-only device based on PFTPA; Figure S13: The $J - V$ curve of slot-die coated perovskite device based on PFTPA; Table S1: The synthetic cost analysis of PFTPA in this study; Table S2: The calculation results of TD-DFT.

Author Contributions: Conceptualization, B.Z., H.H. and Z.L.; Funding acquisition, B.Z., H.H.; Investigation, M.T., X.C., P.X., J.Y. and P.H.; Methodology, B.Z., H.H. and Z.L.; Project administration, B.Z.; Resources, H.H.; Supervision, B.Z.; Validation, H.H. and Z.L.; Writing—original draft, B.Z. and Z.L.; Writing—review and editing, B.Z., H.H. and Z.L. All authors have read and agreed to the published version of the manuscript.

Funding: This research was financially supported by the National Natural Science Foundation of China (No. 62204123), the Foundation of Key Laboratory of Low-grade Energy Utilization Technologies and Systems (No. LLEUTS-201704), the Project of State Key Laboratory of Organic Electronics and Information Displays (No. ZS030ZR22038), the Foundation of Key Laboratory of Flexible Electronics of Zhejiang Province (No. 2022FE001), NJUPT Culturing Project (No. NY218056 and NY219061).

Institutional Review Board Statement: Not applicable.

Informed Consent Statement: Not applicable.

Data Availability Statement: The data presented in this study are available on request from the corresponding author.

Acknowledgments: The authors are grateful for the support from the National Natural Science Foundation of China (No. 62204123), the Foundation of Key Laboratory of Low-grade Energy Utilization Technologies and Systems (No. LLEUTS-201704), the Project of State Key Laboratory of Organic Electronics and Information Displays (No. ZS030ZR22038), the Foundation of Key Laboratory of Flexible Electronics of Zhejiang Province (No. 2022FE001), NJUPT Culturing Project (No. NY218056 and NY219061).We are grateful to the High Performance Computing Center in Nanjing Tech University for supporting the computational resources.

Conflicts of Interest: The authors declare no conflict of interest.

References

1. Kojima, A.; Teshima, K.; Shirai, Y.; Miyasaka, T. Organometal halide perovskites as visible-light sensitizers for photovoltaic cells. *J. Am. Chem. Soc.* **2009**, *131*, 6050–6051. [CrossRef] [PubMed]
2. Etgar, L.; Gao, P.; Xue, Z.; Peng, Q.; Chandiran, A.K.; Liu, B.; Grätzel, M. Mesoscopic $CH_3NH_3PbI_3/TiO_2$ heterojunction solar cells. *J. Am. Chem. Soc.* **2012**, *134*, 17396–17399. [CrossRef]
3. Meng, D.; Xue, J.J.; Zhao, Y.P.; Zhang, E.; Zheng, R.; Yang, Y. Configurable organic charge carriers toward stable perovskite photovoltaics. *Chem. Rev.* **2022**, *122*, 14954–14986. [CrossRef] [PubMed]
4. Burschka, J.; Pellet, N.; Moon, S.J.; Humphry-Baker, R.; Gao, P.; Nazeeruddin, M.K.; Grätzel, M. Sequential deposition as a route to high-performance perovskite-sensitized solar cells. *Nature* **2013**, *499*, 316–319. [CrossRef] [PubMed]
5. Liu, M.; Johnston, M.B.; Snaith, H.J. Efficient planar heterojunction perovskite solar cells by vapour deposition. *Nature* **2013**, *501*, 395–398. [CrossRef] [PubMed]
6. Luo, D.; Yang, W.; Wang, Z.; Sadhanala, A.; Hu, Q.; Su, R.; Zhu, R. Enhanced photovoltage for inverted planar heterojunction perovskite solar cells. *Science* **2018**, *360*, 1442–1446. [CrossRef] [PubMed]
7. Jeong, J.; Kim, M.; Seo, J.; Lu, H.; Ahlawat, P.; Mishra, A.; Kim, J.Y. Pseudo-halide anion engineering for α-FAPbI$_3$ perovskite solar cells. *Nature* **2021**, *592*, 381–385. [CrossRef]
8. Zhou, H.; Liang, L.; Guo, Z.; Li, L. Anti-corrosion strategy to improve the stability of perovskite solar cells. *Nanoscale* **2023**, *15*, 8473.
9. Khare, S.; Gohel, J.V. Performance enhancement of cost-effective mixed cationic perovskite solar cell with MgCl$_2$ and n-BAI as surface passivating agents. *Opt. Mater.* **2022**, *132*, 112845. [CrossRef]
10. Green, M.A.; Ho-Baillie, A.; Snaith, H.J. The emergence of perovskite solar cells. *Nat. Photonics* **2014**, *8*, 506–514. [CrossRef]
11. Correa-Baena, J.P.; Saliba, M.; Buonassisi, T.; Grätzel, M.; Abate, A.; Tress, W.; Hagfeldt, A. Promises and challenges of perovskite solar cells. *Science* **2017**, *358*, 739–744. [CrossRef] [PubMed]
12. Rong, Y.; Hu, Y.; Mei, A.; Tan, H.; Saidaminov, M.I.; Seok, S.I.; Han, H. Challenges for commercializing perovskite solar cells. *Science* **2018**, *361*, eaat8235. [CrossRef]
13. Kim, J.Y.; Lee, J.W.; Jung, H.S.; Shin, H.; Park, N.G. High-efficiency perovskite solar cells. *Chem. Rev.* **2020**, *120*, 7867–7918. [CrossRef] [PubMed]

14. Isikgor, F.H.; Zhumagali, S.T.; Merino, L.V.; De Bastiani, M.; McCulloch, I.; De Wolf, S. Molecular engineering of contact interfaces for high-performance perovskite solar cells. *Nat. Rev. Mater.* **2023**, *8*, 89–108. [CrossRef]
15. Liu, T.; Chen, K.; Hu, Q.; Zhu, R.; Gong, Q. Inverted perovskite solar cells: Progresses and perspectives. *Adv. Energy Mater.* **2016**, *6*, 1600457. [CrossRef]
16. Lin, X.; Cui, D.; Luo, X.; Zhang, C.; Han, Q.; Wang, Y.; Han, L. Efficiency progress of inverted perovskite solar cells. *Energy Environ. Sci.* **2020**, *13*, 3823–3847. [CrossRef]
17. Yao, Y.; Cheng, C.; Zhang, C.; Hu, H.; Wang, K.; De Wolf, S. Organic Hole-Transport Layers for Efficient, Stable, and Scalable Inverted Perovskite Solar Cells. *Adv. Mater.* **2022**, *34*, 2203794. [CrossRef]
18. Li, X.; Zhang, W.; Guo, X.; Lu, C.; Wei, J.; Fang, J. Constructing heterojunctions by surface sulfidation for efficient inverted perovskite solar cells. *Science* **2022**, *375*, 434–437. [CrossRef]
19. Al-Ashouri, A.; Magomedov, A.; Roß, M.; Jošt, M.; Talaikis, M.; Chistiakova, G.; Albrecht, S. Conformal monolayer contacts with lossless interfaces for perovskite single junction and monolithic tandem solar cells. *Energy Environ. Sci.* **2019**, *12*, 3356–3369. [CrossRef]
20. Ullah, A.; Park, K.H.; Nguyen, H.D.; Siddique, Y.; Shah, S.F.A.; Tran, H.; Hong, S. Novel phenothiazine-based self-assembled monolayer as a hole selective contact for highly efficient and stable p-i-n perovskite solar cells. *Adv. Energy Mater.* **2022**, *12*, 2103175. [CrossRef]
21. Li, L.; Wang, Y.; Wang, X.; Lin, R.; Luo, X.; Liu, Z.; Tan, H. Flexible all-perovskite tandem solar cells approaching 25% efficiency with molecule-bridged hole-selective contact. *Nat. Energy* **2022**, *7*, 708–717. [CrossRef]
22. Li, Z.; Tan, Q.; Chen, G.; Gao, H.; Wang, J.; Zhang, X.; He, Z. Simple and robust phenoxazine phosphonic acid molecules as self-assembled hole selective contacts for high-performance inverted perovskite solar cells. *Nanoscale* **2023**, *15*, 1676–1686. [CrossRef] [PubMed]
23. Zhang, S.; Ye, F.; Wang, X.; Chen, R.; Zhang, H.; Zhan, L.; Wu, Y. Minimizing buried interfacial defects for efficient inverted perovskite solar cells. *Science* **2023**, *380*, 404–409. [CrossRef] [PubMed]
24. Al-Ashouri, A.; Köhnen, E.; Li, B.; Magomedov, A.; Hempel, H.; Caprioglio, P.; Albrecht, S. Monolithic perovskite/silicon tandem solar cell with> 29% efficiency by enhanced hole extraction. *Science* **2020**, *370*, 1300–1309. [CrossRef]
25. Tong, J.; Song, Z.; Kim, D.H.; Chen, X.; Chen, C.; Palmstrom, A.F.; Zhu, K. Carrier lifetimes of> 1 µs in Sn-Pb perovskites enable efficient all-perovskite tandem solar cells. *Science* **2019**, *364*, 475–479. [CrossRef]
26. Urieta-Mora, J.; García-Benito, I.; Molina-Ontoria, A.; Martín, N. Hole transporting materials for perovskite solar cells: A chemical approach. *Chem. Soc. Rev.* **2018**, *47*, 8541–8571. [CrossRef]
27. Pham, H.D.; Yang, T.C.J.; Jain, S.M.; Wilson, G.J.; Sonar, P. Development of dopant-free organic hole transporting materials for perovskite solar cells. *Adv. Energy Mater.* **2020**, *10*, 1903326. [CrossRef]
28. Murugan, P.; Hu, T.; Hu, X.; Chen, Y. Advancements in organic small molecule hole-transporting materials for perovskite solar cells: Past and future. *J. Mater. Chem. A* **2022**, *10*, 5044–5081. [CrossRef]
29. Cho, K.T.; Paek, S.; Grancini, G.; Roldán-Carmona, C.; Gao, P.; Lee, Y.; Nazeeruddin, M.K. Highly efficient perovskite solar cells with a compositionally engineered perovskite/hole transporting material interface. *Energy Environ. Sci.* **2017**, *10*, 621–627. [CrossRef]
30. Wu, X.; Gao, D.; Sun, X.; Zhang, S.; Wang, Q.; Li, B.; Zhu, Z. Backbone Engineering Enables Highly Efficient Polymer Hole-Transporting Materials for Inverted Perovskite Solar Cells. *Adv. Mater.* **2023**, *35*, 2208431. [CrossRef]
31. Desoky, M.M.H.; Bonomo, M.; Barbero, N.; Viscardi, G.; Barolo, C.; Quagliotto, P. Polymeric dopant-free hole transporting materials for perovskite solar cells: Structures and concepts towards better performances. *Polymers* **2021**, *13*, 1652. [CrossRef] [PubMed]
32. Kim, G.W.; Kang, G.; Kim, J.; Lee, G.Y.; Kim, H.I.; Pyeon, L.; Park, T. Dopant-free polymeric hole transport materials for highly efficient and stable perovskite solar cells. *Energy Environ. Sci.* **2016**, *9*, 2326–2333. [CrossRef]
33. Sun, X.; Yu, X.; Li, Z.A. Recent advances of dopant-free polymer hole-transporting materials for perovskite solar cells. *ACS Appl. Energy Mater.* **2020**, *3*, 10282–10302. [CrossRef]
34. Wang, L.; Zhuang, Q.; You, G.; Lin, X.; Li, K.; Lin, Z.; Ling, Q. Donor-Acceptor Type Polymers Containing Fused-Ring Units as Dopant-Free, Hole-Transporting Materials for High-Performance Perovskite Solar Cells. *ACS Appl. Energy Mater.* **2020**, *3*, 12475–12483. [CrossRef]
35. Kim, D.W.; Choi, M.W.; Yoon, W.S.; Hong, S.H.; Park, S.; Kwon, J.E.; Park, S.Y. A dopant-free donor-acceptor type semi-crystalline polymeric hole transporting material for superdurable perovskite solar cells. *J. Mater. Chem. A* **2022**, *10*, 12187–12195. [CrossRef]
36. You, G.; Zhuang, Q.; Wang, L.; Lin, X.; Zou, D.; Lin, Z.; Ling, Q. Dopant-free, donor–acceptor-type polymeric hole-transporting materials for the perovskite solar cells with power conversion efficiencies over 20%. *Adv. Energy Mater.* **2020**, *10*, 1903146. [CrossRef]
37. Rombach, F.M.; Haque, S.A.; Macdonald, T.J. Lessons learned from spiro-OMeTAD and PTAA in perovskite solar cells. *Energy Environ. Sci.* **2021**, *14*, 5161–5190. [CrossRef]
38. Reza, K.M.; Mabrouk, S.; Qiao, Q. A review on tailoring PEDOT: PSS layer for improved performance of perovskite solar cells. *Proc. Nat. Res. Soc.* **2018**, *2*, 02004. [CrossRef]
39. Erazo, E.A.; Ortiz, P.; Cortés, M.T. Tailoring the PEDOT: PSS hole transport layer by electrodeposition method to improve perovskite solar cells. *Electrochim. Acta* **2023**, *439*, 141573. [CrossRef]

40. Duan, C.; Zhao, M.; Zhao, C.; Wang, Y.; Li, J.; Han, W.; Jiu, T. Inverted $CH_3NH_3PbI_3$ perovskite solar cells based on solution-processed V_2O_5 film combined with P3CT salt as hole transport layer. *Mater. Today Energy* **2018**, *9*, 487–495. [CrossRef]
41. Zuo, C.; Vak, D.; Angmo, D.; Ding, L.; Gao, M. One-step roll-to-roll air processed high efficiency perovskite solar cells. *Nano Energy* **2018**, *46*, 185–192. [CrossRef]
42. Shao, J.Y.; Yu, B.C.; Wang, Y.D.; Lan, Z.R.; Li, D.M.; Meng, Q.B.; Zhang, Y.W. In-situ electropolymerized polyamines as dopant-free hole transporting materials for efficient and stable inverted perovskite solar cells. *ACS Appl. Energy Mater.* **2020**, *3*, 5058–5066. [CrossRef]
43. Niu, T.; Zhu, W.; Zhang, Y.; Xue, Q.; Jiao, X.; Wang, Z.; Cao, Y. DA-π-AD-type dopant-free hole transport material for low-cost, efficient, and stable perovskite solar cells. *Joule* **2021**, *5*, 249–269. [CrossRef]
44. Lee, K.; Yu, H.; Lee, J.W.; Oh, J.; Bae, S.; Kim, S.K.; Jang, J. Efficient and moisture-resistant hole transport layer for inverted perovskite solar cells using solution-processed polyaniline. *J. Mater. Chem. C* **2018**, *6*, 6250–6256. [CrossRef]

Disclaimer/Publisher's Note: The statements, opinions and data contained in all publications are solely those of the individual author(s) and contributor(s) and not of MDPI and/or the editor(s). MDPI and/or the editor(s) disclaim responsibility for any injury to people or property resulting from any ideas, methods, instructions or products referred to in the content.

Article

Determination of Mechanical Properties of Sand-Coated Carbon Fiber Reinforced Polymer (CFRP) Rebar

Hyun-Do Yun [1], Sun-Hee Kim [2] and Wonchang Choi [2,*]

[1] Department of Architectural Engineering, Chungnam National University, Daejeon 305-764, Republic of Korea
[2] Department of Architectural Engineering, Gachon University, Seongnam-si 13120, Republic of Korea
* Correspondence: wchoi@gachon.ac.kr; Tel.: +82-31-5355

Abstract: This experimental study investigates the fundamental mechanical characteristics of the carbon fiber-reinforced polymer (CFRP) bars, including the tensile strength, compressive strength, shear strength, and modulus of elasticity of the CFRP bar. The properties need to be accurately determined to understand the behavior of the concrete structures reinforced with CFRP rebars. The CFRP rebar was coated with sand to enhance the adhesive strength of the concrete. Three diameters of CFRP rebar (D10, D12, and D16) were considered in accordance with ASTM provisions. A coefficient, i.e., the ratio of shear strength to tensile strength, was employed to predict the tensile strength of the CFRP rebar specimens. The test results confirm that the tensile strength of CFRP rebar is dependent on its diameter due to the shear lag effect. A coefficient in the range of 0.17 to 0.2 can be used to predict the tensile strength of CFRP rebar using shear strength.

Keywords: tensile strength; compressive strength; elastic modulus; shear strength; sand-coated carbon fiber-reinforced polymer (CFRP)

Citation: Yun, H.-D.; Kim, S.-H.; Choi, W. Determination of Mechanical Properties of Sand-Coated Carbon Fiber Reinforced Polymer (CFRP) Rebar. *Polymers* **2023**, *15*, 2186. https://doi.org/10.3390/polym15092186

Academic Editor: Yang Li

Received: 22 February 2023
Revised: 19 April 2023
Accepted: 1 May 2023
Published: 4 May 2023

Copyright: © 2023 by the authors. Licensee MDPI, Basel, Switzerland. This article is an open access article distributed under the terms and conditions of the Creative Commons Attribution (CC BY) license (https://creativecommons.org/licenses/by/4.0/).

1. Introduction

For decades, research into composite materials has explored the feasibility of a replacement material for conventional reinforcement in concrete structures. In Japan, Europe, Canada, and the United States, research for alternative materials for conventional steel rebar has been actively conducted, among them, fiber-reinforced polymer (FRP) composites are employed as an alternative. FRP materials are widely used in the construction industry due to their superior mechanical and physical advantages such as high chemical resistance, high corrosion resistance, lightweight, non-conductivity, etc. FRP rebar is anisotropic material that can be manufactured by either a pultrusion process or braiding technique. The pultrusion process is inexpensive and can rapidly produce a member with a constant cross-section. Composites produced by the pultrusion process are good in structural applications due to their continuous mass production with homogeneous mechanical properties. However, because members generated by the pultrusion process have a smooth surface, an additional step of digging or excavating with a machine and protruding or coating the surface is needed to increase the strength of its bond with the concrete interface. The braiding technique is a modification of the pultrusion process that creates protrusions via a weaving process prior to the hardening stage. Although this braiding method is difficult in practice and the fiber content is less than in the pultrusion process, external protrusions can be created easily. The braiding technique was used to manufacture the carbon fiber-reinforced polymer (CFRP) rebar specimens in this study. In order to improve the adhesion performance of the FRP rebar, surface treatment of the protrusion was preferred. Among FRPs, the CFRP has higher tensile strength and elastic modulus than steel rebar. Based on the previous results, the mechanical performance of the FRP rebar used in construction and civil engineering was verified, and the CFRP rebar was confirmed to be used as a steel rebar substitute [1,2]. Additional studies are needed to better understand the structural behavior of FRP concrete structures. First, the mechanical properties, which include the tensile strength, elastic modulus, and shear strength, of FRP rebar must be determined. Test standards for FRP rebar already have been established in the United States and Canada. In 2006, the American Concrete Institute (ACI) 440 Committee presented the specification 440.1R-06 and published a revised version of ACI 440.1R-15 in 2015 [2,3]. The ACI 440.3R-12 [4] specifies a test method for FRP rebar applied to structures and concrete. The test methods include those for mechanical properties, such as tensile strength, adhesion, creep, and flexure properties. In Canada, a specification referred to as the Design and Construction of Building Structures with Fibre-Reinforced Polymers (CHBDC CAN/CSA-S6-06) was proposed in 2006, and CHBDC CAN/CSA-S6-14 and CAN/CSA S806-12 have been proposed more recently [5–7]. In Korea, KS F ISO 10406-1 was enacted in 2017 as a test standard for FRP rebar [8]. However, the existing standards have not been fully developed for large diameters of CFRP rebars. Several studies have been conducted over the past decades and available findings are reviewed herein.

Benmokrane et al. [9] successfully completed the tensile test for CFRP rebar (less than $\phi 8$ mm). He reported that the pull-out behavior is affected by the surface geometry of FRP rods, the properties of the grout, and the stiffness of the anchoring tube. Khan et al. determined the mechanical properties of glass fiber-reinforced polymer (GFRP) rebar ($\phi 15.9$ mm) and CFRP rebar ($\phi 15$ mm) in accordance with ASTM D7205 (tension test) and ASTM D695 (compression test) [10]. Khan et al. reported that the modulus of elasticity of CFRP rebar is greater than that of GFRP rebar, although the tensile strength of CFRP rebar is less than that of GFRP rebar due to the lower percentage of CFRP fibers by volume than the GFRP rebars [11]. Plevkov et al. determined the tensile and compressive strength values of CFRP rebar ($\phi 10$ mm and GFRP rebar ($\phi 10$ mm) and reported that CFRP rebar has greater tensile strength and a higher elasticity modulus value than GFRP rebar. Plevkov et al. also reported crushing failure at the end tips, so they devised a compressive strength test that involves fitting steel caps to the ends of GFRP rebar and then filling them with concrete [12]. Koosha and Pedram [13] introduced a new test method to determine the compressive properties of GFRP rebar. GFRP rebar with diameters of 13 mm, 16 mm, and 19 mm were used. Steel caps were attached to both ends of the specimen to prevent alignment of the specimen and premature failure at the end of the specimen. In addition, two gauges were installed on the specimen to conduct a compression test. AlAjarmeh et al. [14] proposed a new test for compressive properties by mounting a hollow steel cap on both ends of a GFRP rebar and filling it with cement grout because there was no test method for compressive strength test of GFRP. GFRP rebars with diameters of 9.5 mm, 15.9 mm, and 19.1 mm were used, and the length of the specimen was set to diameter ratios of 2, 4, 8, and 16. Only a few test results are available for the CFRP rebars.

In short, as addressed above, the fundamental mechanical characteristics of the CFRP bars, including the tensile strength, compressive strength, shear strength, and modulus of elasticity of the CFRP bar, need to be accurately determined to understand the behavior of the concrete structures reinforced with CFRP rebars. In addition, the bar size effects on the mechanical properties of CFRP rebars are investigated in this study.

2. Testing of Mechanical Properties of CFRP Rebar Specimens

2.1. Materials

The FRP rebar has lower adhesion to concrete than steel rebar [15]. The CFRP rebar specimens used in this study were sand-coated to improve the bond strength of the concrete. Figure 1 shows the sand-coated CFRP rebar specimens with diameters of 10 mm, 12 mm, and 16 mm. The manufacturer (SK chemical, Seoungnamsi, Korea) provided the following information regarding the properties: ultimate stress > 2850 MPa (based on ASTM D3039M [16]), modulus of elasticity > 158 GPa, and ultimate strain > 1.8 percent. The ratio of carbon fiber (CF) to total area is about 42%.

(a)   (b)

Figure 1. Sand-coated carbon fiber reinforced polymer rebar specimens with three different diameters. (**a**) Various sizes of CFRP rebar; (**b**) Surface of sand-coated CFRP rebar.

2.2. Tensile Strength Testing of CFRP Rebar Specimens

The tensile strength tests of the CFRP rebar specimens were conducted in accordance with ASTM D7205 [17]. Five CFRP rebars for each of the three diameters, D10, D12, and D16 were prepared for the tensile strength test. The ASTM D7205 standard specifies the tensile strength test method suggested in ACI 440.3R-12 [4]. This test can determine the tensile strength of FRP matrix composite rebar that typically is used as a tensile element in rebar and prestressed post-tension concrete. Table 1 presents the dimensions of the tensile strength test CFRP rebar specimens used in this study. A steel tube (thickness: 2 mm) filled with epoxy at both ends of the specimen was fabricated in accordance with ASTM D7205. Due to the length limitation of testing equipment, the specimen for D16 was designed with a grip length of 660 mm.

Table 1. Dimensions of CFRP rebar specimens used for tensile strength tests.

Diameter (mm)	Grip Length (mm)	Free Length (mm)	Total Length (mm)
10	550	400	1500
12	660	480	1800
16	660	480	1800

Figure 2a schematically presents the CFRP D10 rebar specimen fabricated for the tensile strength test. Figure 2a presents the CFRP D10 rebar specimens fabricated for the tensile strength test with a grip length of 550 mm, free length of 400 mm, and total length of 1500 mm. Figure 2b schematically presents the CFRP D12 and D16 rebar specimens fabricated for the tensile strength test with a grip length of 660 mm, free length of 480 mm, and total length of 1800 mm. Figure 2c presents the tensile strength test set-up whereby the load is applied in displacement control mode using a universal testing machine (UTM) with a capacity of 1200 kN at the rate of 3 mm/min.

2.3. Compressive Strength Testing of CFRP Rebar Specimens

The compressive strength tests were conducted using CFRP rebar specimens with their lengths set to two times the diameter of the specimen in accordance with the compressive strength test method specified in ASTM D695 [18]. Five specimens for each of the three diameters, D10, D12, and D16, were prepared for each compressive strength test. Modulus of elasticity tests were conducted using specimens with their lengths set to four times the diameter of the specimen, as shown in Figure 3. Table 2 provides the dimensions of the compressive strength test CFRP rebar specimens. Note that '2D' and '4D' refer to two times and four times the diameter, respectively.

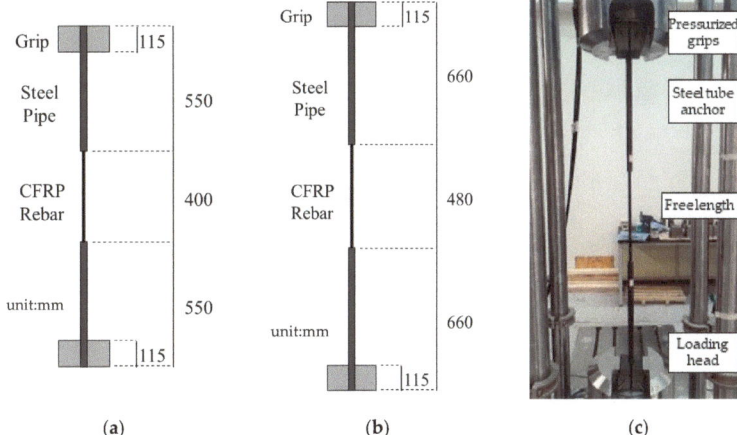

Figure 2. Tensile strength test of CFRP rebar based on ASTM. (**a**) dimensions for CFRP D10 rebar specimen; (**b**) dimensions for CFRP D12 and D16 rebar specimens; (**c**) tensile strength test set-up D7205 (at Intelligent Construction System Core-Support Center, Keimyung University).

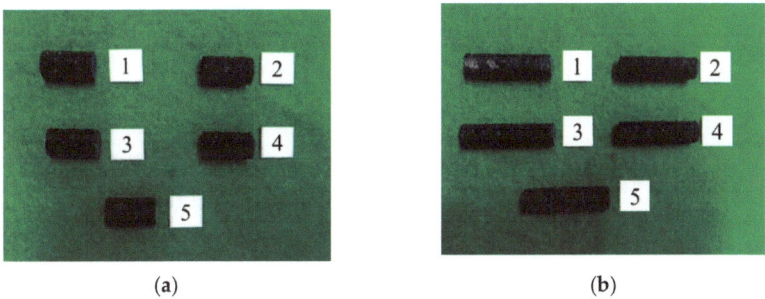

Figure 3. Compressive strength test specimens (D12). (**a**) Specimen length two times its diameter; (**b**) specimen length four times its diameter (for modulus of elasticity tests).

Table 2. Dimensions of CFRP rebar specimens used for compressive strength tests.

Specimen	Diameter (mm)	Height (mm)
D10 (2D)	10	20
D10 (4D)	10	40
D12 (2D)	12	24
D12 (4D)	12	48
D16 (2D)	16	32
D16 (4D)	16	64

Figure 4a,b show identical compressive strength test set-ups for the specimens with their lengths two times and four times their diameters, respectively. The load was applied in displacement control mode at a rate of 1 mm/min using a 100-kN UTM.

2.4. Shear Strength Testing of CFRP Rebar Specimens

The shear strength tests were conducted using five specimens for each of the three diameters, D10, D12, and D16.

Figure 5 shows the test specimens that were fabricated with the length of 225 mm in accordance with ASTM D7617 [19]. The set-up required for this test was designed specifically to fit the specimen for each diameter according to the ASTM D7617 standard.

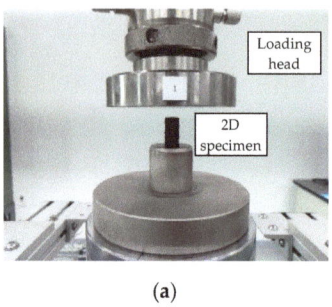 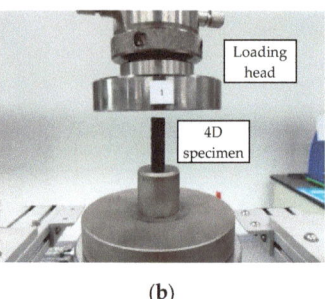

(a) (b)

Figure 4. Compressive strength test set-up based on ASTM D695. (**a**) D12 (length 2 times diameter) test set-up; (**b**) D12 (length 4 times diameter).

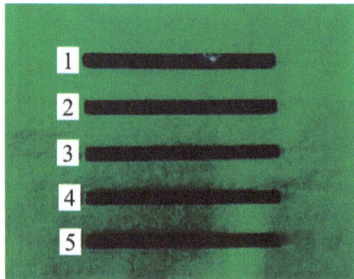

Figure 5. Five specimens used for shear strength tests.

Figure 6a shows the shear jig used for the shear strength test and Figure 6b shows the shear strength test set-up. The load was applied in displacement control mode at the rate of 1 mm/mm using a 100-kN UTM.

 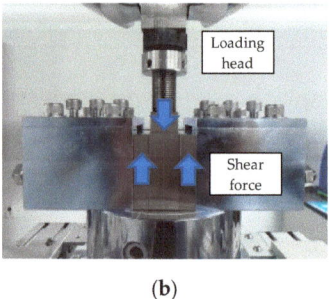

(a) (b)

Figure 6. Shear strength test of CFRP rebar based on ASTM D7617. (**a**) shear jig; (**b**) test set-up.

3. Test Results and Discussion

3.1. Tensile Strength Test Results

Figure 7 shows the tensile strength test specimens and the location and mode of failure. Figure 7a,c show that the CFRP rebar specimens fractured at the center and grip of the specimens, respectively. Figure 7b shows the fracture of the CFRP fiber at the center and that fiber weave is unidirectional. As the fiber weave is generated in one direction, the fibers of the CFRP rebar break sequentially, thus resulting in a brittle fracture. Figure 7d shows a fracture at the grip of the specimen.

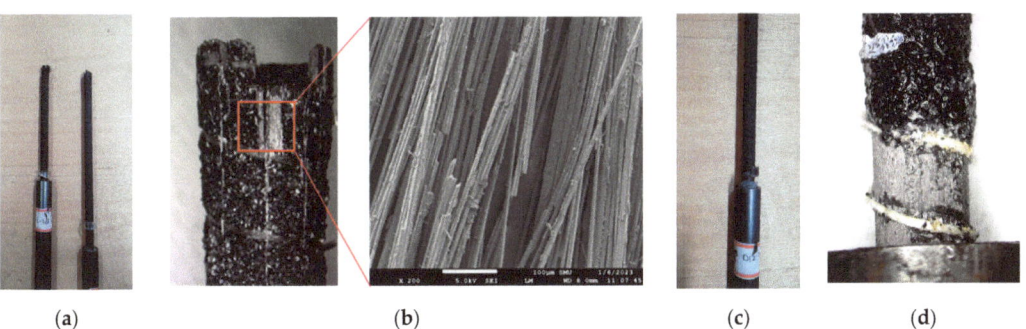

Figure 7. Tensile strength test specimens and failure location or mode. (**a**) center; (**b**) central failure mode; (**c**) grip; (**d**) grip failure mode.

Table 3 presents the tensile strength test results for the CFRP rebar specimens in terms of tensile strength value, modulus of elasticity value, and failure mode. Table 3 also provides the average values of the tensile strength and modulus of elasticity.

Table 3. Tensile test results.

Specimen	Tensile Strength (MPa)	Tensile Modulus of Elasticity (GPa)	Failure Mode
D10-1	2107	161	Grip failure
D10-2	2229	144	Grip failure
D10-3	2083	153	Grip failure
D10-4	2100	158	Grip failure
D10-5	2062	133	Grip failure
Average	2116 ± 58.50	150 ± 10.19	
D12-1	1854	152	Grip failure
D12-2	1811	173	Grip failure
D12-3	1732	144	Center failure
D12-4	1762	162	Grip failure
D12-5	1758	158	Grip failure
Average	1784 ± 43.57	158 ± 9.72	
D16-1	1859	136	Grip failure
D16-2	1799	135	Specimen end
D16-3	1839	145	Grip failure
D16-4	1871	136	Grip failure
D16-5	1786	131	Grip failure
Average	1831 ± 33.16	136 ± 4.59	

Figure 8a–c present the stress–displacement relationship of the D10, D12, and D16 CFRP rebar specimens, respectively. The results clearly indicate that CFRP rebar has no yield point. For the D16 specimens, an unexpected bilinear relationship was observed due to the slippage at the grip area. The average tensile strength value of the five D12 specimens tested is 1784 MPa. This value satisfies the standard tensile strength range of 600 MPa to 3690 MPa for CFRP rebar specified in ACI 440.1R-15 [2]. The average modulus of elasticity value is 158 GPa, which is also within the standard modulus of elasticity range of 120 GPa to 580 GPa for CFRP rebar specified in ACI 440.1R-15. Moreover, these results also satisfy the standard (KS F ISO 10406-1) modulus of elasticity value for CFRP rebar because the design guidelines (KS F ISO 10406-1) specify 140 GPa for FRP rebar. In addition, tensile strength tests of the five D10 and five D16 specimens show tensile strength values of 2116.7 MPa ± 58.47 MPa for D10 and 1831.3 MPa ± 33.19 MPa for D16.

3.2. Compressive Strength Test Results

Figure 9a,b present photos of compressive strength test D12 specimens with lengths that are two times and four times their diameters, respectively, and their failure modes. All the CFRP D12 rebar specimens were crushed at the point where the load was applied, as shown in the figures. The fibers became separated from each other due to the failure of the resin rather than buckling.

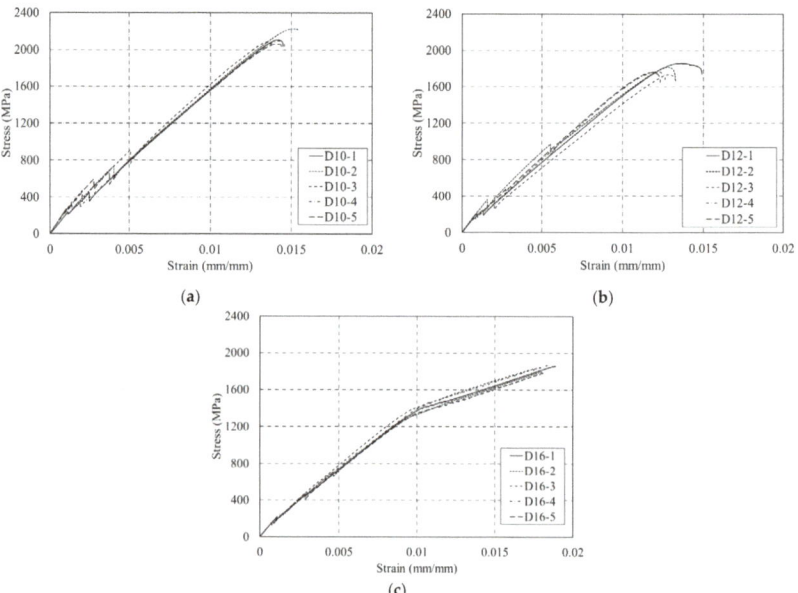

Figure 8. Stress–displacement relationship of five sand-coated CFRP rebar specimens. (**a**) D10; (**b**) D12; (**c**) D16.

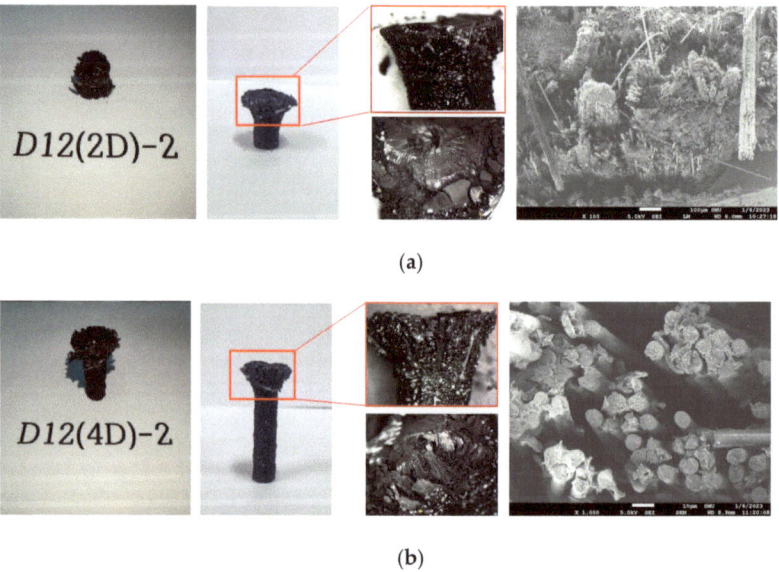

Figure 9. Compressive strength test specimens and their failure modes. (**a**) D12 (length 2 times diameter); (**b**) D12 (length 4 times diameter).

Table 4 presents the results of the compressive strength tests for each of the five specimens with each of the three diameters (D10, D12, and D16) and for each of the lengths (two times and four times the diameter, respectively). The compressive strength test results of the CFRP rebar indicate an average compressive strength of 357 MPa in the case of the D12 CFRP rebar. The compressive strength of the CFRP rebar specimen with a length that is twice the diameter is approximately 7% greater than that of the CFRP rebar specimen with a length that is four times the diameter. In addition, the compressive strength values of the CFRP rebar with lengths that are twice and four times the diameter are 79.9% and 81.2% lower than the tensile strength, respectively. The results of the compressive strength tests of the CFRP rebar indicate that the average compressive strength values of D10 and D16 CFRP rebar are 399 MPa and 360 MPa, respectively. The compressive strength of the CFRP D10 rebar specimen with a length that is twice the diameter is approximately 12% greater than that of the CFRP D10 rebar specimen with the length that is four times the diameter, whereas the compressive strength of the CFRP D16 rebar specimen with the length that is twice the diameter is approximately 9% smaller than that of the CFRP D 16 rebar specimen with the length that is four times the diameter.

Table 4. Compressive test results.

		Description	1	2	3	4	5	Average
D10	(2D)	$f_{comp.}$ (MPa)	402	452	427	314	402	399
	(2D)	Compressive modulus of elasticity E_{comp} (GPa)	14	17	15	13	15	15
	(4D)	$f_{comp.}$ (MPa)	369	355	314	333	408	356
	(4D)	Compressive modulus of elasticity E_{comp} (GPa)	29	26	23	23	30	26
D12	(2D)	$f_{comp.}$ (MPa)	350	392	420	344	277	357
	(2D)	Compressive modulus of elasticity E_{comp} (GPa)	16	16	16	13	13	15
	(4D)	$f_{comp.}$ (MPa)	356	311	335	398	273	335
	(4D)	Compressive modulus of elasticity E_{comp} (GPa)	31	30	28	36	26	30
D16	(2D)	$f_{comp.}$ (MPa)	326	348	332	410	382	360
	(2D)	Compressive modulus of elasticity E_{comp} (GPa)	13	18	17	19	18	17
	(4D)	$f_{comp.}$ (MPa)	453	429	321	403	368	395
	(4D)	Compressive modulus of elasticity E_{comp} (GPa)	42	38	34	46	41	40

3.3. Shear Strength Test Results

Shear strength tests of five specimens for each of the three diameters (D10, D12, and D16) were conducted, and Figure 10 shows a photo of the typical shear failure observed for the D12 specimens.

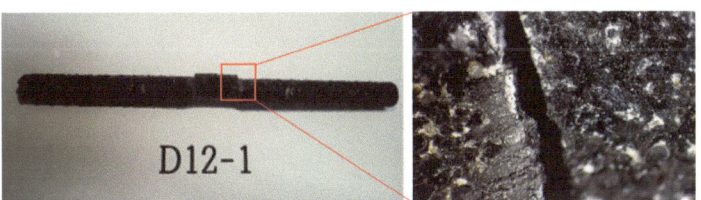

Figure 10. Typical failure mode for shear strength test of D12 Specimen.

Table 5 presents a summary of the shear strength test results for the five specimens for each of the three diameters. The results of the shear strength tests of the CFRP rebar indicate that the average shear strength values of D10, D12, and D16 CFRP rebar are 371 MPa, 360 MPa, and 283 MPa, respectively.

Table 5. Shear strength test results.

Description	D10	D12	D16
Shear Strength (MPa)	350	353	282
	368	360	256
	368	376	314
	375	353	275
	393	357	290
Average	371	360	283

Figure 11 shows the stress–displacement relationships of five shear strength test specimens for each of the three dimensions. Figure 11a shows that the stress increased up to 371 MPa for the CFRP rebar D10. Figure 11b shows that the stress of the CFRP rebar D12 test specimens continuously increased until failure. The CFRP rebar D16 had smaller shear strength than CFRP rebars D10 and D12, as shown in Figure 11c. According to the results of shear strength tests of FRP in previous studies [20], CFRP rebar maintains constant stress before failure and exhibits failure in terms of horizontal and vertical cracks. However, by contrast, the shear strength test results obtained in this study indicate that CFRP rebar shows a tendency to fracture immediately without resistance to a constant load. The reason for this outcome appears to be due to the FRP weaving method. The CFRP specimens used in this study showed significant resistance to loading in the longitudinal direction because they were fabricated in one direction and thus were vulnerable to shear. Therefore, specimens should be fabricated based on three dimensions instead of using single-directional weaving methods to improve the shear performance of CFRP rebar.

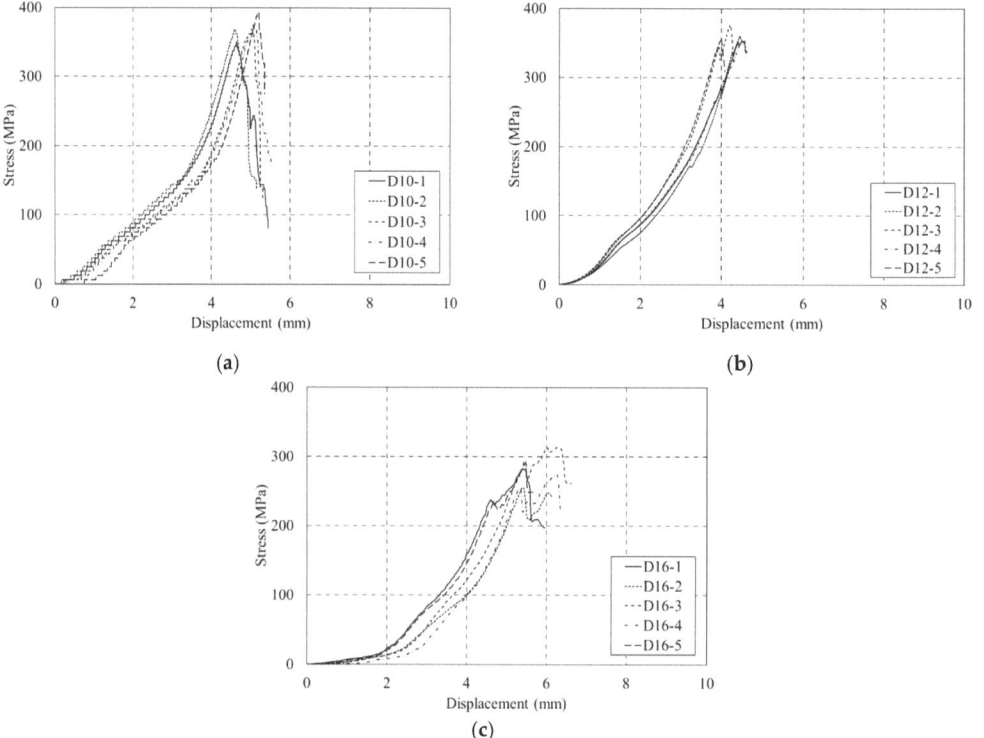

Figure 11. Stress–displacement relationship of five shear strength test specimens. (a) D10; (b) D12; (c) D16.

4. Discussion
4.1. Effect of Size of CFRP Rebar

With an increase in the diameter of CFRP rebar, the tensile strength tends to decrease due to an uneven tensile stress distribution throughout the cross-section. This result is matched well with the experimental results in the literature [21]. Furthermore, unlike steel rebar, CFRP rebar has an orientation due to its fiber inclusion, and both the strength and stiffness of CFRP rebar vary according to the fiber content and resin used. Therefore, the tensile strength tests should be conducted using many specimens in order to ensure the reliability of the material's mechanical performance. Similarly, the shear strength is significantly affected by the size of CFRP rebar due to the fiber matrix and resin. The shear strength is reduced as an increase in the CFRP rebar size as shown in Figure 12. Regardless of the CFRP rebar size, the tensile modulus of elasticity of CFRP rebars was higher than their compressive modulus of elasticity. The results are matched well with the findings in the literature [10].

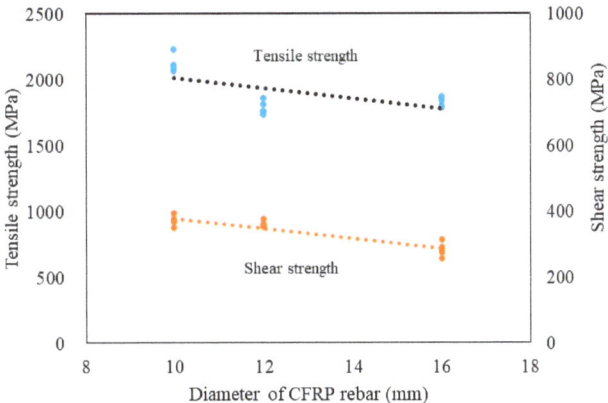

Figure 12. Effect of bar size for tensile strength and shear strength.

Jung et al. [22] reported the modulus of elasticity of hybrid rebar to be approximately 100 GPa. The CFRP D12 rebar investigated in this study showed a modulus of elasticity value that is 1.58 times higher than that of the hybrid rebar but lower than that of steel rebar (200 GPa). Plevkov et al. [12] found that the modulus of elasticity value of general CFRP rebar is 144 GPa, which is 0.91 times lower than that of the CFRP D12 rebar used in this study. Therefore, the CFRP rebar produced in the future as steel rebar replacement should have enhanced strength, which can be accomplished by conducting additional research studies using different fiber arrangements/orientations and different resin contents.

4.2. Relationship between Shear Strength and Tensile Strength

As addressed in the literature [10], the tensile test for the CFRP rebar needs high attention to avoid premature failure in tension. As increasing CFRP rebar size, the test setup required a large free length, enough steel pipe anchor, and a high-capacity loading machine. Simply, a correlation between tensile strength and shear strength can be employed to predict the tensile strength of CFRP rebar according to the rebar's diameter. Equation (1) can be used to calculate the ratio of shear strength to tensile strength using the results of shear and tensile strength tests of CFRP rebar with a diameter up to 16 mm [21].

$$k_d = \frac{\sum \frac{f_s}{f_t}}{n_t}, \tag{1}$$

where k_d is the coefficient; f_s is the shear strength; f_t is the tensile strength.

Figure 13 shows the computed coefficients (k_d). The literature shows that these coefficients (k_d) range from 0.163 to 0.207 for basalt FRP rebar and basalt/CFRP hybrid rebar depending on the diameter of the rebar [21]. The values obtained in this study match well with the results from the literature. Therefore, the calculated coefficients (k_d) can be used to predict the tensile strength of CFRP D10 to D16 rebar.

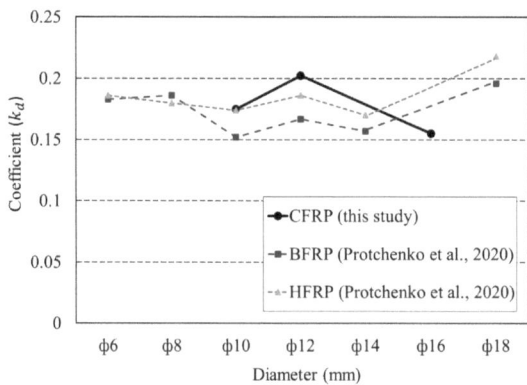

Figure 13. Calculated coefficients with respect to rebar diameter for sand-coated CFRP, basalt FRP, and hybrid FRP [21].

5. Conclusions

Tensile, compressive, and shear-strength tests were conducted in this study in accordance with the test methods specified in ASTM and other international standards to evaluate the mechanical performance of CFRP rebar. As a result of the tests and quantification of the mechanical performance of the CFRP rebar specimens, the following conclusions can be drawn.

The results of the tensile strength tests of CFRP rebar conducted in accordance with ASTM D7205 indicate that the average tensile strength value is 1784 MPa and the average modulus of elasticity value is 158 GPa. This value satisfies the standard modulus of elasticity values specified in ACI 440.1R-15 (in the range of 120 GPa to 580 GPa) and the minimum modulus of elasticity value of 140 GPa specified in guidelines for the structural design of FRP rebar.

For the compressive strength and modulus of elasticity tests, the test results indicate that the average compressive strength value is 357 MPa and the average modulus of elasticity value is 30 GPa. The tensile modulus of elasticity of CFRP rebars was higher than their compressive modulus of elasticity.

As the results of the shear strength test, all specimens are continuously increased until failure. In short, the shear strength and tensile strength of CFRP rebar are affected by the CFRP rebar size. In addition, as a result of investigating the relationship between the shear strength and tensile strength of CFRP, it was possible to predict the coefficient according to the diameter of the CFRP rebar.

Based on the mechanical performance testing, the sand-coated CFRP rebar was determined in this study. The following study will use this information for further structural tests and analyses. To achieve the target tensile strength value of 2100 MPa in future research, the resistance to tension should be increased throughout the polymer section by changing the fiber arrangement/orientation of the CFRP rebar to improve its brittle-resistant properties. In addition, to prevent grip failure and slippage, an adequate bond system along with ASTM specification is needed for future tests.

Author Contributions: Conceptualization, S.-H.K. and W.C.; methodology, S.-H.K. and H.-D.Y.; validation, S.-H.K. and W.C.; investigation, S.-H.K. and H.-D.Y.; data curation, S.-H.K.; writing—original draft preparation, S.-H.K.; writing—review and editing, H.-D.Y. and W.C.; project administration, W.C.; funding acquisition, H.-D.Y. All authors have read and agreed to the published version of the manuscript.

Funding: The author(s) disclosed receipt of the following financial support for the research, authorship, and/or publication of this article: the research was supported by a grant (RS-2021-KA163381) from Construction Technology Research Project Funded by the Ministry of Land, Infrastructure and Transport of Korea Government.

Institutional Review Board Statement: Not applicable.

Informed Consent Statement: Not applicable.

Data Availability Statement: Not applicable.

Conflicts of Interest: The authors declare no conflict of interest.

References

1. Cho, J.R.; Park, Y.H.; Park, S.Y.; Park, C.W. Development of Design Guidelines for FRP Reinforced Concrete Structure and Application of FRP Reinforcement in Korea. *Mag. Korea Concr. Inst.* **2018**, *30*, 81–86. Available online: https://www.dbpia.co.kr/journal/articleDetail?nodeId=NODE07571350 (accessed on 19 April 2023).
2. *ACI 440.1R-15*; Guide for the Design and Construction of Structural Concrete Reinforced with Fiber-Reinforced Polymer (FRP) bars. American Concrete Institute: Farmington Hills, MI, USA, 2015.
3. *ACI 440.1R-06*; Guide for the the Design and Construction of Structural Concrete Reinforced with fiber-Reinforced Polymer (FRP) bars. American Concrete Institute: Farmington Hills, MI, USA, 2006.
4. *ACI 440.3R*; Guide for the Design and Construction of Structural Concrete Reinforced with Fiber-Reinforced Polymer (FRP) Bars. American Concrete Institute: Farmington Hills, MI, USA, 2012.
5. *CAN/CSA-S6-14*; Canadian Highway Bridge Design Code. Canadian Standards Authority (CSA): Toronto, ON, Canada, 2006.
6. *CAN/CSA-S6:19*; Canadian Highway Bridge Design Code. Canadian Standards Authority (CSA): Toronto, ON, Canada, 2014.
7. *CAN/CSA S806*; Design and Construction of Building Structures with Fibre-reinforced Polymers. Canadian Standards Authority (CSA): Toronto, ON, Canada, 2012.
8. *KS F ISO 10406-1*; Fibre-Reinforced Polymer (FRP) Reinforcement of Concrete-Test Methods-Part 1: FRP Bars and Grids. Korean Standards Association: Seoul, Republic of Korea, 2017.
9. Benmokrane, B.; Zhang, B.; Chennouf, A. Tensile Properties and Pull out Behaviour of AFRP and CFRP rods for grouted anchor applications. *Constr. Build. Mater.* **2000**, *14*, 157–170. [CrossRef]
10. Khan, Q.S.; Sheikh, M.N.; Hadi, M.N.S. Tension and Compression Testing of Fibre Reinforced Polymer (FRP) Bars. In Proceedings of the 12th International Symposium on Fiber Reinforced Polymers for Reinforced Concrete Structures (FRPRCS-12) & the 5th Asia-Pacific Conference on Fiber Reinforced Polymers in Structures (APFIS-2015) Joint Conference, Nanjing, China, 14–16 December 2015.
11. Khan, Q.S.; Sheikh, M.N.; Hadi, M.N.S. Tensile Testing of Carbon FRP (CFRP) and Glass FRP (GFRP) Bars: An Experimental Study. *J. Test. Eval.* **2021**, *49*, 2035–2050. [CrossRef]
12. Plevkov, V.; Baldin, I.; Kudyakov, K.; Nevskii, A. Mechanical Properties of Composite Rebar under Static and Short-term Dynamic Loading. *AIP Conf. Proc.* **2017**, *1800*, 040018.
13. Koosha, K.; Pedram, S. New Testing Method of GFRP Bars in Compression. In Proceedings of the CSCE Annual Conference 2018, Frederiction, NB, Canada, 13–16 June 2018.
14. Ashrafi, H.; Bazli, M.; Najafabadi, E.P.; Oskouei, A.V. The Effect of Mechanical and Thermal Properties of FRP bars on their Tensile Performance under Elevated Temperatures. *Constr. Build. Mater.* **2017**, *157*, 1001–1010. [CrossRef]
15. Lee, S.T.; Park, K.P.; Park, K.T.; You, Y.J.; Seo, D.W. A study on the Application of FRP Hybrid Bar to Prevent Corrosion of Reinforcing Bar in Concrete Structure. *J. Korea Acad. Ind. Coop. Soc.* **2019**, *20*, 559–568.
16. *ASTM D3039/D3039M-14*; Standard Test Method for Tensile Properties of Polymer Matrix Composite Materials. ASTM International: West Conshohocken, PA, USA, 2014.
17. *ASTM D7205/D7205M*; Standard Test Method for Tensile Properties of Fiber Reinforced Polymer Matrix Composite Bars. American Society for Testing and Materials (ASTM): West Conshohocken, PA, USA, 2016.
18. *ASTM D695*; Standard Test Method for Compressive Properties of Rigid Plastics. American Society for Testing and Materials (ASTM): West Conshohocken, PA, USA, 2010.
19. *ASTM D7617/D7617M*; Standard Test Method for Transverse Shear Strength of Fiber-reinforced Polymer Matrix Compo-site Bars. American Society for Testing and Materials (ASTM): West Conshohocken, PA, USA, 2017.
20. Park, C.G.; Won, J.P. Mechanical Properties of Hybrid FRP Rebar. *J. Korean Soc. Agric. Eng.* **2003**, *45*, 58–67.

21. Protchenko, K.; Zayoud, F.; Urba´nski, M.; Szmigiera, E. Tensile and Shear Testing of Basalt Fiber Reinforced Polymer (BFRP) and Hybrid Basalt/Carbon Fiber Reinforced Polymer (HFRP) Bars. *Materials* **2020**, *13*, 5839. [CrossRef] [PubMed]
22. Jung, K.S.; Park, K.T.; Seo, D.W.; Kim, B.C.; Park, J.S. Prediction of Tensile Behavior for FRP Hybrid Bar. *J. Korean Soc. Adv. Compos. Struct.* **2017**, *8*, 25–33. [CrossRef]

Disclaimer/Publisher's Note: The statements, opinions and data contained in all publications are solely those of the individual author(s) and contributor(s) and not of MDPI and/or the editor(s). MDPI and/or the editor(s) disclaim responsibility for any injury to people or property resulting from any ideas, methods, instructions or products referred to in the content.

Article

Dual-Function Smart Windows Using Polymer Stabilized Cholesteric Liquid Crystal Driven with Interdigitated Electrodes

Xiaoyu Jin [1], Yuning Hao [1], Zhuo Su [2], Ming Li [3,*], Guofu Zhou [2] and Xiaowen Hu [2,*]

1. College of Physics and Electronic Information, Yunnan Normal University, Kunming 650500, China
2. SCNU-TUE Joint Research Lab of Device Integrated Responsive Materials (DIRM), National Center for International Research on Green Optoelectronics, South China Normal University, No. 378, West Waihuan Road, Guangzhou Higher Education Mega Center, Guangzhou 510006, China
3. Solar Energy Research Institute, Yunnan Normal University, Kunming 650500, China
* Correspondence: lmllldy@126.com (M.L.); xwhu@m.scnu.edu.cn (X.H.)

Citation: Jin, X.; Hao, Y.; Su, Z.; Li, M.; Zhou, G.; Hu, X. Dual-Function Smart Windows Using Polymer Stabilized Cholesteric Liquid Crystal Driven with Interdigitated Electrodes. *Polymers* **2023**, *15*, 1734. https://doi.org/10.3390/polym15071734

Academic Editor: Beom Soo Kim

Received: 26 February 2023
Revised: 27 March 2023
Accepted: 30 March 2023
Published: 31 March 2023

Copyright: © 2023 by the authors. Licensee MDPI, Basel, Switzerland. This article is an open access article distributed under the terms and conditions of the Creative Commons Attribution (CC BY) license (https:// creativecommons.org/licenses/by/ 4.0/).

Abstract: In this study, we present an electrically switchable window that can dynamically transmit both visible light and infrared (IR) light. The window is based on polymer stabilized cholesteric liquid crystals (PSCLCs), which are placed between a top plate electrode substrate and a bottom interdigitated electrode substrate. By applying a vertical alternating current electric field between the top and bottom substrates, the transmittance of the entire visible light can be adjusted. The cholesteric liquid crystals (CLC) texture will switch to a scattering focal conic state. The corresponding transmittance decreases from 90% to less than 15% in the whole visible region. The reflection bandwidth in the IR region can be tuned by applying an in-plane interdigital direct current (DC) electric field. The non-uniform distribution of the in-plane electric field will lead to helix pitch distortion of the CLC, resulting in a broadband reflection. The IR reflection bandwidth can be dynamically adjusted from 158 to 478 nm. The electric field strength can be varied to regulate both the transmittance in the visible range and the IR reflection bandwidth. After removing the electric field, both features can be restored to their initial states. This appealing feature of the window enables on-demand indoor light and heat management, making it a promising addition to the current smart windows available. This technology has considerable potential for practical applications in green buildings and automobiles.

Keywords: cholesteric liquid crystal; interdigitated electrodes; transmittance; reflection bandwidth; smart window

1. Introduction

Smart windows have gained popularity in recent decades due to their ability to control indoor solar irradiation, reduce air-conditioning energy consumption and provide and maintain a comfortable visual environment indoors [1]. Liquid crystals (LCs) offer an excellent material system for developing smart windows, and applications such as switchable privacy glass [2], solar heat rejection [3] and switchable color [4] have been demonstrated. Various LC technologies based on light scattering or reflection have been examined to create smart windows for regulating indoor solar irradiation, especially for visible light and infrared (IR) light. Visible light is used to maintain interior illumination levels, while IR light is responsible for indoor temperature control.

Various technologies are available for controlling visible light, including LC molecules mixed with ionic dopants or a guest LC material, which can be used to switch between the haze-free transparent and high-haze opaque states owing to the electrohydrodynamic effect [5–7]. Furthermore, technologies based on light scattering using LC/polymer composites, such as polymer stabilized LCs (PSLC) and polymer dispersed LCs (PDLC), have been developed to control visible light [8–12]. Smart windows have been developed using smectic/cholesteric liquid crystal (CLC) phases, where the window can be reversibly switched between transparent (planar and homeotropic) and scattering (focal conic) states [13,14]. For technologies controlling IR light, polymer stabilized CLC (PSCLC) has been widely studied [15–17]. CLC are nematic liquid crystals (NLC) that contain a chiral component that generates a helical twist between LC layers, thus allowing the selective reflection of circularly polarized incident light of the same handedness as its helix. The wavelength of light reflected by the CLC is determined by its helical pitch (p). The bandwidth of the reflected light is defined as $\Delta\lambda \approx \Delta n \times p = (n_e - n_o) \times p$, where Δn is the birefringence of the host LC, and n_e and n_o are extraordinary and ordinary refractive indices, respectively [18,19]. For PSCLC-based IR reflectors, the IR reflection band can be electric-regulated dynamically, which is attributed to pitch distortion caused by the displacement of the polymer network under an external E-field [15,17,20]. We previously reported an IR reflector that can reflect a broad band of IR light from 725 to 1435 nm dynamically upon application of the E-field, while maintaining high transmittance in the visible region [17]. An IR reflector of this type could be used as a smart window, allowing solar IR energy to enter during the winter while reflecting it during the summer, reducing energy consumption for heating and cooling energy demands in the built environment.

Although much effort has been dedicated to developing LC-based smart windows for regulating visible and IR light, the ability to regulate both types of light is extremely desirable. Du et al. demonstrated an LC based window that can be switched between transparent and opaque for visible light through a voltage pulse while maintaining a 220 nm reflection bandwidth for near IR light [21]. However, the fabrication process is complicated, involving the construction of a hybrid structure using CLC and chiral polymer film. Moreover, the reflection bandwidth of the IR light is narrow and cannot be dynamically regulated, indicating that it will always reject solar heat, regardless of whether it is winter or summer.

In this work, we demonstrate an electrically switchable window that can dynamically regulate both visible and IR light. The window is based on PSCLC sandwiched between two transparent glass substrates. The PSCLC has three terminal electrodes—one plate electrode on the top substrate, and the other two interdigitated electrodes on the bottom substrate. When an alternating current (AC) E-field is applied between the top and bottom electrodes, the PSCLC switches from a transparent cholesteric state to scattering focal conic state, reducing transmittance from 90% to less than 15% in the whole visible region. When a direct current (DC) E-field is applied between the interdigitated electrodes, the non-uniform E-field distribution leads to helix pitch distortion of the CLC, thus resulting in a broadband reflection. The IR reflection bandwidth can be dynamically tuned from 158 to 478 nm. By adjusting the E-field intensity, the transmittance in the visible range and the IR reflection bandwidth can both be controlled, and after the E-field is removed, both can be returned to initial conditions. This attractive feature of the window makes it possible to perform indoor light and heat management on demand. Such windows would be an extension of the currently available smart windows, with considerable potential for practical applications in green buildings and automobiles.

2. Materials and Methods

2.1. Materials

The PSCLC mixture was prepared by mixing 81.4 wt% of a nematic LC E7 (The Δn of E7 is 0.217 at λ = 589 nm and T = 25 °C, and the $\Delta\varepsilon$ of E7 is 11.4 at 1 kHz and T = 25 °C; Merck, Darmstadt, Germany), 13.8 wt% of a right-handed chiral dopant S811 (Merck), 3.0 wt% of a diacrylate monomer (RM257, Merck, Darmstadt, Germany) and 1.0 wt% of photo-initiator Irgacure-651 (Ciba Specialty Chemicals (China) Ltd., Shanghai, China). All ingredients were used without further purification. Figure 1A shows the chemical structures of S811, RM257 and Irgacure-651. To enable the reflection of IR light, the chiral dopant concentration was selected such that the reflection notch of the PSCLC cells was centered at 1050 nm.

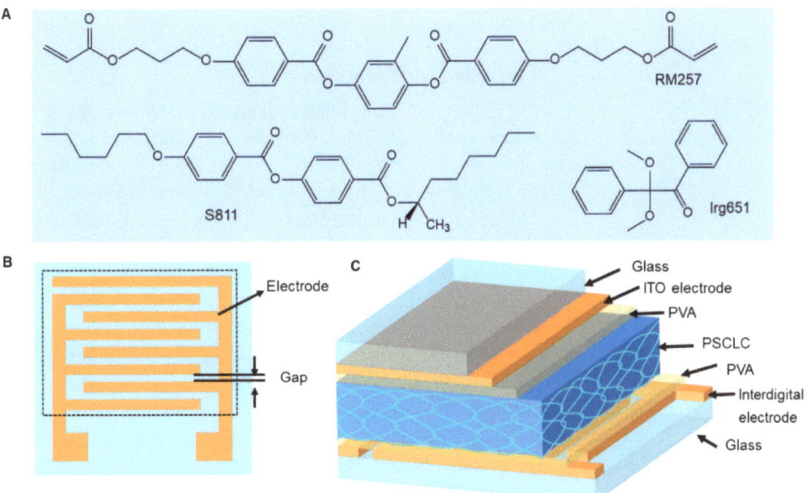

Figure 1. (**A**) The chemical structures of S811, RM257 and Irgacure-651. (**B**) Top view of the bottom glass substrate and the interdigitated electrodes pattern (the analyzed responsive region is indicated by the dashed square). (**C**) Three-dimensional structure side view of the device.

2.2. Sample Preparation

The interdigitated electrodes (Wuhu Changxin Technology Co., Ltd., Anhui, China) and the plate electrode glass were used as the bottom and top substrates to make the cell. Figure 1B shows the two-dimensional top view of the interdigitated electrodes pattern. The width of the electrode is 95 μm, and the gap between the interdigitated electrodes is 25 μm. The electrode and gap are periodically distributed. Both the top and bottom substrates underwent a 10 min ultrasonic cleaning process in acetone, isopropanol, and deionized water, followed by a 20 min UV-ozone treatment (UV Products, BZS250GF-TC, Shenzhen, China) to make them more hydrophilic. The polyvinyl alcohol (PVA) alignment layer was spin-coated on both substrates. Then, the substrates with the PVA layer were placed parallel to each other in the rubbing direction and glued together with UV curable glue containing a 30 μm spacer (SiO_2). After 1 min of UV irradiation, the LC cell with a cell thickness of 30 μm was obtained, as shown in Figure 1C. To examine the impact of interdigitated electrodes with different gaps on the performance of the sample, interdigitated electrodes with a 95 μm electrode width and various gap sizes (25, 45, 65, 85 and 105 μm) were used. Moreover, for investigating the impact of interdigitated electrodes with different electrodes on the performance of the sample, interdigitated electrodes with a gap width of 25 μm and different electrode widths (35, 55, 75, 95 and 115 μm) were used. The LC mixture was stirred and filled into the cell by capillary force. The cells were then exposed to UV light of 365 nm (28 mW/cm^2) to polymerize. The whole preparation process was conducted in the yellow-light area.

2.3. Characterization

The optical characterization of the sample cells was conducted using unpolarized spectrophotometry (Lambda 950, PerkinElmer, Shanghai, China) in transmission mode at normal incidence to obtain the spectra of the sample cells [16]. A bare ITO glass substrate was used for the baseline correction of the instrument prior to the measurement of the photoelectric properties. Optical microscope photographs of PSCLC sample cells were observed at room temperature using a polarizing light microscope (POM, LEICA DM2700P,

Leica, Solms, Germany) to analyze changes in the liquid crystal texture of the sample cells [16]. The device was driven by a single-phase AC power source with a frequency of 50 Hz, which generates voltages from 0 to 300 V with a power factor of 0.8, and a DC power source that generates voltages from 0 to 210 V.

3. Results and Discussion

Firstly, an AC E-field was applied between the top and bottom electrodes. Because of the small gap (25 μm) of the interdigitated electrodes on the bottom substrate, an approximated vertical E-field was created in the sample cell (Figure 2(Ai)). In the gap region, the E-field was slightly inclined. Under this E-field, the alignment of the LC molecules changed from a cholesteric texture (Figure 2(Aii)) to a focal conic texture (Figure 2(Aiii)), which is evidenced by the POM images shown in Figure 2(Bi,Bii). Accordingly, the sample switched from transparent to opaque because of the realignment of the LC molecules. As shown in Figure 2(Biii,Biv), under the vertical E-field, the background patterns under the sample could not be observed, thus confirming a scattering opaque state. Figure 2C shows the transmittance spectra of the sample under different AC voltages. At the off state (0 V/μm), the transmittance in the entire visible region was greater than 90%, and there was a reflection band centered at 1050 nm due to the Bragg reflection of the cholesteric phase. In the on state, the transmittance decreased as the applied voltage increased. For example, under the voltage of 3.34 V/μm, the transmittance dropped to less than 15% in the visible region (from 400 to 760 nm). Moreover, the reflection band disappeared under high voltage, which was due to the randomly distributed helical axis of the focal conic state [13]. It could be noticed that the operational voltage was high, which is attributed to two reasons. One is that the inclined E-field in the gap regions of the interdigitated electrodes cannot completely act on rotating the LC molecules with positive dielectric anisotropy (E7, $\Delta\varepsilon = 11.4$). The horizontal component of the inclined E-field actually prevents the LC molecules from rotating. The other reason is the anchoring force of the polymer networks among the LC molecules. When removing the AC voltage, the alignment of the LC molecules goes back to cholesteric texture because of the anchoring force of the polymer network. Therefore, the sample will revert to its initial transparent state.

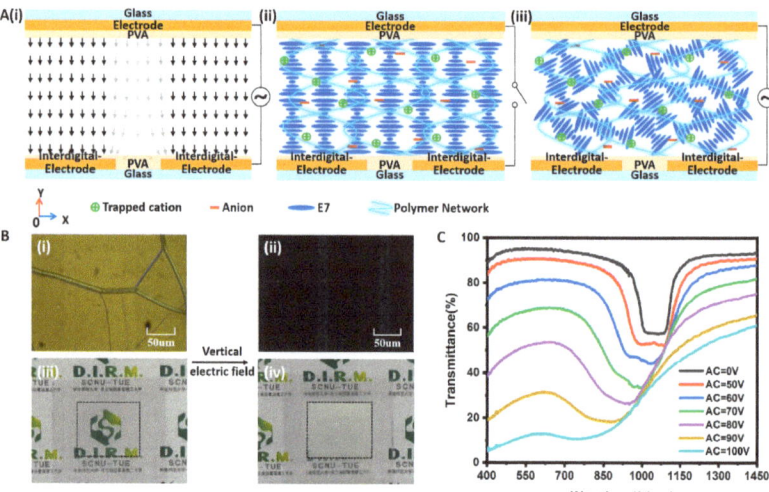

Figure 2. (**A**) Schematic diagram of (**i**) the vertical AC E-field between the top and bottom electrodes, (**ii**) the initial CLC cholesteric texture before applying the E-field, (**iii**) the focal conic texture after applying the E-field. (**B**) The POM images of the sample before (**i**) and after (**ii**) applying the E-field. The pictures of the sample before (**iii**) and after (**iv**) applying the E-field. (**C**) Transmittance spectra of the sample under different AC voltages.

When a DC voltage is applied between the interdigitated electrodes, the sample remains transparent, as shown in Figure 3(Ai,Aii). Regardless of whether the voltage is applied, the background patterns under the sample are clearly visible, indicating a good transparency state. The POM images (Figure 3(Aiii,Aiv)) indicate that the alignment of CLC molecules remains in the cholesteric texture under the DC voltage. Figure 3B shows the transmittance spectra of the sample under different DC voltages, thus exhibiting an electrically tunable bandwidth broadening. The reflection band is centered at 1050 nm with an initial bandwidth of 158 nm. When a DC voltage is applied, the bandwidth broadening becomes almost symmetric, extending simultaneously towards the red and blue sides compared to the original reflection notch wavelength. At the DC voltage of 2.76 V/μm, the sample shows a broad band of IR light from 800 to 1278 nm (with a reflection bandwidth of 478 nm), while predominantly remaining transparent in the visible region with high transmittance in the 400–760 nm range. When the DC interdigital E-field is removed, the alignment of the LC molecules goes back to cholesteric texture because of the anchoring force of the polymer network. Therefore, the reflection bandwidth will revert to its initial value of 158 nm.

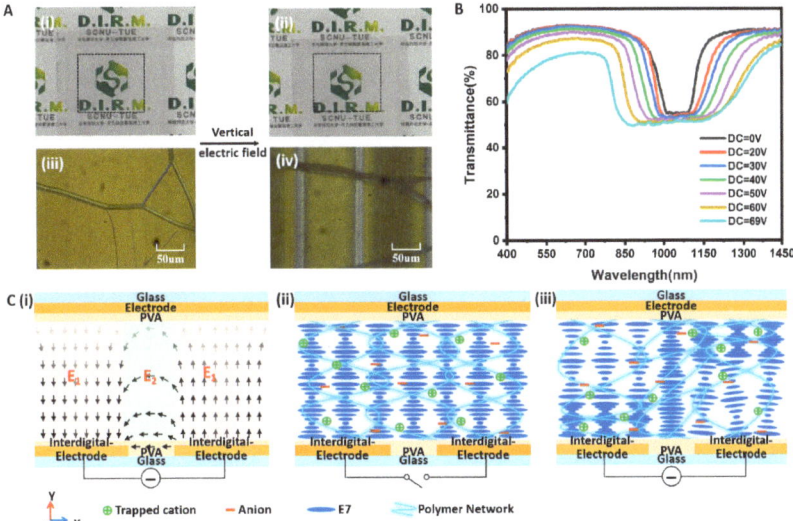

Figure 3. (**A**) The pictures of the sample before (**i**) and after (**ii**) applying the in-plane interdigital E-field. The POM images of the sample before (**iii**) and after (**iv**) applying the in-plane interdigital E-field. (**B**) Transmittance spectra of the sample under different in-plane interdigital voltages. (**C**) Schematic diagram of (**i**) the in-plane interdigital E-field between the neighboring electrodes, (**ii**) the initial CLC cholesteric texture before applying the E-field and (**iii**) the non-uniform pitch texture after applying the E-field.

The bandwidth broadening of the CLC can be attributed to the distortion of its helix pitch under an in-plane E-field between the interdigitated electrodes. Typically, applying a DC voltage between the interdigitated electrodes generates a non-uniformly distributed in-plane E-field in the sample cell [22]. Figure 3(Ci) shows the in-plane E-field configuration over a spatial period. In the region of the interdigitated electrodes, the E-field is perpendicular to the substrate (E_1 in Figure 3(Ci)), while in the gap region between the neighboring electrodes, the E-field is parallel to the substrate (E_2 in Figure 3(Ci)). The parabola-like distribution of the in-plane E-field induces a complicated orientation change of the CLC molecules. In particular, in the gap region, the E_2 is perpendicular to the helical axis of the CLC, which results in the unwinding of the CLC helix. Therefore, the helical pitch p will be elongated [22]. Moreover, E_2 is not homogeneous in the gap region, where the E-field close to the bottom substrate is larger than that close to the top substrate. As a result, the unwinding of the CLC helix is not uniform, and the pitch varies along the cell thickness in the samples. Accordingly, the reflection peak of the CLC will be shifted to the red side with band broadening. In the region of the electrodes, the DC E-field (E_1) is parallel to the helix axis of the CLC. The polymer network traps cations by electrostatic force [23–25]. Upon applying the DC E-field (E_1), an electromechanical force is exerted on the polymer network because of the trapped ions, leading to the deformation of the polymer network. Because of the aligning effect of the polymer network on the CLC molecules, the deformation results in a non-uniform pitch distribution by stretching the pitch near the anode side and compressing the pitch near the cathode side [26], as shown in Figure 3(Ciii). Consequently, the reflection band is broadened to include both the blue and red sides (Figure 3B). It is noted that, with increases in the applied DC voltage, the transmittance of the sample in the visible region slightly decreases. For instance, without applying the DC voltage, the transmittance in the visible region is above 90%, and it decreases to less than 80% at a high voltage of 2.76 V/μm. The decrease in the transmittance in the visible region is likely attributed to the scattering resulting from the focal conic state at such a high E-field.

Next, we examined the effect of the interdigitated electrodes with different gaps on the device's performance. We created the cell using interdigitated electrodes with a fixed electrode width of 95 μm, but with various gaps ranging from 25 to 105 μm.

After applying an AC voltage of 80 V between the top and bottom electrodes, we measured the transmittance spectra of the samples created using interdigitated electrodes with different gaps. Figure 4A displays the transmittance of the samples at 600 nm. The sample with a small interdigitated electrodes gap shows low transmittance. For example, when the gap was 25 μm, the transmittance was 40%. As the gap increased, the transmittance increased, rising to more than 60% when the gap was magnified to 105 μm. As previously discussed, when an AC E-field is applied between the top and bottom electrodes, the E-field is vertical to the substrate in the area of the interdigitated electrodes and inclined in the gap region. As the gap increases, the inclination angle of the E-field becomes larger, resulting in a smaller vertical component of the inclined E-field. Thus, a weaker vertical field acts on the CLC molecules in the gap region, making it more challenging to realign the CLC molecules from cholesteric texture to focal conic texture. Consequently, under the same applied AC voltage, the sample with the larger interdigitated electrodes gap results in a higher transmittance.

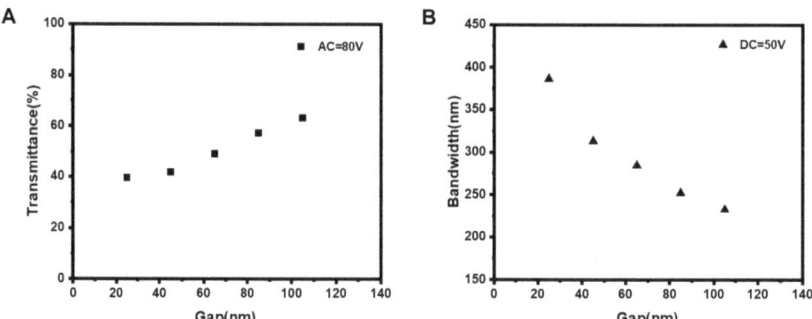

Figure 4. Samples are fabricated by using interdigitated electrodes with different gaps. (**A**) The transmittance (@600 nm) of the samples when applying an AC voltage of 80 V between the top and bottom electrodes. (**B**) The reflection bandwidth of the samples in the infrared region when applying a DC voltage of 50 V between the interdigitated electrodes.

As mentioned earlier, a DC voltage applied between the interdigitated electrodes creates a non-uniformly distributed in-plane E-field that distorts the CLC helix pitch, thus resulting in broadband reflection. The different gaps of the interdigitated electrodes intrinsically affect the intensity of the in-plane E-field, which in turn influences the reflection bandwidth broadening of the sample. Hence, we measured the reflection band of the samples fabricated with various interdigitated electrodes gaps. Figure 4B shows the reflection bandwidth of samples under a DC interdigital voltage of 50 V. The reflection bandwidth decreases as the gap increases. For instance, at a gap of 25 μm, the reflection bandwidth of the sample is 387 nm, which gradually decreases to 313, 285, 252 and 233 nm when the gap increases from 25 to 45, 65, 85 and 105 μm. We attributed this decrease to the reduction in interdigital E-field intensity as the gap widens. In particular, in the sample with a small gap, the stronger in-plane E-field made it easier for the CLC helix in the gap region to be unwound. Furthermore, in the interdigitated electrodes region, the stronger E-field, which is parallel to the CLC helix, caused a larger displacement of the polymer network, creating a larger pitch gradient and resulting in a broader reflection bandwidth.

Furthermore, we investigated the effect of the interdigitated electrodes with different electrode widths on the device performance. We made the cell by using the interdigitated electrodes with the gap fixed at 25 μm, and various electrode widths ranging from 35 to 115 μm.

We measured the transmittance spectra of the samples under an AC voltage of 80 V applied between the top and bottom electrodes. The transmittance of the samples at a wavelength of 600 nm is shown in Figure 5A. The sample with the small electrode width showed low transmittance. For example, when the electrode width was 35 μm, the transmittance was 35%. As the electrode width increased, the transmittance increased and rose to 49% when the electrode width was increased to 115 μm. As we discussed above, when an AC E-field is applied between the top and bottom electrodes, an approximated vertical E-field is created in the sample cell. The electric field intensity between the capacitor's two plates can be shown as $E = Q \div (\varepsilon_0 \times S)$, where Q is the amount of charge in the capacitor's plate, ε_0 is the dielectric constant of the material between the capacitor's two plates (in this case, ε_0 is the dielectric constant of the PSCLC material), and S is the relative area of the two plates [27,28]. We can regard the cell as a capacitor, with the top substrate's plate electrode and the bottom substrate's interdigitated electrodes serving as its two plates. The top and bottom electrodes of the capacitor contain an equal quantity of charge when an AC field is applied between them. As the electrode width increases, the relative area of the capacitor's two plates increases. Since the AC power supply and the PSCLC material are fixed, the samples created with interdigitated electrodes of various electrode widths have the same charge Q and dielectric constant ε_0. Thus, the E-field of the capacitor decreases as the relative area of the two plates increases. In other words, when the AC voltage is held constant, the E-field intensity in the cell decreases as the electrode width increases. In this sense, a weaker E-field acts on the CLC molecules in the cell, which makes it harder to realign the CLC molecules from cholesteric texture to focal conic texture. As a result, the sample with a larger electrode width produces a higher transmittance with the same applied AC voltage.

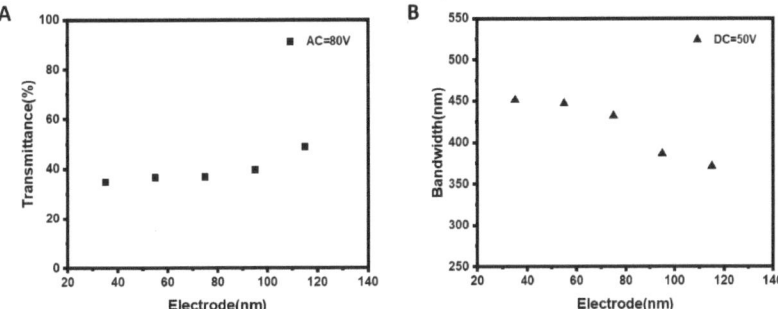

Figure 5. Samples are fabricated by using the interdigitated electrodes with different electrode widths. (**A**) The transmittance (@600 nm) of the samples when applying an AC voltage of 80V between the top and bottom electrodes. (**B**) The reflection bandwidth of the samples in the infrared region when applying a DC voltage of 50V between the interdigitated electrodes.

As we discussed above, a DC voltage between the interdigitated electrodes creates a non-uniformly distributed E-field, which causes a distortion of the CLC helix pitch, resulting in a broadband reflection. Different electrode widths of the interdigitated electrodes intrinsically influence the E-field intensity, which in turn affects the reflection bandwidth broadening of the sample. Therefore, we measured the reflection band of the samples fabricated with various electrodes. Figure 5B shows the reflection bandwidth of the samples under a DC interdigital voltage of 50 V. We can observe that the reflection bandwidth decreases as the electrode width increases. For example, at an electrode width of 35 μm, the reflection bandwidth of the sample was 451 nm, which gradually decreased to 447, 432, 386 and 371 nm when the electrode width enlarged from 35 to 55, 75, 95 and 115 μm. We attributed this to the reduction in interdigital E-field intensity when the electrode width increases. According to Gauss's theorem, the electric field intensity generated by a charged plate can be calculated as $E = \sigma \div 2\varepsilon_0 = (Q \div S) \div 2\varepsilon_0$, where σ is the charge surface density, ε_0 is the dielectric constant, Q is the amount of charge and S is the area of the plate [29,30]. When a DC E-field is applied to interdigitated electrodes, we can regard the interdigitated electrodes as a charged plate. The equivalent charged plate area increases as the electrode width increases in the unit area. The DC power supply and the PSCLC material are fixed, so the amount of charge Q and dielectric constant ε_0 of the samples are the same, both of which were fabricated by using interdigitated electrodes with different electrode widths. Therefore, the charge surface density σ of the sample decreases as the electrode width increases. Furthermore, the electric field intensity decreases as the charge surface density decreases. Specifically, in the sample with small electrode width, the total electrode area per unit area was small and the E-field intensity was strong. The stronger E-field makes it easier for the CLC helix to have a non-uniform pitch distribution, resulting in a broader reflection bandwidth.

For the practical use of electrically responsive dual-function glass as a smart window, an important factor that needs to be considered is its energy consumption [17]. The electric power needed to operate the PSCLC was calculated to evaluate its power consumption. Figure 6A shows exponential decays in the current flow when operation AC voltages of 60, 70, 80, 90 and 100 V were applied. The current of the device reaches its steady state after nearly 200 s. The steady current increases as the operational voltage increases. Driven by an AC voltage of 100 V, the steady current is 9.076 μA. The power consumption can be calculated as $P = V \times I$, where V is the applied voltage and I is the current at steady state [17]. The effective area of the PSCLC is 1.6×2.0 cm^2. The power factor of the single-phase AC power source is 0.8. Accordingly, at 100 V AC voltage, the total power consumption to switch and maintain the PSCLC in an opaque state was around 2.3×10^3 mW m^{-2}. Figure 6B shows exponential decays in the current flow when DC voltages of 30, 40, 50, 60 and 69 V were applied. The current of the device reaches its steady state after nearly 240 s. Driven by a DC voltage of 69 V, the steady current is 0.227 μA. The corresponding power consumption to switch and maintain the device in a broad reflection state is calculated to be only 48.9 mW m^{-2}.

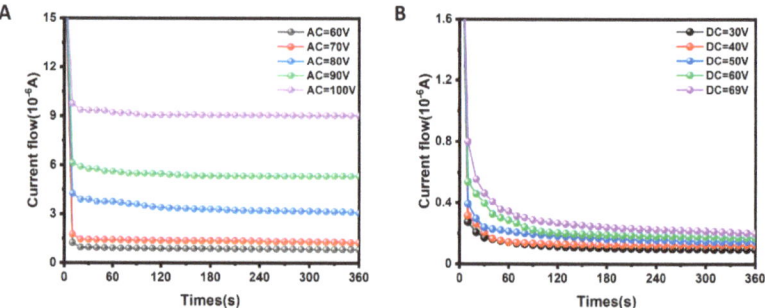

Figure 6. Samples fabricated by using the interdigitated electrodes with electrode width 95 μm and gap width 25 μm. (**A**) Current flow in the visible light scattering in the presence of different operation voltages. (**B**) Current flow in the IR broadband reflection in the presence of different operation voltages.

4. Conclusions

After combining a top plate electrode substrate with a bottom interdigitated electrode substrate, we successfully fabricated a PSCLC smart window capable of regulating both visible light and infrared (IR) light by applying different electric fields. The interdigitated electrodes design enables the window to achieve dual functionality by creating different electric fields. In particular, a vertical alternating current (AC) electric field between the top and bottom substrate switches the CLC texture to a focal conic texture, thus resulting in a decrease in transmittance throughout the visible region from 90% to less than 15%. Furthermore, an in-plane interdigital direct current (DC) electric field leads to helix pitch distortion of the CLC, thus resulting in a broadband reflection. The reflection bandwidth in the IR region can be dynamically tuned from 158 to 478 nm. The electric field strength can be adjusted to regulate both the transmittance in the visible range and the IR reflection bandwidth. Once the electric field is removed, the window is restored to its initial state. This attractive feature of the window makes it possible to perform indoor light and heat management on demand. Such a window would be an extension of the currently available smart windows. However, the practical application of the LC smart window is limited by its stability under a strong electric field and under strong light flux. If this problem is solved or optimized, smart windows have great potential for practical applications in green buildings and automobiles.

Author Contributions: Funding acquisition, M.L. and G.Z.; Resources, M.L., X.H. and G.Z.; Investigation, X.J., Z.S. and Y.H.; Formal analysis, X.J. and X.H.; Validation, X.J. and Y.H.; Writing—original draft, X.J., M.L. and X.H. All authors have read and agreed to the published version of the manuscript.

Funding: This research was funded by the National Key R&D Program of China [No. 2020YFE0100200], the Science and Technology Program of Guangzhou [No. 202201010344, 20202030148, 2019050001], the Natural Science Foundation of Guangdong Province under Grant [No. 2021A1515010653, 2021A1515011388], the Innovation Research Project of Foshan Education Bureau under Grant [No. 2021XCL06], Guangdong Provincial Key Laboratory of Optical Information Materials and Technology [No. 2017B030301007], MOE International Laboratory for Optical Information Technologies and the 111 Project, Yunnan expert workstation (No. 202005AF150028) and the Graduate Research and Innovation Foundation of Yunnan Normal University (YJSJJ22-B80).

Institutional Review Board Statement: Not applicable.

Informed Consent Statement: Not applicable.

Data Availability Statement: Data presented in this study are available on request from the corresponding author.

Acknowledgments: The authors are grateful for the support from the National Key R&D Program of China [No. 2020YFE0100200], the Science and Technology Program of Guangzhou [No. 202201010344, 20202030148, 2019050001], the Natural Science Foundation of Guangdong Province under Grant [No. 2021A1515010653, 2021A1515011388], the Innovation Research Project of Foshan Education Bureau under Grant [No. 2021XCL06], Guangdong Provincial Key Laboratory of Optical Information Materials and Technology [No. 2017B030301007], MOE International Laboratory for Optical Information Technologies and the 111 Project, Yunnan expert workstation (No. 202005AF150028) and the Graduate Research and Innovation Foundation of Yunnan Normal University (YJSJJ22-B80).

Conflicts of Interest: The authors declare no conflict of interest.

References

1. Cannavale, A.; Ayr, U.; Fiorito, F.; Martellotta, F. Smart electrochromic windows to enhance building energy efficiency and visual comfort. *Energies* **2020**, *13*, 1449. [CrossRef]
2. Han, C.-H.; Eo, H.; Choi, T.-H.; Kim, W.-S.; Oh, S.-W. A simulation of diffractive liquid crystal smart window for privacy application. *Sci. Rep.* **2022**, *12*, 11384. [CrossRef] [PubMed]
3. Khandelwal, H.; Schenning, A.P.; Debije, M.G. Infrared regulating smart window based on organic materials. *Adv. Energy Mater.* **2017**, *7*, 1602209. [CrossRef]

4. Ma, L.-L.; Li, C.-Y.; Pan, J.-T.; Ji, Y.-E.; Jiang, C.; Zheng, R.; Wang, Z.-Y.; Wang, Y.; Li, B.-X.; Lu, Y.-Q. Self-assembled liquid crystal architectures for soft matter photonics. *Light Sci. Appl.* **2022**, *11*, 1–24.
5. Jo, Y.-S.; Choi, T.-H.; Ji, S.-M.; Yoon, T.-H. Control of haze value by dynamic scattering in a liquid crystal mixture without ion dopants. *AIP Adv.* **2018**, *8*, 085004. [CrossRef]
6. Heilmeier, G.H.; Zanoni, L.A.; Barton, L.A. Dynamic scattering: A new electrooptic effect in certain classes of nematic liquid crystals. *Proc. IEEE* **1968**, *56*, 1162–1171. [CrossRef]
7. Zhang, Y.; Yang, X.; Zhan, Y.; Zhang, Y.; He, J.; Lv, P.; Yuan, D.; Hu, X.; Liu, D.; Broer, D.J. Electroconvection in zwitterion-doped nematic liquid crystals and application as smart windows. *Adv. Opt. Mater.* **2021**, *9*, 2001465. [CrossRef]
8. Natarajan, L.V.; Brown, D.P.; Wofford, J.M.; Tondiglia, V.P.; Sutherland, R.L.; Lloyd, P.F.; Bunning, T.J. Holographic polymer dispersed liquid crystal reflection gratings formed by visible light initiated thiol-ene photopolymerization. *Polymer* **2006**, *47*, 4411–4420. [CrossRef]
9. Hu, X.; Zhang, X.; Yang, W.; Jiang, X.F.; Jiang, X.; de Haan, L.T.; Yuan, D.; Zhao, W.; Zheng, N.; Jin, M. Stable and scalable smart window based on polymer stabilized liquid crystals. *J. Appl. Polym. Sci.* **2020**, *137*, 48917. [CrossRef]
10. Sun, H.; Xie, Z.; Ju, C.; Hu, X.; Yuan, D.; Zhao, W.; Shui, L.; Zhou, G. Dye-doped electrically smart windows based on polymer-stabilized liquid crystal. *Polymers* **2019**, *11*, 694. [CrossRef]
11. Coates, D. Polymer-dispersed liquid crystals. *J. Mater. Chem.* **1995**, *5*, 2063–2072. [CrossRef]
12. Liu, Y.; Sun, X. Holographic polymer-dispersed liquid crystals: Materials, formation, and applications. *Adv. OptoElectronics* **2008**, *2008*, 1–52. [CrossRef]
13. Mo, L.; Sun, H.; Liang, A.; Jiang, X.; Shui, L.; Zhou, G.; de Haan, L.T.; Hu, X. Multi-stable cholesteric liquid crystal windows with four optical states. *Liq. Cryst.* **2022**, *49*, 289–296. [CrossRef]
14. Kim, K.-H.; Jin, H.-J.; Park, K.-H.; Lee, J.-H.; Kim, J.C.; Yoon, T.-H. Long-pitch cholesteric liquid crystal cell for switchable achromatic reflection. *Opt. Express* **2010**, *18*, 16745–16750. [CrossRef] [PubMed]
15. Hu, X.; de Haan, L.T.; Khandelwal, H.; Schenning, A.P.; Nian, L.; Zhou, G. Cell thickness dependence of electrically tunable infrared reflectors based on polymer stabilized cholesteric liquid crystals. *Sci. China Mater.* **2018**, *61*, 745–751. [CrossRef]
16. Hu, X.; Zeng, W.; Zhang, X.; Wang, K.; Liao, X.; Jiang, X.; Jiang, X.-F.; Jin, M.; Shui, L.; Zhou, G. Pitch gradation by ion-dragging effect in polymer-stabilized cholesteric liquid crystal reflector device. *Polymers* **2020**, *12*, 96. [CrossRef] [PubMed]
17. Hu, X.; Zeng, W.; Yang, W.; Xiao, L.; De Haan, L.T.; Zhao, W.; Li, N.; Shui, L.; Zhou, G. Effective electrically tunable infrared reflectors based on polymer stabilised cholesteric liquid crystals. *Liq. Cryst.* **2019**, *46*, 185–192. [CrossRef]
18. Huang, Y.; Zhou, Y.; Doyle, C.; Wu, S.-T. Tuning the photonic band gap in cholesteric liquid crystals by temperature-dependent dopant solubility. *Opt. Express* **2006**, *14*, 1236–1242. [CrossRef]
19. Chen, L.; Li, Y.; Fan, J.; Bisoyi, H.K.; Weitz, D.A.; Li, Q. Photoresponsive monodisperctional cholesteric liquid crystalline microshells for tunable omnidirectional lasing enabled by a visible light-driven chrial molecular switch. *Adv. Opt. Mater.* **2014**, *2*, 845–848. [CrossRef]
20. Khandelwal, H.; Loonen, R.C.; Hensen, J.L.; Debije, M.G.; Schenning, A.P. Electrically switchable polymer stabilised broadband infrared reflectors and their potential as smart windows for energy saving in buildings. *Sci. Rep.* **2015**, *5*, 11773. [CrossRef]
21. Du, X.; Li, Y.; Liu, Y.; Wang, F.; Luo, D. Electrically switchable bistable dual frequency liquid crystal light shutter with hyper-reflection in near infrared. *Liq. Cryst.* **2019**, *46*, 1727–1733. [CrossRef]
22. Xianyu, H.; Faris, S.; Crawford, G.P. In-plane switching of cholesteric liquid crystals for visible and near-infrared applications. *Appl. Opt.* **2004**, *43*, 5006–5015. [CrossRef] [PubMed]
23. Lu, L.; Sergan, V.; Bos, P.J. Mechanism of electric-field-induced segregation of additives in a liquid-crystal host. *Phys. Rev. E* **2012**, *86*, 051706. [CrossRef] [PubMed]
24. Bremer, M.; Naemura, S.; Tarumi, K. Model of ion solvation in liquid crystal cells. *Jpn. J. Appl. Phys.* **1998**, *37*, L88. [CrossRef]
25. Son, J.-H.; Park, S.B.; Zin, W.-C.; Song, J.-K. Ionic impurity control by a photopolymerisation process of reactive mesogen. *Liq. Cryst.* **2013**, *40*, 458–467. [CrossRef]
26. Nemati, H.; Liu, S.; Zola, R.S.; Tondiglia, V.P.; Lee, K.M.; White, T.; Bunning, T.; Yang, D.-K. Mechanism of electrically induced photonic band gap broadening in polymer stabilized cholesteric liquid crystals with negative dielectric anisotropies. *Soft Matter* **2015**, *11*, 1208–1213. [CrossRef]
27. Nawaka, K.; Putson, C. Enhanced electric field induced strain in electrostrictive polyurethane composites fibers with polyaniline (emeraldine salt) spider-web network. *Compos. Sci. Technol.* **2020**, *198*, 108293. [CrossRef]
28. SaeidNahaei, S.; Jo, H.-J.; Lee, S.J.; Kim, J.S.; Lee, S.J.; Kim, Y. Investigation of the Carrier Movement through the Tunneling Junction in the InGaP/GaAs Dual Junction Solar Cell Using the Electrically and Optically Biased Photoreflectance Spectroscopy. *Energies* **2021**, *14*, 638. [CrossRef]
29. Lv, H.; Chen, X.; Li, X.; Ma, Y.; Zhang, D. Finding the optimal design of a Cantor fractal-based AC electric micromixer with film heating sheet by a three-objective optimization approach. *Int. Commun. Heat Mass Transf.* **2022**, *131*, 105867. [CrossRef]
30. Gatzia, D.E.; Ramsier, R.D. Dimensionality, symmetry and the inverse square law. *Notes Rec.* **2021**, *75*, 333–348. [CrossRef]

Disclaimer/Publisher's Note: The statements, opinions and data contained in all publications are solely those of the individual author(s) and contributor(s) and not of MDPI and/or the editor(s). MDPI and/or the editor(s) disclaim responsibility for any injury to people or property resulting from any ideas, methods, instructions or products referred to in the content.

Article

3D Printing of Solar Crystallizer with Polylactic Acid/Carbon Composites for Zero Liquid Discharge of High-Salinity Brine

Qing Yin [1], Fangong Kong [1], Shoujuan Wang [1], Jinbao Du [1], Ling Pan [2], Yubo Tao [1,*] and Peng Li [1,2,*]

[1] State Key Laboratory of Biobased Material and Green Papermaking, Qilu University of Technology, Shandong Academy of Sciences, Jinan 250353, China; qluyinqing@163.com (Q.Y.); kfgwsj1566@163.com (F.K.); nancy5921@163.com (S.W.); zyq19980609@163.com (J.D.)

[2] College of Material Science and Engineering, Northeast Forestry University, Harbin 150040, China; panling@nefu.edu.cn

* Correspondence: taoyubo@qlu.edu.cn (Y.T.); lipeng@qlu.edu.cn (P.L.)

Abstract: Zero liquid discharge (ZLD) is a technique for treating high-salinity brine to obtain freshwater and/or salt using a solar interface evaporator. However, salt accumulation on the surface of the evaporator is a big challenge to maintaining stable water evaporation. In this study, a simple and easy-to-manufacture evaporator, also called a crystallizer, was designed and fabricated by 3D printing. The photothermal layer printed with polylactic acid/carbon composites had acceptable light absorption (93%) within the wavelength zone of 250 nm–2500 nm. The micron-sized voids formed during 3D printing provided abundant water transportation channels inside the crystallizer. After surface hydrophilic modification, the crystallizer had an ultra-hydrophilic channel structure and gravity-assisted salt recovery function. The results revealed that the angles between the photothermal layers affected the efficacy of solar evaporation and the yield of solid salt. The crystallizer with the angle of 90° between two photothermal layers could collect more solid salt than the three other designs with angles of 30°, 60°, and 120°, respectively. The crystallizer has high evaporation and salt crystallization efficiency in a high-salinity brine environment, which is expected to have application potentials in the zero liquid discharge of wastewater and valuable salt recovery.

Keywords: high-salinity brine treatment; solar crystallizer; 3D printing; salt collection; zero liquid discharge

Citation: Yin, Q.; Kong, F.; Wang, S.; Du, J.; Pan, L.; Tao, Y.; Li, P. 3D Printing of Solar Crystallizer with Polylactic Acid/Carbon Composites for Zero Liquid Discharge of High-Salinity Brine. *Polymers* **2023**, *15*, 1656. https://doi.org/10.3390/polym15071656

Academic Editors: Giorgio Luciano and Maurizio Vignolo

Received: 20 February 2023
Revised: 16 March 2023
Accepted: 24 March 2023
Published: 27 March 2023

Copyright: © 2023 by the authors. Licensee MDPI, Basel, Switzerland. This article is an open access article distributed under the terms and conditions of the Creative Commons Attribution (CC BY) license (https:// creativecommons.org/licenses/by/ 4.0/).

1. Introduction

According to the United Nations Environment Programme, more than half of the world population will face freshwater shortage by 2025, and water scarcity and water security should be taken seriously [1,2]. In order to alleviate the shortage of freshwater resources, many countries have attached great importance to desalinations [3,4]. However, traditional desalination plants could produce a large amount of high-salinity brine in the desalination process [5,6]. This results in a large amount of untreated brine discharged directly into surface water, deep-well evaporation ponds, or the ocean, which causes water pollution and land salinization [7,8]. Therefore, zero liquid discharge (ZLD) technology was developed, which is a circular economy-based method for treating high-salinity brine, involving brine concentration and brine crystallization [9,10]. Using the ZLD technique, the dissolved minerals in the brine can be concentrated into solids for collection [11,12]. Nevertheless, traditional ZLD has shortcomings, such as low treatment efficiency, energy consumption, and high cost [13]. Reverse osmosis (RO) membrane filtration is a crucial component of traditional ZLD. The RO membrane pores could be contaminated by fouling after long periods of operation. The fouling comes from the accumulation of bacteria, sludge, metal ions, and other suspended matter. This results in a significant decline in membrane filtration performance. Therefore, the brine requires pretreatment before passing through the RO membrane for desalination [14].

In recent years, the solar-driven interfacial evaporator has been widely studied in the desalination research due to its low cost, effective use of clean energy, no carbon emission, suitability for high-salinity brine, and other benefits [15–17]. At present, the evaporator is mainly used for the treatment of low-salinity brine. By designing the photothermal conversion materials and structure of the evaporator, the salt inside the evaporator can quickly diffuse into water bodies [18,19]. Such an evaporator can maintain water evaporation without being affected by salt crystallization [20,21]. Although the aforementioned studies can prevent the degradation of evaporator performance caused by salt accumulation, the direct diffusion of salt will cause the salinity increase in the original water body, resulting in secondary contamination. As we know, the evaporator is also suitable for salt recovery from high-salinity brine. However, it is very easy for salt to crystallize on the evaporator surface and within the water channel while treating high-salinity brine. The efficiency of photothermal conversion and the water transport performance of evaporators will decrease due to salt accumulation [22,23]. Therefore, enabling an evaporator to have durable water evaporation performance is key to research during salt recovery.

Some innovative interface evaporators have been designed to achieve efficient water evaporation while also enabling solid salt collection. Those ideas also realized ZLD goals of pollution-free, zero energy consumption for high-salinity brine treatment. As shown in Figure 1, the interface evaporator is able to simplify the traditional ZLD treatment process by replacing the expensive, contamination-prone RO membrane.

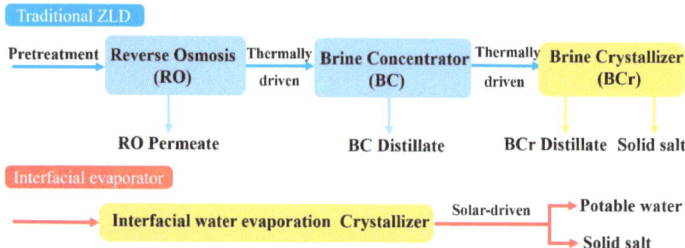

Figure 1. A schematic diagram of traditional ZLD and interfacial evaporator. (Reprinted with permission from Ref. [12]. 2022, Elsevier).

The research of interface evaporators focuses on material and structure design. In terms of evaporator material research, the surface of the evaporator is commonly hydrophobically modified to avoid salt crystallization, and the other part remains hydrophilic for water transmission. Gu et al. [24] modified hydrophobic carbon felt to make it hydrophilic through acid treatment, and then coated the top surface with hydrophobic poly(1,1-difluoroethylene) (PVDF). After the treatment, salt could only crystallize on the edge of evaporator, without salt accumulation on the top surface. Similarly, Dong et al. [25] first prepared hydrophilic cellulose acetate film by electrospinning. Then, through hydrophobic modification of the top surface with polydimethylsiloxane (PDMS) coating, they ensured that salt could only crystallize in the hydrophilic part of the evaporator.

In terms of evaporator structure design, some studies have realized selective salt crystallization at the edge of the evaporator by designing the evaporator's water transport channel. Li et al. [26] prepared a water transport structure with a different bending degree using polyvinyl acetate (PVA) hydrogel. Salt could crystallize along a designated water flow direction. Bionic structural design is another route to achieve selective salt crystallization. Being inspired by the salt dilution and salt secretion mechanism of halophyte, Zhang et al. [27] prepared an evaporator capable of surface salt secretion and edge salt dilution using glass fiber felt and Polypyrrole (PPy). Sun et al. [28] 3D-printed conical PLA structures with concave grooves by mimicking mussel seashells and then coated the top layer of the evaporator with PPy to realize freshwater collection.

In the above research, although the constructed evaporators achieved the ZLD treatment goal of zero energy consumption, the coating materials or the preparation process used are costly. In addition, when the water delivery channel in the evaporator is nanoscale, Mg^{2+} and Ca^{2+} will block the channel during the desalination of high-salinity brine, which requires the addition of salt crystallization inhibitors to pretreat high-salinity brine. This also increases the cost of desalination.

With the aim of free control of the outer shape and inner water transport channel of the evaporator, seawater evaporation crystallizers were constructed by 3D-printing technology in this study. The crystallizer structure consists of two parts: (1) the top photothermal conversion layer, which seems like a sloping roof; and (2) the bottom cube for water transfer. In addition, the supporting buoyancy layer structure was designed to provide buoyancy for the crystallizer body and the solid salt collected. The photothermal conversion part was printed with Polylactic acid (PLA)/bio carbon composite filaments. The water transfer cube was printed with pure PLA filaments. Due to the hydrophobicity of PLA, the 3D-printed parts were subjected to hydrophilic modification. Considering that the slope of the photothermal conversion layer is the crucial factor that determines the effective photothermal conversion area, four types of crystallizers were constructed, with the photothermal layer containing angles of 30°, 60°, 90°, and 120°, respectively. To evaluate the salt crystallization performance of the crystallizers, the salt crystallization process was run under an extreme salt concentration environment (20 wt%). The experimental results showed that the micron-sized water transport channel and superhydrophilicity of the crystallizer produced continuous water evaporation and optimal salt crystallization performance.

2. Materials and Methods

2.1. Materials

Sodium chloride (NaCl, AR) and Tannic acid (TA, AR) were supplied by Macklin Biochemical Co., Ltd. (Shanghai, China). Polyvinylpyrrolidone (PVP, Mw~8000) was supplied by Aladdin Biochemical Technology Co., Ltd. (Shanghai, China). Iron(III) chloride ($FeCl_3$, AR) was supplied by Damao Chemical Co., Ltd. (Tianjin, China). PLA (4032D) was supplied by Nature Works LLC (Minneapolis, MN, USA). Hazelnut shells were from Benxi Agricultural Development Co., LTD. (Benxi, China).

The PLA-C filament for 3D printing was prepared by melt extrusion as follows: 98 g of PLA and 2 g of the dried hazelnut shell carbon. The preparation process of hazelnut shell carbon was as follows: in the first step, hazelnut shells were pulverized into particles. Then, hazelnut shell particles were pyrolyzed for 2 h at 700 °C under an N_2 atmosphere in a tube furnace. The heating rate in the furnace was 5 °C per minute (OTF-1200X, KEJING, Hefei, China). Finally, further grinding was run in a ball mill machine (QM 0.4 L, Chishun LLC, Nanjing, China) and then the ground carbon powder was passed through a 120-mesh sieve and collected. The PLA was fused and blended with hazelnut shell carbon in a double roll mixer (BP-8175-A, Baopin precise instrument, Dongguan, China). The blending procedure was run at 180 °C, with the rollers' rotation speed at 1.5 r min^{-1}. Then, the blends were smashed into particles smaller than 2 mm by a disintegrator (FW135, Taisite instruments, Tianjin, China). Finally, a single-screw plastic extruder was used to extrude PLA-C filament (C2 model, Wellzoom LLC, Shenzhen, China). The extrusion temperature was in the range of 178–183 °C. Pure PLA filament was extruded under the same conditions.

2.2. Design and 3D Printing of the Crystallizer Structure

The model of crystallizer was designed using AutoCAD software (2022 Student edition, Autodesk software, San Rafael, CA, USA). The T-shaped crystallizer consists of two parts: the photothermal layers and the water transport layer. The detailed crystallizer design is presented in Figure 2a and the CAD drawing is provided as a supplementary document (Figure S1). The photothermal layer of the crystallizer has two rectangular water evaporation surfaces of 36 mm × 40 mm with a thickness of 6 mm. The angles between two photothermal layers were designed as 30°, 60°, 90°, and 120°, respectively. The horizontal water transport layer has a cross-section of 6 mm × 40 mm. The water transport layer has the same volume in contact with brine and a height of 15 mm. The Cura software (Cura 15.04, Ultimaker, Waltham, MA, USA) was used to slice the crystallizer model. The model slice program was set as without the exterior of the model and an infill density of 60%. An FFF (Fused Filament Fabrication) double-head printer (Inker 334, Wuhan Allcct Technology Co., Ltd., Wuhan, China) was used to construct the crystallizer. As shown in Figure 2b, the photothermal layers and the water transport layer were printed using PLA-C and PLA filament, respectively. The printing parameters were set as: a nozzle temperature of 200 °C, a hot bed temperature of 60 °C, a printing speed of 50 mm s^{-1}, a printing thickness of 0.2 mm, and a nozzle diameter of 400 μm, respectively.

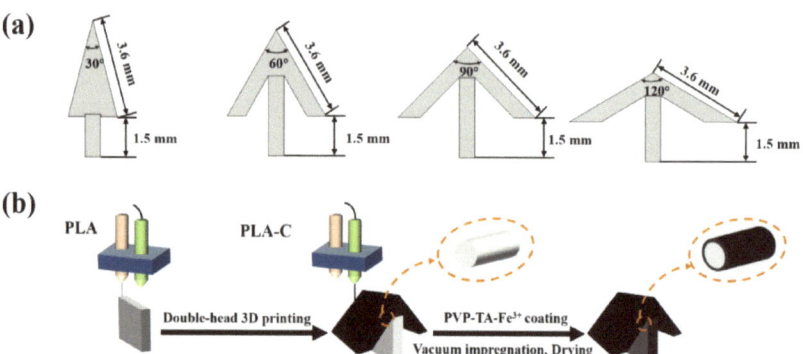

Figure 2. (**a**) numerical design of angle between photothermal layer of the crystallizer; (**b**) schematic diagram of crystallizer preparation process.

2.3. Surface Hydrophilic Modification of Crystallizer

PVP-TA-Fe^{3+} coating was synthesized to improve the hydrophilicity of the crystallizer. The preparation steps of PVP-TA-Fe^{3+} coating are as follows: PVP (0.32 g) and TA (0.32 g) were mixed in 100 mL deionized water by magnetic stirring at room temperature for 5 min. Then, FeCl$_3$ (0.69 g) was added into the PVP-TA solution by magnetic stirring. The 3D-printed structure was immersed into the PVP-TA-Fe^{3+} solution and then placed in a vacuum oven for vacuum impregnation. After 12 h impregnation, the structure was dried at 50 ± 2 °C for 6 h. Eventually, the dried structure was rinsed with distilled water under ultrasound treatment for 2 h to remove the unattached PVP-TA-Fe^{3+} coating.

2.4. Design and 3D Printing of the Buoyant Structure

The buoyancy layer can ensure that the crystallizer system floats on the brine and reduces the heat transfer between the crystallizer and brine. In this work, the buoyancy layer provided a buoyancy force greater than the total weight of the crystallizer itself, its water content, and salt crystals. The design of buoyancy layer is presented in Figure 3 and the CAD model is provided as a supplementary document (Figure S2). According to Archimedes' theorem, the mass of the water body occupied by the buoyancy layer is equal to the buoyancy exerted on it. The following formula (1) is used to calculate the buoyancy provided by the buoyant structure [29].

$$F_{buo} = \rho \times g \times V_{disp} \tag{1}$$

where ρ is brine density, 1035 kg × m^{-3}, g is radio of gravity to mass, 9.8 N × kg^{-1}, V_{disp} is volume of displacement, m^3.

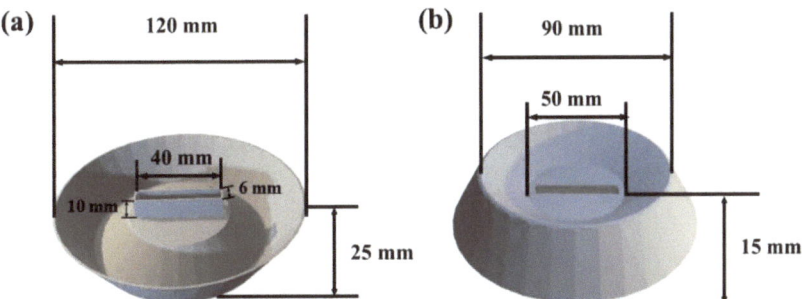

Figure 3. Buoyancy layer model: (**a**) front view; (**b**) back view.

2.5. Characterization

The microstructure of the crystallizer was observed by scanning electron microscopy at 5 kV (SEM, JSM-7401F, Hitachi, Tokyo, Japan). The contact angle was recorded to compare the wettability of the crystallizer surface (the optical contact angle, Dataphysics OCA20, Filderstadt, Germany). The tensile specimens were designed according to ASTM D638 Type IV [30]. All samples were 3D-printed as described in Section 2.2. Tensile tests were conducted by a universal tensile machine (TSE104B, Wance, Shenzhen, China) with the tensile rate of 5 mm/min^{-1}. The reflectance (R%) and transmittance (T%) of the crystallizer at 250 nm–2500 nm were tested by Ultraviolet-visible-near-infrared spectrophotometer (UV-VIS-NIR, Cary 7000, Santa Clara, CA, USA). The light absorbance (A%) of the crystallizer was calculated by the equation: A% = 1 − (R% + T%).

2.6. Salt Crystallization Test

The photothermal conversion and crystallization capabilities of the crystallizer were evaluated under a customized solar simulation test system (Xenon light source, CME-SL500, Micro energy, Beijing, China). A solar power meter (TES132, ShenZhen, China) was used to maintain the surface light intensity of the crystallizer at 1000 W m^{-2}. An electronic balance equipped with a computer was adopted to record the water mass change continuously. The accuracy of the balance was 0.0001 g (BSA224S, Sartorius, Gottingen, Germany). The surface temperature of the crystallizer was monitored with an infrared thermal camera (FLIR E8, Wilsonville, OR, USA). During the water evaporation test, the crystallizer was embedded in the expandable polyethylene foam (EPF) to isolate the natural evaporation of bulk water. This also ensured that the photothermal layer of the crystallizer can convert light to heat and promote the evaporation of brine to generate steam at the interface. To test the water evaporation of the crystallizer, a 50 mL quartz beaker was used to hold the crystallizer, high-salinity brine, and EPF. To test the continuous salt crystallization of the crystallizer, a glass flume was used to hold the crystallizer, high-salinity brine, and buoyant structure. The test was conducted at an ambient temperature (20 ± 2 °C) and relative humidity (50 ± 5%).

2.7. COMSOL Simulation Test

The mass transfer processes of water and salt include advection, diffusion, and evaporation. The laminar flow and dilute material transfer were used for evaluating the ion diffusion in the crystallizer. The distribution of ion concentration was studied by solving the following equation:

$$\rho \times \partial u/\partial t + \rho \times (u \times \nabla u) \times u = \nabla[-pI + K] + F + \rho g \qquad (2)$$

$$\rho \times \nabla u = 0 \qquad (3)$$

$$\partial c/\partial t + \nabla J + u \times \nabla c = R \qquad (4)$$

$$J = -D \times \nabla c \tag{5}$$

where u is the velocity field, set as 4×10^{-7} m × s^{-1}; I is the constitutive relation coefficient; K is the viscosity, set as 8.9×10^{-4} Pa·s; ρ is the brine density, set as 1 035 kg × m^{-3}; F is the surface tension at the interface between air and liquid, set as 0. R is the reaction on the ion surface, set as 0; c is the ion concentration, set as 3589 mol × L^{-1}, corresponding to 20 wt% brine; t is the time, and J is the concentration gradient.

3. Results and Discussion

3.1. Preparation and Surface Characterization of Crystallizer

Bio-based carbon composites are ideal photothermal materials due to their advantages such as being robust and inexpensive, and their capability to absorb full-spectrum solar wavelengths [31–33]. In this study, 3D printing was combined with bio-based photothermal materials to construct controllable crystallizer structure. As shown in Figure 4a, the crystallizer structure built by 3D printing consists of the black part printed with PLA-C filament and the white part printed with PLA filament. PLA-C and PLA filaments were used to construct the photothermal layer and water transfer layer of the crystallizer, respectively. After being printed, these two parts were integrated together to form the main structure of the crystallizer. Then, the crystallizer structure was treated with a PVP-TA-Fe^{3+} coating. After PVP-TA-Fe^{3+} treatment, the whole structure became significantly black. The PVP-TA-Fe^{3+} system utilized its hydrophilic as well as strong adhesive property to form a stable coating [34–36]. As shown in Figure 4c, the surface of micron-sized water transport channels without coating was smooth. In comparison, the surface of the structure became rough after coating (Figure 4d). The contact angle results in Figure 4c, d proved that the hydrophobic water transfer channels were modified to be ultra-hydrophilic after the coating treatment. Furthermore, the mechanical property of the crystallizer is crucial for stable operation of the evaporator system. The tensile experiments demonstrated that all specimens constructed by 3D printing had a tensile strength above 45 MPa, as shown in Figure 4b.

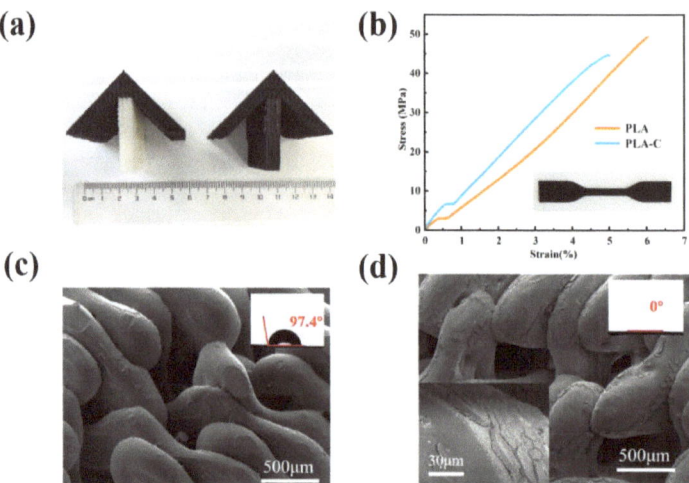

Figure 4. (**a**) digital photograph of the 3D-printed crystallizer structure without and with coating; (**b**) tensile properties of the standard tensile specimen; (**c**) Microstructure and wettability of crystallizer without PVP-TA-Fe^{3+} coating; (**d**) Microstructure and wettability of crystallizer with PVP-TA-Fe^{3+} coating.

3.2. Buoyant Structure

The hollow cone-shaped buoyant layer was designed parametrically by CAD. The center of the buoyancy layer holds the crystallizer, and the concave groove of the buoyancy layer collects the dislodged salt crystals. The buoyancy layer also acts as an insulator, reducing the heat transfer between the crystallizer and the brine. According to the buoyancy formula (1) and the specific design data of the buoyancy layer, it can be calculated that the volume of brine discharged by the buoyancy layer was 158.57 cm^3. Therefore, the buoyancy layer can enable an object weighing 168.375 g to float on the brine. In this experiment, the buoyancy layer itself weighed 21.64 g, and the weight of crystallizer containing brine was 25 ± 5 g. Therefore, the buoyancy layer can float an object weighing 120 ± 5 g on the brine, which indicates that it can provide a stable salt crystallization environment for the crystallizer.

3.3. Salt Crystallization Performance of the Crystallizer

The hydrophilic coating could ensure efficient water delivery in the micron-sized channels of the crystallizer. However, salt crystals may accumulate on the photothermal layer during continuous water evaporation, which affects its evaporation efficiency. The schematic illustration of the assembly of crystallizer device is shown in Figure 5. The whole crystallizing system consists of the photothermal layer, water transport layer, and buoyancy layer. The function of the water transport layer is to transfer brine from the water body to the photothermal layer. The function of the photothermal layer is to absorb solar radiation and convert solar energy into heat, warming and evaporating brine to obtain salt crystals. The buoyancy layer collects the salt crystals produced during the stable operation of the entire system. By regulating the difference in surface area between the photothermal layer and the water transport layer, the salt can crystallize on the surface of the photothermal layer. Meanwhile, the angle between two photothermal layers was designed to ensure that the salt crystals can fall off on their own by gravity, which supports the stable operation of the crystallizer.

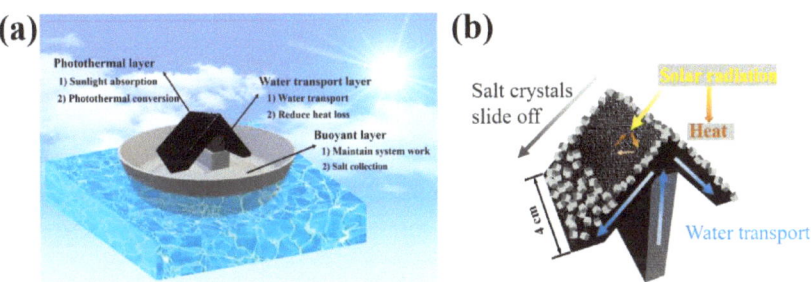

Figure 5. (a,b) schematic illustration of crystallizer based on 3D printing.

Excellent photothermal conversion performance can promote water evaporation and salt crystallization. The solar absorption of the crystallizer at the full-spectrum wavelength is shown in Figure 6a, reaching 93%. It indicated that the photothermal layer of the crystallizer had good light-absorption capacity. As shown in Figure 6b, an indoor solar simulation device was used to evaluate the water evaporation performance. As shown in Figure 6c, the water evaporation rate of the crystallizer increased as the angle between the photothermal layers increased. This is due to the height of the crystallizer itself becoming smaller as the angle between the photothermal layers becomes larger, and then the brine was able to reach the photothermal layer more quickly. After the crystallizer had been running for a specified period, the solid salt was collected from the crystallizer surface as well as from the buoyancy layer. The solid salt collected was dried at 55 °C for 6 h to a constant weight. As shown in Figure 6c, d, the angle between the photothermal layers played a key role in the evaporating rate of the crystallizer. During a specific operation period, brine mass decreased faster as the angle of photothermal layer inclusion increased. Correspondingly, the amount of solid salt collected followed the opposite trend. However, a larger angle between the photothermal layers does not mean better salt crystallization. When the system had been operating for 8 h, the crystallizer with an inclusion angle of 120° collected less solid salt than the one with a 90° inclusion angle. This is due to the fact that the solid salt on the photothermal surface of the crystallizer with an inclusion angle of 90° was subjected to a larger gravitational component parallel to the slope than the one with a 120° inclusion angle. As the photothermal conversion process continued, the mass of salt crystals on the slope increased. When the gravitational component parallel to the slope overcomes the friction between the salt and the photothermal surface, the salt will drop off. However, at an inclusion angle of 120°, it was more difficult for solid salt falling off from the evaporation surface, resulting in more salt accumulation on the slope, which reduced its salt collecting efficacy.

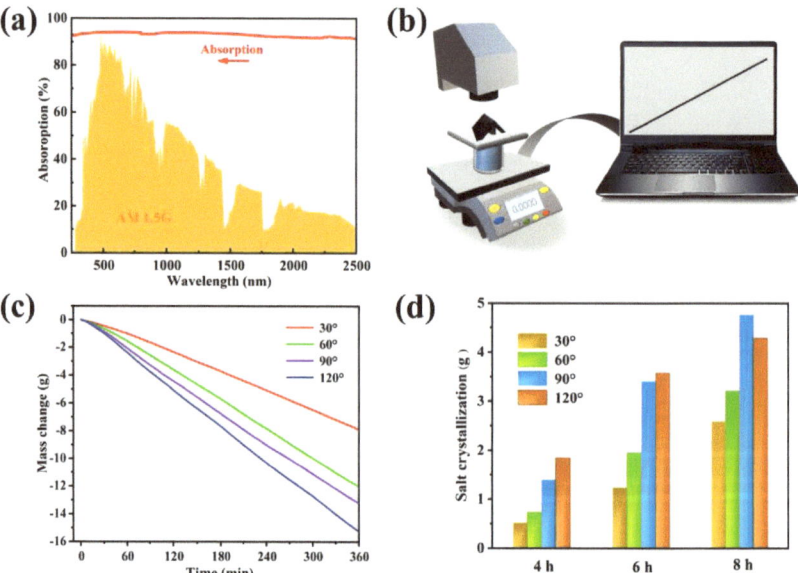

Figure 6. (**a**) Light absorption of crystallizer at the 250–2500 nm wavelength spectrum; (**b**) schematic of an indoor solar evaporation setup; (**c**) mass change of the brine over time at different crystallizer angles; (**d**) salt collection of different crystallizers within three durations of test.

Figure 7 demonstrated the distribution of salt crystals on the surface of the crystallizers at different angles after 1 h and 6 h of operation. After 1 h of operation, salt crystals appeared on the surface of all crystallizers. After 6 h of operation, the top edge of the crystallizer at 30° and 60° were covered with salt crystals, while the top edge of the crystallizers at 90° and 120° did not show this phenomenon. The reason for this is that the brine was not replenished after evaporation, resulting in salt crystal accumulation on the surface of the photothermal layer. The salt crystalline encrustation could reduce the photothermal efficiency of the crystallizer, which is detrimental to the long-term operation of the crystallizer. In contrast, the top edges of the crystallizers with inclusion angles of 90° and 120° showed no salt crystals wrapped around them, proving that their photothermal layer enabled rapid brine replenishment along with water evaporation. Among four types of crystallizers, the water transfer distance from the brine to the photothermal layer of the crystallizers with inclusion angles of 30° and 60° is longer, and the water transfer efficiency is lowered.

Figure 7. Digital photograph of salt crystallization: (**a**) 1 h operation of the crystallizer; (**b**) 6 h operation of the crystallizer.

3.4. Simulation of Salt Ion Distribution on the Crystallizer Surface

To perceive the water transportation and the ion concentration distribution during the operation of the crystallizer, the salt ion distribution in the crystallizer was simulated. The profile structure of the crystallizers was simulated using COMSOL Multiphysics software. The results of the water transport simulation are shown in Figure 8a. It is clear that the brine was quickly transported from the water transport layer to the photothermal layer. During brine transportation, brine backflow as well as diffusion occurred simultaneously. As the inclusion angle between the photothermal layers became bigger, the diffusion effect of brine became stronger at the intersection of the photothermal layer and the water transport layer. At the same time, the backflow effect of brine became more pronounced. The results of the salt concentration distribution simulation are shown in Figure 8b. The salt concentration within the water transport layer maintained a bottom-up increase, whereas the salt concentration within the photothermal layer changed from the center to the edges. The salt concentration in the photothermal layer were higher in the crystallizers with inclusion angles of 60° and 90° than in those with 30° and 120° inclusion angles. A higher salt concentration indicated a trend towards better salt crystallization on the photothermal layer.

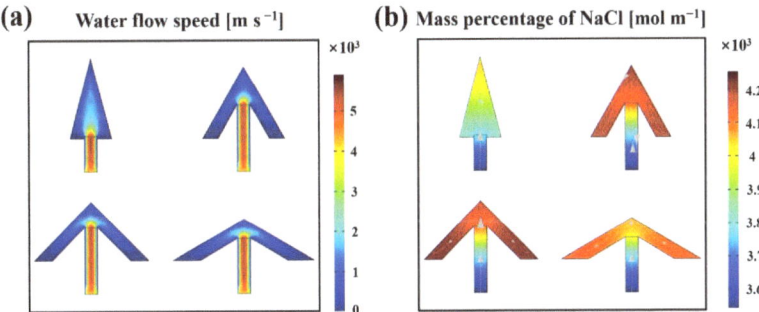

Figure 8. (**a**) Simulated water flow speed; (**b**) simulated salt concentration distribution(arrows is the direction of water flow).

3.5. Long-Term Stability of the Crystallizer at 90°

Through experiments and simulations, it can be seen that the crystallizer with an inclusion angle of 90° had higher salt crystallization efficiency. To evaluate the long-term stability of the crystallizer at 90°, a 12 h run was recorded and analyzed. As shown in Figure 9a, it is obvious that salt crystals continuously generated on the surface of the crystallizer with the increase of time. Though the salt mass rose with time, the pores in the salt crust ensured successful water transportation [37]. During the evaporation process, the salt ion concentration could keep in a dynamic equilibrium. Salt crystals would slide off the photothermal layer when the mass of the salt crystals was large enough [38]. The shedding of salt crystals ensured that the upper of the photothermal layer was not blocked by salt crystals, allowing the crystallizer to operate for a long time to produce salt crystals. Figure 9b shows that under the irradiation of one-sun illumination, the surface temperature on the upper and middle parts of the photothermal layer was maintained at 42 ± 2 °C, and the lower part of it was maintained at 38 ± 2 °C. Although salt accumulated at the lower part of the photothermal layer, which led to the temperature drop, the overall temperature of the photothermal layer still showed a high temperature. This was due to the upper and middle parts of the photothermal layer having a continuous photothermal conversion ability, which allowed the evaporation surface to transfer heat quickly and ensured the continuous salt crystallization capacity of the photothermal layer.

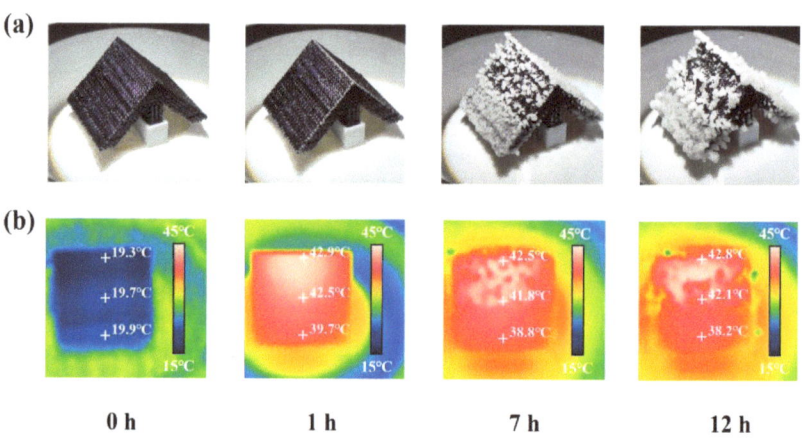

Figure 9. During 12 h operation of the crystallizer: (**a**) Digital photograph of salt crystallization; (**b**) infrared thermal images.

4. Conclusions

In this work, a seawater evaporation crystallizer with designated geometry and an inner channel was constructed through 3D printing. This crystallizer was capable of producing salt crystals during the treatment of high-salinity brine. The buoyancy layer in the design not only provided buoyancy but also acted as salt collector and could reduce the overall heat loss of the crystallizer. Through experiment and simulation analyses, it was determined that the salt crystallizer with an angle of 90° between the photothermal layers could maintain continuous salt crystallization and a higher efficiency of salt harvesting in a high-salinity brine environment. This crystallization system has long-term stability in the treatment of high-salinity brine and would have practical application in achieving the zero liquid discharge of high-salinity water and the recovery of valuable salts.

Supplementary Materials: The following supporting information can be downloaded at: https://www.mdpi.com/article/10.3390/polym15071656/s1, Figure S1: parameters for designing crystallizers; Figure S2: model of buoyancy layer.

Author Contributions: Conceptualization, P.L. and Y.T.; methodology, Q.Y. and P.L.; software, Q.Y. and Y.T.; validation, F.K. and S.W.; formal analysis, Q.Y. and Y.T.; investigation, Q.Y., J.D. and L.P.; writing—original draft preparation, Q.Y.; writing—review and editing, Y.T.; supervision, Y.T. and P.L.; project administration, P.L. and Y.T.; funding acquisition, F.K., S.W., Y.T. and P.L. All authors have read and agreed to the published version of the manuscript.

Funding: This research was funded by the Start-up Funding from Qilu University of Technology (Shandong Academy of Sciences), grant number 81110696 and 81110592, the Program for New Century Excellent Talents in University of China, grant number NCET-13-0711, the National Natural Science Foundation of China, grant number 31971605, the QUTJBZ Program, grant number 2022JBZ01-05.

Institutional Review Board Statement: Not applicable.

Informed Consent Statement: Not applicable.

Data Availability Statement: The data presented in this study are available on request from the corresponding author.

Acknowledgments: The authors thank Huiying Liu for technical assistance.

Conflicts of Interest: The authors declare no conflict of interest.

References

1. Prisciandaro, M.; Capocelli, M.; Piemonte, V.; Barba, D. Process analysis applied to water reuse for a "closed water cycle" approach. *Chem. Eng. J.* **2016**, *304*, 602–608. [CrossRef]
2. Shatat, M.; Worall, M.; Riffat, S. Opportunities for solar water desalination worldwide: Review. *Sustain. Cities Soc.* **2013**, *9*, 67–80. [CrossRef]
3. Chen, K.; Li, L.; Zhang, J. Design of a Separated Solar Interfacial Evaporation System for Simultaneous Water and Salt Collection. *ACS Appl. Mater. Interfaces* **2021**, *13*, 59518–59526. [CrossRef] [PubMed]
4. Gu, R.; Yu, Z.; Sun, Y.; Xie, P.; Li, Y.; Cheng, S. Enhancing stability of interfacial solar evaporator in high-salinity solutions by managing salt precipitation with Janus-based directional salt transfer structure. *Desalination* **2022**, *524*, 115470. [CrossRef]
5. Xie, W.; Duan, J.; Li, J.; Qi, B.; Liu, R.; Yu, B.; Wang, H.; Zhuang, X.; Xu, M.; Zhou, J. Charge-Gradient Hydrogels Enable Direct Zero Liquid Discharge for Hypersaline Wastewater Management. *Adv. Mater.* **2021**, *33*, 2100141. [CrossRef]
6. Chen, Q.; Burhan, M.; Shahzad, M.W.; Ybyraiymkul, D.; Akhtar, F.H.; Li, Y.; Ng, K.C. A zero liquid discharge system integrating multi-effect distillation and evaporative crystallization for desalination brine treatment. *Desalination* **2021**, *502*, 114928. [CrossRef]
7. Panagopoulos, A. Brine management (saline water & wastewater effluents): Sustainable utilization and resource recovery strategy through Minimal and Zero Liquid Discharge (MLD & ZLD) desalination systems. *Chem. Eng. Process.-Process Intensif.* **2022**, *176*, 108944. [CrossRef]
8. Tsai, J.-H.; Macedonio, F.; Drioli, E.; Giorno, L.; Chou, C.-Y.; Hu, F.-C.; Li, C.-L.; Chuang, C.-J.; Tung, K.-L. Membrane-based zero liquid discharge: Myth or reality? *J. Taiwan Inst. Chem. Eng.* **2017**, *80*, 192–202. [CrossRef]
9. Panagopoulos, A.; Giannika, V. Comparative techno-economic and environmental analysis of minimal liquid discharge (MLD) and zero liquid discharge (ZLD) desalination systems for seawater brine treatment and valorization. *Sustain. Energy Technol. Assess.* **2022**, *53*, 102477. [CrossRef]
10. Farahbod, F.; Mowla, D.; Jafari Nasr, M.R.; Soltanieh, M. Experimental study of forced circulation evaporator in zero discharge desalination process. *Desalination* **2012**, *285*, 352–358. [CrossRef]
11. Chen, F.; Zhang, Z.; Zeng, F.; Yang, Y.; Li, X. Pilot-scale treatment of hypersaline coal chemical wastewater with zero liquid discharge. *Desalination* **2021**, *518*, 115303. [CrossRef]
12. Panagopoulos, A. Techno-economic assessment of zero liquid discharge (ZLD) systems for sustainable treatment, minimization and valorization of seawater brine. *J. Environ. Manag.* **2022**, *306*, 114488. [CrossRef] [PubMed]
13. Wang, R.; Lin, S. Thermodynamics and Energy Efficiency of Zero Liquid Discharge. *ACS EST Eng.* **2022**, *2*, 1491–1503. [CrossRef]
14. Maiti, S.; Kane, P.; Pandit, P.; Singha, K.; Maity, S. Chapter Nine-Zero liquid discharge wastewater treatment technologies. In *Sustainable Technologies for Textile Wastewater Treatments*; Muthu, S.S., Ed.; Woodhead Publishing: Soston, UK, 2021; pp. 209–234. [CrossRef]
15. Xu, N.; Li, J.; Wang, Y.; Fang, C.; Li, X.; Wang, Y.; Zhou, L.; Zhu, B.; Wu, Z.; Zhu, S.; et al. A water lily–inspired hierarchical design for stable and efficient solar evaporation of high-salinity brine. *Sci. Adv.* **2022**, *5*, eaaw7013. [CrossRef]
16. Wang, Y.; Wu, X.; Wu, P.; Yu, H.; Zhao, J.; Yang, X.; Li, Q.; Zhang, Z.; Zhang, D.; Owens, G.; et al. Salt isolation from waste brine enabled by interfacial solar evaporation with zero liquid discharge. *J. Mater. Chem. A* **2022**, *10*, 14470–14478. [CrossRef]
17. Yu, Z.; Li, S.; Chen, Y.; Zhang, X.; Chu, J.; Zhang, Y.; Tan, S.C. Intensifying the co-production of vapor and salts by a one-way brine-flowing structure driven by solar irradiation or waste heat. *Desalination* **2022**, *539*, 115942. [CrossRef]
18. Zhou, Q.; Li, H.; Li, D.; Wang, B.; Wang, H.; Bai, J.; Ma, S.; Wang, G. A graphene assembled porous fiber-based Janus membrane for highly effective solar steam generation. *J. Colloid Interface Sci.* **2021**, *592*, 77–86. [CrossRef]
19. Shi, Y.; Zhang, C.; Li, R.; Zhuo, S.; Jin, Y.; Shi, L.; Hong, S.; Chang, J.; Ong, C.; Wang, P. Solar Evaporator with Controlled Salt Precipitation for Zero Liquid Discharge Desalination. *Environ. Sci. Technol.* **2018**, *52*, 11822–11830. [CrossRef]
20. Ni, G.; Zandavi, S.H.; Javid, S.M.; Boriskina, S.V.; Cooper, T.A.; Chen, G. A salt-rejecting floating solar still for low-cost desalination. *Energy Environ. Sci.* **2018**, *11*, 1510–1519. [CrossRef]
21. Xia, Y.; Li, Y.; Yuan, S.; Kang, Y.; Jian, M.; Hou, Q.; Gao, L.; Wang, H.; Zhang, X. A self-rotating solar evaporator for continuous and efficient desalination of hypersaline brine. *J. Mater. Chem. A* **2020**, *8*, 16212–16217. [CrossRef]

22. Menon, A.K.; Haechler, I.; Kaur, S.; Lubner, S.; Prasher, R.S. Enhanced solar evaporation using a photo-thermal umbrella for wastewater management. *Nat. Sustain.* **2020**, *3*, 144–151. [CrossRef]
23. Cooper, T.A.; Zandavi, S.H.; Ni, G.W.; Tsurimaki, Y.; Huang, Y.; Boriskina, S.V.; Chen, G. Contactless steam generation and superheating under one sun illumination. *Nat. Commun.* **2018**, *9*, 5086. [CrossRef]
24. Gu, R.; Yu, Z.; Sun, Y.; Su, Y.; Wu, W.; Cheng, S. Janus 3D solar crystallizer enabling an eco-friendly zero liquid discharge of high-salinity concentrated seawater with antiscalant. *Desalination* **2022**, *537*, 115862. [CrossRef]
25. Dong, X.; Li, H.; Gao, L.; Chen, C.; Shi, X.; Du, Y.; Deng, H. Janus Fibrous Mats Based Suspended Type Evaporator for Salt Resistant Solar Desalination and Salt Recovery. *Small* **2022**, *18*, 2107156. [CrossRef] [PubMed]
26. Li, L.; He, N.; Jiang, B.; Yu, K.; Zhang, Q.; Zhang, H.; Tang, D.; Song, Y. Highly Salt-Resistant 3D Hydrogel Evaporator for Continuous Solar Desalination via Localized Crystallization. *Adv. Funct. Mater.* **2021**, *31*, 2104380. [CrossRef]
27. Zhang, S.; Yuan, Y.; Zhang, W.; Song, F.; Li, J.; Liu, Q.; Gu, J.; Zhang, D. A bioinspired solar evaporator for continuous and efficient desalination by salt dilution and secretion. *J. Mater. Chem. A* **2021**, *9*, 17985–17993. [CrossRef]
28. Sun, S.; Shi, C.; Kuang, Y.; Li, M.; Li, S.; Chan, H.; Zhang, S.; Chen, G.; Nilghaz, A.; Cao, R.; et al. 3D-printed solar evaporator with seashell ornamentation-inspired structure for zero liquid discharge desalination. *Water Res.* **2022**, *226*, 119279. [CrossRef] [PubMed]
29. Loverude, M.E.; Kautz, C.H.; Heron, P.R.L. Helping students develop an understanding of Archimedes' principle. I. Research on student understanding. *Am. J. Phys.* **2003**, *71*, 1178–1187. [CrossRef]
30. Tao, Y.; Liu, M.; Han, W.; Li, P. Waste office paper filled polylactic acid composite filaments for 3D printing. *Compos. Part B Eng.* **2021**, *221*, 108998. [CrossRef]
31. Yin, Q.; Zhang, J.; Tao, Y.; Kong, F.; Li, P. The emerging development of solar evaporators in materials and structures. *Chemosphere* **2022**, *289*, 133210. [CrossRef] [PubMed]
32. Finnerty, C.; Zhang, L.; Sedlak, D.L.; Nelson, K.L.; Mi, B. Synthetic Graphene Oxide Leaf for Solar Desalination with Zero Liquid Discharge. *Environ. Sci. Technol.* **2017**, *51*, 11701–11709. [CrossRef]
33. Guan, W.; Guo, Y.; Yu, G. Carbon Materials for Solar Water Evaporation and Desalination. *Small* **2021**, *17*, 2007176. [CrossRef] [PubMed]
34. Li, M.; Wu, L.; Zhang, C.; Chen, W.; Liu, C. Hydrophilic and antifouling modification of PVDF membranes by one-step assembly of tannic acid and polyvinylpyrrolidone. *Appl. Surf. Sci.* **2019**, *483*, 967–978. [CrossRef]
35. Fan, H.; Wang, L.; Feng, X.; Bu, Y.; Wu, D.; Jin, Z. Supramolecular Hydrogel Formation Based on Tannic Acid. *Macromolecules* **2017**, *50*, 666–676. [CrossRef]
36. Nam, H.G.; Nam, M.G.; Yoo, P.J.; Kim, J.-H. Hydrogen bonding-based strongly adhesive coacervate hydrogels synthesized using poly(N-vinylpyrrolidone) and tannic acid. *Soft Matter* **2019**, *15*, 785–791. [CrossRef]
37. Zhang, C.; Shi, Y.; Shi, L.; Li, H.; Li, R.; Hong, S.; Zhuo, S.; Zhang, T.; Wang, P. Designing a next generation solar crystallizer for real seawater brine treatment with zero liquid discharge. *Nat. Commun.* **2021**, *12*, 998. [CrossRef] [PubMed]
38. Xia, Y.; Hou, Q.; Jubaer, H.; Li, Y.; Kang, Y.; Yuan, S.; Liu, H.; Woo, M.W.; Zhang, L.; Gao, L.; et al. Spatially isolating salt crystallisation from water evaporation for continuous solar steam generation and salt harvesting. *Energy Environ. Sci.* **2019**, *12*, 1840–1847. [CrossRef]

Disclaimer/Publisher's Note: The statements, opinions and data contained in all publications are solely those of the individual author(s) and contributor(s) and not of MDPI and/or the editor(s). MDPI and/or the editor(s) disclaim responsibility for any injury to people or property resulting from any ideas, methods, instructions or products referred to in the content.

Article

A Flexible Bi-Stable Composite Antenna with Reconfigurable Performance and Light-Responsive Behavior

Yaoli Huang [1], Cong Zheng [1], Jinhua Jiang [1], Huiqi Shao [2,3,*] and Nanliang Chen [1,*]

1. Engineering Research Center of Technical Textiles, Ministry of Education, College of Textiles, Donghua University, Shanghai 201620, China
2. Engineering Research Center of Digitalized Textile and Fashion Technology, Ministry of Education, Shanghai 201620, China
3. Innovation Center for Textile Science and Technology, Donghua University, Shanghai 200051, China
* Correspondence: hqshao@dhu.edu.cn (H.S.); nlch@dhu.edu.cn (N.C.)

Abstract: An integrated solution providing a bi-stable antenna with reconfigurable performance and light-responsive behavior is presented in this paper for the first time. The proposed antenna includes a radiation layer with conductivity, which is integrated onto the bi-stable substrate. First, the effect of the radiation layer material and substrate layer parameters on antenna performance was studied. The experiment showed that an antenna with CNTF has a wider impedance bandwidth than one with CSP, namely 10.37% versus 3.29%, respectively. The resonance frequency increases gradually with the increase in fiber laying density and fiber linear density. Second, the influence of state change of the substrate layer on the antenna radiation pattern was studied. The measured results showed that the maximum radiation angle and gain of states I and II are at 90°, 1.21 dB and 225°, 1.53 dB, respectively. The gain non-circularities of the antenna at states I and II are 4.48 dB and 8.35 dB, respectively, which shows that the antenna has good omnidirectional radiation performance in state I. The display of the array antenna, which shows that the array antenna has good omnidirectional radiation performance in state A, with gain non-circularities of 4.20 dB, proves the feasibility of this bi-stable substrate in reconfigurable antennas. Finally, the antenna deforms from state I to state II when the illumination stimulus reaches 22 s, showing good light-responsive behavior. Moreover, the bi-stable composite antenna has the characteristics of small size, light weight, high flexibility, and excellent integration.

Keywords: reconfigurable antenna; bi-stable substrate; radiation pattern; carbon nanotube film (CNTF); conductive silver paste (CSP)

Citation: Huang, Y.; Zheng, C.; Jiang, J.; Shao, H.; Chen, N. A Flexible Bi-Stable Composite Antenna with Reconfigurable Performance and Light-Responsive Behavior. *Polymers* **2023**, *15*, 1585. https://doi.org/10.3390/polym15061585

Academic Editors: Giorgio Luciano and Maurizio Vignolo

Received: 8 February 2023
Revised: 15 March 2023
Accepted: 17 March 2023
Published: 22 March 2023

Copyright: © 2023 by the authors. Licensee MDPI, Basel, Switzerland. This article is an open access article distributed under the terms and conditions of the Creative Commons Attribution (CC BY) license (https:// creativecommons.org/licenses/by/ 4.0/).

1. Introduction

Recently, the proposal and research of flexible, multifunctional antennas have greatly expanded the application of integrated antennas in flexible wearable devices such as sensors and actuators [1–3]. At the same time, the development of fifth-generation (5G) wireless necessitates the development of a flexible antenna with integrated structure and function. The concept of a reconfigurable antenna is a perfect demonstration of structural and functional integration [4,5]. As mentioned in the previous literature, the reconfigurable performance of an antenna is mainly reflected by the change in its radiation pattern [6]. At present, the reconfigurable performance of antennas has been realized by researchers through certain technical means. For example, Alfred et al. [7,8] prepared a reconfigurable antenna that can achieve reconfigurable changes in the antenna pattern through MEMS switching in its operating frequency. However, these antennas need an additional MEMS switch to control the antenna in addition to the antenna device itself, which greatly increases the weight of the antenna and limits its application. At the same time, these methods require continuous electrical stimulation to maintain the deformation. After removing the stimulation, the antenna structure immediately changes back to its initial state, which causes energy waste. Therefore, it is necessary to develop a composite antenna with reconfigurable structure and antenna function.

The bi-stable structure is used to solve the energy-waste problem, by achieving reconfigurable antenna performance. "Bi-stable" structure refers to two stable configurations that can be maintained in one of the stable states without the maintenance of external energy. A short stimulus can make it jump from one stable state to another stable state, and removing the stimulus leaves the steady state shape unchanged [9]. Researchers have prepared a flexible film with bi-stable characteristics, owing to its structural design as well as silicone oil dissolved in organic solvents [10–12]. However, this bi-stable structure has low stiffness, and the dissolution process requires a large number of organic solvents, which is not environmentally friendly and not suitable for mass production. At present, the external actuating forces for the response of intelligent structures are mainly smart-actuating [13–17] and mechanical-force-actuating [18,19]. Among them, light is a common, natural source of light and clean energy, which can be turned on and off at long distances [20,21]. This simple actuating method can play a great role in the shape transformation of bi-stable structures.

According to the existing reports, it is known that the bending angle of antennas has a significant impact on their working frequency and pattern [22]. Hu et al. [23] prepared the reconfigurable antenna by layering, which laid the foundation for the feasibility of the research in this paper. The results showed that the antenna has omnidirectional radiation when it is curled, with a gain of −1.52 dB, and directional radiation when it is unfolded, with a gain of 9.77 dB. However, these antennas require shape memory alloys and electricity as external actuating sources during the state-conversion process, and the antennas are too stiff to be widely used in smart wearable devices. In addition, researchers prepared a bi-layer encapsulated PCL-TPU shape memory composite structure by 4D printing [24]. The successful implementation of this technology will help people to expand the structural diversity of shape memory composites. Perhaps in the future, bi-stable structures with intelligent responses could be prepared directly by 4D printing [25]. Zheng Zhang et al. [26] prepared a bi-stable laminate using commercial silicone rubber-reinforced carbon fiber ply, and the state transition was made via gas actuation. However, the gas drive generally requires substantial air pressure to change its state. Therefore, to meet the needs of space detection and smart wearable devices, it is crucial and practical to design a light-driven, flexible, reconfigurable antenna with a simple fabrication method, deformable structure, low price, low energy consumption, and integrated structure and function. It has been reported that polydimethylsiloxane (PDMS) can be used as a good antenna substrate layer material due to its excellent optical properties and good hydrophobic and high dielectric constants [27–29]. Similarly, as reported in the References, polyimide (PI) fiber is widely used due to its high strength, low coefficient of thermal expansion (CTE), and high rigidity [30,31].

In this study, a bi-stable composite antenna is composed of a substrate layer and a radiation layer. First, the substrate layer, containing a functional layer and middle layer, is fabricated by antisymmetric laying and a high-temperature curing method. Next, the radiation layer is prepared with conductive material using screen-printing or laser-cutting technologies. Next, the antenna performance is studied by experimental methods. The reflection coefficient (S11) and the radiation pattern in state I and state II are tested. Lastly, the effects of the radiation layer material and substrate layer parameters on the antenna reflection coefficient are compared and analyzed.

2. Experimental Section
2.1. Materials

Carbon nanotube film (CNTF) with a thickness of 10–20 μm was purchased from Zhongke Times Nano Co., Chengdu, China. Conductive silver paste (CSP) (DJ002) was purchased from Maintenance Co., HongKong, China. PDMS material was commercially purchased (Sylgard 184 silicone elastomer) Dow Corning., Midland, MI, USA. with a viscosity of 5500 mPA·s. PI fibers with different linear densities were purchased from

Jiangsu Xiannuo New Material Technology Co., Ltd., China. PI electro-spun membranes with thicknesses of 80 µm were supplied by the Hunan Institute of Engineering.

2.2. Preparation of the Bi-Stable Substrate Layer

As shown in Figure 1a, flexible thin-film antennas have important applications in interpersonal communication, satellite communication, Bluetooth, etc. Figure 1b demonstrates a bi-stable composite antenna with light-responsive and reconfigurable performance, which provides new reference and theoretical data for the long-term application of reconfigurable flexible electronic devices. As shown in Figure 1b, a complete antenna is composed of a substrate layer and a radiation layer. In particular, the substrate layer has a direct influence on the reconfigurable performance of the antenna. Therefore, we first discuss the preparation method for the reconfigurable substrate with bi-stable characteristics.

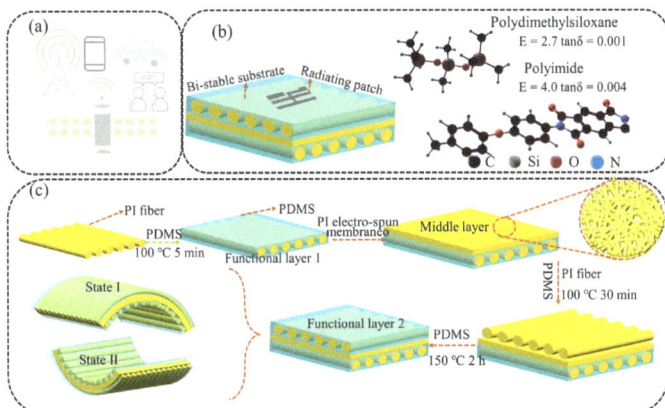

Figure 1. (a) Schematic of the potential applications of flexible film antennas. (b) Structural schematic of a micro-strip antenna. (c) Design and preparation of a reconfigurable substrate with bi-stable characteristics.

Figure 1c shows a schematic for designing and fabricating a bi-stable substrate layer for the antenna. It can be seen that the bi-stable substrate layer consists of a middle layer and a functional layer. The functional layer includes Functional layer 1 and Functional layer 2, which are arranged in an anti-symmetric method. The bi-stable substrate layer is pre-cured using the residual thermal stress caused by the large CTE difference between the PI fiber and PDMS during high-temperature curing and the arrangement of antisymmetric layers.

It is worth noting that the PI electro-spun membrane should be pasted at the PDMS curing temperature of 100 °C and a curing time of 5 min. At this time, the PDMS is pre-cured, but it is not completely cross-linked. The PI electro-spun membrane can be pasted onto Functional layer 1 by the viscosity of PDMS, to prevent slipping. The bi-stable substrate layer will maintain a stable structure when the sample is taken out at room temperature (state I). It can reach another stable structure (state II) through external stimulation. Significantly, in previous studies, we have proved that the oriented-PI-fiber-reinforced PDMS film has bi-stable characteristics and demonstrates a close relationship between fiber laying densities and fiber linear densities [32].

2.3. Preparation of the Radiation Layer

The antenna radiation layer was prepared by screen printing and laser-cutting, as shown in Figure 2. First, the CNTF radiation layer was cut from the commercially available CNTF film with laser cutting technology and then pasted on the bi-stable substrate during the curing process. The process used a cutting power of 3 W and 5 cutting times.

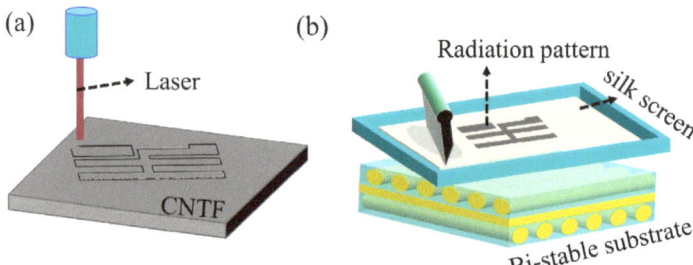

Figure 2. (a) Schematic diagram of laser cutting CNTF. (b) Schematic diagram of the screen-printing process.

Next, the CSP radiation layer was printed directly onto the cured bi-stable substrate using screen printing technology. At this time, 300 mesh screen cloth was used to penetrate the conductive silver paste.

2.4. Characterization

The morphological features of the radiation layer were investigated by a field-emission scanning electron microscope (FE-SEM 7500F, JEOL Technology and Trade Co., Ltd., Beijing, China). A UV light source with a wavelength of 395 nm was used to actuate the soft actuator, by means of a UV curing machine (SJUV3D-104, Shengju Machinery Automation Co., Ltd., Shanghai, China). A thermogravimetric analyzer (TGA8000, PerkinElmer Enterprise Management Co., Ltd., Shanghai, China) was used to test the high-temperature thermal stability of the radiation layer. Carbon nanotube films were cut by a portable micro-engraving machine (K6PRO, Shanghai Diaotu Industrial Co., Ltd., Shanghai, China).

The Tru EBox Workstation (01RC, LinkZill Corporation, Shanghai, China) was used to test the capacitance of the substrate layer. The reflection coefficient (S11) of all antennas was tested using a network analyzer (Keysight E5071C, Keysight Technology Co., Ltd., Shanghai, China) with open-short-load calibration. The radiation pattern made by the rotation angle position of the horn antenna and the testing antenna was tested using the network analyzer.

3. Results and Discussion

3.1. Effect of Radiation Layer Material on Antenna Performance

The development of fifth-generation (5G) wireless will meet the basic requirements of small size, light weight, flexibility, and low price. At the same time, 2.1 GHz and 3.5 GHz are receiving more and more attention as the mainstream frequency bands for 5G network deployment. The pattern and conductivity of an antenna's radiation layer play a decisive role in its electromagnetic performance. The radiation layer pattern and dimensions used in this paper are shown in Figure 3a. The bi-stable composite prepared in this section was cut to a size of 40 mm × 40 mm × 0.35 mm to obtain a reconfigurable substrate of the first state (state I) and the second state (state II).

Carbon nanotube film (CNTF) is a relatively mature conductive film material which, owing to the in-depth study and maturity of its production process, is becoming more and more popular [33,34]. In addition, conductive silver paste (CSP) is also very common in various electronic devices, due to its low price and mature formula [35]. To compare these, this paper investigates and analyzes the effect of using CNTF and CSP as radiation layer materials on antenna performance. Figure 3b shows physical pictures of the CSP-based and CNTF-based bi-stable composite antennas in states I and II. Figure 3c shows the physical diagram of a CNTF-based bi-stable antenna with a coaxial line feed. One of the enlarged images is of the connection between the antenna radiation layer and the coaxial line and the other is of the SMA interface connected to the test end.

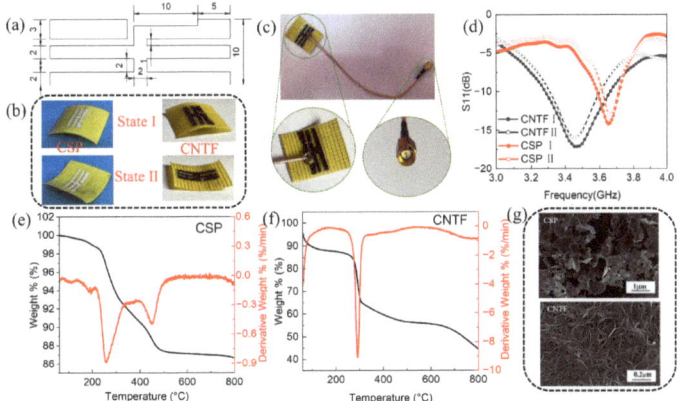

Figure 3. (**a**) Digital image of the geometry and size of as-fabricated reconfigurable antennas. (**b**) The physical images of the CSP-based and CNTF-based bi-stable antennas in state I and state II, respectively. (**c**) Physical diagram of the CNTF-based bi-stable antenna with a coaxial line feed. (**d**) Reflection coefficient (S11) of bi-stable antennas with different radiation layer materials in two states. (**e**) TG curves of conductive silver paste. (**f**) TG curves of carbon nanotube film. (**g**) SEM image of the radiation layer material.

Figure 3d shows the reflection coefficients of bi-stable antennas with different radiation layer materials in two states. First, it can be seen that the S11 curves of the antennas with the same radiation layer material coincide in state I and state II. This indicates that the stable-state changes do not have a large impact on the reflection coefficient and resonant frequency. However, it can be seen from Figure 3d that the change in radiation layer material does affect the resonant frequency and reflection coefficient of the antennas. It can be seen that the resonant frequency and reflection coefficient of the bi-stable antenna with CNTF as the radiation layer material are 3.47 GHz and −17 dB, respectively. In comparison, the resonant frequency and reflection coefficient of the bi-stable antenna with CSP as the radiation layer material are 3.65 GHz and −15 dB, respectively. This is mainly because the CNTF radiation layer material has better conductivity. The resistivity of the CSP radiation layer is 5×10^{-5} $\Omega \cdot$cm, while the resistivity of the CNTF radiation layer is 5×10^{-4} $\Omega \cdot$cm.

At the same time, impedance matching efficiency and impedance bandwidths are also important indexes for evaluating antenna performance, and are thus worth discussing and analyzing. When the impedance matching efficiency is closer to 100%, the antenna is in an ideal state. Similarly, when the impedance matching efficiency is closer to 0%, the antenna impedance is completely mismatched.

As shown in Figure 3d, all antennas with different radiation layer materials at two states have a reflection coefficient of <−10 dB, indicating the impedance matching efficiency of all antennas is greater than 90%, which meets the application requirements of antennas. However, the impedance matching efficiency of the antenna with the CNTF radiation layer is closer to 100%, which is closer to the ideal state.

The measured characteristics indicate that the resonant frequency of the antenna with the CNTF radiation layer is f = 3.47 GHz, and the impedance bandwidths (S11 ≤ −10 dB) are 10.37% (3.29–3.65 GHz). The resonant frequency of the antenna with the CSP radiation layer is f = 3.65 GHz, and the impedance bandwidths (S11 ≤ −10 dB) are 3.29% (3.59–3.71 GHz). This indicates that the antenna with the CNTF radiation layer has wider impedance bandwidths than the one with CSP. In summary, the bi-stable antenna with the CNTF radiation layer has a smaller reflection coefficient, higher impedance matching efficiency, and higher impedance bandwidths. Meanwhile, the resonant frequency of the bi-stable antenna with the CNTF radiation layer is closer to 3.5 GHz, which is more in-line with the use range of a 5G network than that of the antenna with CSP. Since the radiation layer is a light-driven heat transfer layer, it will directly contact the heat generated by the light source during the subsequent light response stimulation process. The study of thermal stability is therefore very necessary. Figure 3e,f shows the TG curves of CSP and CNTF. It can be seen that the mass loss rates of CSP and CNTF are 1.71% and 12.98%, respectively, at 50–230 °C, which indicates that they have good thermal stability. In addition, Figure 3g shows the SEM morphology of the CSP and CNTF layers, from which the sheet structure of the CSP and the tubular structure of the CNTF can be observed clearly. This shows that the radiation materials themselves have no defects and will not produce errors in the experimental results, nor affect the testing performance.

3.2. Effect of Substrate Layer Parameters on Antenna Performance

As mentioned earlier, a complete antenna is composed of the substrate layer and radiation layer, so the impact of changes to the substrate layer parameters on the antenna performance is also well worth studying and exploring. To make a comparative test, this paper prepared a bi-stable substrate with different fiber laying densities and fiber linear densities. The radiation material used for all of the contrast antennas in this section was CSP. Fiber laying densities refer to the number of fibers per unit length of the functional layer. Linear density is a textile term that means the fineness of the fiber. Its unit is denier/D, which refers to the mass (in grams) of fibers with lengths of 9000 m. Figure 4a,b shows the reflection coefficient of bi-stable antennas with different fiber laying densities and fiber linear densities in state I. It can be seen that the resonance frequency of the antenna increases gradually with the increase in fiber laying density and fiber linear density. Conversely, the impedance bandwidth decreases gradually with the increase in fiber laying density and fiber linear density. We speculate that this is caused by the change in the dielectric constant of the substrate layer, because the substrate is generally required to be non-conductive in antenna design, and the dielectric constant is often used to characterize the degree of non-conductivity of the material. To verify this conjecture, we looked at the dielectric constant of the substrate layer with different fiber laying densities and fiber linear densities by inferring it from the capacitance values.

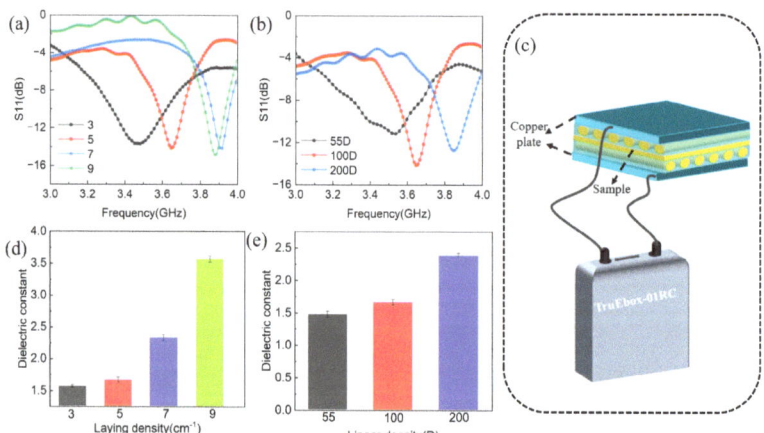

Figure 4. (**a**) Reflection coefficients (S11) of reconfigurable antennas with different fiber laying densities in state I. (**b**) Reflection coefficients (S11) of reconfigurable antennas with different fiber linear densities in state I. (**c**) Schematic diagram of the capacitance testing method. (**d**) Dielectric constants of substrate layers with different laying densities. (**e**) Dielectric constants of substrate layers with different linear densities.

First, a schematic diagram of the substrate layer capacitance testing method is shown in Figure 4c. Second, because the substrate layer is not conductive, the following formula can be used to derive the dielectric constant of the substrate layer.

$$\varepsilon_r = \frac{C \times d}{\varepsilon_0 \times S} \quad (1)$$

where ε_0 is the vacuum dielectric constant, d is the thickness of the substrate layer, S is the contact area, ε_r is the substrate layer dielectric constant, and C is the capacitance. The dielectric constants of the substrate layers with different fiber laying densities and fiber linear densities are shown in Figure 4d,e. It can be seen from here that the dielectric constant increased with the increase in fiber laying density and fiber linear density. Combined with Figure 4a,b, it can be found that the increase in the dielectric constant of the substrate layer drives the resonant frequency of the antenna to be larger. This phenomenon can be explained and verified by the following formula:

$$f_0 = \frac{c}{\lambda_0}\sqrt{\varepsilon_r} \quad (2)$$

where c is the speed of light, λ_0 is the wavelength of free space, and ε_r is the substrate layer dielectric constant. Therefore, the resonant frequency of the bi-stable antenna can be adjusted by the dielectric properties of the substrate layer in theoretical practice. However, it can be seen that the resonant frequency of the laying densities of 9 cm^{-1} and 7 cm^{-1} does not change much, which may be because the influence on the resonant frequency of the antenna is not obvious when the dielectric constant is too large. In addition, the change in laying density has little effect on the reflection coefficient of the antenna, but the fiber linear density seems to have a great influence on the reflection coefficient of the antenna. This is because the increase in fiber linear density increases the thickness of the bi-stable antenna's substrate layer.

3.3. Reconfigurable Performance Analysis of Bi-Stable Antennas

The normalized radiation patterns and gain of an antenna are also important indexes to measure its electromagnetic performance, especially its reconfigurable performance. Therefore, it is very important to study the changes to the bi-stable antenna's radiation pattern in different states. Figure 5a shows the testing scheme of the normalized radiation pattern. It can be seen that the whole testing system is composed of a network analyzer, standard horn antenna, rotary table, experimental antenna, and microwave anechoic chamber. The experimental antenna is placed on the rotary table to measure the gain values at different angles. The actual gain of the measured antenna is calculated from the standard gain antenna and its calibration gain, as shown below:

$$G = G_s + (G_x - G_w) \quad (3)$$

where G is the actual gain of the measured antenna, G_s is the gain calibration value (7.5 dB) of the standard gain antenna at 3.5 GHz, G_x is the receiving gain of the experimental antenna at 3.5 GHz, and G_w is the receiving gain (-27.39 dB) of the calibrated standard gain antenna at 3.5 GHz. According to the relationship between the experimental antenna and the electromagnetic field direction of the test environment, the expression methods of E-plane and H-plane come into being. Among them, the E-plane is also called the electric plane, which refers to the direction plane parallel to the direction of the electric field. The H-plane, also known as the magnetic plane, refers to the direction plane parallel to the direction of the magnetic field. Figure 5b shows the E-plane and H-plane radiation patterns of state I at 3.5 GHz. It can be seen from Figure 5b that the radiation pattern of state I in the electric field plane (E-plane) is quasi-octagonal, and the radiation pattern in the magnetic field plane (H-plane) is closer to a circle, which obtains good omnidirectional radiation. The experimental results show that the maximum radiation angles of the E-plane and H-plane are at 135° and 90°, respectively. The measured gains of the antenna are 1.1 dB and 1.21 dB, respectively.

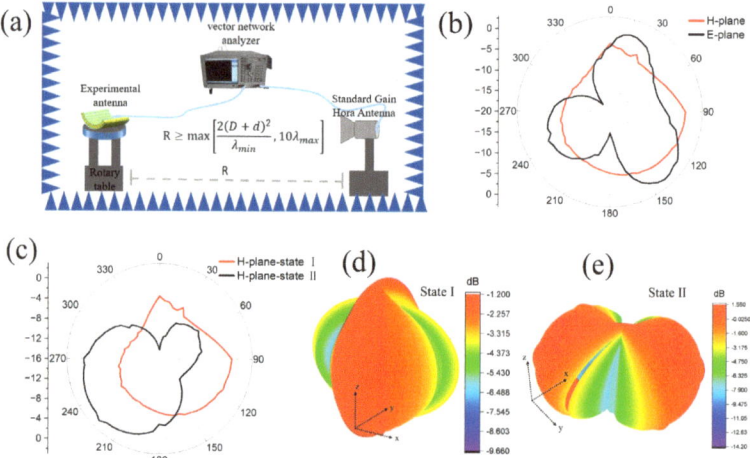

Figure 5. (**a**) Schematic diagram of the normalized radiation patterns. (**b**) Radiation patterns of the E- and H-planes of the bi-stable antenna at state I. (**c**) Radiation patterns of the H-plane of the bi-stable antenna at states I and II, respectively. (**d**) 3D radiation pattern of state I. (**e**) 3D radiation pattern of state II at 3.5 GHz.

Figure 5c shows the H-plane radiation pattern of the bi-stable antenna in states I and II at 3.5 GHz. It can be found that the radiation pattern of the bi-stable antenna also changes accordingly with the change between state I and II. The experimental results show that the maximum radiation angles of states I and II are at 90° and 225°, respectively. The measured gains for states I and II are 1.21 dB and 1.53 dB, respectively. Gain non-circularity is often used to describe the omnidirectional radiation performance of the antenna. The following formula is a method for calculating gain non-circularity.

$$G_R = \max(|G_{max} - G_{avg}|, |G_{min} - G_{avg}|) \quad (4)$$

where G_R represents the antenna gain non-circularity and G_{max}, G_{min}, and G_{avg} represent the maximum gain, minimum gain, and average gain of the antenna in the H-plane.

The gain non-circularities of the antenna at states I and II are 4.48 dB and 8.35 dB, respectively. This shows that the bi-stable antenna has good omnidirectional radiation performance in state I. This is because the change to antenna structure causes a change in radiation direction and intensity of the radiation layer. We speculate that this is because the antenna radiation layer in state I is flat, while the antenna radiation layer in state II is bent, and the corresponding current radiation intensity is weakened in a bending direction. As a result, the main current transmission direction of the antenna is deflected; that is, the pattern of the same antenna is different under different structures, so it can be considered reconfigurable. At the same time, it can be seen from the 3D radiation pattern of Figure 5d,e that the gain of state I and II also changes due to the deformation of the antenna structure, but both have good gain performance.

To improve the transmission capacity of the antenna, reduce the cost of building a network in high-traffic areas, and meet the applications of various occasions, several single antennas with the same resonant frequency can together be used as the basic unit and, with reasonable planning, be combined in space to form an array antenna. The main function of the antenna array is to strengthen and improve the directivity of the radiation field and to strengthen the intensity of the radiation field. The bi-stable composite material prepared in this section was cut to a size of 150 mm × 80 mm × 0.35 mm to obtain a reconfigurable substrate of the first state (curled) and the second state (unfolded). Figure 6a shows the specific size and shape of the antenna radiation layer array.

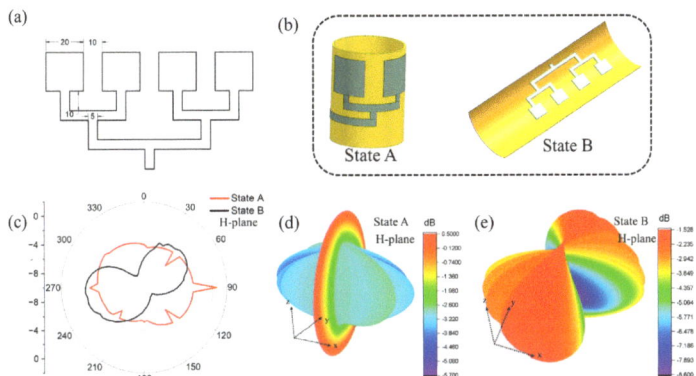

Figure 6. (**a**) Digital image of the geometry and size of the antenna array's radiation layer. (**b**) Schematic diagram of the pattern of the bi-stable antenna array. (**c**) Radiation patterns of the H-plane of the bi-stable antenna array at states A and B. (**d**) 3D radiation pattern of state A. (**e**) 3D radiation pattern of state B.

As shown in Figure 6a, the distance between each single antenna element is 10 mm. Figure 6b shows the schematic diagram of the pattern of the reconfigurable antenna array. As shown in Figure 6b, the size of the radiation layer is just one cycle around the first stable-state substrate layer, which is defined as state A; the second stable-state is defined as state B. Figure 6c shows the radiation patterns of the H-plane of the reconfigurable antenna array at states A and B

It can be seen that in state A, the gain value of the reconfigurable array antenna in each direction does not change much, while in state B, the reconfigurable array antenna shows strong directional radiation characteristics in the 90° direction. The gain non-circularities of the array antenna at states A and B are 3.47 dB and 4.20 dB, respectively. This shows that the array antenna has good omnidirectional radiation performance in state A. This further proves the possibility for reasonable application of bi-stable substrates in the reconfigurable performance of array antennas.

3.4. Analysis of Light Response Behavior

Recently, with the introduction and implementation of energy-saving and emission-reduction measures, the application of light response actuation methods has become more and more popular. The experimental device established to analyze the light response performance of the bi-stable antenna is shown in Figure 7a. Two cameras were mounted on the xz and yz planes, respectively, to observe the curvature changes of the xz and yz planes. The reconfigurable composite antenna used to demonstrate light response behavior was 40 mm × 40 mm × 0.35 mm. The radiation layer material of the antenna used to test the light response behavior was CNTF, and the fiber linear density and fiber laying density were 100D and 5 cm^{-1}, respectively. In the actuation process of the bi-stable composite antenna, the CNTF radiation layer was used as a photothermal conversion layer for the bi-stable composite antenna to convert the light response between state I and state II. The oriented-PI-fiber-reinforced PDMS layer acted as an actuating layer with photothermal sensitivity, and the electro-spun polyimide film acted as an inert layer. The wavelength of the light source for the light response experiment was 395 nm.

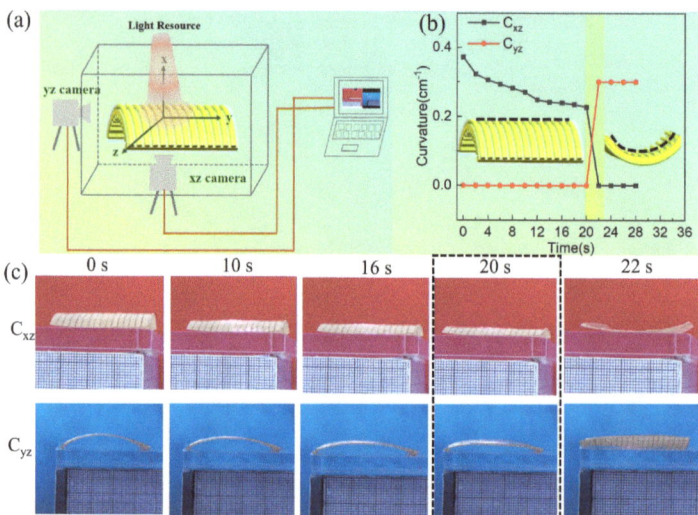

Figure 7. (a) Schematic of the experimental device established for testing the light response of the bi-stable composite antenna. (b) The relationship between the bending of the bi-stable composite antenna with the light-actuated time. (c) Photograph of the bi-stable composite antenna changing with light-actuated time.

Figure 7b shows the relationship between the bending of the bi-stable composite antenna and the light-actuated time. It can be seen that the curvature change curves can be divided into state I stage, state II stage, and transition stage according to the state shape.

In addition, a photograph of the bi-stable composite antenna changing with light-actuated time is shown in Figure 7c. These photos show that the reconfigurable antenna exhibits significant snapping-through movements under the light. The bi-stable antenna remains in state II when the light source is removed at 22 s, and the snapping-through occurs at 20 s, indicating that the bi-stable antenna has a fast response speed and good stable-structure retention.

4. Conclusions

The reconfigurable antenna represents a new concept for designing antenna structures by using structurally effective materials. In this paper, a novel bi-stable composite antenna with reconfigurable performance and light-responsive behavior is realized. First, it is found that the conductivity of the antenna's radiation layer material and the dielectric properties of the substrate layer can significantly change the antenna's electromagnetic properties. However, the reflection coefficient curves of state I and II are coincident, indicating that the structural change has little effect on the reflection coefficient curve for antennas with the same radiation layer material. Second, the state change has a great influence on the radiation patterns; the experimental results show that the maximum radiation angles of states I and II are at 90° and 225°, respectively. The experimental gains for states I and II are 1.21 and 1.53 dB, respectively. The gain non-circularities of the antenna at states I and II are 4.48 dB and 8.35 dB, respectively. This shows that the bi-stable antenna has good omnidirectional radiation performance in state I. Finally, the bi-stable antenna is converted from state I to state II when the 395 nm ultraviolet light irradiation time is 20 s, and then the shape of state II remains unchanged. This indicates that the bi-stable antenna has fast response speed and good stable-structure retention. In summary, the work also shows that the flexible bi-stable composite structure has the characteristics of light weight, low profile, large deformation, low energy input, and multiple electromagnetic functions. It therefore has broad application prospects in deformable and multifunctional structures.

Author Contributions: Conceptualization, H.S.; software, Y.H.; investigation, C.Z. and N.C.; data curation, Y.H.; writing—original draft, Y.H.; supervision, J.J. and N.C.; funding acquisition, J.J. and H.S. All authors have read and agreed to the published version of the manuscript.

Funding: This work was sponsored by the Shanghai Natural Science Foundation of the Shanghai Municipal Science and Technology Commission (Grant No. 20ZR1401600, 20ZR1400600) and the Fundamental Research Funds for the Central Universities (Grant No. 2232020D-09, 2232020G-06).

Institutional Review Board Statement: Not applicable.

Informed Consent Statement: Not applicable.

Data Availability Statement: Not applicable.

Conflicts of Interest: The authors declare no conflict of interest.

References

1. Zhang, K.; Zhao, D.; Chen, W.; Zheng, L.; Yao, L.; Qiu, Y.; Xu, F. Three-dimensional woven structural glass fiber/polytetrafluoroethylene (PTFE) composite antenna with superb integrity and electromagnetic performance. *Compos. Struct.* **2022**, *281*, 115096–115103. [CrossRef]
2. Zhao, Q.; Liu, J.; Yang, H.; Liu, H.; Zeng, G.; Huang, B. High Birefringence D-Shaped Germanium-Doped Photonic Crystal Fiber Sensor. *Micromachines* **2022**, *13*, 826–836. [CrossRef] [PubMed]
3. Lu, C.; Zhou, H.; Li, Y.; Yang, A.; Xu, C.; Ou, Z.; Wang, J.; Wang, X.; Tian, F. Split-core magnetoelectric current sensor and wireless current measurement application. *Measurement* **2022**, *188*, 110527–110535. [CrossRef]
4. Joseph, S.D.; Manoj, S.; Waghmare, C.; Nandakumar, K.; Kothari, A. UWB Sensing Antenna, Reconfigurable Transceiver and Reconfigurable Antenna Based Cognitive Radio Test Bed. *Wirel. Pers. Commun.* **2017**, *96*, 3435–3462. [CrossRef]
5. Anagnostou, D.E.; Chryssomallis, M.T.; Goudos, S. Reconfigurable Antennas. *Electronics* **2021**, *10*, 897–900. [CrossRef]
6. Isa, S.R.; Jusoh, M.; Sabapathy, T.; Nebhen, J.; Kamarudin, M.R.; Osman, M.N.; Abbasi, Q.H.; Rahim, H.A.; Yasin, M.N.M.; Soh, P.J. Reconfigurable Pattern Patch Antenna for Mid-Band 5G: A Review. *CMC Comput. Mater. Contin.* **2022**, *70*, 2699–2725.
7. Besoli, A.G.; De Flaviis, F. A Multifunctional Reconfigurable Pixeled Antenna Using MEMS Technology on Printed Circuit Board. *IEEE Trans. Antennas Propag.* **2011**, *59*, 4413–4424. [CrossRef]
8. Won Jung, C.; Lee, M.J.; Li, G.P.; De Flaviis, F. Reconfigurable scan-beam single-arm spiral antenna integrated with RF-MEMS switches. *IEEE Trans. Antennas Propag.* **2006**, *54*, 455–463. [CrossRef]
9. Arrieta, A.F.; Gemmeren, V.V.; Anderson, A.J.; Weaver, P.M. Dynamics and control of twisting bi-stable structures. *Smart Mater. Struct.* **2018**, *27*, 025006–025021. [CrossRef]
10. Chen, Z.; Kong, S.; He, Y.; Yi, S.; Liu, G.; Mao, Z.; Huo, M.; Chan, C.H.; Lu, J. Soft, Bistable Actuators for Reconfigurable 3D Electronics. *ACS Appl. Mater. Interfaces* **2021**, *13*, 41968–41977. [CrossRef]
11. Pezzulla, M.; Shillig, S.A.; Nardinocchi, P.; Holmes, D.P. Morphing of geometric composites via residual swelling. *Soft Matter* **2015**, *11*, 5812–5820. [CrossRef] [PubMed]
12. Egunov, A.I.; Korvink, J.G.; Luchnikov, V.A. Polydimethylsiloxane bilayer films with an embedded spontaneous curvature. *Soft Matter* **2016**, *12*, 45–52. [CrossRef] [PubMed]
13. Li, H.; Dai, F.; Du, S. Numerical and experimental study on morphing bi-stable composite laminates actuated by a heating method. *Compos. Sci. Technol.* **2012**, *72*, 1767–1773. [CrossRef]
14. Epstein, E.; Yoon, J.; Madhukar, A.; Hsia, K.J.; Braun, P.V. Colloidal Particles that Rapidly Change Shape via Elastic Instabilities. *Small* **2015**, *11*, 6051–6057. [CrossRef]
15. Ishii, H.; Ting, K.-L. SMA actuated compliant bistable mechanisms. *Mechatronics* **2004**, *14*, 421–437. [CrossRef]
16. Novelino, L.S.; Ze, Q.; Wu, S.; Paulino, G.H.; Zhao, R. Untethered control of functional origami microrobots with distributed actuation. *Proc. Natl. Acad. Sci. USA* **2020**, *117*, 24096–24101. [CrossRef]
17. Zhang, Z.; Li, Y.; Yu, X.; Li, X.; Wu, H.; Wu, H.; Jiang, S.; Chai, G. Bistable morphing composite structures: A review. *Thin Walled Struct.* **2019**, *142*, 74–97. [CrossRef]
18. Wang, B.; Fancey, K.S. A bistable morphing composite using viscoelastically generated prestress. *Mater. Lett.* **2015**, *158*, 108–110. [CrossRef]
19. Lin, J.; Guo, Q.; Dou, S.; Hua, N.; Zheng, C.; Pan, Y.; Huang, Y.; Chen, Z.; Chen, W. Bistable structures with controllable wrinkled surface. *Extrem. Mech. Lett.* **2020**, *36*, 100653. [CrossRef]
20. Li, Q. Photochromism into nanosystems: Towards lighting up the future nanoworld. *Chem. Soc. Rev.* **2018**, *47*, 1044–1097.
21. Yoshida, M. Photoinduced Directional Motions of Microparticles at Air-Liquid-Crystal Interfaces of Azobenzene-Doped Liquid-Crystal Films with Homeotropic or Homogeneous Alignment Structures. *Appl. Phys. Express* **2012**, *5*, 1701–1704.
22. Shao, L.; Tang, X.; Yang, Y.; Wei, D.; Lin, Y.; He, G.; Wei, D. Flexible force sensitive frequency reconfigurable antenna base on stretchable conductive fabric. *J. Phys. D Appl. Phys.* **2022**, *55*, 195301–195310. [CrossRef]
23. Hu, J.; Pan, D.; Dai, F. Microstrip Patch Array Antenna with Reconfigurable Omnidirectional and Directional Patterns Using Bistable Composite Laminates. *IEEE Antennas Wirel. Propag. Lett.* **2017**, *16*, 2485–2488. [CrossRef]

24. Rahmatabadi, D.; Aberoumand, M.; Soltanmohammadi, K.; Soleyman, E.; Ghasemi, I.; Baniassadi, M.; Abrinia, K.; Bodaghi, M.; Baghani, M. 4D Printing-Encapsulated Polycaprolactone–Thermoplastic Polyurethane with High Shape Memory Performances. *Adv. Eng. Mater.* **2022**, *25*, 2201309–2201318. [CrossRef]
25. Aberoumand, M.; Soltanmohammadi, K.; Soleyman, E.; Rahmatabadi, D.; Ghasemi, I.; Baniassadi, M.; Abrinia, K.; Baghani, M. A comprehensive experimental investigation on 4D printing of PET-G under bending. *J. Mater. Res. Technol.* **2022**, *18*, 2552–2569. [CrossRef]
26. Zhang, Z.; Ni, X.; Wu, H.; Sun, M.; Bao, G.; Wu, H.; Jiang, S. Pneumatically Actuated Soft Gripper with Bistable Structures. *Soft Robot* **2021**, *9*, 57–71. [CrossRef] [PubMed]
27. Wei, L.; Wang, J.W.; Gao, X.H.; Wang, H.Q.; Ren, H. Enhanced Dielectric Properties of Poly(dimethyl siloxane) Bimodal Network Percolative Composite with MXene. *ACS Appl. Mater. Interfaces* **2020**, *12*, 16805–16814. [CrossRef]
28. Zhang, G.; Sun, Y.; Qian, B.; Gao, H.; Zuo, D. Experimental study on mechanical performance of polydimethylsiloxane (PDMS) at various temperatures. *Polym. Test.* **2020**, *90*, 106670–106680. [CrossRef]
29. Wang, F.; Lei, S.; Ou, J.; Li, W. Effect of PDMS on the waterproofing performance and corrosion resistance of cement mortar. *Appl. Surf. Sci.* **2020**, *507*, 145016–145026. [CrossRef]
30. Chen, L.; Xu, Z.; Wang, F.; Duan, G.; Xu, W.; Zhang, G.; Yang, H.; Liu, J.; Jiang, S. A flame-retardant and transparent wood/polyimide composite with excellent mechanical strength. *Compos. Commun.* **2020**, *20*, 100355–100360. [CrossRef]
31. Li, X.; Zhang, B.; Wu, Z.; Liu, Y.; Hu, J.; Zhang, C.; Cao, G.; Zhang, K.; Sun, J.; Liu, X.; et al. Highly flexible, large scaled and electrical insulating polyimide composite paper with nanoscale polyimide fibers. *Compos. Commun.* **2023**, *38*, 101463–101469. [CrossRef]
32. Huang, Y.; Jiang, J.; Li, J.; Su, C.; Yu, Q.; Wang, Z.; Chen, N.; Shao, H. Light-driven Bi-stable actuator with oriented polyimide fiber reinforced structure. *Compos. Commun.* **2022**, *31*, 101128–101134. [CrossRef]
33. Xu, J.; Gong, X.; Yong, Z.; Ramakrishna, S. Construction of various nanostructures on carbon nanotube films. *Mater. Today Chem.* **2020**, *16*, 100253–100259. [CrossRef]
34. Wang, S.; Zhao, J.; Wang, Q.; Zhang, D. Preparation and Recycling of High-Performance Carbon Nanotube Films. *ACS Sustain. Chem. Eng.* **2022**, *10*, 3851–3861. [CrossRef]
35. Donley, G.J.; Hyde, W.W.; Rogers, S.A.; Nettesheim, F. Yielding and recovery of conductive pastes for screen printing. *Rheol. Acta* **2019**, *58*, 361–382. [CrossRef]

Disclaimer/Publisher's Note: The statements, opinions and data contained in all publications are solely those of the individual author(s) and contributor(s) and not of MDPI and/or the editor(s). MDPI and/or the editor(s) disclaim responsibility for any injury to people or property resulting from any ideas, methods, instructions or products referred to in the content.

Article

Structure and Thermomechanical Properties of Polyvinylidene Fluoride Film with Transparent Indium Tin Oxide Electrodes

Vitaliy Solodilov [1], Valentin Kochervinskii [1,*], Alexey Osipkov [1,*], Mstislav Makeev [1], Aleksandr Maltsev [2], Gleb Yurkov [3,*], Boris Lokshin [4], Sergey Bedin [5], Maria Shapetina [5], Ilya Tretyakov [3] and Tuyara Petrova [3]

1 Laboratory of Ferroelectric Polymers, Bauman Moscow State Technical University, 105005 Moscow, Russia
2 Department of Electronics of Organic Materials and Nanostructures, N.M. Emanuel Institute of Biochemical Physics (IBCP), Russian Academy of Science (RAS), 119334 Moscow, Russia
3 N.N. Semenov Federal Research Center of Chemical Physics, Russian Academy of Sciences, 119991 Moscow, Russia
4 Department of Physical and Physico-Chemical Methods for Studying the Structure of Substances, A.N. Nesmeyanov Institute of Organoelement Compounds, Russian Academy of Sciences, 119334 Moscow, Russia
5 Laboratory of Physics of Advanced Materials and Nanostructures, Moscow Pedagogical State University, 119991 Moscow, Russia
* Correspondence: kochval@mail.ru (V.K.); osipkov@bmstu.ru (A.O.); ygy76@mail.ru (G.Y.)

Citation: Solodilov, V.; Kochervinskii, V.; Osipkov, A.; Makeev, M.; Maltsev, A.; Yurkov, G.; Lokshin, B.; Bedin, S.; Shapetina, M.; Tretyakov, I.; et al. Structure and Thermomechanical Properties of Polyvinylidene Fluoride Film with Transparent Indium Tin Oxide Electrodes. Polymers 2023, 15, 1483. https://doi.org/10.3390/polym15061483

Academic Editors: Giorgio Luciano and Maurizio Vignolo

Received: 12 February 2023
Revised: 6 March 2023
Accepted: 13 March 2023
Published: 16 March 2023

Copyright: © 2023 by the authors. Licensee MDPI, Basel, Switzerland. This article is an open access article distributed under the terms and conditions of the Creative Commons Attribution (CC BY) license (https:// creativecommons.org/licenses/by/ 4.0/).

Abstract: This paper is devoted to the study of the structure and thermomechanical properties of PVDF-based ferroelectric polymer film. Transparent electrically conductive ITO coatings are applied to both sides of such a film. In this case, such material acquires additional functional properties due to piezoelectric and pyroelectric effects, forming, in fact, a full-fledged flexible transparent device, which, for example, will emit a sound when an acoustic signal is applied, and under various external influences can generate an electrical signal. The use of such structures is associated with the influence of various external influences on them: thermomechanical loads associated with mechanical deformations and temperature effects during operation, or when applying conductive layers to the film. The article presents structure investigation and its change during high-temperature annealing using IR spectroscopy and comparative results of testing a PVDF film before and after deposition of ITO layers for uniaxial stretching, its dynamic mechanical analysis, DSC, as well as measurements of the transparency and piezoelectric properties of such structure. It is shown that the temperature-time mode of deposition of ITO layers has little effect on the thermal and mechanical properties of PVDF films, taking into account their work in the elastic region, slightly reducing the piezoelectric properties. At the same time, the possibility of chemical interactions at the polymer–ITO interface is shown.

Keywords: PVDF; ITO; ferroelectric polymers; thermomechanical testing; structure; transparency; piezoelectric properties

1. Introduction

Over the past few years, flexible electronics have been increasingly incorporated into our daily lives. The first smartphone with a flexible display was officially presented at the CES 2018 conference, and in 3 years almost all major players in this market have similar phone models in their lineup. At the same conference, LG also announced the first TV screen which can be rolled up. Today's rapid development of flexible electronic devices is primarily due to significant progress in organic crystalline materials (OCMs) over the past 10 years. In contrast to the traditional complementary metal-oxide-semiconductor (CMOS) structures mounted on rigid substrates in modern electronics, OCMs, which offer flexibility, transparency, and cost and weight reduction, have shown great potential in optoelectronics applications including organic field-effect transistors, LEDs, organic photovoltaics, and various types of sensors [1,2].

One such class of promising OCMs for next-generation flexible electronics is ferroelectric polymers (a relatively new class of electroactive materials based on copolymers of PVDF and some nylons) [3]. At present, these materials are being actively studied and find application in various fields of science and technology: as piezo sensors [4,5], biocompatible prosthetic materials, nanogenerators/sensors [6–8], and adaptive optical system elements [8–11]. In contrast to traditional piezoelectrics based on oxide ceramics (PZT, quartz, lithium niobate, etc.), ferroelectric polymers have mechanical and technological flexibility and high impact toughness allowing their effective use as large area piezoelectric generators, high breakdown fields, transparency [12]. Application of transparent electrodes, such as indium tin oxide (ITO) [13] to these films makes it possible to create flexible transparent sound sources [14] or sub-screen sensors, such as fingerprint scanners [15], acousto-optical transducers (modulators) and other devices. The creation of structures based on flexible transparent ferroelectric polymers, which also have high pyroactivity [16] and exhibit high electrocaloric effect, will significantly increase the potential of this class of polymeric materials in flexible electronics devices; these structures can be obtained by adding nanoscale particles to the melted or dissolved PVDF matrix [17]. In contrast, layered structures also can be useful for various applications. For example, an aluminum electrode deposited on poly (ethylene-2,6 naphtalate) film was used as a substrate for three-layer hybrid material consisting of vacuum-deposited barium-strontium titanate coated on both sides with a spray-deposited PVDF-TrFE copolymer film [18]; this material has a significant piezoelectric response with low degradation after 340,000 mechanical bending cycles and can potentially be used in wearable energy collection devices. In opposition to layered structures, powder-filled PVDF films usually show significant light absorption. For example, filled with 15% $BaTiO_3$ PVDF-PMMA composite, film with 0.012 cm thickness has an optical density of 0.7–1.0 in the visible light wavelength band [19].

Optical characteristics of PVDF-based composites also are being thoroughly investigated at the present time. Rare earth ions doped fibrous PVDF material, for example [20] can be used for converting near-infrared light to visible light. Owing to its high reflectance in the wide part of the IR optical band, PVDF copolymer fibers may be used as radiative cooling material [21].

This research is devoted to the study of some characteristics of hybrid materials obtained by depositing transparent ITO electrodes on a polarized PVDF film, with which, for example, acoustic waves can be generated when an alternating electric field is applied to it. The use of such a structure in flexible electronic devices is subject to a number of external factors. First of all, these are thermomechanical stresses associated with mechanical deformations and temperature effects during operation or during the deposition of conductive layers to the film. The paper presents a study of the structure by infrared spectroscopy and comparative results of tests of PVDF film before and after the deposition of ITO layers in uniaxial tension, its dynamic mechanical analysis, DSC, as well as measurements of transparency and piezoelectric properties of such a structure.

2. Materials and Methods

The objects of the study are samples of 50 μm-thick polarized PVDF-P00050 film produced by PolyK (State College, PA, USA) with a declared piezoelectric constant value of d_{33} over 30 pC/N. Some samples were coated on both sides of the film with 98 nm thick optically transparent electrically conductive ITO coating with a specific surface resistivity of 190 Ohm/□ (PVDF + ITO samples). The ITO was deposited by the reactive magnetron method from 2 metal (Sn and In) targets in an oxygen atmosphere on the QUADRA series quadrupole magnetron sputtering unit (NPF "Elan-Practic" Ltd., Russia). The magnetrons of the quadrupole system are evenly spaced around the carousel device, which ensures high plasma homogeneity during deposition and practically excludes the zones with low ionization degrees and low atom flux density of the deposited material [22]. In the deposition process, the polymer film is slowly heated for 10 min from 60 to 100 °C, stays at this temperature for 7 min in the process during the ITO layer deposition, and slowly cools down.

IR-spectra were recorded on a VERTEX 70v Fourier-transform IR-spectrometer using the attenuated total reflection (ATR) method using a Pike Glady ATR adapter with a diamond crystal in the 4000–400 cm^{-1} range and spectral resolution of 4 cm^{-1}. Measured ATR spectra were corrected using OPUS 7 software to allow for the wavelength dependence of radiation penetration depth into the sample. The absorption spectra of the polymer films were measured on a TENSOR 37 Fourier-transform spectrometer with a spectral resolution of 2 cm^{-1}. The structure of PVDF films was investigated by infrared spectroscopy in two stages. In the first stage, the role of high-temperature annealing (200 °C) on volumetric characteristics of the pure PVDF film was tested. In the second stage, PVDF + ITO samples were investigated before and after high-temperature annealing.

Thermal and physical-mechanical studies of PVDF film were performed by analogy with the studies of polylactic acid films, which we conducted earlier in [23].

Thermal properties of the film materials were determined using a differential scanning calorimeter NETZSCH DSC 204F1 Phoenix (Selb, Germany) in an inert atmosphere at a flow rate of argon 100 mL/min. Film samples (raw and coated) were heated in the temperature range from 20 °C to 200 °C at a rate of 10 °C/min. After the first heating cycle, the samples were kept at 200 °C for 5 min and cooled to room temperature, and then reheated to 200 °C at a rate of 10 °C/min to record the second DSC curve. These curves were used to determine the melting heat effect (ΔH_m) and then to calculate the percentage of crystallinity (χ_c) using the following expression:

$$\chi_c(\%) = \frac{\Delta H_m}{\Delta H_m^*} \times 100\% \tag{1}$$

where ΔH_m^* = 105 (thermal melting effect for 100% crystalline PVDF [24])

The film samples were tested by the DMA method under tension on a NETZSCH DMA 242 E Artemis dynamic mechanical analyzer (Selb, Germany). The samples were coated and uncoated strips with a width of 5 mm and a length of the working part of 10 mm. Samples were cut out along the material orientation axis (direction 0) and transversely (direction 90). The cut specimens were tested under isothermal conditions at 30 °C The frequency of external load application was varied from 0.25 to 100 Hz.

Tensile mechanical characteristics of the films (original and coated) were determined on a Zwick Z100 testing machine (Ulm, Germany) at room temperature and a loading rate of 1 mm/min. The samples were also cut in two directions: 0 and 90 (Figure 1).

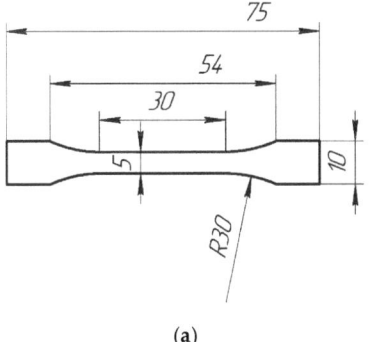
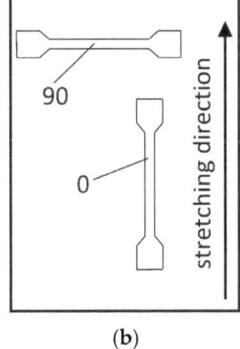

(a) (b)

Figure 1. Dimensions of the film samples tested in tension (**a**) and the scheme of their cutting from the film (**b**).

The given values (fracture strength σ_f, maximum strength σ_m, modulus of elasticity E, elongation at failure ε_f, and elongation at maximum strength ε_m) represent the arithmetic mean of the measurements for 5 samples for each series of specimens.

The piezoelectric coefficient d_{33} was measured according to the Berlincourt method [25] using a PKD3-2000 device (PolyK, State College, PA, USA) for 2×2 cm^2 samples at 16 points of each sample. Measurements were performed at a calibrated force of 0.25 N on a frequency of 110 Hz.

The light transmission in the visible and IR wavelength range from 380 to 2600 nm was determined on a Shimadzu UV-3600i Plus spectrophotometer with a resolution of 1 nm at normal incidence of light on the sample using the integrating sphere.

Average light transmission and color rendering coefficients were calculated in accordance with European Standard EN 410:2011 Glass in building—Determination of luminous and solar characteristics of glazing.

3. Results

3.1. Structure of PVDF Films by DSC and IR Spectroscopy

Figure 2 shows the DSC curves obtained for the PVDF and PVDF + ITO films.

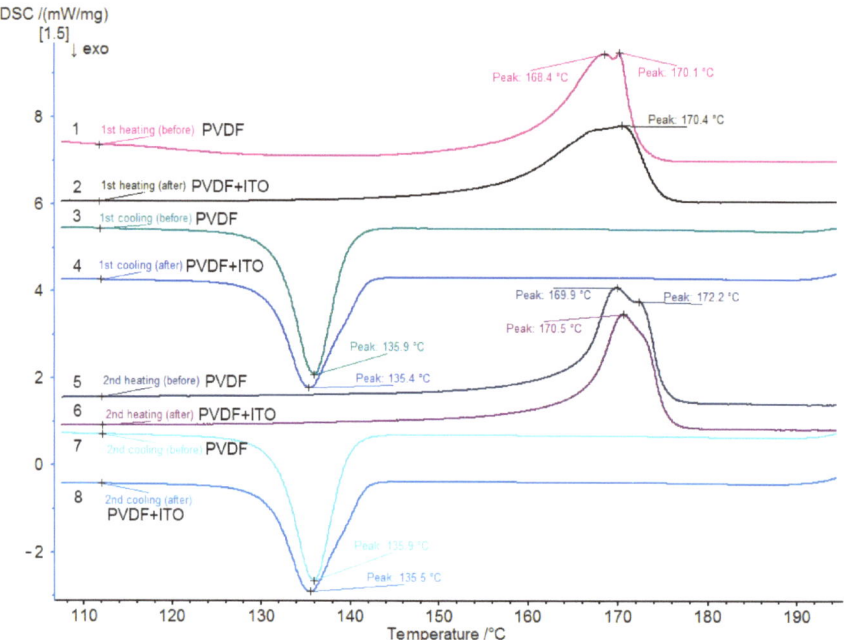

Figure 2. DSC curves for the original PVDF films without ITO (curves 1, 3, 5, 7) and with ITO (curves 2, 4, 6, 8).

The DSC curves obtained by heating the pure PVDF film show a double endothermic peak (T_{m1} and T_{m2}, Figure 2) which corresponds to the characteristic melting temperature of the polymer. When the film is reheated, the double endothermic peak is also preserved, but the intensity of the peak changes (curve 5, Figure 2). The T_{m1} peak becomes more intense which indicates a change in the supramolecular structure. Based on the data obtained, we can assume that the supramolecular structure of PVDF is formed by two types of crystal structures. It is known [26–28] that PVDF macromolecules can have at least three types of crystal structures (α-, β- and γ- phases) depending on the macromolecule conformation. The double peak on the DSC curve describing PVDF melting indicates the simultaneous presence of α- and β-crystalline phases in the samples, which is confirmed by infrared spectroscopy data (Figures 3 and 4). The double peak in the sample curve is preserved after deposition of ITO coating on its surface (Figure 2, curves 2, 6). It can be assumed that the coating conditions do not affect the supramolecular structure of PVDF film.

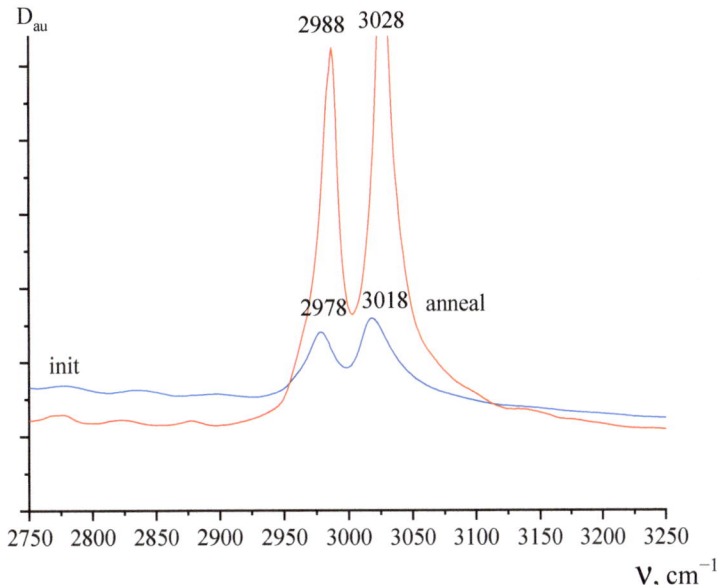

Figure 3. IR absorption spectrum in the range of 2750–3250 cm^{-1} for the original PVDF film (initial) and after heating to 200 °C (annealing).

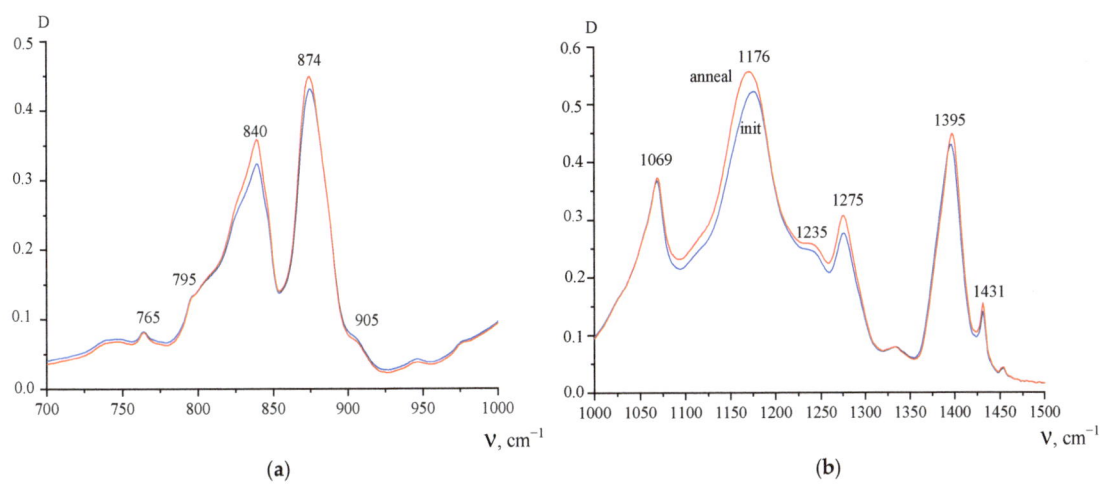

Figure 4. ATR IR absorption spectrum in the 700–1000 cm^{-1} (**a**) and 1000–1500 cm^{-1} (**b**) range for the original PVDF film (blue curve) and after heating to 200 °C (red curve).

Thermograms of the cooling of PVDF samples (coated and uncoated) (Figure 2, curves 3, 4, 7, 8) are described by a single endothermic peak corresponding to PVDF crystallization. The peaks' positions and shapes remain constant for PVDF and PVDF + ITO. It may be noted that the width of the crystallization peak in the PVDF films with ITO is noticeably wider than in the pure PVDF films.

The crystallization and melting temperatures as well as the thermal effects of melting and percent crystallinity are more clearly shown in Table 1.

Table 1. Melting and crystallization temperatures, melting thermal effects, and percent crystallinity for PVDF films with and without ITO.

Sample	T_{m1}, °C	T_{m2}, °C	T_{cr}, °C	ΔH_m, J/g	χ_c, %
PVDF, heating 1	168	170	136	50	48
PVDF, heating 2	170	172	136		
PVDF + ITO heating 1	168	170	135	46	44
PVDF + ITO, heating 2	170	172	136		

The results of the study of the pure PVDF film by infrared spectroscopy before and after annealing are shown in Figures 3 and 4. Figure 3 shows that doublet absorption spectrum bands (due to the symmetric antisymmetric component) in the region of valence vibrations of methyl groups noticeably shift toward higher frequencies after annealing. As follows from the DSC data, the melting peak is characterized by a doublet. One reason for this may be related to the polymorphism of PVDF crystals [1–3]. Since α-, β- and γ-phases have different chain conformations, they have their own absorption bands in the vibrational spectra. Figure 4a shows that in the initial state of the film, there is a large fraction of the polar β-phase observed, for which an intense absorption band of 840 cm^{-1} is responsible [1–3].

Since PVDF crystallization from the melt under normal conditions occurs in the α- phase, its conversion to the polar β- modification is usually performed by uniaxial stretching at low temperatures [1–3]. It is obvious that this also takes place for the studied film. Confirmation of this can be found in mechanical tests, where a large difference in the deformation behavior in two mutually perpendicular directions can be seen. However, along with the 840 cm^{-1} band, weak 765 and 795 cm^{-1} bands characteristic of the TGTG-chain conformation are observed in the spectrum, which indicate the presence of small amounts of α-phase [1–3]. It is the presence of the latter that can lead to the low-temperature endotherm. If this hypothesis is correct, then an increase in the α-phase fraction in the film should lead to an increase in its intensity. As can be seen from the DSC data, this is noted in the second heating cycle of the studied film. As follows from Figure 3, the absorption peaks in the film after annealing are shifted towards higher frequencies. As follows from the results of [3], this indicates an increase in the α-phase fraction during the annealing, which is confirmed by an increase in the low-temperature peak on the DSC during the second heating as compared to the first.

Other information can be obtained from Figure 4, namely the change in the microstructure of the film surface after annealing, since the micron-sized layer is investigated by ATR imaging. As follows from the above figures, the intensity of the 840 and 1275 cm^{-1} bands are practically unchanged after annealing. Since these bands are sensitive to the presence of long chain regions in the planar zigzag conformation [1–3], the reason for the observed change may be the pre-polarization of the film. As noted above, the initial film has a piezo effect.

Figure 5 shows that PVDF film with ITO coatings in the 700–1500 cm^{-1} range has strong absorption, as the bands characteristic of the polymer are almost absent in the coated film. This may be due to the fact that the thickness of the applied ITO turns out to be greater than the thickness of the probed layer. The possibility of chemical interactions at the PVDF–ITO interface can be judged from Figure 5c,d. This is indicated by the appearance of two wide maxima in the region of 1500 cm^{-1} and 3500 cm^{-1}. They are visible only because, as can be seen, the polymer itself practically does not absorb in these spectral regions. Comparison with the spectrophotometer data gives, as can be observed, a qualitative correspondence.

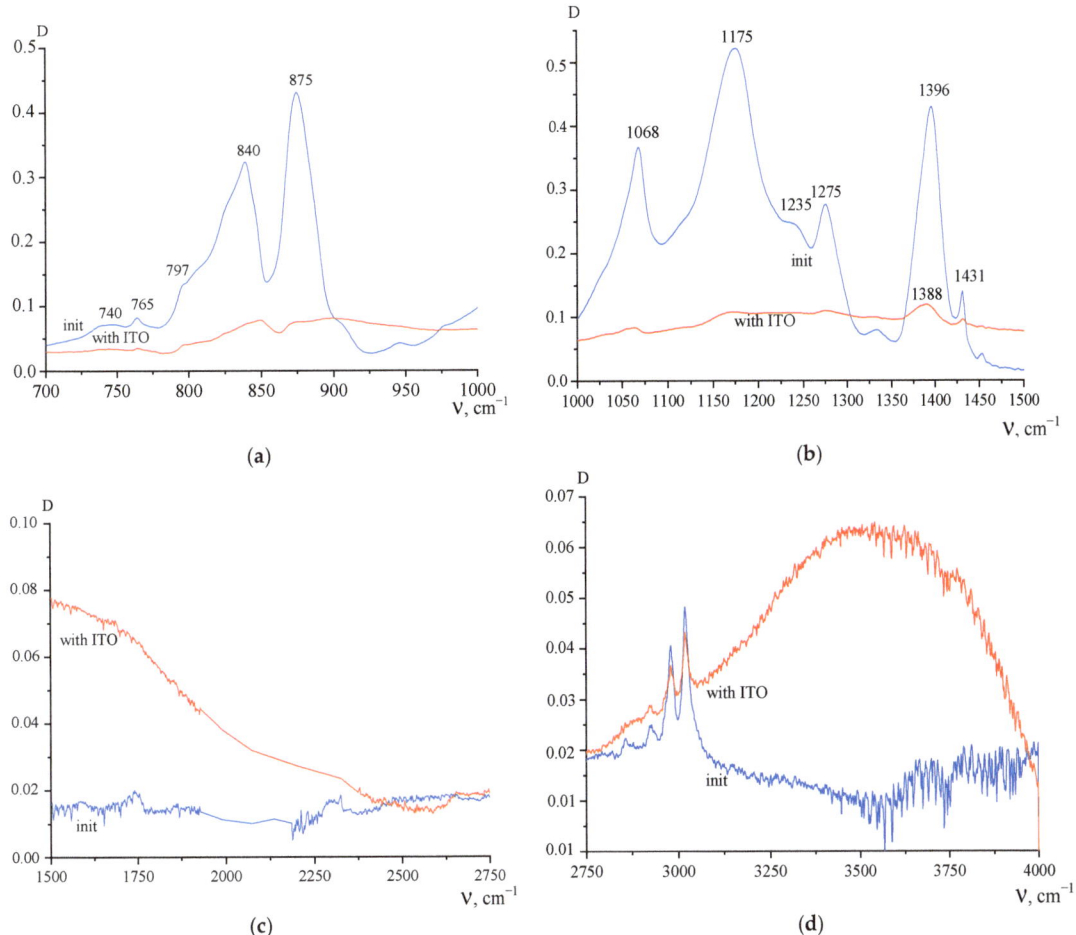

Figure 5. ATR IR absorption spectrum in the 700–1000 cm^{-1} (**a**), 1000–1500 cm^{-1} (**b**), 1500–2750 cm^{-1} (**c**) and 2750–4000 cm^{-1} (**d**) range for the original PVDF film (init) and with applied ITO electrodes.

Thermal treatment of films with ITO electrodes (Figure 6) leads to an unusual effect: bands characteristic of the polymer appear in the spectral region after annealing. The reason for this phenomenon is not completely clear, but one reason may be related to thermally activated chemical reactions between ITO molecules and PVDF surface molecules. An indication of the possibility of such a reaction follows from Figure 6a. It shows that the wide ITO 3500 cm^{-1} absorption band after annealing is strongly shifted toward low frequencies.

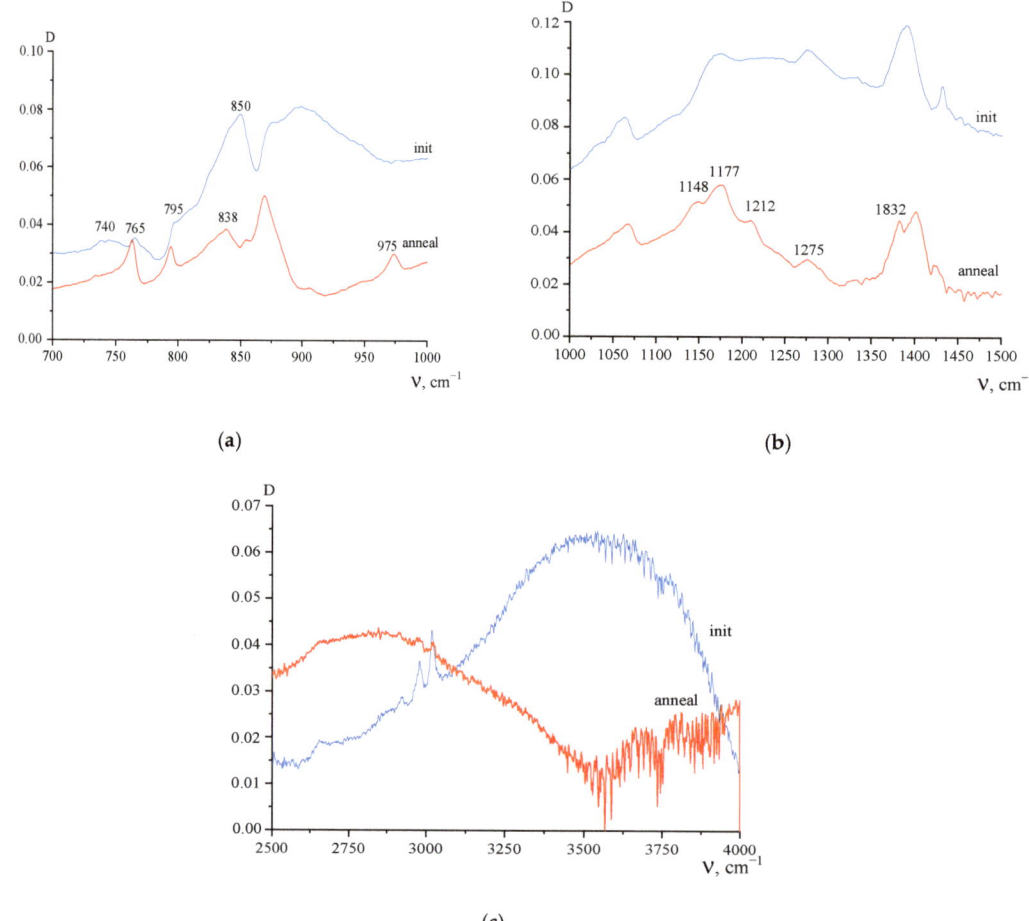

Figure 6. ATR IR absorption spectrum in the 700–1000 cm^{-1} (**a**), 1000–1500 cm^{-1} (**b**), and 2500–4000 cm^{-1} (**c**) range for PVDF and PVDF + ITO films with ITO before (init) and after annealing (aneal).

The possibility of intensification of the noted chemical interactions at the polymer–ITO interface during annealing indirectly follows from the frequency dependences of the components of the complex permittivity of three-layer capacitors, metal-polymer-ITO-metal, which are shown in Figure 7a,b. For comparison, the curves for a similar copolymer with a conductive layer of graphene-containing material [29] are shown in the graph. It can be seen that in the case of the inorganic conductive coating, two new dispersion regions arise in the material at low frequencies.

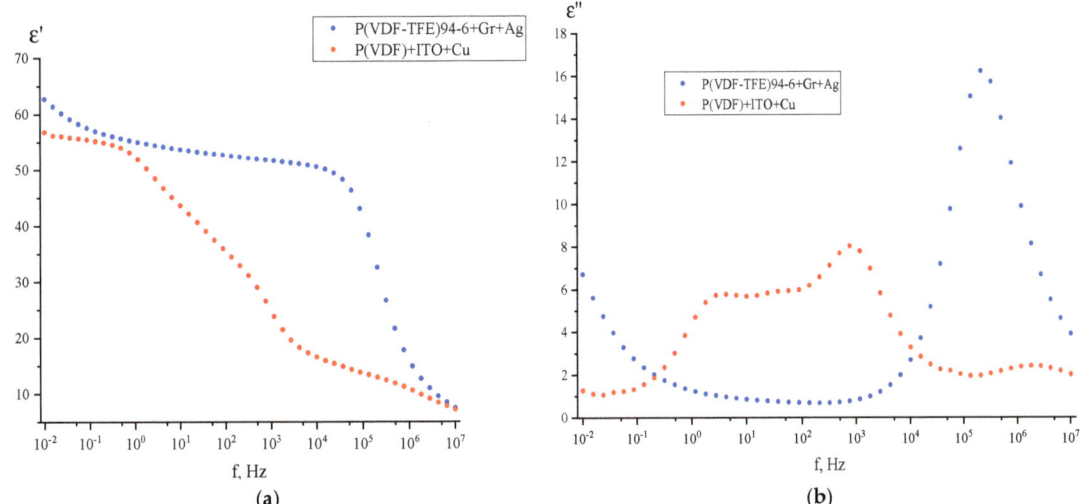

Figure 7. Frequency dependences of the real (**a**) and imaginary (**b**) components of the complex permittivity of PVDF with ITO.

3.2. Thermomechanical Test Results

The change in elastic modulus of PVDF films with increasing frequency of load application is shown in Figure 8. It can be seen that in all cases, the elastic modulus changes little in the frequency range of 0.25 to 100 Hz. At the same time, the change in E' values is about 5%. It should be noted that the curves describing the change in elastic modulus of PVDF and PVDF + ITO films with frequency are at a different level. The difference between the modulus E' values is about 15%.

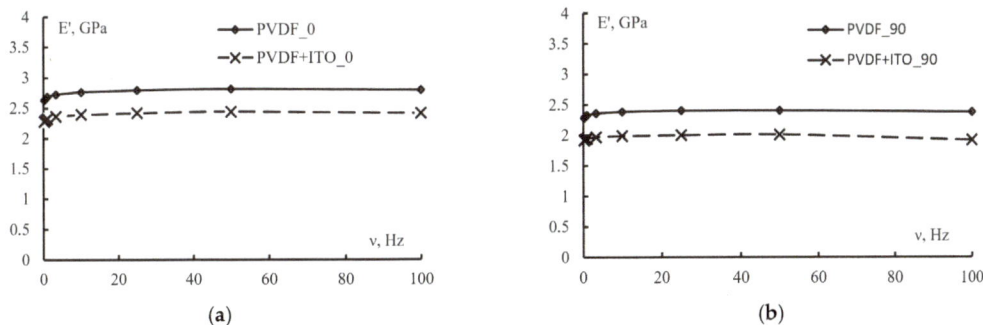

Figure 8. Variation of the elastic modulus E' in direction 0 (curve 1) and in direction 90 (curve 2) with increasing frequency ν for the original and coated PVDF films in directions 0 (**a**) and 90 (**b**).

Table 2 (E' at a fixed frequency (50 Hz)) shows that elastic modulus differs by about 15% for the specimens tested in and across the orientational tensile direction. The coating decreased the elastic modulus by about 15%. At the same time, the difference in the E' values for directions 0 and 90 remained at the same level.

Table 2. Elastic moduli E' of coated and uncoated PVDF films at a load frequency of 50 Hz.

Cutting Direction	E', GPa (PVDF)	E', GPa (PVDF + ITO)
0	2.8	2.4
90	2.4	2.0

Figure 9 shows typical load diagrams of coated and uncoated PVDF film. In direction 0, the studied samples (coated and uncoated) have the same load diagrams. In both cases, stresses grow elastically at small deformations. After reaching a strain of 1.5%, the specimens begin to deform irreversibly. At about 3% strain (stress of about 50 MPa), a small yield point is observed. With further increase in strain, stresses increase until the materials fracture.

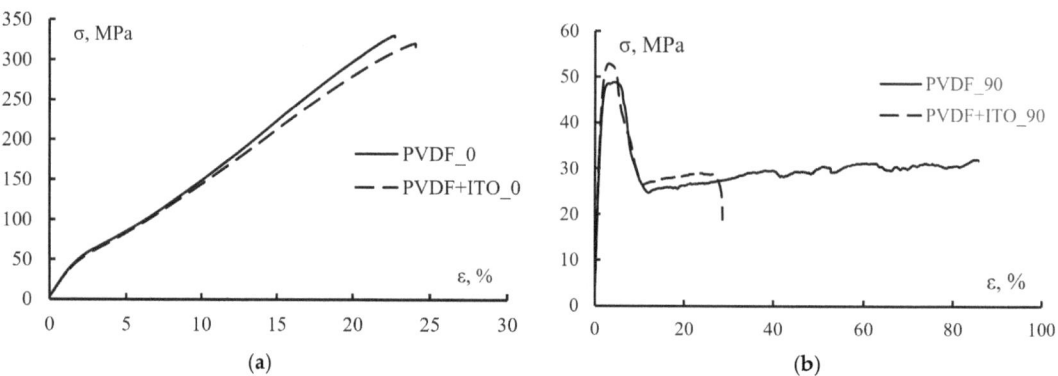

Figure 9. Typical load diagrams for the original and coated PVDF films in directions 0 (a) and 90 (b).

The tensile diagrams of the samples cut in the 90 direction are significantly different from the diagrams described above. As in the previous case, the treated and untreated films deform elastically until the relative elongation reaches 1.5%. With further increase in strain, a maximum of stresses corresponding to the temporary strength of the material is observed. It should be noted that the maximum strength value for specimens cut in the 90 direction approximately corresponds to the "yield strength" of specimens cut in the 0 direction. Furthermore, there is a decrease in stresses to the level of 20–25 MPa and an increase in elongation up to the point of specimen failure. It should be noted that the coating deposition influenced the film deformability in direction 90.

The change in the elastic-strength properties during coating is shown more clearly in Table 3.

Table 3. Elastic strength properties of PVDF film.

Sample	E, GPa	σ_m, MPa	ε_m, %	σ_p, MPa	ε_f, %
PVDF_0	3.10 ± 0.10	334 ± 15	22.5 ± 1	330 ± 18	22.4 ± 10.0
PVDF_90	3.11 ± 0.13	49 ± 2	4.3 ± 0.3	29 ± 3	100.5 ± 42.1
PVDF + ITO_0	2.87 ± 0.06	305 ± 7	24.0 ± 0.1	289 ± 15	25.7 ± 2.4
PVDF + ITO_90	3.47 ± 0.11	52 ± 2	3.1 ± 0.1	17 ± 5	19.4 ± 7.4

It can be seen that the maximum strength σ_m in direction 0 for the materials with and without treatment is almost the same, but differs significantly for different orientation directions (almost six times greater in direction 0). This difference in maximum strength values indicates a high degree of film orientation. Tensile strength changes similarly. The high degree of orientation of the film is evidenced by its deformation corresponding to the maximum strength and fracture strength. Thus, the fracture strain ε_f for the pure PVDF film differs almost 5 times depending on its orientation (22.4% and 100% respectively for orientation 0 and 90). Additionally, if coating deposition practically does not change ε_f in the case of orientation 0, then in the direction of 90 ε_f decreases by up to 20%. Such a sharp decrease in the deformability in the direction 90 is due to the ITO deposition.

The result of determining the elastic modulus was somewhat unexpected, which for the untreated PVDF film in both directions left about 3.1 GPa, and after coating deposition it decreased to 2.87 GPa in the 0 direction and increased to 3.47 GPa in the 90 direction.

It should be noted that the modulus determined by the DMA method differs markedly from the elastic modulus measured under quasi-static loading conditions. Thus, the elastic moduli in directions 0 and 90 differ by 10% and 20%, respectively.

3.3. Optical and Piezoelectric Properties

The value of the piezoelectric coefficient d_{33} for the initial films is 28 ± 3 pC/N (corresponds to the value declared by the manufacturer), and after the deposition of ITO, respectively, 25 ± 6 pC/N. It can be seen that the average value of the piezoelectric modulus d_{33} after ITO application slightly decreases, although the error increases.

The dependences of the transmittance coefficients of the experimental samples on the wavelength of the visible and IR ranges are shown in Figure 10.

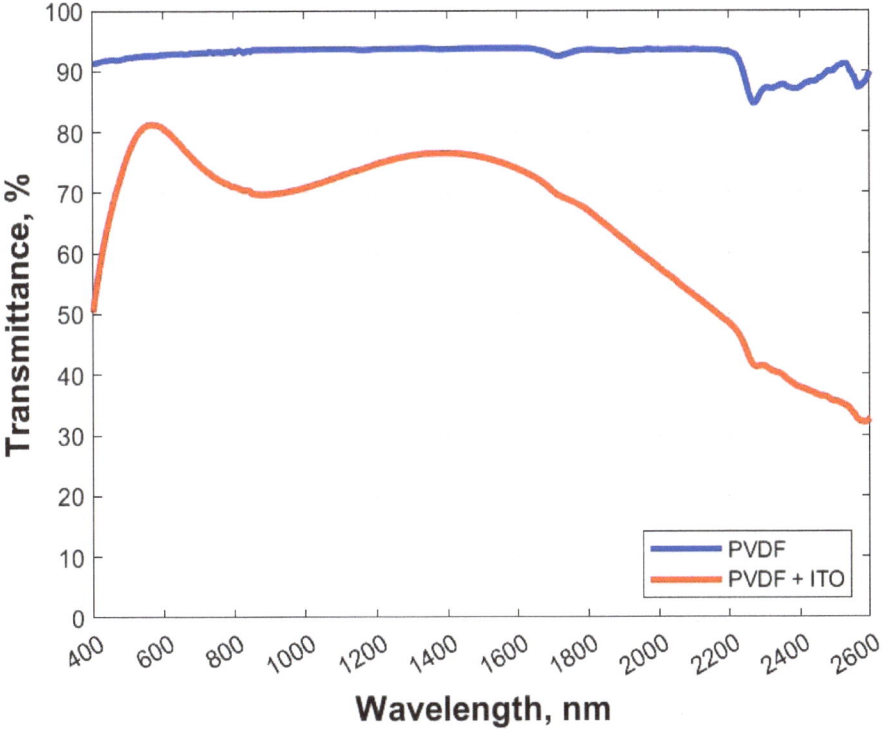

Figure 10. Transmittance for PVDF and PVDF + ITO films.

The determined values of light transmission coefficients, the average value of transmittance coefficients in the IR wavelength range, and color rendering coefficients of experimental samples are presented in Table 4.

Table 4. Optical properties of experimental samples.

Sample	Average Transmittance in the Visible Range, %	Average Transmittance in the IR Range from 780 to 2600 nm, %	Color Rendering Index
PVDF	92.6	92.4	81.0
PVDF_ITO	79.6	62.0	84.6

The transparency of the ITO-coated samples decreased by 13% in the visible wavelength range and by almost 30% in the IR wavelength range from 780 to 2600 nm. The color rendering of the ITO-coated sample even improved slightly compared to the original sample.

4. Discussion

As expected, the results of the studies revealed that the viscoelastic behavior of the films under tension along and across the orientation is described by different loading diagrams because the initial PVDF film was uniaxial stretched. Therefore, the strength of uncoated and coated PVDF films differs by an order of magnitude depending on the test direction.

PVDF is characterized by two melting temperatures, which indicates the presence of two types of crystal structures (α- and β- phases) in the polymer supramolecular structure. IR spectroscopy indicates the predominance of the β-phases, which is confirmed by the high piezoelectric constant of the initial PVDF film.

Heating the PVDF film to 200 °C (above the melting point) leads to a slight increase in the α-, which is confirmed by DSC (increasing T_{m1} peak intensity) and IR spectroscopy. Despite the fact that PVDF crystallization usually takes place in the α-phase, the β-phase is still present after heating the films above the melting temperature, which can be explained by the preliminary polarization of the film. This indicates that the polarized PVDF film retains its ferroelectric properties even after heating to the melting temperature.

The temperature-time mode of ITO deposition has little effect on the structure, thermophysical and mechanical properties of the polymer. The maximum strength σ_m in direction 0 and the shape of the DSC curves during heating for the samples with and without ITO are almost the same. The piezoelectric constant d_{33} also changed slightly (from 28 to 25 pC/N). This allows us to conclude that the ITO deposition process had practically no effect on the internal structure of the polymer material.

At the same time, the results of the annealing of PVDF films with ITO indicate the possibility of chemical interactions at the polymer–ITO interface that is indirectly confirmed by the frequency dependences of the components of the complex permittivity of three-layer metal-polymer-ITO-metal capacitors.

The possibility of such interactions is confirmed by the broadening of the crystallization peaks on the DSC curve after ITO deposition. The chemical interaction between ITO and the surface layer of the polymer can interfere with the crystallization process during cooling, and the transition of the melt into a crystal will occur in a wider temperature range. The slight decrease in crystallinity from 48% to 44% can be explained by structural disturbance in the polymer surface layer due to the interaction of ITO with polymer chains.

This interaction can also affect the mechanical properties of the ITO-PVDF-ITO structure. This can explain why the deposition of ITO films affected the film stretch in the 90 direction decreasing the film strain by a factor of 5 times or the unexpected elastic modulus results noted above.

5. Conclusions

In this paper, we demonstrated flexible transparent film ITO-PVDF-ITO with transparency of 79.6%, a color rendering index of 84, and a piezoelectric coefficient d_{33} 25 ± 6 pC/N and investigated the effect of the ITO deposition technological process on the properties of the PVDF film. The manifestation of memory effects was noted for the initial film.

It is shown by the fact that the characteristics of conformationally sensitive absorption bands of the film change weakly when the temperature rises even above the melting point. Hypotheses for this phenomenon are proposed. Infrared spectroscopy in the ATR variant demonstrates chemical interactions of ITO molecules with the polymer surface layer where an increased content of defects of both chemical and physical nature is observed for PVDF films with an applied ITO layer. It can be concluded that the deposition process has almost no effect on the mechanical, structural, and piezoelectric properties of initial PVDF film taking into account the operation of such structures in the elastic region, decreasing the transparency of pure PVDF film by 14% in the visible range. The discovered possible chemical interactions between ITO and PVDF can provide good adhesion of ITO layers to the polymer and expand the application potential of such films.

Author Contributions: Conceptualization, A.O., G.Y. and V.K.; Funding acquisition, M.M. and G.Y.; Investigation, V.K., V.S., A.M., S.B., M.S., I.T., T.P. and B.L.; Methodology, A.O., V.K. and V.S.; Project administration, M.M.; Resources, M.M.; Supervision, G.Y.; Validation, V.K.; Writing—original draft, A.O., V.S. and V.K.; Writing—review & editing, A.O., M.M., V.K. and G.Y. All authors have read and agreed to the published version of the manuscript.

Funding: The research was carried out within the state assignment of Ministry of Science and Higher Education of the Russian Federation (theme No. FSFN-2022-0007).

Institutional Review Board Statement: Not applicable.

Informed Consent Statement: Not applicable.

Data Availability Statement: The data presented in this study are available on request from the corresponding author.

Acknowledgments: This work was performed employing the equipment of Center for molecular composition studies of INEOS RAS. The authors also express their gratitude to the students of Bauman Moscow State Technical University and to Bogdan Basov and Nikita Ryshkov for their assistance with experiments and data processing.

Conflicts of Interest: The authors declare no conflict of interest.

References

1. Wang, Y.; Sun, L.; Wang, C.; Yang, F.; Ren, X.; Zhang, X.; Hu, W. Organic crystalline materials in flexible electronics. *Chem. Soc. Rev.* **2019**, *48*, 1492–1530. [CrossRef] [PubMed]
2. Wang, C.; Dong, H.; Jiang, L.; Hu, W. Organic semiconductor crystals. *Chem. Soc. Rev.* **2018**, *47*, 422–500. [CrossRef] [PubMed]
3. Lang, S.B.; Muensit, S. Review of some lesser-known applications of piezoelectric and pyroelectric polymers. *Appl. Phys. A* **2006**, *85*, 125–134. [CrossRef]
4. Molodets, A.M.; Eremchenko, E.N. Electrical response of a piezoelectric polymer film to shock compression. *Combust. Explos. Shock. Waves* **1994**, *30*, 149–154. [CrossRef]
5. Lin, B.; Giurgiutiu, V. Modeling and testing of PZT and PVDF piezoelectric wafer active sensors. *Smart Mater. Struct.* **2006**, *15*, 1085–1093. [CrossRef]
6. Yu, Y.; Sun, H.; Orbay, H.; Chen, F.; England, C.G.; Cai, W.; Wang, X. Biocompatibility and in vivo operation of implantable mesoporous PVDF-based nanogenerators. *Nano Energy* **2016**, *27*, 275–281. [CrossRef] [PubMed]
7. Netyaga, A.A.; Parfenov, A.O.; Nutfullina, G.M.; Zhukovskiy, V.A. Comparative experimental study of the biomechanical properties of the standard, light and composite meshes for abdominal wall after implantation. *Mod. Probl. Sci. Educ.* **2013**, *5*, 370.
8. Gradov, O.V.; Gradova, M.A.; Kochervinskii, V.V. Biomimetic biocompatible ferroelectric polymer materials with an active response for implantology and regenerative medicine. In *Woodhead Publishing Series in Electronic and Optical Materials, Organic Ferroelectric Materials and Applications*; Asadi, K., Ed.; Woodhead Publishing: Sawston, UK, 2022; pp. 571–619.
9. Makeev, M.O.; Osipkov, A.S.; Batshev, V.I.; Polschikova, O.V.; Ryshkov, N.S.; Kochervinskii, V.V.; Yurkov, G.Y. Investigation of the phase delay of radiation by a transparent ferroelectric polymer film. *J. Phys. Conf. Ser.* **2021**, *2127*, 012048. [CrossRef]
10. Sato, T.; Ishida, H.; Ikeda, O. Adaptive PVDF piezoelectric deformable mirror system. *Appl. Opt.* **1980**, *19*, 1430–1434. [CrossRef]
11. Tahtali, M.; Lambert, A.J.; Zhitao, Z.; Fraser, D. Using the Modeshapes of a Vibrating PVDF Film as a Deformable Mirror Surface. *CANEUS: MNT Aerosp. Appl.* **2006**, *42541*, 31–36.
12. Kochervinskii, V.V.; Gradova, M.A.; Gradov, O.V.; Kiselev, D.A.; Ilina, T.S.; Kalabukhova, A.V.; Kozlova, N.V.; Shmakova, N.A.; Bedin, S.A. Structural, optical, and electrical properties of ferroelectric copolymer of vinylidenefluoride doped with Rhodamine 6G dye. *J. Appl. Phys.* **2019**, *125*, 044103. [CrossRef]

13. Minami, T.; Sonohara, H.; Kakumu, T.; Takata, S. Physics of very thin ITO conducting films with high transparency prepared by DC magnetron sputtering. *Thin Solid Film.* **1995**, *270*, 37–42. [CrossRef]
14. Sugimoto, T.; Ono, K.; Ando, A.; Kurozumi, K.; Hara, A.; Morita, Y.; Miura, A. PVDF-driven flexible and transparent loudspeaker. *Appl. Acoust.* **2009**, *70*, 1021–1028. [CrossRef]
15. Xin, Y.; Li, X.; Tian, H.; Guo, C.; Sun, H.; Jiang, Y.; Wang, C. A fingerprint sensor based on PVDF film for a manipulator. *Integr. Ferroelectr.* **2017**, *183*, 91–99. [CrossRef]
16. Zhang, B.; Guo, C.; Cao, X.; Yuan, X.; Li, X.; Huang, H.; Dong, S.; Wang, Z.L.; Ren, K. Flexoelectricity on the photovoltaic and pyroelectric effect and ferroelectric memory of 3D-printed $BaTiO_3$/PVDF nanocomposite. *Nano Energy* **2022**, *104*, 107897. [CrossRef]
17. Panicker, S.S.; Rajeev, S.P.; Thomas, V. Impact of PVDF and its copolymer-based nanocomposites for flexible and wearable energy harvesters. *Nano-Struct. Nano-Objects* **2023**, *34*, 100949. [CrossRef]
18. Aleksandrova, M.; Tudzharska, L.; Nedelchev, K.; Kralov, I. Hybrid Organic/Inorganic Piezoelectric Device for Energy. Harvesting and Sensing Application. *Coatings* **2023**, *13*, 464. [CrossRef]
19. Sengwa, R.J.; Kumar, N.; Saraswat, M. Morphological, structural, optical, broadband frequency range dielectric and electrical properties of PVDF/PMMA/$BaTiO_3$ nanocomposites for futuristic microelectronic and optoelectronic technologies. *Mater. Today Commun.* **2023**, *35*, 105625. [CrossRef]
20. Bose, S.; Summers, J.R.; Srivastava, B.B.; Padilla-Gainza, V.; Peredo, M.; Trevino De Leo, C.M.; Hoke, B.; Gupta, K.S.; Lozano, K. Efficient near infrared to visible light upconversion from Er/Yb codoped PVDF fibrous mats synthesized using a direct polymer doping technique. *Opt. Mater.* **2022**, *123*, 111866. [CrossRef]
21. Cheng, N.; Miao, D.; Wang, C.; Lin, Y.; Babar, A.A.; Wang, X.; Wang, Z.; Yu, J.; Ding, B. Nanosphere-structured hierarchically porous PVDF-HFP fabric for passive daytime radiative cooling via one-step water vapor-induced phase separation. *Chem. Eng. J.* **2023**, *460*, 141581. [CrossRef]
22. Nartsev, V.M.; Ageeva, M.S.; Prokhorenkov, D.S.; Zaitsev, S.V.; Karatsupa, S.V.; Vashchilin, V.S. Influence of deposition conditions of high-quality AlN and SiC on the coating characteristics. *Bull. BSTU Named After V.G. Shukhov.* **2013**, *6*, 68–172.
23. Alexeeva, O.; Olkhov, A.; Konstantinova, M.; Podmasterev, V.; Tretyakov, I.; Petrova, T.; Koryagina, O.; Lomakin, S.; Siracusa, V.; Iordanskii, A.L. Improvement of the Structure and Physicochemical Properties of Polylactic Acid Films by Addition of Glycero-(9,10-trioxolane)-Trialeate. *Polymers* **2022**, *14*, 3478. [CrossRef] [PubMed]
24. Figoli, A.; Simone, S.; Criscuoli, A.; Aljlil, S.A.; Shabouna, F.S.A.; AlRomaih, H.; Nicolò, E.; Al-Harbi, O.; Drioli, E. Hollow fibers for seawater desalination from blends of PVDF with different molecular weights: Morphology, properties and VMD performance. *Polymer* **2014**, *55*, 1296–1306. [CrossRef]
25. Stewart, M. *Characterisation of ferroelectric bulk materials and thin films*; Springer: Berlin/Heidelberg, Germany, 2014; p. 290.
26. Ruan, L.; Yao, X.; Chang, Y.; Zhou, L.; Qin, G.; Zhang, X. Properties and applications of the β phase poly (vinylidene fluoride). *Polymers* **2018**, *10*, 228. [CrossRef] [PubMed]
27. Gregorio, L.; Cestari, M. Effect of crystallization temperature on the crystalline phase content and morphology of poly (vinylidene fluoride). *J. Polym. Sci. Part B Polym. Phys.* **1994**, *32*, 859–870. [CrossRef]
28. Sharma, M.; Madras, G.; Bose, S. Process induced electroactive β-polymorph in PVDF: Effect on dielectric and ferroelectric properties. *Phys. Chem. Chem. Phys.* **2014**, *16*, 14792–14799. [CrossRef]
29. Kochervinskii, V.V.; Baskakov, S.A.; Malyshkina, I.A.; Kiselev, D.A.; Ilina, T.S.; Rybin, M.G.; Bedin, S.A.; Chubunova, E.V.; Shulga, Y.M. The application of organic graphene-based electrodes forstudies of electrophysical properties of polymer dielectrics and ferroelectrics. *Ferroelectrics* **2022**, *600*, 59–72. [CrossRef]

Disclaimer/Publisher's Note: The statements, opinions and data contained in all publications are solely those of the individual author(s) and contributor(s) and not of MDPI and/or the editor(s). MDPI and/or the editor(s) disclaim responsibility for any injury to people or property resulting from any ideas, methods, instructions or products referred to in the content.

Article

Synthesis of KH550-Modified Hexagonal Boron Nitride Nanofillers for Improving Thermal Conductivity of Epoxy Nanocomposites

Bolin Tang [1,2], Miao Cao [2], Yaru Yang [2], Jipeng Guan [2], Yongbo Yao [2], Jie Yi [2], Jun Dong [2], Tianle Wang [3,4,*] and Luxiang Wang [1,*]

[1] State Key Laboratory of Chemistry and Utilization of Carbon Based Energy Resources, College of Chemistry, Xinjiang University, Urumqi 830017, China
[2] Nanotechnology Research Institute, School of Materials and Textile Engineering, Jiaxing University, Jiaxing 314001, China
[3] Zhejiang Provincial Key Laboratory for Cutting Tools, School of Materials Science and Engineering, Taizhou University, Taizhou 318000, China
[4] College of Chemical and Materials Engineering, Zhejiang A&F University, Hangzhou 311300, China
* Correspondence: wtl0203@tzc.edu.cn (T.W.); wangluxiangxju@163.com (L.W.)

Citation: Tang, B.; Cao, M.; Yang, Y.; Guan, J.; Yao, Y.; Yi, J.; Dong, J.; Wang, T.; Wang, L. Synthesis of KH550-Modified Hexagonal Boron Nitride Nanofillers for Improving Thermal Conductivity of Epoxy Nanocomposites. *Polymers* **2023**, *15*, 1415. https://doi.org/10.3390/polym15061415

Academic Editor: Maurizio Vignolo

Received: 15 February 2023
Revised: 3 March 2023
Accepted: 11 March 2023
Published: 13 March 2023

Copyright: © 2023 by the authors. Licensee MDPI, Basel, Switzerland. This article is an open access article distributed under the terms and conditions of the Creative Commons Attribution (CC BY) license (https://creativecommons.org/licenses/by/4.0/).

Abstract: In this work, KH550 (γ-aminopropyl triethoxy silane)-modified hexagonal boron nitride (BN) nanofillers were synthesized through a one-step ball-milling route. Results show that the KH550-modified BN nanofillers synthesized by one-step ball-milling (BM@KH550-BN) exhibit excellent dispersion stability and a high yield of BN nanosheets. Using BM@KH550-BN as fillers for epoxy resin, the thermal conductivity of epoxy nanocomposites increased by 195.7% at 10 wt%, compared to neat epoxy resin. Simultaneously, the storage modulus and glass transition temperature (Tg) of the BM@KH550-BN/epoxy nanocomposite at 10 wt% also increased by 35.6% and 12.4 °C, respectively. The data calculated from the dynamical mechanical analysis show that the BM@KH550-BN nanofillers have a better filler effectiveness and a higher volume fraction of constrained region. The morphology of the fracture surface of the epoxy nanocomposites indicate that the BM@KH550-BN presents a uniform distribution in the epoxy matrix even at 10 wt%. This work guides the convenient preparation of high thermally conductive BN nanofillers, presenting a great application potential in the field of thermally conductive epoxy nanocomposites, which will promote the development of electronic packaging materials.

Keywords: hexagonal boron nitride; epoxy; surface modification; polymer matrix nanocomposites; thermal conductivity

1. Introduction

The rapid development of modern electronic devices toward high integration, high power density, and miniaturization presents an increasing requirement for heat dissipation, which makes it urgent to develop high thermally conductive materials for thermal management applications [1–4]. Currently, the most commonly used thermal management materials (TMMs) are mainly some types of thermosetting polymers, such as epoxy resin [5], organosilicone [6], polyurethane [7,8], etc. Among them, epoxy resin receives extensive research as a TMM, by virtue of its high adhesion strength, chemical durability, and excellent mechanical strength [9–11]. However, the low thermal conductivity of epoxy resins (~0.2 W m^{-1}K^{-1}) is far from meeting the need for efficient heat dissipation of modern electronic devices. Therefore, developing TMMs with high thermal conductivity is of great importance for the electronic and electrical industry.

Incorporating high-thermal-conductivity filler into epoxy matrix is a widely used method for improving the thermal conductivity of epoxy resin [12,13]. Typical thermal conductive fillers include metals (Ag, Al, etc.) [14], carbon-based materials (carbon nanotube, graphene, etc.) [15–17], and ceramics (Al$_2$O$_3$, BN, etc.) [18,19]. Compared with metals or carbon-based materials, ceramic fillers usually exhibit good electrical insulating characteristic besides high thermal conductivity, which is quite suitable for use as thermally conductive fillers of TMMs. As a ceramic filler with a similar two-dimensional (2D) structure of graphene, hexagonal BN has excellent thermal conductivity and electrical insulating property, especially the BN nanosheets exfoliated from BN that possess a theoretical thermal conductivity as high as ~2000 W m^{-1}K^{-1} [20,21]. Nevertheless, the strong atom interactions between neighboring planar B and N atoms, commonly referred to as "lip-lip" interactions, make exfoliating BN significantly more difficult [22].

To date, different means have been explored for the preparation of BN nanosheets, such as mechanical ball-milling [23], chemical vapor deposition [24], and thermal exfoliation [25]. However, these exfoliation methods still face three main challenges to practical application: the low yield (Y) of BN nanosheets (<20%), difficult separation of BN nanosheets, and poor interfacial compatibility between BN nanosheets and polymer matrix. Usually, the obtained BN nanosheets need to be further surface-modified with silane coupling agents or other organic molecules in the application of polymer composites [26].

Recently, Chen et al. [27] developed a simple and efficient one-step method for the preparation of functionalized few-layer BNNPs by solid-state ball milling of commercially available h-BN and urea powder. Ren et al. [28] reported that the yield and dispersibility of BN nanosheets can be effectively enhanced via a tannic acid (TA)-assisted liquid-phase exfoliation, due to the intermolecular interaction. Among these methods, the interaction between organic molecules and BN, such as Lewis acid–base interactions, was considered to play a critical role in the exfoliation of BN. Moreover, Agrawal's group [29] and Liu's group [11] demonstrated that by using aminosilane-coupling-agent-modified BN as fillers, the thermal conductivity of BN/epoxy composites can be improved effectively. Herein, the KH550-modified BN nanopowders with high content of nanosheets were synthesized via a one-step ball-milling process for improving the thermal conductivity and mechanical property of epoxy nanocomposites. As shown in Scheme 1, the BN particles were directly added into a mixed solution of deionized water (H$_2$O) and absolute ethyl alcohol (CH$_3$CH$_2$OH) containing KH550, followed by performing the ball-milling process. On one hand, the -NH$_2$ group (lewis base) of KH550 can interact with B atom (Lewis acid) of BN; on the other hand, the mechanical impact and shear forces generated by grinding balls could make KH550 molecules easy to insert the gap of BN layers. As a result, a high yield of BN nanosheets was obtained, and the KH550 were also grafted onto the surface of BN nanosheets under the action of ball-milling internal heat. Using ball-milled KH550-modified BN as fillers, epoxy nanocomposites were prepared after blending and curing, and the thermal conductivity and dynamic thermomechanical property of epoxy nanocomposites were investigated.

Scheme 1. Illustration for preparation of BN/epoxy nanocomposites (**A**) and modification process of BN fillers (**B**).

2. Materials and Methods

2.1. Materials

BN powders (hexagonal, 99.9%, ~3 μm), γ-aminopropyl triethoxy silane (KH550, 97%), and absolute ethyl alcohol (99.7%) were purchased from McLean Biochemical Technology Co., Ltd. (Shanghai, China). Epoxy resin (E-51), methylhexahydrophthalic anhydride (MHHPA), and 2.4.6-Tri(dimethylaminomethyl)Phenol (DMP-30) were provided by Haining Hailong Chemical Co., Ltd. (Jiaxing, China). All raw materials or reagents were not further purified before use.

2.2. Preparation of BN Nanofillers

Raw BN (R-BN) powders were added into a stainless steel ball grinding tank containing 50 stainless steel grinding balls, followed by pouring a mixed solution of H_2O and CH_3CH_2OH (volume ratio: 3:7) containing 1% KH550 into the ball-grinding tank at a ratio of 0.05 g·mL^{-1} [23]. Subsequently, the ball-grinding tanks were loaded onto the planetary ball mill, and the ball-milling process was performed at a speed of 500 r/min for 8 h at room temperature. After this, the BN powders in mixed solution were collected by vacuum filtration, and washed three times with deionized water to remove the unreacted KH550 molecules. After they were dried for 12 h at 60 °C in a vacuum-drying oven, the KH550-modified BN nanofillers synthesized by one-step ball-milling (BM@KH550-BN) were obtained. As a contrast, another BN nanofiller, denoted as BM-BN, was prepared according to the same procedure as BM@KH550-BN, except no KH550 was used in the mixed solution of H_2O and CH_3CH_2OH. In addition, the BM-BN were also post-modified with 1% KH550 solution at 60 °C for 2 h to obtain the conventional KH550-modified BN nanofillers (BM-KH550-BN).

2.3. Preparation of BN/Epoxy Nanocomposites

The desired ratios of BN nanofillers were ultrasonically dispersed into absolute ethyl alcohol at a concentration of 0.05 g·mL^{-1}, and then the epoxy resin was added into the BN dispersion. With vigorous stirring, the mixed solution of BN/epoxy was reduced-pressure distilled at 60 °C until the absolute ethyl alcohol was fully evaporated. Afterwards, the MHHPA and DMP-30 were added into the mixed solution in sequence, followed by stirring for 30 min at room temperature in a vacuum. Finally, the BN/epoxy nanocomposites were obtained after BN/epoxy mixture was cured at 120 °C for 2 h in stainless steel mold. Herein, the weight ratio of epoxy resin, MHHPA, and DMP-30 was set to 100:85:0.8, and the weight content of BN nanofillers was set to 1%, 4%, 7%, and 10%.

2.4. Characterization and Measurement

The micromorphology of BN nanofillers and BN/epoxy nanocomposites were investigated by field-emission scanning electron microscope (FE-SEM, Apero 2, a resolution of 1 nm, Thermo Scientific, Waltham, USA) equipped with an energy-dispersive spectrum (EDS) component and transmission electron microscope (TEM, Talos F200X, Thermo Scientific, Waltham, MA, USA). The samples for EDS were prepared by natural deposition BN nanofillers on a polished Ti substrate. The fracture surfaces of the epoxy nanocomposites were prepared by cryo fractured in liquid nitrogen. The optical photographs of BN nanofillers solution (2 mg·mL^{-1}) were taken on a white paper by a smartphone (Mate 40, Huawei, Shenzhen, China). Fourier-transform infrared (FTIR) spectra of BN nanofillers were recorded at room temperature by an infrared spectrometer (Bruker, Vertex 70) with the range from 400 cm^{-1} to 4000 cm^{-1}. The thermogravimetric analysis (TGA) was performed by a thermal gravimetric analyzer (TA Instruments (New Castle, DE, USA), Q50) at a heating rate of 5 °C/min from 30 to 800 °C under a nitrogen atmosphere according to the ASTM E1131 standard. Thermal conductivity of BN/epoxy nanocomposites (50 mm in diameter and 20 mm in height) was tested with a quick thermal conductivity meter (Xiangyi instrument, DRE-III). The dynamic thermomechanical properties of BN/epoxy nanocomposites were measured with DMA 850 (TA Instruments) according to the ASTM standard D4065-94. The nanocomposite samples were cut into strip-shaped specimens ($60 \times 10 \times 4$ mm^3) to match with machine. A temperature scan was conducted from 50 to 200 °C with a heating rate of 5 °C/min at a frequency of 1 Hz and a strain of 0.05% in a double cantilever mode. The storage modulus and tan δ (the ratio of loss modulus to storage modulus) were obtained from DMA analysis, and the glass transition temperature (Tg) was gained from the peak value of tan δ.

3. Results and Discussion

3.1. Morphology and Element Mapping of BN Nanofillers

Using raw BN (R-BN) powders as materials, three different kinds of BN nanofillers were prepared by ball-milling method, i.e., BM-BN, BM-KH550-BN, and BM@KH550-BN. As shown in Figure 1A, the R-BN particles present the typical stacked layered structure and uneven sizes, with an approximate size range from 1 to 8 μm. After ball-milling treatment with an aqueous solution (Figure 1B), the obtained BN particles seem to have a smaller size and thinner thickness than R-BN, and some BN nanosheets appear in BM-BN. This can be attributed to the impact and shear effect of the ball-milling process [30]. It is also found that the post-modification with KH550 coupling agents does not change the morphology of BM-BN nanofillers (Figure 1C). When aqueous solution is substituted with KH550 solution in the ball-milling process, a large number of BN nanosheets are observed in the obtained BM@KH550-BN nanofillers (Figure 1D), suggesting the successful exfoliation of R-BN particles in the KH550-assisted ball-milling process.

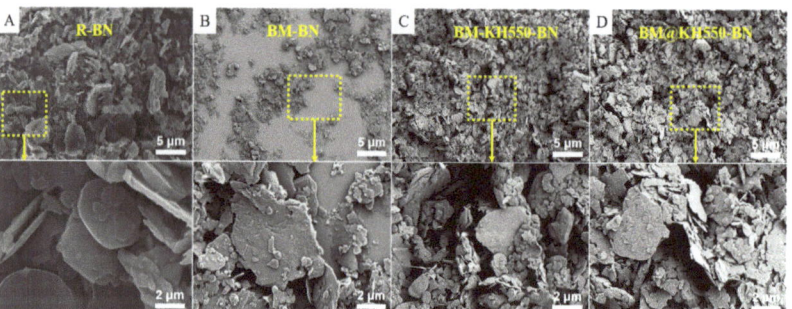

Figure 1. SEM images of different BN nanofillers (**A–D**).

To estimate the ratio of BN nanosheets (exfoliated BN) in different nanofillers, a centrifugal separation method was performed at 3000 rpm. Results indicate that the BM-BN and BM-KH550-BN exhibit a similar yield of BN nanosheets (~38%), indicating that the surface modification with KH550 does not change the amount of BN nanosheets. However, a yield of BN nanosheets as high as 73.3% is gained in BM@KH550-BN nanofillers (Figure 2A). This may be ascribed to the fact that the electron donor -NH$_2$ in KH550 can readily insert into the layered structure of BN with the assistance of mechanical force and weaken the atom interactions between electron acceptor B atom and neighboring planar N atom, resulting in the higher yield of BN nanosheets [31]. For the BM@KH550-BN nanofillers, the unambiguous BN nanosheets structure are observed in TEM images (Figure 2B). The dispersibility of BN nanofillers were further evaluated by observing the stability of BN ethanol dispersion. It can be seen from Figure 2C that the R-BN dispersion almost becomes clear, and BM-BN dispersion is a little turbid after standing for 24 h. It is noteworthy that the BM-KH550-BN and BM@KH550-BN dispersion are still relatively turbid after 24 h, especially BM@KH550-BN, which is ascribed to the fact that the KH550 silane coupling agent can improve the dispersibility of nanofillers, and, simultaneously, the high ratio of BN nanosheets in BM@KH550-BN is also conducive to the dispersibility of nanofillers.

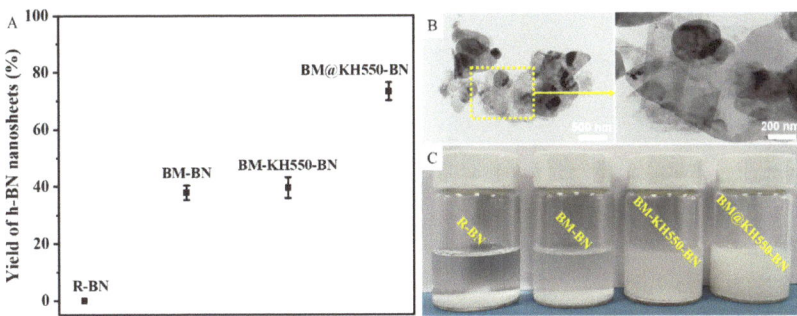

Figure 2. The yield of BN nanosheets in different BN nanofillers (**A**); TEM images of BM@KH550-BN nanofillers (**B**); optical photographs of different BN nanofillers ethanol dispersion after standing for 24 h (**C**).

The key element distribution of different BN nanofillers were detected by EDS (BN nanofillers deposited on polished Ti substrate). The element mapping images in Figure 3 show that all the BN nanofillers exhibit a strong nitrogen (N) element signal (green) with uniform distribution. For the silicon (Si) element (red), it is found that no Si element signal is detected in R-BN and BM-BN samples. However, the apparent Si element signals are observed in the BM-KH550-BN and BM@KH550-BN nanofillers, which implies that the KH550 molecules are successfully bound to the surface of BN nanofillers during the ball-milling process, since the Si element only exists in KH550 molecules.

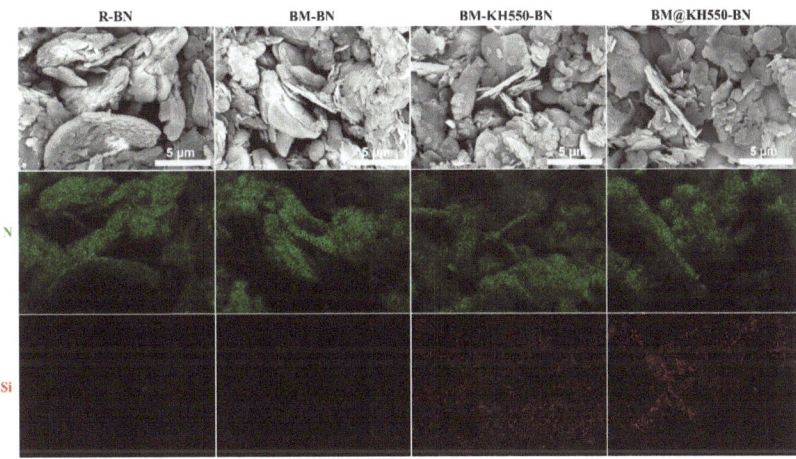

Figure 3. Nitrogen (N) and silicon (Si) element mapping images of different BN nanofillers.

3.2. FTIR Analysis of BN Nanofillers

In order to further investigate the chemical compositions of BN nanofillers, the FTIR spectra of BN nanofillers were recorded at room temperature. As shown in Figure 4, the two strong adsorption bands at 1380 and 810 cm^{-1} are indexed to the in-plane stretching vibration and out-of-plane bending vibration of B-N bond, respectively [32,33]. It is found that the adsorption band of typical -OH stretching vibration at 3431 cm^{-1} is hardly found in R-BN, but an obvious adsorption band is observed in ball-milling-treated BN nanofillers (BM-BN, BM-KH550-BN, and BM@KH550-BN), which indicates that the ball-milling treatment enhances the degree of hydroxylation of R-BN [34]. It is noteworthy that a new adsorption band at 1078 cm^{-1}, which is indexed to the stretching vibration of the Si-O bond of KH550, emerges in BM-KH550-BN and BM@KH550-BN, implying that KH550 is successfully grafted onto the surface of BN nanofillers through the ball-milling process [35].

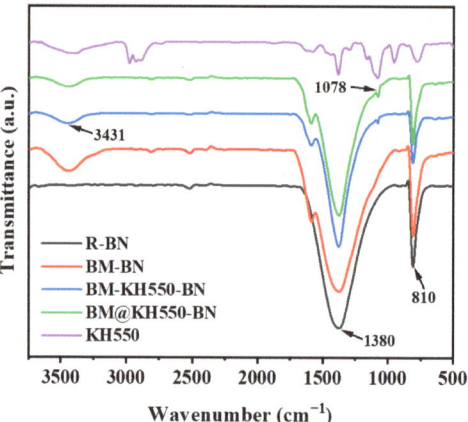

Figure 4. FTIR spectra of different BN nanofillers.

3.3. TGA Analysis of BN Nanofillers

The thermal stability of BN nanofillers was analyzed by TGA. It can be seen from Figure 5 that the R-BN shows little weight loss (only 0.25%) over the whole temperature range, but the weight loss of BM-BN is estimated to 3.26%, which is mainly ascribed to the removal of adsorbed water and -OH groups on the surface of BM-BN [36]. In addition, the weight loss of BM-KH550-BN and BM@KH550-BN are determined to 5.96% and 9.97%, respectively, implying the higher content of KH550 molecules on the BN surface. The weight loss of BM-KH550-BN and BM@KH550-BN is mainly attributed to the removal of adsorbed water and -OH groups, and the decomposition of KH550 molecules [37]. The weight loss of all the BN nanofillers mainly takes place between 100 °C and 400 °C. It is noteworthy that the initial weight loss (below 200 °C) of BM-BN is larger than other BM-KH550-BN and BM@KH550-BN, which may be due to the higher level of -OH groups, as confirmed by FTIR spectra. Compared to BM-KH550-BN, the higher content of KH550 in BM@KH550-BN usually indicates the better interfacial compatibility between nanofillers and epoxy resin.

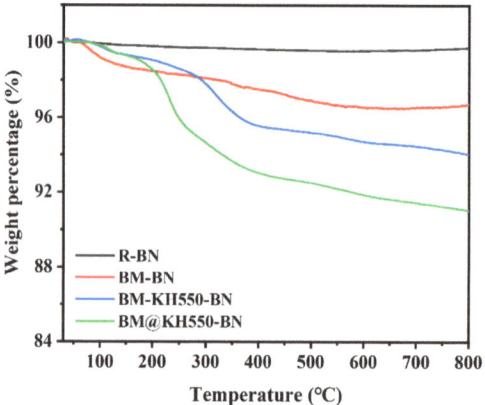

Figure 5. TGA curves of different BN nanofillers.

3.4. Thermal Conductivity Analysis of BN Nanofillers

Using the as-prepared BN nanoparticles as fillers, the BN/epoxy nanocomposites with different content of BN nanofillers were prepared, and the thermal conductivity property of BN/epoxy nanocomposites was measured by thermal conductivity meter. The testing result in Figure 6A shows that the thermal conductivity of the neat epoxy is only 0.208 W m^{-1}K^{-1}, which is attributed to the fact that the highly cross-linked network structure of epoxy resin presents a low crystallinity and orderliness, which is not enough to make phonons propagate quickly [38]. When the four different BN nanofillers are added into epoxy matrix, the thermal conductivity of epoxy resin has an insignificant enhancement at 1 wt%. Although the weight content of BN nanofillers increases to 4%, the highest thermal conductivity with a value of 0.328 W m^{-1}K^{-1} is obtained in the BM@KH550/epoxy nanocomposite, which is only 57.7% higher than neat epoxy. This can be understood by the fact that the small amount of thermally conductive BN nanofillers make it difficult to form an effective heat conduction path inside the epoxy resin. After the content of BN nanofillers is greater than 4%, the thermal conductivity of all the BN/epoxy nanocomposites enhances significantly with the increase in BN nanofillers, due to the gradual contact with each other among BN nanofillers. It is found that the thermal conductivity of R-BN/epoxy, BM-BN/epoxy, BM-KH550/epoxy, and BM@KH550/epoxy nanocomposites at 7 wt% increase to 0.307, 0.354, 0.391, and 0.431 W m^{-1}K^{-1}, respectively. When the content of BN nanofillers is further increased to 10 wt%, the enhancement of thermal conductivity of four epoxy nanocomposites presents relatively large differences. For R-BN/epoxy nanocomposites, the thermal conductivity is only enhanced to 0.325 W m^{-1}K^{-1} from 0.307 W m^{-1}K^{-1}, which may be ascribed to the fact that the poor dispersion of R-BN nanofillers and the worse interfacial compatibility between R-BN nanofillers and epoxy matrix produces a lot of R-BN aggregation formations inside the R-BN/epoxy nanocomposite, hindering the conduction of heat. However, the BM-BN/epoxy, BM-KH550/epoxy, and BM@KH550/epoxy nanocomposites are determined to be 0.433, 0.507, and 0.615 W m^{-1}K^{-1}, which is increased by 108.2%, 143.8%, and 195.7%, respectively (Figure 6B), compared to the thermal conductivity of neat epoxy resin. The highest thermal conductivity in BM@KH550/epoxy nanocomposites may be ascribed to the good interfacial compatibility between the BM@KH550-BN nanofillers and epoxy matrix and the high content of BN nanosheets in BM@KH550-BN nanofillers, forming more efficient thermal conductive paths in BM@KH550/epoxy nanocomposite.

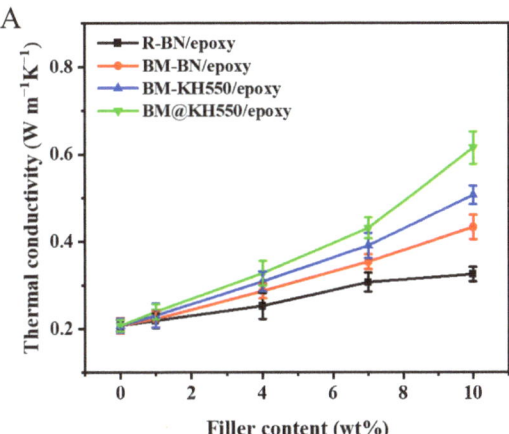

Figure 6. *Cont.*

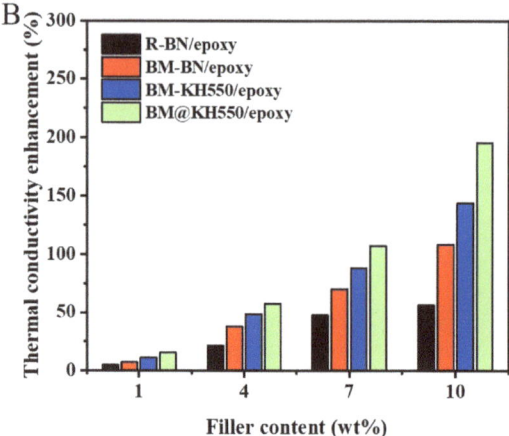

Figure 6. Thermal conductivity (**A**) and thermal conductivity enhancement (**B**) of different BN/epoxy nanocomposites.

3.5. Thermomechanical Properties of the Epoxy Nanocomposites

The dynamic thermomechanical properties of BN/epoxy nanocomposites were further tested. The test results show that the storage modulus of neat epoxy resin is 2.28 GPa (Figure 7A). After incorporation of the 10 wt% BN nanofillers, the storage modulus of neat epoxy resin enhances in different degrees. For R-BN/epoxy, the storage modulus only increases from 2.28 GPa to 2.49 GPa, mainly ascribed to the easy aggregations of raw BN particles at high content. For BM-BN/epoxy, a storage modulus of 2.63 GPa is achieved, due to the existence of some BN nanosheets in the BM-BN nanofillers. It is found that the storage moduli of BM-KH550/epoxy and BM@KH550/epoxy nanocomposites at 10 wt% reaches 2.85 and 3.09 GPa at 50 °C, respectively, which is 25% and 35.6% higher than that of neat epoxy resin, respectively. This is mainly ascribed to the fact that the reactive amino groups on the surface of the BM-KH550 and BM@KH550 nanofillers enhance the interfacial bonding strength between nanofillers and epoxy matrix, and the surface of BM@KH550 has more abundant reactive amino groups [39]. In addition, the glass transition temperature (Tg) of all the epoxy nanocomposites at 10 wt% were analyzed by the peak value of tan δ curves. As shown in Figure 7B, the Tg of neat epoxy resin is determined to be 143.9 °C, and an insignificant enhancement in Tg is found in the R-BN/epoxy (145.6 °C). For the BM-BN/epoxy and BM-KH550-BN/epoxy nanocomposites, the Tg is determined to be 148.9 °C and 153 °C, respectively, which is higher than the Tg of neat epoxy. Expectedly, the BM@KH550-BN/epoxy nanocomposite presents the highest Tg, with a value reaching up to 156.3 °C, which is 12.4 °C higher than that of neat epoxy resin. The enhanced Tg in the BM@KH550-BN/epoxy nanocomposite is mainly ascribed to the fact that the covalent connection between amino groups in BM@KH550-BN and epoxy groups in the epoxy molecule chain restrain the mobility of the local epoxy molecules around the BM@KH550-BN nanofillers, enhancing the degree of cross-linking [40]. It is worth noting that the height of tan δ peak reduces gradually from neat epoxy to the four kinds of epoxy nanocomposites (R-BN/epoxy, BM-BN/epoxy, BM-KH550/epoxy, and BM@KH550/epoxy nanocomposites), implying a decrease in the amount of mobile epoxy chains in epoxy nanocomposites [41].

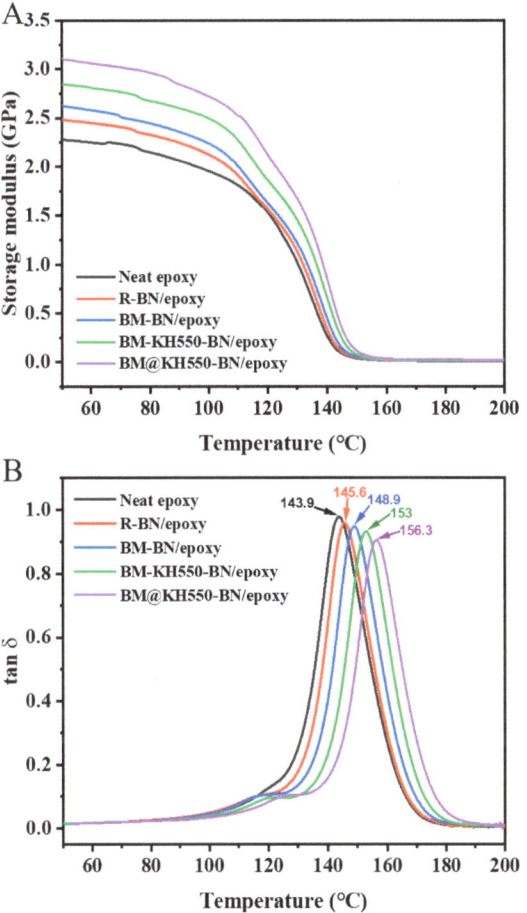

Figure 7. Temperature-dependent storage modulus (**A**) and tan δ (**B**) curves of different BN/epoxy nanocomposites at 10 wt%.

3.6. Effectiveness of the BN Nanofillers and Constrained Region of the Epoxy Nanocomposites

To explore the effect of different BN nanofillers on the dynamic thermomechanical properties of epoxy matrix, the effectiveness of BN nanofillers and constrained region of the epoxy nanocomposites were analyzed. The factor β_f represents the effectiveness of fillers on the moduli of composites, and can be given by the equation:

$$\beta_f = \frac{(G_g'/G_r')\ \text{composite}}{(G_g'/G_r')\ \text{matrix}} \quad (1)$$

where G_g' and G_r' are the storage modulus in the glassy region and rubbery region, respectively [42]. The G_g', G_r', and β_f of epoxy nanocomposites are listed in Table 1. The lower β_f value represents the higher effectiveness of the filler. Clearly, all the epoxy nanocomposites have a lower β_f value than neat epoxy. Also, the BM@KH550-BN/epoxy nanocomposite exhibits the lowest β_f value (79.2) among the four types of BN nanofillers (R-BN/epoxy, BM-BN/epoxy, BM-KH550/epoxy, and BM@KH550/epoxy nanocomposites), indicating the strongest reinforcing effect of BM@KH550 nanofillers on the epoxy matrix. In addition, the incorporation of fillers in polymer matrix can also affect the entanglement dynamics and mobility of the polymer chains around the fillers. Herein, the volume fraction of the constrained region can be quantitatively calculated by the tan δ value. The relationship between tan δ and energy loss fraction (W) of the polymer nanocomposites is given by the equation [43]:

$$W = \frac{\pi \tan \delta}{\pi \tan \delta + 1} \tag{2}$$

where the energy loss fraction W at the tan δ peak can be expressed by the dynamic viscoelastic data with the following equation:

$$W = \frac{(1-C) W_0}{1 - C_0} \tag{3}$$

where C is the volume fraction of the constrained region of the composite, and W_0 and C_0 are the energy loss fraction and volume fraction of the constrained region of neat epoxy, respectively [44]. Equation (3) can be rearranged as follows:

$$C = 1 - \frac{(1-C_0) W}{W_0} \tag{4}$$

Table 1. Some characteristic parameters of epoxy nanocomposites at 10 wt% derived from DMA.

Samples	G_g' (GPa)	G_r' (GPa)	β_f	W	C
Neat epoxy	2.28	0.011	207.3	0.754	0
R-BN/epoxy	2.49	0.013	191.5	0.751	0.004
BM-BN/epoxy	2.63	0.018	146.1	0.747	0.009
BM-KH550-BN/epoxy	2.85	0.026	109.6	0.746	0.011
BM@KH550-BN/epoxy	3.09	0.039	79.2	0.739	0.02

Herein, C_0 is taken as zero for neat epoxy, and the C value of epoxy nanocomposites are calculated by Equation (4). As listed in Table 1, the volume fraction of the constrained region in the R-BN/epoxy nanocomposite is only 0.004, indicating the weak effect of R-BN on the epoxy chains. However, the volume fraction of the constrained region in the BM-BN/epoxy and BM-KH550/epoxy is higher than that of the R-BN/epoxy nanocomposite, implying the stronger effect of BM-BN and BM-KH550 nanofillers on the epoxy chains than R-BN nanofillers. For BM@KH550-BN/epoxy nanocomposite at 10 wt%, the volume fraction of the constrained region reaches 0.02, which is higher than that of the other epoxy nanocomposites. This may be explained by the fact that a stronger interfacial attraction exists between BM@KH550-BN nanofillers and epoxy matrix, better restricting the mobility of the epoxy chains [45].

3.7. Fracture Surface Analysis of the Epoxy Nanocomposites

As is well known, the distribution of fillers in polymer matrix plays an important role in the thermal conductivity and thermomechanical properties of epoxy nanocomposites. Since the BM@KH550-BN/epoxy nanocomposite achieves the optimal thermal conductivity and thermomechanical properties in all the epoxy nanocomposites, it is necessary to investigate the distribution of BM@KH550-BN nanofillers at different loading content in epoxy matrix. Thus, the morphology of fracture surfaces of BM@KH550-BN/epoxy nanocomposites was analyzed. As shown in Figure 8A, the neat epoxy displays a flat and clean fracture surface, which is a typical characteristic of epoxy resin [46,47]. After the incorporation of 1% BM@KH550-BN nanofillers into the epoxy matrix, an uneven fracture surface and little BN particles (yellow arrow) embedded in epoxy matrix are observed, and the small area of irregular cracks and slightly rougher fracture surface for the composites are also observed (Figure 8B). It is found that the unevenness of the fracture surface of the epoxy nanocomposite and the amount of BN particles increase obviously with the increase in BM@KH550-BN nanofillers, and the large area of irregular cracks and the roughness of the fracture surface is also enhanced greatly (Figure 8C–E). When the weight content of BM@KH550-BN nanofillers reaches 10 wt%, a large number of BN particles emerge on the fracture surface of the epoxy nanocomposite. It should be noted that the BM@KH550-BN nanofillers still display a uniform distribution in the epoxy matrix even if the content of BM@KH550-BN is 10%, which indicates the good dispersion of BM¬@KH550-BN nanofillers in epoxy matrix.

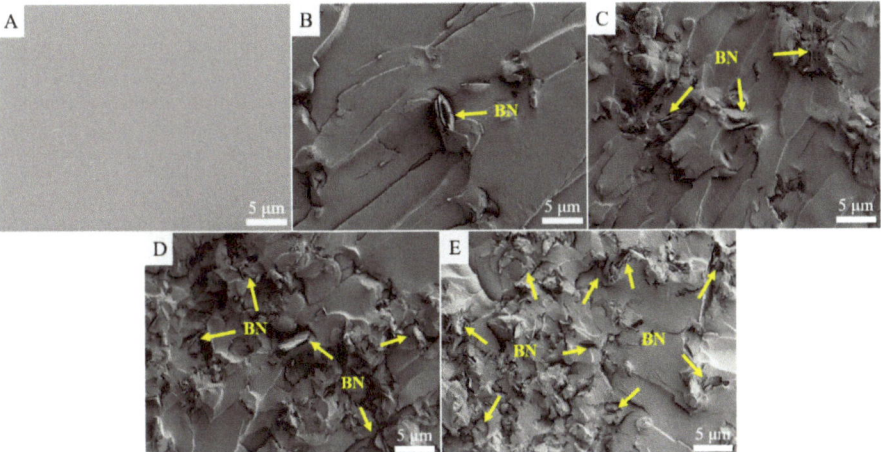

Figure 8. The SEM images of fracture surface of neat epoxy (**A**) and BM@KH550-BN/epoxy nanocomposites with different content: 1% (**B**), 4% (**C**), 7% (**D**), 10% (**E**).

4. Conclusions

In summary, the KH550-modified BN (BM@KH550-BN) nanofillers were successfully prepared via a one-step ball-milling route, and a high proportion of BN nanosheets and excellent dispersion stability were confirmed in BM@KH550-BN nanofillers. As a result, the BM@KH550-BN nanofillers can enhance the thermal conductivity of epoxy nanocomposites significantly, with an enhancement of 195.7% at 10 wt%, compared to neat epoxy resin. In addition, the storage modulus and glass transition temperature (Tg) of BM@KH550-BN/epoxy nanocomposites also increase by 35.6% and 12.4 °C, respectively. The calculated results from the dynamical mechanical analysis show that the BM@KH550-BN nanofillers have the strongest reinforcing effect and the higher volume fraction of constrained region in all the epoxy nanocomposites. This is mainly attributed to the high ratio of BN nanosheets in BM@KH550-BN nanofillers and the good interfacial compatibility between nanofillers and matrix. This work develops a simple way to prepare high thermally conductive BN nanofillers, showing a good application prospect in the field of thermally conductive polymer nanocomposites.

Author Contributions: Conceptualization, L.W.; methodology, B.T. and M.C.; software, J.G.; validation, B.T.; formal analysis, Y.Y. (Yaru Yang); investigation, B.T.; resources, L.W.; data curation, Y.Y. (Yongbo Yao) and J.Y.; writing—original draft B.T.; writing—review and editing, L.W. and T.W.; visualization, J.D.; supervision, L.W. and T.W.; project administration, Y.Y. (Yongbo Yao) and J.Y.; funding acquisition, B.T., L.W. and T.W. All authors have read and agreed to the published version of the manuscript.

Funding: This work is financially supported by Zhejiang Provincial Natural Science Foundation of China (No. LY22E020013), the Zhejiang Public Welfare Technology Application Research Project (No. LGC20E010003), and the Foundation of Zhejiang Educational Committee (No. Y201738304).

Institutional Review Board Statement: Not applicable.

Informed Consent Statement: Not applicable.

Data Availability Statement: The data presented in this study are available on request from the corresponding author.

Conflicts of Interest: The authors declare no conflict of interest. The funders had no role in the design of the study; in the collection, analyses, or interpretation of data; in the writing of the manuscript; or in the decision to publish the results.

References

1. Chen, H.Y.; Ginzburg, V.V.; Yang, J.; Yang, Y.F.; Liu, W.; Huang, Y.; Du, L.B.; Chen, B. Thermal conductivity of polymer-based composites: Fundamentals and applications. *Prog. Polym. Sci.* **2016**, *59*, 41–85. [CrossRef]
2. Li, L.; Zhou, B.; Han, G.J.; Feng, Y.Z.; He, C.G.; Su, F.M.; Ma, J.M.; Liu, C.T. Understanding the effect of interfacial engineering on interfacial thermal resistance in nacre-like cellulose nanofiber/graphene film. *Compos. Sci. Technol.* **2020**, *197*, 108229. [CrossRef]
3. Moore, A.L.; Shi, L. Emerging challenges and materials for thermal management of electronics. *Mater. Today* **2014**, *17*, 163–174. [CrossRef]
4. Lin, Z.Y.; Mcnamara, A.; Liu, Y.; Moon, K.S.; Wong, C.P. Exfoliated hexagonal boron nitride-based polymer nanocomposite with enhanced thermal conductivity for electronic encapsulation. *Compos. Sci. Technol.* **2014**, *90*, 123–128. [CrossRef]
5. Song, S.H.; Park, K.H.; Kim, B.H.; Choi, Y.W.; Jun, G.H.; Lee, D.J.; Kong, B.S.; Paik, K.W.; Jeon, S. Enhanced thermal conductivity of epoxy-graphene composites by using non-oxidized graphene flakes with non-covalent functionalization. *Adv. Mater.* **2013**, *25*, 732–737. [CrossRef]
6. Niu, H.Y.; Guo, H.C.; Ren, Y.J.; Ren, L.C.; Lv, R.C.; Kang, L.; Bashir, A.; Bai, S.L. Spherical aggregated BN/AlN filled silicone composites with enhanced through-plane thermal conductivity assisted by vortex flow. *Chem. Eng. J.* **2022**, *430*, 133155. [CrossRef]
7. Ji, Z.M.; Liu, W.Y.; Ouyang, C.G.; Li, Y.B. High thermal conductivity thermoplastic polyurethane/boron nitride/liquid metal composites: The role of the liquid bridge at the filler/filler interface. *Mater. Adv.* **2021**, *2*, 5977–5985. [CrossRef]
8. Zhu, Z.Z.; Li, C.W.; E, S.F.; Xie, L.Y.; Geng, R.J.; Lin, C.T.; Li, L.Q.; Yao, Y.G. Enhanced thermal conductivity of polyurethane composites via engineering small/large sizes Interconnected boron nitride nanosheets. *Compos. Sci. Technol.* **2019**, *170*, 93–100. [CrossRef]
9. Ganguli, S.; Roy, A.K.; Anderson, D.P. Improved thermal conductivity for chemically functionalized exfoliated graphite/epoxy composites. *Carbon* **2008**, *46*, 806–817. [CrossRef]
10. Wang, F.Z.; Drzal, L.T.; Qin, Y.; Huang, Z.X. Mechanical properties and thermal conductivity of graphene nanoplatelet/epoxy composites. *J. Mater. Sci.* **2015**, *50*, 1082–1093. [CrossRef]
11. Liu, Z.; Li, J.H.; Liu, X.H. Novel functionalized BN nanosheets/epoxy composites with advanced thermal conductivity and mechanical properties. *ACS Appl. Mater. Inter.* **2020**, *12*, 6503–6515. [CrossRef]
12. Du, P.Y.; Wang, Z.X.; Ren, J.W.; Zhao, L.H.; Jia, S.L.; Jia, L.C. Scalable polymer-infiltrated boron nitride nanoplatelet films with high thermal conductivity and electrical insulation for thermal management. *ACS Appl. Electron. Mater.* **2022**, *4*, 4622–4631. [CrossRef]
13. Meng, Q.S.; Han, S.S.; Liu, T.Q.; Ma, J.; Ji, S.D.; Dai, J.B.; Kang, H.L.; Ma, J. Noncovalent modification of boron nitrite nanosheets for thermally conductive, mechanically resilient epoxy nanocomposites. *Ind. Eng. Chem. Res.* **2020**, *59*, 20701–20710. [CrossRef]
14. Li, Y.; Gan, G.; Huang, Y.; Yu, X.; Cheng, J.; Liu, C. Ag-NPs/MWCNT composite-modified silver-epoxy paste with improved thermal conductivity. *RSC Adv.* **2019**, *9*, 20663–20669. [CrossRef] [PubMed]
15. Bao, D.; Gao, Y.Y.; Cui, Y.X.; Xu, F.; Shen, X.S.; Geng, H.L.; Zhang, X.G.; Lin, D.Y.; Zhu, J.; Wang, H.Y. A novel modified expanded graphite/epoxy 3D composite with ultrahigh thermal conductivity. *Chem. Eng. J.* **2022**, *433*, 133519. [CrossRef]
16. Yu, A.P.; Ramesh, P.; Sun, X.B.; Bekyarova, E.; Itkis, M.E.; Haddon, R. Enhanced thermal conductivity in a hybrid graphite nanoplatelet-carbon nanotube filler for epoxy composites. *Adv. Mater.* **2008**, *20*, 4740–4744. [CrossRef]
17. Bryning, M.B.; Milkie, D.E.; Islam, M.F.; Kikkawa, J.M.; Yodh, A.G. Thermal conductivity and interfacial resistance in single-wall carbon nanotube epoxy composites. *Appl. Phys. Lett.* **2005**, *87*, 161909. [CrossRef]
18. Liu, C.; Wu, W.; Drummer, D.; Shen, W.T.; Wang, Y.; Schneider, K.; Tomiak, F. ZnO nanowire-decorated Al_2O_3 hybrids for improving the thermal conductivity of polymer composites. *J. Mater. Chem. C* **2020**, *8*, 5380–5388. [CrossRef]
19. Hou, J.; Li, G.; Yang, N.; Qin, L.; Grami, M.E.; Zhang, Q.; Wang, N.; Qu, X. Preparation and characterization of surface modified boron nitride epoxy composites with enhanced thermal conductivity. *RSC Adv.* **2014**, *4*, 44282–44290. [CrossRef]
20. Zhang, K.; Feng, Y.; Wang, F.; Yang, Z.; Wang, J. Two dimensional hexagonal boron nitride (2D-hBN): Synthesis, properties and applications. *J. Mater. Chem. C* **2017**, *5*, 11992–12022. [CrossRef]
21. Shi, X.; Wang, K.; Tian, J.; Yin, X.; Guo, B.; Xi, G.; Wang, W.; Wu, W. Few-layer hydroxyl-functionalized boron nitride nanosheets for nanoscale thermal management. *ACS Appl. Nano Mater.* **2020**, *3*, 2310–2321. [CrossRef]
22. Zhi, C.; Bando, Y.; Tang, C.; Kuwahara, H.; Golberg, D. Large-scale fabrication of boron nitride nanosheets and their utilization in polymeric composites with improved thermal and mechanical properties. *Adv. Mater.* **2009**, *21*, 2889–2893. [CrossRef]
23. Li, L.H.; Chen, Y.; Behan, G.; Zhang, H.; Petravic, M.; Glushenkov, A.M. Large-scale mechanical peeling of boron nitride nanosheets by low-energy ball milling. *J. Mater. Chem.* **2011**, *21*, 11862–11866. [CrossRef]

24. Song, L.; Ci, L.; Lu, H.; Sorokin, P.B.; Jin, C.; Ni, J.; Kvashnin, A.G.; Kvashnin, D.G.; Lou, J.; Yakobson, B.I.; et al. Large scale growth and characterization of atomic hexagonal boron nitride layers. *Nano Lett.* **2010**, *10*, 3209–3215. [CrossRef] [PubMed]
25. Cui, Z.; Oyer, A.J.; Glover, A.J.; Schniepp, H.C.; Adamson, D.H. Large scale thermal exfoliation and functionalization of boron nitride. *Small* **2014**, *10*, 2352–2355. [CrossRef]
26. Jang, I.; Shin, K.H.; Yang, I.; Kim, H.; Kim, J.; Kim, W.H.; Jeon, S.W.; Kim, J.P. Enhancement of thermal conductivity of BN/epoxy composite through surface modification with silane coupling agents. *Colloid. Surface A* **2017**, *518*, 64–72. [CrossRef]
27. Lei, W.; Mochalin, V.N.; Liu, D.; Qin, S.; Gogotsi, Y.; Chen, Y. Boron nitride colloidal solutions, ultralight aerogels and freestanding membranes through one-step exfoliation and functionalization. *Nat. Commun.* **2015**, *6*, 8849. [CrossRef]
28. Zhao, L.; Yan, L.; Wei, C.; Wang, Z.; Jia, L.; Ran, Q.; Huang, X.; Ren, J. Aqueous-phase exfoliation and functionalization of boron nitride nanosheets using tannic acid for thermal management applications. *Ind. Eng. Chem. Res.* **2020**, *59*, 16273–16282. [CrossRef]
29. Agrawal, A.; Chandrakar, S. Influence of particulate surface treatment on physical, mechanical, thermal, and dielectric behavior of epoxy/hexagonal boron nitride composites. *Polym. Compos.* **2020**, *41*, 1574–1583. [CrossRef]
30. Chen, S.; Xu, R.; Liu, J.; Zou, X.; Qiu, L.; Kang, F.; Liu, B.; Cheng, H.M. Simultaneous production and functionalization of boron nitride nanosheets by sugar-assisted mechanochemical exfoliation. *Adv. Mater.* **2019**, *31*, 1804810. [CrossRef]
31. Yang, N.; Ji, H.F.; Jiang, X.X.; Qu, X.W.; Zhang, X.J.; Zhang, Y.; Liu, B.Y. Preparation of boron nitride nanoplatelets via amino acid assisted ball milling: Towards thermal conductivity application. *Nanomaterials* **2020**, *10*, 1652. [CrossRef]
32. Xiao, Q.; Zhan, C.; You, Y.; Tong, L.; Wei, R.; Liu, X. Preparation and thermal conductivity of copper phthalocyanine grafted boron nitride nanosheets. *Mater. Lett.* **2018**, *227*, 33–36. [CrossRef]
33. Xiao, F.; Naficy, S.; Casillas, G.; Khan, M.H.; Katkus, T.; Jiang, L.; Liu, H.; Li, H.; Huang, Z. Edge-hydroxylated boron nitride nanosheets as an effective additive to improve the thermal response of hydrogels. *Adv. Mater.* **2015**, *27*, 7196–7203. [CrossRef]
34. Wu, K.; Liao, P.; Du, R.N.; Zhang, Q.; Chen, F.; Fu, Q. Preparation of a thermally conductive biodegradable cellulose nanofiber/hydroxylated boron nitride nanosheet film: The critical role of edge-hydroxylation. *J. Mater. Chem. A* **2018**, *6*, 11863–11873. [CrossRef]
35. Zhou, W.Y. Effect of coupling agents on the thermal conductivity of aluminum particle/epoxy resin composites. *J. Mater. Sci.* **2011**, *46*, 3883–3889. [CrossRef]
36. Shang, X.J.; Zhu, Y.M.; Li, Z.H. Surface modification of silicon carbide with silane coupling agent and hexadecyl iodiele. *Appl. Surf. Sci.* **2017**, *394*, 169–177. [CrossRef]
37. Hao, L.F.; Gao, T.T.; Xu, W.; Wang, X.C.; Yang, S.Q.; Liu, X.G. Preparation of crosslinked polysiloxane/SiO$_2$ nanocomposite via in-situ condensation and its surface modification on cotton fabrics. *Appl. Surf. Sci.* **2016**, *371*, 281–288. [CrossRef]
38. Jiang, Y.; Shi, X.; Feng, Y.; Li, S.; Zhou, X.; Xie, X. Enhanced thermal conductivity and ideal dielectric properties of epoxy composites containing polymer modified hexagonal boron nitride. *Compos. Part A Appl. Sci. Manuf.* **2018**, *107*, 657–664. [CrossRef]
39. Zhang, Y.L.; He, X.Z.; Cao, M.; Shen, X.J.; Yang, Y.R.; Yi, J.; Guan, J.P.; Shen, J.X.; Xi, M.; Zhang, Y.J.; et al. Tribological and thermo-mechanical properties of TiO$_2$ nanodot-decorated Ti$_3$C$_2$/epoxy nanocomposites. *Materials* **2021**, *14*, 2509. [CrossRef]
40. Cao, Y.; Deng, Q.H.; Liu, Z.D.; Shen, D.Y.; Wang, T.; Huang, Q.; Du, S.Y.; Jiang, N.; Lin, T.C.; Yu, J.H. Enhanced thermal properties of poly (vinylidene fluoride) composites with ultrathin nanosheets of MXene. *RSC Adv.* **2017**, *7*, 20494–20501. [CrossRef]
41. Joy, J.; George, E.; Thomas, S.; Anas, S. Effect of filler loading on polymer chain confinement and thermomechanical properties of epoxy/boron nitride (h-BN) nanocomposites. *New J. Chem.* **2020**, *44*, 4494–4503. [CrossRef]
42. Sunny, A.; Vijayan, P.; Adhikari, R.; Mathew, S.; Thomas, S. Copper oxide nanoparticles in an epoxy network: Microstructure, chain confinement and mechanical behaviour. *Phys. Chem. Chem. Phys.* **2016**, *18*, 19655–19667. [CrossRef] [PubMed]
43. Zhang, X.; Loo, L. Study of glass transition and reinforcement mechanism in polymer/layered silicate nanocomposites. *Macromolecules* **2009**, *42*, 5196–5207. [CrossRef]
44. Vijayan, P.; Puglia, D.; Kenny, J.; Thomas, S. Effect of organically modified nanoclay on the miscibility, rheology, morphology and properties of epoxy/carboxyl-terminated (butadiene-co-acrylonitrile) blend. *Soft Matter* **2013**, *9*, 2899–2911. [CrossRef]
45. Chirayil, C.; Joy, J.; Mathew, L.; Koetz, J.; Thomas, S. Nanofibril reinforced unsaturated polyester nanocomposites: Morphology, mechanical and barrier properties, viscoelastic behavior and polymer chain confinement. *Ind. Crop. Prod.* **2014**, *56*, 246–254. [CrossRef]
46. Zhao, R.G.; Luo, W.B. Fracture surface analysis on nano-SiO$_2$/epoxy composite. *Mat. Sci. Eng. A-Struct.* **2008**, *483*, 313–315. [CrossRef]
47. Kim, B.C.; Park, S.W.; Lee, D.G. Fracture toughness of the nano-particle reinforced epoxy composite. *Compos. Struct.* **2008**, *86*, 69–77. [CrossRef]

Disclaimer/Publisher's Note: The statements, opinions and data contained in all publications are solely those of the individual author(s) and contributor(s) and not of MDPI and/or the editor(s). MDPI and/or the editor(s) disclaim responsibility for any injury to people or property resulting from any ideas, methods, instructions or products referred to in the content.

Article

Investigation on the Effect of Calcium on the Properties of Geopolymer Prepared from Uncalcined Coal Gangue

Qingping Wang [1,2,3,*], Longtao Zhu [1], Chunyang Lu [1], Yuxin Liu [1], Qingbo Yu [1] and Shuai Chen [1]

1 School of Materials Science and Engineering, Anhui University of Science and Technology, Huainan 232001, China
2 State Key Laboratory of Mining Response and Disaster Prevention and Control in Deep Coal Mines, Anhui University of Science and Technology, Huainan 232001, China
3 Anhui Generic Technology Research Center for New Materials from Coal-Based Solid Wastes, Anhui University of Science and Technology, Huainan 232001, China
* Correspondence: wqp.507@163.com

Abstract: In this paper, the influence of calcium on coal gangue and fly ash geopolymer is explored, and the problem of low utilization of unburned coal gangue is analyzed and solved. The experiment took uncalcined coal gangue and fly ash as raw materials, and a regression model was developed with the response surface methodology. The independent variables were the CG content, alkali activator concentration, and $Ca(OH)_2$ to NaOH ratio (CH/SH). The response target value was the coal gangue and fly-ash geopolymer compressive strength. The compressive strength tests and the regression model obtained by the response surface methodology showed that the coal gangue and fly ash geopolymer prepared with the content of uncalcined coal gangue is 30%, alkali activator content of 15%, and the value of CH/SH is 1.727 had a dense structure and better performance. The microscopic results demonstrated that the uncalcined coal gangue structure is destroyed under an alkali activator's action, and a dense microstructure is formed based on C(N)-A-S-H and C-S-H gel, which provides a reasonable basis for the preparation of geopolymers from the uncalcined coal gangue.

Keywords: calcium hydroxide; coal gangue; geopolymer; response surface method; alkali activator

Citation: Wang, Q.; Zhu, L.; Lu, C.; Liu, Y.; Yu, Q.; Chen, S. Investigation on the Effect of Calcium on the Properties of Geopolymer Prepared from Uncalcined Coal Gangue. *Polymers* 2023, 15, 1241. https://doi.org/10.3390/polym15051241

Academic Editor: Yung-Sheng Yen

Received: 7 January 2023
Revised: 23 February 2023
Accepted: 24 February 2023
Published: 28 February 2023

Copyright: © 2023 by the authors. Licensee MDPI, Basel, Switzerland. This article is an open access article distributed under the terms and conditions of the Creative Commons Attribution (CC BY) license (https:// creativecommons.org/licenses/by/ 4.0/).

1. Introduction

China is one of the countries with the largest coal reserves and produces the highest amount of coal annually. Coal will be the country's primary energy pillar in the future, but its mining process yields large quantities of solid waste. China currently has over 7 billion tons of coal gangue (CG) and more than 2600 large-scale CG hills [1]. Due to these severe pollution problems, many researchers are searching for materials that can effectively utilize CG; geopolymers have been proposed as a suitable inorganic binder because of their excellent durability, mechanical strength [2], and large consumption of CG [3,4]. Alumina and silica are the main elemental components of CG, and the main mineral components are quartz, kaolinite, and muscovite. Compared with the high pozzolanic property of fly ash [5], CG without special treatment has a stable structure and low activity. In the use of CG, its pozzolanic properties should be maximized through activator. The internal crystal phase composition of CG can be transformed via calcination at 700–900 °C [6,7]. Experimental findings obtained under different reaction conditions [8] have shown that an appropriate calcination temperature and time are needed to destroy the structures of kaolinite and quartz and improve the hydration reactivity of CG [9,10]. Raw CG is calcined at 550 °C to convert the kaolinite in its structure into metakaolin, which is then converted into mullite at 950 °C [11].

According to its calcium content, CG can be classified as high-, medium-, or low-calcium gangue. Compared with alkali-activated slag, geopolymers prepared by calcining

CG have poor strength and other properties, with their differences in calcium content being the main factor affecting the performance discrepancy between the two materials [12]. Lime or red mud and other materials can be added during the calcination of CG to increase its calcium content and thus enhance its activation efficiency. Li et al. [6] effectively improved the compressive strength of geopolymers by adding $CaSO_4$ and CaO as activation additives during CG calcination. Mineralizers, such as fluorite and gypsum, can also promote CG activation during calcination [13]. Compared with the commonly used high-calcium mineral admixtures, geopolymers exhibit denser structures, higher bulk densities, and better compressive strength [14,15]. Chen et al. [16] used CaO and SO_3 as admixtures to promote the geopolymer reaction, thereby increasing the number of gel products and compensating for the shrinkage of the geopolymer.

Due to considerable research data and activation processes, high-calcium additives can be added during the preparation of CG-based geopolymers to obtain ideal properties. However, researchers should focus on the energy consumption and environmental impact of geopolymers, a new gel material, before using them to replace traditional cement and other gel materials. CG calcination consumes large amounts of energy, so uncalcined CG (UCG) should be used to prepare geopolymers. Through experiments, Geng et al. [17] found that UCG can be mixed with red mud to prepare geopolymers with excellent development strength. Guo et al. [18] prepared suitable geopolymer grouting materials by compounding UCG, fly ash (FA), and slag. Preparing geopolymers from UCG is significant for energy consumption reduction and the environment.

To sum up, high-strength geopolymers can be prepared by mixing calcined coal gangue with fly ash and other raw materials. However, the energy consumption in the process of coal gangue calcination is an important factor limiting the utilization of coal gangue. In this study, UCG was used as the mineral raw material and mixed with FA to prepare geopolymer (CG–FA geopolymer (CFG)); the effect of calcium on the mechanical properties and microstructure of geopolymer was investigated by using the mixture of calcium hydroxide and sodium hydroxide as an activator. Desulfurization gypsum and a water-reducing agent were used as admixtures to adjust the compressive strength of the geopolymer. The response surface methodology (RSM) was chosen as the experimental design to optimize the factors that influence the optimization of geopolymer properties, and a multifactor, multiresponse collaborative optimization method was adopted. The CG content, alkali activator content, and $Ca(OH)_2$ to NaOH ratio (CH/SH) were the independent variables, and the compressive strengths for different curing periods were the target values. Relevant models were established to analyze the other, different experimental conditions. The CFG microstructure was examined via XRD and SEM–EDS. FTIR and ^{29}Si nuclear magnetic resonance (NMR) was used to characterize the changes in the chemical bonds and degrees of polymerization within the CFG structure.

2. Experimental
2.1. Materials Selection and Pretreatment

The chemical composition of UCG and FA is shown in Table 1. It can be seen from the table that UCG and FA contain a large amount of SiO_2 and Al_2O_3. Therefore, UCG and FA were selected as the silicon and aluminum raw materials. The coal gangue used in the experiment was obtained from Xingtai (Hebei, China). The X-ray diffraction (XRD) analysis and microstructure analysis is shown in Figures 1 and 2, which demonstrates that there are a large number of mineral structures such as kaolinite and quartz with stable structures in the original coal gangue. The fly ash was obtained from the Henan Datang Power Plant. Table 1 shows that the UCG and FA contain a large amount of SiO_2 and Al_2O_3, but the calcium oxide content of coal gangue is less than 3%, which belongs to low calcium coal gangue. Additive selection desulfurization gypsum (purity $\geq$ 93%). The alkali activator is prepared by blending sodium hydroxide and calcium hydroxide (analytical grade, purity $\geq$ 98%). The activator must be mixed with water and cooled to room temperature before use.

Table 1. Chemical composition of raw material (%).

Materials	SiO$_2$	Al$_2$O$_3$	Fe$_2$O$_3$	CaO	MgO	K$_2$O	TiO$_2$	Na$_2$O	MnO$_2$	P$_2$O$_5$
FA	59.61	28.85	3.82	3.03	1.02	1.77	1.77	0.78	0.06	0.13
CG	61.72	25.74	4.13	1.18	0.80	2.36	0.93	0.40	0.06	0.08

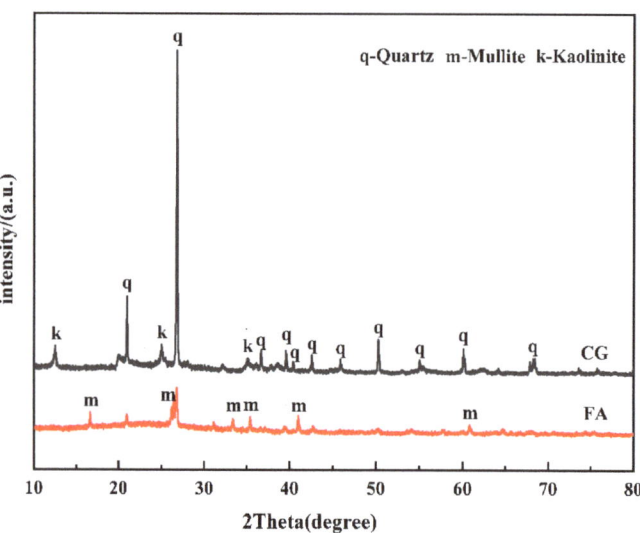

Figure 1. The XRD patterns of raw material.

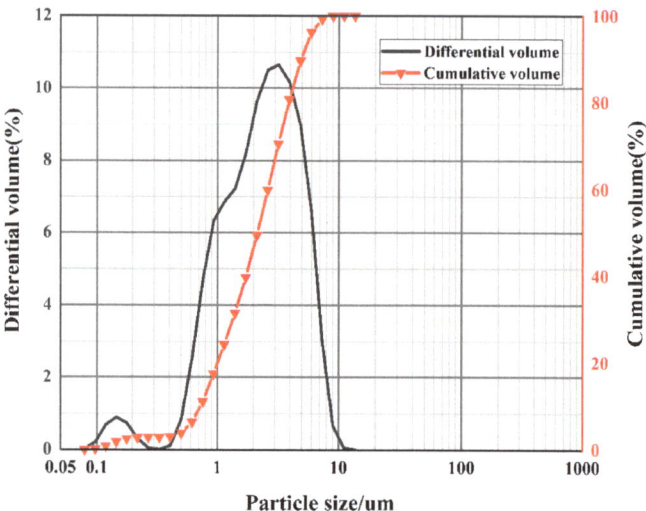

Figure 2. Particle size distributions of coal gangue.

The massive coal gangue shall be pretreated before use. First, use a hammer to crush the large pieces of raw coal gangue, and then put the lumpy coal gangue into the roller ball mill to crush it into coarse aggregate. After 120 min, grind the material with a planetary mill and sieve it to prepare fine coal gangue particles. The particle size distribution of coal gangue after ball milling is shown in Figure 2. The ground coal gangue can increase the specific surface area of the particles, significantly improve the gel activity, and facilitate the dissolution of the reaction process, which is the basis for the preparation of geopolymers.

2.2. Specimen Preparation and Experimental Design

In this study, based on the response surface design experiment, three main factors were selected to control the performance of geopolymer, and a total of 15 experiments were carried out. The different levels of the three independent variables are shown in Table 2. The content of coal gangue was selected as factor A, the activator which has an essential influence on the properties of silicon-alumina raw materials was chosen as factor B, and $Ca(OH)_2$ to NaOH ratio (CH/SH) was selected as factor C. The solid–liquid ratio was fixed at 0.7, and the additional amount of desulfurization gypsum was 12% of the solid waste silica-alumina material.

Table 2. Factors and levels in Box–Behnken design.

Independent Variable Factor	Coding and Level		
	−1	0	1
A, Coal gangue content	30%	40%	50%
B, Alkali activator content	15%	22.5%	30%
C, CH/SH	0.5	1.25	2

According to the experimental conditions, an appropriate amount of sodium hydroxide was dissolved in water with calcium hydroxide powder and allowed to stand for 12 h. The activator and raw materials were mixed in a pure slurry mixer at 1000~1200 r/min for 15 min. The mixed slurry was poured into a 40 mm × 40 mm × 40 mm six-joint mold. Then, the mold wrapped with plastic film was placed into a high temperature curing box at 80 °C for 24 h. The sample taken from the mold was placed in a curing box at 25 ± 1 °C and 95% relative humidity.

2.3. Macroscopic Test and Microstructure Characterization

The CFG macroscopic properties were tested in compressive strength: the experimental data were tested according to GB/T 17671-2020 by an automatic cement constant-force test machine.

The chemical bonds of the geopolymer were characterized in the reaction process by Fourier Transform Infrared Spectroscopy (FT-IR) (Nicolet380). The type and composition of the polymer product was determined by SEM (FlexSEM1000, Hitachi, Hong Kong, China) by observing the microstructure and morphology of the material. An X-ray diffractometer examined the mineral composition. The sample scanning speed was controlled at 5°/min, and the scanning range was 5~80° (2θ°). The chemical shifts of ^{29}Si NMR samples were tested using a spectrometer. The Gaussian linear peak was fitted by PeakFit v4.0 software to obtain the relevant result.

3. Results and Discussion
3.1. Validation and Analysis of ANOVA Model

Table 3 lists the experimental compressive strengths of the geopolymer samples for different curing periods and analysis results obtained from the Design-Expert software. An analysis of the nonlinear fit of models shows that the second-order model is the most effective. The specific regression model for compressive strength (Y) is shown in Equations (1) and (2).

$$Y_{7d} = 5.47 - 1.16A + 0.77B + 0.96C - 0.72AB - 0.15AC - 3.23BC + 1.82A^2 + 1.84B^2 - 0.13C^2 \tag{1}$$

$$Y_{28d} = 11.87 - 3.44A - 0.94B + 1.3C + 2.7AB + 0.42AC - 3.18BC + 2.09A^2 + 0.19B^2 - 3.18C^2 \qquad (2)$$

where A is the CG content, B is the alkali activator content, and C is the CH/SH value.

Table 3. Compressive strength test results of geopolymer.

Run	A (%)	B (%)	C	7d Compressive Strength (MPa)	28d Compressive Strength (MPa)
1	30	15	1.25	8.9	22
2	50	15	1.25	7.5	8.6
3	30	30	1.25	12.2	14.3
4	50	30	1.25	7.9	11.7
5	30	22.5	0.5	7.2	12.8
6	50	22.5	0.5	5.7	6.2
7	30	22.5	2	8.9	14.5
8	50	22.5	2	6.8	9.6
9	40	15	0.5	2.1	5.1
10	40	30	0.5	9.8	10
11	40	15	2	11	14.1
12	40	30	2	5.8	6.3
13	40	22.5	1.25	5.1	13.1
14	40	22.5	1.25	5.8	11
15	40	22.5	1.25	5.5	11.5

The relationship between the 7- and 28-day-curing compressive strength results and the independent variables were analyzed using RSM. Their coefficients of variation were 7.53% and 9.05% (<10%) [19], and the feasibility of the equation analysis was verified. The p value can be used to express the effectiveness of the hypothesis and mismatch test analysis during the analysis; the p value between 0.05 and 0.1 is significant, and that below 0.05 is very substantial [20]. Table 4 lists the results of the ANOVA analysis of the regression model. The p values of the factors are less than 0.05, showing that the regression effect is significant [21], whereas the p values of the interaction terms are all less than 0.05, indicating that the partial p values are significant. The test reliability of the polynomial equations is tested using R^2 values. As seen in Table 5, the R^2 values of the 7- and 28-day-curing compressive strength models are 0.9834 and 0.9787, respectively.

Table 4. Response surface test results of geopolymer.

Response	7 d Compressive Strength		28 d Compressive Strength	
	F-Value	p-Value	F-Value	p-Value
Model	32.82	0.0006	25.48	0.0012
A	35.35	0.0019	89.11	0.0002
B	15.71	0.0107	6.63	0.0498
C	24.23	0.0044	12.74	0.0160
AB	6.87	0.0470	27.49	0.0033
AC	0.29	0.6108	0.68	0.4468
BC	136.03	<0.0001	38.01	0.0016
A^2	39.84	0.0015	15.23	0.0114
B^2	40.95	0.0014	0.13	0.7353
C^2	0.21	0.6626	35.27	0.0019
Lack of Fit		0.2319		0.5963

Table 5. Model reliability test analysis.

Group	Std. Dev./Mpa	R^2	Adj R^2	Pred R^2	C.V./%	Adeq Precisior
Model Y_{7d}	0.55	0.9834	0.9534	0.7706	7.53	21.205
Model Y_{28d}	1.03	0.9787	0.9402	0.7917	9.05	18.892

3.2. Influence of Various Factors on Compressive Strength

The influence of A, B, and C on the 7-day-curing compressive strength is shown in Figure 3. Figure 3a reveals an interaction between A and B. When the value of A is 30, a large amount of FA enhances the reactivity of the raw material. As the B value increases, the concentration of the alkali activator increases, thereby accelerating the dissolution of the raw material structure. The relationship between A and C in Figure 3b shows that as the value of C increases, the dissolution of the raw material structure accelerates, and the calcium ions in the reaction process react with the silicon–oxygen tetrahedra to form a gel structure, which improves the 7d compressive strength. As shown in Figure 3c, with a decrease in the B content, the slurry strength decreases and then increases. As C gradually decreases, the compressive strength increases significantly. However, when C is 2, the 7-day-curing compressive strength decreases slightly with the B value. This is because when C is high, with the increase in the B value, more calcium ions react with the silica tetrahedron to improve the compressive strength and carbonization occurs at the same time. By contrast, excess calcium ions will carbonize with carbon dioxide in the air to form carbides, such as calcite, the reduction in calcium content will reduce the reaction of active silicon–alumina materials and decrease the 28d strength growth.

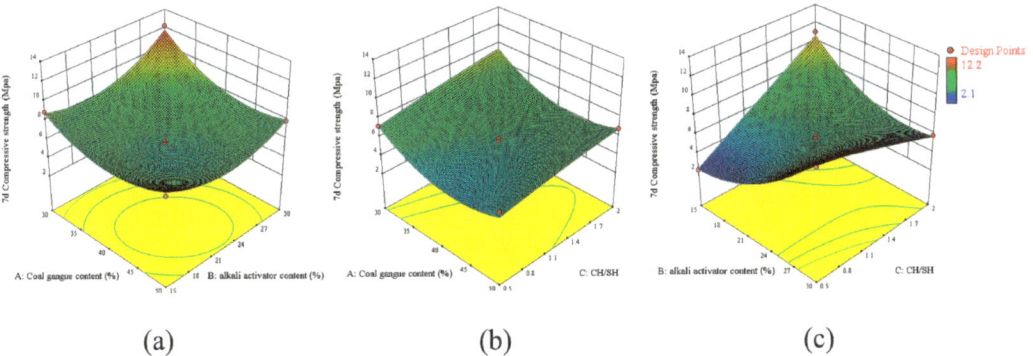

(a)　　　　　　　　　　(b)　　　　　　　　　　(c)

Figure 3. Response surface model of 7 day curing compressive strength, (**a**) CG content vs. alkali activator content, (**b**) CG content vs. CH/SH, (**c**) alkali activator content vs. CH/SH.

Figure 4 shows the response surface of the effects of A, B, and C on the 28-day-curing compressive strengths of the specimens. Figure 4a shows that the experimental results of the slurry decrease with an increase in A because the large quantity of impurities and more structurally stable quartz and kaolinite structures in the UCG reduce the structural compactness of the geopolymer. The correspondence between A and B in Figure 4a suggests that the slurry strength is high when B is about 10%. With a gradual increase in alkali activator concentration, more unreacted sulfate radicals and alkaline cations will remain in the structure in the later stages of the reaction, which will corrode the material structure and reduce the structural strength. As shown in Figure 4c, compressive strength increases with C when the latter is below 1.1. However, when the C value is higher than 1.1, the compressive strength decreases. This is because with an increase in calcium concentration, the gel phase increases and the polymerization degree of the material increases. Still, an excessively high calcium concentration will accelerate the reaction early, thus reducing the calcium content in the later reaction stages and decreasing the gelation in the subsequent curing process. Part of the calcite structure generated in the early stage will also dissolve in the later reaction stages, thereby reducing the gel structure and strength.

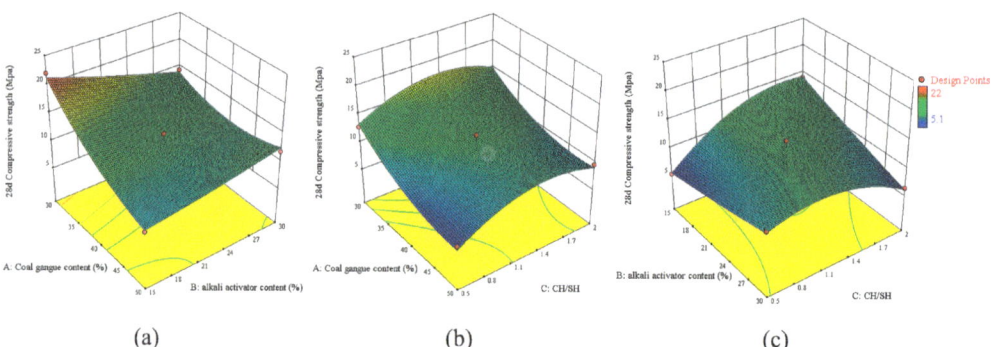

Figure 4. Response surface model of 28 day curing compressive strength, (**a**) CG content vs. alkali activator content, (**b**) CG content vs. CH/SH, (**c**) alkali activator content vs. CH/SH.

After the influence of each independent variable on compressive strength was evaluated, the maximum 28-day curing compressive strengths of the geopolymer were regarded as the optimal values. The optimized conditions are as follows: 30% CG content, 15% alkali activator content, 1.727 CH/SH value, and the results of compressive strength predictions for 7-day curing (11.494 MPa) and 28-day curing (22.513 MPa). Experiments were performed using these optimal ratios, and the experimental 7- and 28-day curing compressive strengths are 11 MPa and 20.5 MPa, respectively. The error between the specimens' compressive strengths in the experiments and the model's predicted values is less than 10%, which means that the numerical model has high accuracy.

3.3. FTIR Spectroscopic and Mineral Morphology Analysis

Figure 5 shows the XRD results for the geopolymer paste samples in Table 3. According to Figure 6, the absorption peak of kaolinite in CG almost disappears. The diffraction peak of quartz still exists but has a reduced intensity, indicating that the structure of the crystal phase of CG has been destroyed under the action of the composite alkali activator. A broad hump is observed in the 2θ range of 22–35°, indicating the presence of C–S–H and C–(N)A–S–H gels [22]. As the A value increases, the diffraction peak intensities of quartz and kaolinite also increase, mainly because the activation efficiency of the UCG decreases [23]. The unreacted coal gangue particles can build up and destroy the integrity of the structure, resulting in reduced strength. As the B value increases, the increase in calcium content promotes the formation of C–S–H in the early stage and improves the early strength. The 2θ characteristic peaks at 14° and 29° indicate the presence of a nosean. The excessive c value increases the sodium content in the reaction precursor, and the excessive sodium reacts with gypsum to form nosean which accumulates in the material structure, damaging the performance of the geopolymer.

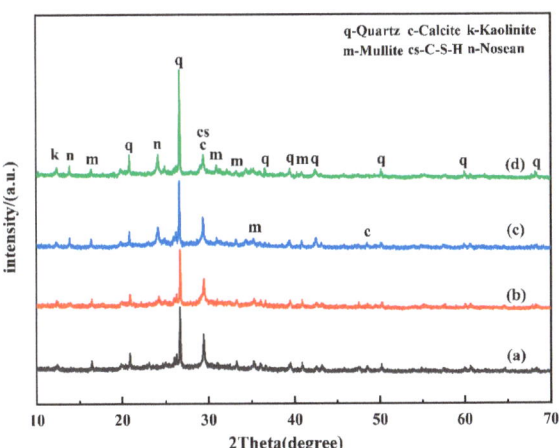

Figure 5. The XRD patterns of CFG at 28d, (**a**) Group1, (**b**) Group3, (**c**) Group11, (**d**) Group6.

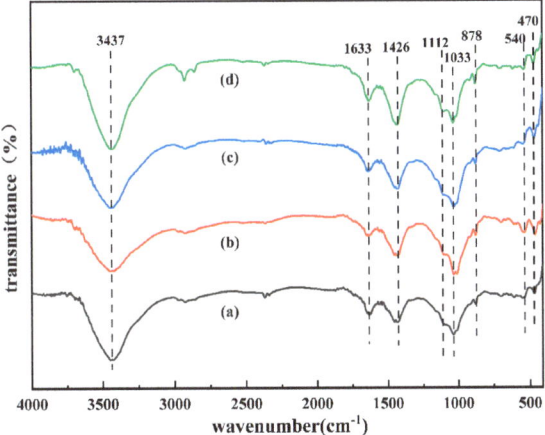

Figure 6. IR analysis of the CFG at 28d, (**a**) Group 1, (**b**) Group 3, (**c**) Group 11, (**d**) Group 6.

Figure 6 depicts the FTIR spectra of different samples on day 28. The bands at 3440 cm^{-1} and 1650 cm^{-1} can be attributed to the tensile and bending vibrations of H–O–H in the molecular water [24]. The stretching vibration peak at 1450 cm^{-1} can be attributed to the tensile vibrations of the sample's O–C–O bonds. The characteristic infrared absorption peaks of geopolymers are usually distributed within 900–1300 cm^{-1}, which is related to Si–O–T (T = Si, Al) asymmetric stretching [25]. The stretching vibration peak from 1009 to 1030 cm^{-1} is related to the asymmetric stretching of the Si–O–Si (Al) bond of the C–(N)A–S–H gel. The vibrational peak at 790 cm^{-1} is associated with quartz [26]. The bands around 460 cm^{-1} are associated with the symmetrical stretching vibrations of Si–O–Si, which may be associated with kaolinite [27]. As the A value decreases, the corresponding absorption peak wave numbers at 460 cm^{-1} and 539 cm^{-1} gradually decrease while moving higher. The absorption peak at 1450 cm^{-1} decreases with increases in the B and C values, indicating that CG does not fully participate in the reaction, which is consistent with the XRD analysis results.

3.4. Microstructure Analysis

The surface morphology of the CFG was observed and analyzed using SEM–EDS. The microstructures of the samples and their EDS spectra under different experimental conditions are shown in Figure 7 and Table 6, respectively. As shown in Figure 7a,d, large numbers of flocculent- and gel-like hydration products are generated in the reaction; the main components are C–N–A–S–H and C–S–H [28,29]. As shown in Figure 7c,d, there is also some calcite and C-S-H in the structure. Microstructure analysis of different samples shows that with a decrease in A, a new C(N)–A–S–H gel forms. This is because of an increase in the active silica–alumina substances participating in the reaction process, which increases the amount of gel product and enhances microstructure density. Moreover, the shrinkage in the geopolymer itself or the mechanical property test may have created microcracks in the sample [30]. Increases in the values of B and C lead to the accumulation of excess alkali cations and residual sulfate in the structure during the reaction, thus compromising the structure's integrity, makes the microstructure loose, increases the pore structure, and reduces the mechanical strength.

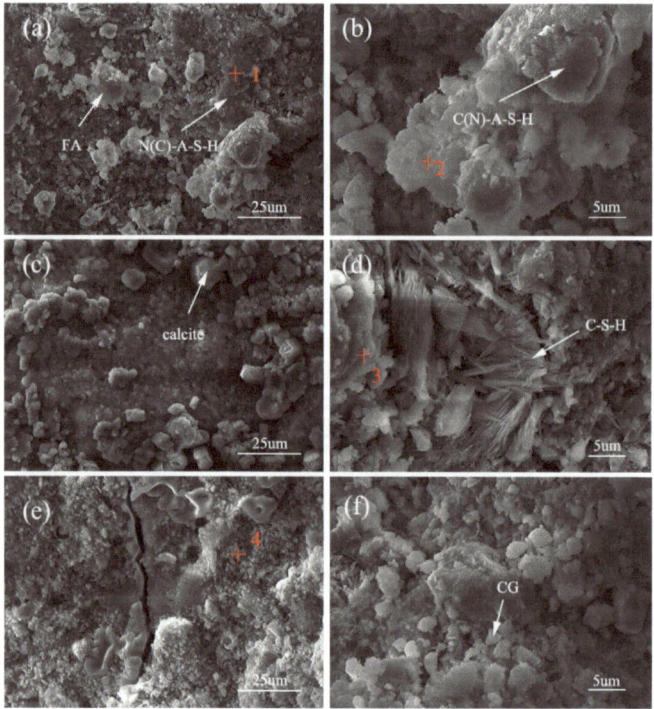

Figure 7. The SEM images of the CFG at 28d of Group 1 (**a**,**b**), Group 11 (**c**,**d**) and Group 6 (**e**,**f**).

Table 6. The atomic percentage of elemental composition at each spot in Figure 7.

Position	O%	Na%	Al%	Si%	S%	Ca%	Si/Al	Description
1	38.494	12.373	16.772	23.939	0	8.422	1.421	N(C)–A–S–H
2	41.284	4.74	10.792	27.743	1.245	14.196	2.7	C(N)–A–S–H
3	40.542	2.526	18.041	18.529	0	15.136	1	C–A–S–H
4	41.456	26.145	2.318	7.253	18.521	4.306	3.14	N–A–S–H

3.5. ^{29}Si NMR Analysis

In the structure of silicate mineral materials, each Si atom is generally surrounded by four O atoms to form a $[SiO_4]^{4-}$ tetrahedron, the basic structural unit of silicate [31]. In the ^{29}Si NMR test analysis, the different ^{29}Si NMR signals corresponding to these five tetrahedral backbones represent the silicon–oxygen tetrahedron's aggregation degree [32]. As seen in Figure 8 and Table 7, the absorption peak in the CG raw material is mainly from the Q_0^3 structural unit from kaolin (at −93.59 ppm) and the Q_0^4 structural unit from quartz (at −110.14 ppm) [33]. Compared with the findings in Figure 8a,b, Q_0^3 and Q_0^4 move to a lower chemical shift when the strength decreases during the reaction between the alkali activator and CG. Under the action of the mixed alkali activator, the quartz and kaolin in CG decrease, and the polymerization degree in the sample increases. As depicted in Figure 8b–d, the peak strengths of quartz and kaolinite at −190 ppm and −109 ppm increase with A. When A increases to 50, the chemical shift at −92 ppm reappears, consistent with the XRD analysis results. The large intensity peak from −83 to −87 ppm comes from the Q^4 structural unit of N–A–S–H or N (C)–A–S–H [34]. As shown in Table 7, with increases in B and C, the relative areas of Q_0^3 and Q_0^4 decrease and then increase. The relative area of Q_2^4 decreases gradually, indicating that the gradual decline in the network polymerization degree of reaction products leads to a reduction in compressive strength.

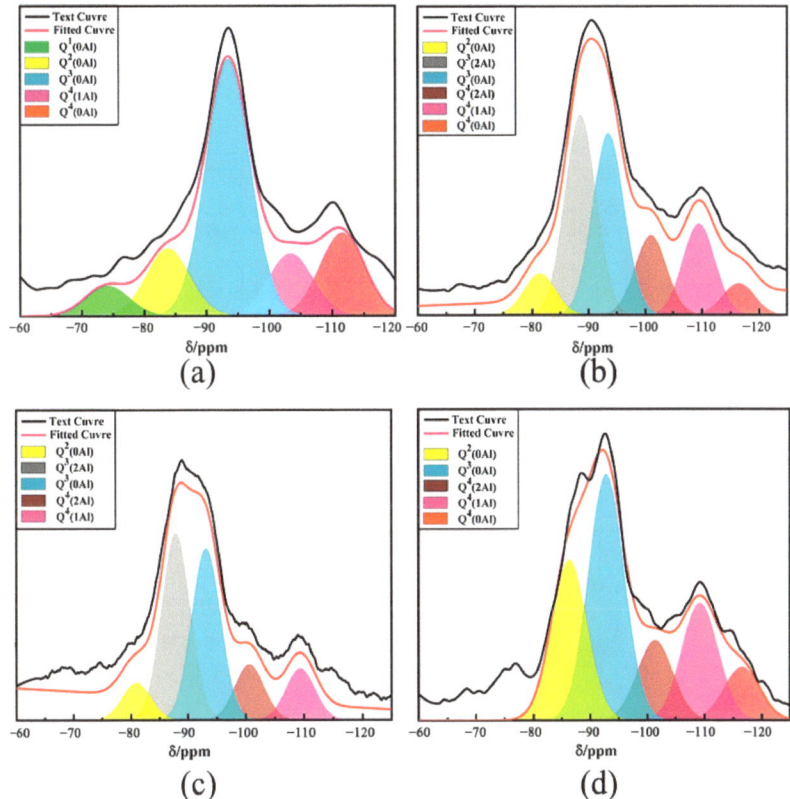

Figure 8. ^{29}Si NMR spectra of CFG at28d, (**a**) CG, (**b**) Group 1, (**c**) Group 11 and (**d**) Group 6.

Table 7. Relative peak area of the fitted curve of ^{29}Si NMR spectra (%).

Sample	Q_0^1	Q_0^2	Q_0^3	Q_2^3	Q_2^4	Q_1^4	Q_0^4
CG	6	13	45	0	0	27	8
1	0	7	29	32	13	14	5
11	0	8	34	37	11	10	0
6	0	24	38	0	12	18	8

3.6. Analysis of Geopolymerization Process

Figure 9 shows the XRD spectra and infrared vibration bands of sample 1 on the 7th and 28th days and the unreacted CG, respectively. The peak of 2θ at 29° corresponds to calcite and C-S-H structure. The calcium content in the structure at the early stage of the reaction is high, which promotes the calcite formation at the early stage and improves the 7d compressive strength [35]. However, as curing progresses, the peak strength of $CaCO_3$ structure decreases [36]. Figure 6 shows that the stretching vibration peak at 1450 cm^{-1} can be attributed to the tensile vibrations of the sample's O-C-O bonds. The absorption peak broadens and shifts higher as the reaction proceeds, indicating that the calcite structure has been decomposed, which is consistent with the XRD analysis findings (calcite ($CaCO_3$)) [16]. The crystal structure and functional group analysis show that the quartz and kaolinite structures were decomposed in an alkaline environment at the early stage of the reaction. At the same time, the high concentration of calcium content was carbonized to generate calcite and C-S-H, which were partially decomposed in the subsequent reaction process. The mechanism diagram of possible bond formation of alkali-activated UCG is shown in Figure 10.

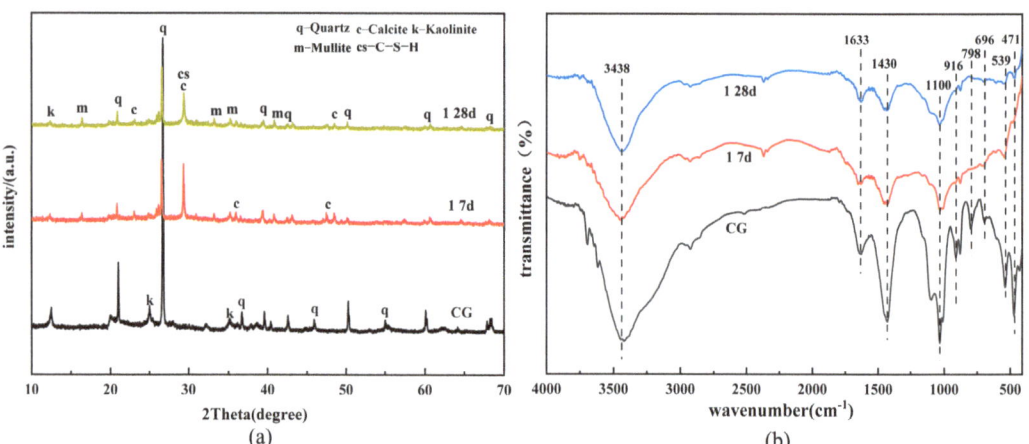

Figure 9. XRD (a) and IR (b) analysis of the CFG.

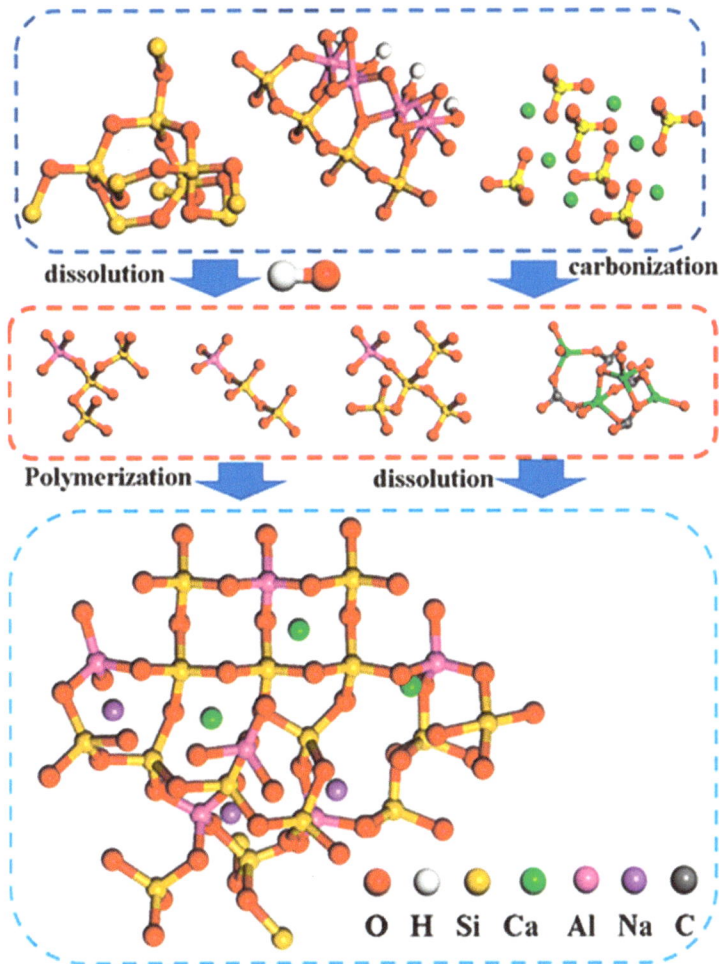

Figure 10. Mechanism diagram of possible bond formation in the reaction between UCG and gypsum.

4. Conclusions

We created a geopolymer called CFG, which is environmentally friendly and has good economic value. UCG was used as the raw material, and desulfurization gypsum and an alkaline activator (essential for CG utilization and recovery) were used as ligands.

The composite alkali activator, prepared by mixing sodium hydroxide and calcium hydroxide while increasing the calcium content, was used to prepare the geopolymer from UCG. It has been found that the addition of unburned coal gangue was an important factor affecting the compressive strength of geopolymer by optimizing the ratio. With the increase in coal gangue content, the compressive strength decreased. When the content of coal gangue was 30%, the maximum 28-day compressive strength was 22 MPa. The microstructure characterization and analysis showed that the inert structure of UCG could be dissolved to a large extent under the action of a mixed activator of sodium hydroxide and calcium hydroxide. The dissolved coal gangue and fly ash formed new structures under the action of activators, such as C-S-H and calcite. Adding the proper $Ca(OH)_2$ and desulfurization gypsum to provide additional calcium content improved the early compressive strength. It could also produce a composite gel structure of N (C)–A–S–H with high strength characteristics.

In this study, coal gangue and fly ash were used as raw materials to prepare geopolymer, and the effect of calcium on the formation of geopolymer was also discussed. The experiment provides an experimental basis for the development of UCG and has good economic value. Future research will continue to simplify the activation process of coal gangue in order to improve the utilization rate of coal gangue and reduce the cost of preparing geopolymer.

Author Contributions: Writing, Q.W. and L.Z.; review & editing, Q.W. and L.Z.; Supervision, Q.W., S.C. and Q.Y.; Investigation, Q.W. and Y.L.; Resources, Q.W.; Formal analysis, Q.W.; Software, C.L. All authors have read and agreed to the published version of the manuscript.

Funding: This research was funded by the National Natural Science Foundation of China [51501002]; the University Synergy Innovation Program of Anhui Province [GXXT-2022-083]; Anhui Provincial Natural Science Foundation [2208085ME103].

Institutional Review Board Statement: Not applicable.

Data Availability Statement: Data obtained as described.

Conflicts of Interest: The authors declare that there is no conflict of interest regarding the publication of this article.

References

1. Jabłońska, B.; Kityk, A.V.; Busch, M.; Huber, P. The structural and surface properties of natural and modified coal gangue. *J. Environ. Manag.* **2017**, *190*, 80–90. [CrossRef]
2. Zhou, S.Q.; Yang, Z.N.; Zhang, R.R.; Li, F. Preparation, characterization and rheological analysis of eco-friendly road geopolymer grouting materials based on volcanic ash and metakaolin. *J. Clean. Prod.* **2021**, *312*, 127822. [CrossRef]
3. Li, Z.F.; You, H.; Gao, Y.F.; Wang, C.; Zhang, J. Effect of ultrafine red mud on the workability and microstructure of blast furnace slag-red mud based geopolymeric grouts. *Powder Technol.* **2021**, *392*, 610–618. [CrossRef]
4. Xu, Z.F.; Zou, X.T.; Chen, J. Preparation of Thermal Activation Sludge and Coal Gangue Polymer. *Integr. Ferroelectr.* **2015**, *160*, 1–9. [CrossRef]
5. Shahedan, N.F.; Abdullah, M.M.A.; Mahmed, N.; Kusbiantoro, A.; TammasWilliams, S.; Li, L.Y.; Aziz, I.H.; Vizureanu, P.; Wysłocki, J.J.; Błoch, K.; et al. Properties of a New Insulation Material Glass Bubble in Geopolymer Concrete. *Materials* **2021**, *14*, 809. [CrossRef]
6. Li, Y.; Yao, Y.; Liu, X.M.; Sun, H.G.; Ni, W. Improvement on pozzolanic reactivity of coal gangue by integrated thermal and chemical activation. *Fuel* **2013**, *109*, 527–533. [CrossRef]
7. Li, C.; Wan, J.H.; Sun, H.H.; Li, L.T. Investigation on the activation of coal gangue by a new compound method. *J. Hazard. Mater.* **2010**, *179*, 515–520. [CrossRef]
8. Guo, Z.H.; Xu, J.J.; Xu, Z.H.; Gao, J.M.; Zhu, X.L. Performance of cement-based materials containing calcined coal gangue with different calcination regimes. *J. Build. Eng.* **2022**, *56*, 104821. [CrossRef]
9. Xie, M.Z.; Liu, F.Q.; Zhao, H.L.; Ke, C.Y.; Xu, Z.Q. Mineral phase transformation in coal gangue by high temperature calcination and high-efficiency separation of alumina and silica minerals. *J. Mater. Res. Technol.* **2021**, *14*, 2281–2288. [CrossRef]
10. Wang, A.G.; Liu, P.; Mo, L.W.; Liu, K.W.; Ma, R.; Guan, Y.M.; Sun, D.S. Mechanism of thermal activation on granular coal gangue and its impact on the performance of cement mortars. *J. Build. Eng.* **2022**, *45*, 103616. [CrossRef]
11. Zhao, Y.B.; Yang, C.Q.; Li, K.F.; Qu, F.; Yan, C.Y.; Wu, Z.R. Toward understanding the activation and hydration mechanisms of composite activated coal gangue geopolymer. *Constr. Build. Mater.* **2022**, *318*, 125999. [CrossRef]
12. Huang, G.D.; Ji, Y.S.; Li, J.; Hou, Z.H.; Dong, Z.C. Improving strength of calcinated coal gangue geopolymer mortars via increasing calcium content. *Constr. Build. Mater.* **2018**, *166*, 760–768. [CrossRef]
13. Wang, R.; Wang, J.S.; Song, Q.C. The effect of Na^+ and H_2O on structural and mechanical properties of coal gangue-based geopolymer: Molecular dynamics simulation and experimental study. *Constr. Build. Mater.* **2020**, *268*, 121081. [CrossRef]
14. Ma, B.; Luo, Y.; Zhou, L.Z.; Shao, Z.Y.; Liang, R.H.; Fu, J.; Wang, Q.; Zang, J.; Hu, Y.Y.; Wang, L.M. The influence of calcium hydroxide on the performance of MK-based geopolymer. *Constr. Build. Mater.* **2022**, *329*, 127224. [CrossRef]
15. Yang, X.Y.; Zhang, Y.; Lin, C. Microstructure Analysis and Effects of Single and Mixed Activators on Setting Time and Strength of Coal Gangue-Based Geopolymers. *Gels* **2022**, *8*, 195. [CrossRef]
16. Chen, X.; Zhang, J.C.; Lu, M.Y.; Chen, B.W.; Gao, S.Q.; Bai, J.W.; Zhang, H.Y.; Yang, Y. Study on the effect of calcium and sulfur content on the properties of fly ash based geopolymer. *Constr. Build. Mater.* **2022**, *314*, 125650. [CrossRef]
17. Geng, J.J.; Zhou, M.; Li, Y.X.; Chen, Y.C.; Han, Y.; Wan, S.; Zhou, X.; Hou, H.B. Comparison of red mud and coal gangue blended geopolymers synthesized through thermal activation and mechanical grinding preactivation. *Constr. Build. Mater.* **2017**, *153*, 185–192. [CrossRef]

18. Guo, L.Z.; Zhou, M.; Wang, X.Y.; Li, C.; Jia, H.Q. Preparation of coal gangue-slag-fly ash geopolymer grouting materials. *Constr. Build. Mater.* **2022**, *328*, 126997. [CrossRef]
19. Quiatchon, P.R.J.; Dollente, I.J.R.; Abulencia, A.B.; Libre, R.G.D.; Villoria, M.B.D.; Guades, E.J.; Promentilla, M.A.B.; Ongpeng, J.M.C. Investigation on the Compressive Strength and Time of Setting of Low-Calcium Fly Ash Geopolymer Paste Using Response Surface Methodology. *Polymers* **2021**, *13*, 3461. [CrossRef]
20. Chen, K.Y.; Wu, D.Z.; Zhang, Z.L.; Pan, C.G.; Shen, X.Y.; Xia, L.L.; Zang, J.W. Modeling and optimization of fly ash–slag-based geopolymer using response surface method and its application in soft soil stabilization. *Constr. Build. Mater.* **2022**, *315*, 125723. [CrossRef]
21. Shi, X.S.; Zhang, C.; Wang, X.Q.; Zhang, T.; Wang, Q.Y. Response surface methodology for multi-objective optimization of fly ash-GGBS based geopolymer mortar. *Constr. Build. Mater.* **2022**, *315*, 125644. [CrossRef]
22. Bai, Y.Y.; Guo, W.C.; Wang, X.L.; Pan, H.M.; Zhao, Q.X.; Wang, D.L. Utilization of municipal solid waste incineration fly ash with red mud-carbide slag for eco-friendly geopolymer preparation. *J. Clean. Prod.* **2022**, *340*, 130820. [CrossRef]
23. Yi, C.; Ma, H.Q.; Chen, H.Y.; Wang, J.X.; Shi, J.; Li, Z.H.; Yu, M.K. Preparation and characterization of coal gangue geopolymers. *Constr. Build. Mater.* **2018**, *187*, 318–326. [CrossRef]
24. Li, Z.F.; Gao, Y.F.; Zhang, J.; Zhang, C.; Chen, J.P.; Liu, C. Effect of particle size and thermal activation on the coal gangue based geopolymer. *Mater. Chem. Phys.* **2021**, *267*, 124657. [CrossRef]
25. Li, Z.F.; Gao, Y.F.; Zhang, M.; Zhang, C.; Zhang, J.; Wang, C.; Zhang, N. The enhancement effect of Ca-bentonite on the working performance of red mud-slag based geopolymeric grout. *Mater. Chem. Phys.* **2022**, *276*, 125311. [CrossRef]
26. de Vargas, A.S.; Dal Molin, D.C.C.; Masuero, A.B.; Vilela, A.C.F.; Castro-Gomes, J.; de Gutierrez, R.M. Strength development of alkali-activated fly ash produced with combined NaOH and Ca(OH)$_2$ activators. *Cem. Concr. Compos.* **2014**, *53*, 341–349. [CrossRef]
27. Koshy, N.; Dondrob, K.; Hu, L.M.; Wen, Q.B.; Meegoda, J.N. Synthesis and characterization of geopolymers derived from coal gangue, fly ash and red mud. *Constr. Build. Mater.* **2019**, *206*, 287–296. [CrossRef]
28. Nie, Y.M.; Ma, H.W.; Liu, S.X.; Niu, F.S. Mechanism of polymerization reaction during the solidification of meta-kaolin based mineral polymer. *Adv. Mater. Res.* **2011**, *177*, 628–635. [CrossRef]
29. Liu, C.J.; Deng, X.W.; Liu, J.; Hui, D. Mechanical properties and microstructures of hypergolic and calcined coal gangue based geopolymer recycled concrete. *Constr. Build. Mater.* **2019**, *221*, 691–708. [CrossRef]
30. Saif, M.S.; El-Hariri, M.O.R.; Sarie-Eldin, A.I.; Tayeh, B.A.; Farag, M.F. Impact of Ca+ content and curing condition on durability performance of metakaolin-based geopolymer mortars. *Case Stud. Constr. Mater.* **2022**, *16*, e00922. [CrossRef]
31. Kunther, W.; Ferreirom, S.; Skibsted, J. Influence of the Ca/Si ratio on the compressive strength of cementitious calcium–silicate–hydrate binders. *J. Mater. Chem. A* **2017**, *5*, 17401–17412. [CrossRef]
32. Singh, P.S.; Bastow, T.; Trigg, M. Structural studies of geopolymers by ^{29}Si and ^{27}Al MAS-NMR. *J. Mater. Sci.* **2005**, *40*, 3951–3961. [CrossRef]
33. Souayfan, F.; Rozière, E.; Paris, M.; Deneele, D.; Loukili, A.; Justino, C. ^{29}Si and ^{27}Al MAS NMR spectroscopic studies of activated metakaolin-slag mixtures. *Constr. Build. Mater.* **2022**, *322*, 126415. [CrossRef]
34. Zhao, X.H.; Liu, C.Y.; Zuo, L.M.; Wang, L.; Zhu, Q.; Wang, M.K. Investigation into the effect of calcium on the existence form of geopolymerized gel product of fly ash based geopolymers *Cem. Concr. Compos.* **2019**, *103*, 279–292. [CrossRef]
35. Li, Y.D.; Li, J.F.; Cui, J.; Shan, Y.; Niu, Y.F. Experimental study on calcium carbide residue as a combined activator for coal gangue geopolymer and feasibility for soil stabilization. *Constr. Build. Mater.* **2021**, *312*, 125465. [CrossRef]
36. Firdous, R.; Hirsch, T.; Klimm, D.; Lothenbach, B.; Stephan, D. Reaction of calcium carbonate minerals in sodium silicate solution and its role in alkali-activated systems. *Miner. Eng.* **2021**, *165*, 106849. [CrossRef]

Disclaimer/Publisher's Note: The statements, opinions and data contained in all publications are solely those of the individual author(s) and contributor(s) and not of MDPI and/or the editor(s). MDPI and/or the editor(s) disclaim responsibility for any injury to people or property resulting from any ideas, methods, instructions or products referred to in the content.

Article

Superparamagnetic Multifunctionalized Chitosan Nanohybrids for Efficient Copper Adsorption: Comparative Performance, Stability, and Mechanism Insights

Ahmed A. Al-Ghamdi [1], Ahmed A. Galhoum [2], Ahmed Alshahrie [1,3], Yusuf A. Al-Turki [4,5], Amal M. Al-Amri [6] and S. Wageh [1,*]

[1] Department of Physics, Faculty of Science, King Abdulaziz University, Jeddah 21589, Saudi Arabia
[2] Nuclear Materials Authority, El-Maadi, Cairo P.O. Box 530, Egypt
[3] Centre of Nanotechnology, King Abdulaziz University, Jeddah 21589, Saudi Arabia
[4] Department of Electrical and Computer Engineering, Faculty of Engineering, King Abdulaziz University, Jeddah 21589, Saudi Arabia
[5] K. A. CARE Energy Research and Innovation Center, King Abdulaziz University, Jeddah 21589, Saudi Arabia
[6] Physics Department, Rabigh College of Science and Arts, King Abdulaziz University, P.O. Box 344, Rabigh 21911, Saudi Arabia
* Correspondence: wswelm@kau.edu.sa

Citation: Al-Ghamdi, A.A.; Galhoum, A.A.; Alshahrie, A.; Al-Turki, Y.A.; Al-Amri, A.M.; Wageh, S. Superparamagnetic Multifunctionalized Chitosan Nanohybrids for Efficient Copper Adsorption: Comparative Performance, Stability, and Mechanism Insights. *Polymers* 2023, 15, 1157. https://doi.org/10.3390/polym15051157

Academic Editors: Giorgio Luciano and Maurizio Vignolo

Received: 21 January 2023
Revised: 11 February 2023
Accepted: 18 February 2023
Published: 24 February 2023

Copyright: © 2023 by the authors. Licensee MDPI, Basel, Switzerland. This article is an open access article distributed under the terms and conditions of the Creative Commons Attribution (CC BY) license (https://creativecommons.org/licenses/by/4.0/).

Abstract: To limit the dangers posed by Cu(II) pollution, chitosan-nanohybrid derivatives were developed for selective and rapid copper adsorption. A magnetic chitosan nanohybrid (r-MCS) was obtained via the co-precipitation nucleation of ferroferric oxide (Fe_3O_4) co-stabilized within chitosan, followed by further multifunctionalization with amine (diethylenetriamine) and amino acid moieties (alanine, cysteine, and serine types) to give the TA-type, A-type, C-type, and S-type, respectively. The physiochemical characteristics of the as-prepared adsorbents were thoroughly elucidated. The superparamagnetic Fe_3O_4 nanoparticles were mono-dispersed spherical shapes with typical sizes (~8.5–14.7 nm). The adsorption properties toward Cu(II) were compared, and the interaction behaviors were explained with XPS and FTIR analysis. The saturation adsorption capacities (in $mmol.Cu.g^{-1}$) have the following order: TA-type (3.29) > C-type (1.92) > S-type (1.75) > A-type(1.70) > r-MCS (0.99) at optimal pH_0 5.0. The adsorption was endothermic with fast kinetics (except TA-type was exothermic). Langmuir and pseudo-second-order equations fit well with the experimental data. The nanohybrids exhibit selective adsorption for Cu(II) from multicomponent solutions. These adsorbents show high durability over multiple cycles with desorption efficiency > 93% over six cycles using acidified thiourea. Ultimately, QSAR tools (quantitative structure-activity relationships) were employed to examine the relationship between essential metal properties and adsorbent sensitivities. Moreover, the adsorption process was described quantitatively, using a novel three-dimensional (3D) nonlinear mathematical model.

Keywords: superparamagnetic nanohybrids; chitosan derivatives; copper adsorption; polyamine and amino acid moieties; selectivity; mathematical modeling

1. Introduction

Water pollution is among the most severe problems facing humanity today, and it has drawn the attention of all scientists. The release of industrial effluent into the aquatic environment is a serious global concern due to the possibility of heavy metal contamination in water reserves [1–3]. The mining, electroplating, automobiles, metal processing, textile, and battery manufacturing sectors all contribute significantly to heavy metal pollution. Heavy-metal-containing effluent is routinely released into water bodies as a result of industrial activity, causing a slew of environmental issues [4–7]. Copper is among the most precious and widely utilized metals in the industry [3,8]. Copper in industrial effluent is particularly dangerous. When ingested in excess, copper accumulation in the liver causes gastro-intestinal problems. Copper poisoning occurs when an excessive amount of copper is consumed, producing nausea, liver and kidney failure, vomiting, and abdominal discomfort [9,10]. The World Health Organization states that the Cu(II) permitted maximum in surface water is 3 mg L^{-1} [9]. Thus, different technologies are frequently applied for removing heavy metals from water, including (i) precipitation, (ii) solvent extraction, (iii) impregnated resins, (iv) ultrafiltration, and (v) adsorption and ion exchange [2,9,11,12]. However, practically all of these techniques are often costly, require multiple stages, are environmentally unfriendly, produce organic wastes, and are ineffective, especially at low metal concentrations [13,14]. Thus, adsorption methods are usually thought to be more effective for treating diluted effluents, due to fast kinetics, reusability, environmental friendliness, selectivity, and high efficiency [1,15,16].

More recently, adsorbent materials such as modified natural materials [1,11], low-cost biopolymers [2,6,17,18], metal–organic-frameworks-based materials [16,19,20], carbonaceous materials (e.g., biochar and activated carbon derived from biomass wastes) [14,21,22], synthetic polymers [23,24], and nanomaterials [18,25] have been used in the field of Cu(II) removal. Biosorption has been proposed as an environmentally acceptable green technique for the elimination of different contaminants [2,26–28]. Sustainable biomass with numerous functional groups (–NH$_2$ and –OH), renewable resources, and good hydrophilicity are promising and economically feasible alternatives to synthetic polymer adsorbents [9,29,30]. Modified biopolymers based on polysaccharides are excellent examples of these adsorbent types [29]. Chitosan (CS, as an amino-polysaccharide) is a powerful biosorbent for eliminating heavy metals because it possesses a distinct set of characteristics (for example, nontoxicity, biocompatibility, biodegradability, and bioactivity) and is nature's most abundant and least expensive biopolymer [2,26,30]. To promote adsorption capacity, selectivity, and mass transfer: multiphase nanohybrids were designed as a development of functionalized organic–inorganic nanohybrids [3,24,31]. CS was selected and adapted due to its chemical compatibility with the Fe$_3$O$_4$ nanoparticles' fabrication [32,33].

The synergism and mixing between two or more individual components were proposed to design and create new multiphase nanomaterials with improved characteristics, such as multifunctionalization, reaction velocity, stability, and the cost-effectiveness [25,29,30]. This is the driving force behind the development of multifunctionalized chitosan-based hybrid magnetic materials. Magnetic nanohybrids using Fe$_3$O$_4$ nanoparticles are an innovative type of composite [24,34]. Magnetic-adsorption technologies have been widely employed for contaminants' removal because it significantly enhances the specific surface area (which reduces restrictions imposed by intraparticle diffusion), and it is rapid, simple, sensitive, and extraordinarily effective. Furthermore, after equilibration is completed, depleted adsorbents can be recovered by utilizing an external magnetic field [31,34–37]. Until now, no published work has been reported containing the extensive and in-depth description required for mechanistic investigations of the synthesis stages and post-chemical modification, adsorption, and desorption mechanisms for copper with such magnetic multifunctionalized chitosan nanohybrids, which pose different active sites that have not been tested before. Herein, we describe the synthesis of four multifunctionalized chitosan adsorbents based on the functionalization of polymer support (by in situ Fe$_3$O$_4$ nanoparticles co-precipitation covered with a thin chitosan layer) with different reactive groups via amine polydentate (diethylenetriamine) and amino acids (e.g., alanine/cysteine/serine) grafting. These adsorbents were extensively characterized to justify the synthesis routes (including the iron oxide and organic synthesis mechanism involved in the chemical modification). These functionalization pathways brought a good opportunity for comparing the impact of grafted groups on the physicochemical properties of the materials (including CHNS/O (elemental analysis), HR-TEM (high-resolution transition microscope), XRD (X-ray diffraction), textural properties (using the Brunauer–Emmett–Teller (BET technique) and Barrett–Joyner–Halenda (BJH) method)), pH$_{ZPC}$ titration (pH zero-point charge), FTIR (Fourier-transform infrared spectroscopy), TGA/DTA (Thermogravimetric analysis/Differential Thermal Analysis), VSM (vibrating sample magnetometer), EDX (Energy dispersive X-ray Spectroscopy), and XPS (X-ray photoelectron spectroscopy) techniques. The batch experiments were used to investigate the structure activity relationship and compare adsorption performance, selectivity test, and the adsorption–desorption cycle was used to investigate the regeneration and reusability of these nanocomposites. Furthermore, the experimental results were fitted with kinetics and isotherm models (with conventional modeling equations). The adsorption and desorption mechanisms were analyzed and studied using spectroscopic techniques (particularly, FTIR and XPS). Moreover, an effective mathematical method for determining adsorption capacity at varying initial concentrations and pHs was developed using a special modeling methodology for quantitative nonlinear description in three dimensions.

2. Materials and Methods

2.1. Materials

Chitosan (CS, CAS: 9012-76-4) was purchased from Acros Organics. Diethylenetriamine (99%), 1,4-dioxane (>99%), epichlorohydrin (99%), cysteine ($\geq$98.5%), $FeCl_2 \cdot 4H_2O$ (>99%), alanine ($\geq$98%), $FeCl_3$ (>99%), Arsenazo III, and serine ($\geq$99%) were provided from Sigma-Aldrich (Saint-Louis, MS, USA). Every reagent was used as received. Standard atomic absorption solutions (Cu, Co, Ni, Cd, and Zn, 1000 mg/L^{-1} nitrate form in nitric acid) were from Scharlau Chemie (Barcelona, Spain). Samples (5 mL) were centrifuged to separate them. The supernatants were taken, and the remaining Cu(II) concentration was analyzed. Total Zn, Cd, Ni, Co, and Cu concentrations in mixed complex solution were measured using AAS (flame atomic absorption spectrophotometer) (GBC Avanta-EGf 3000, Scientific equipment Ltd., Melbourne, Australia).

2.2. Preparation of Adsorbents

The adsorbents employed in this study were synthesized according to the previous article, as a foundation, and with minor modifications [25,32]. A schematic representation for the fabrication of magnetic nanohybrid adsorbents is represented as a general method in Scheme 1.

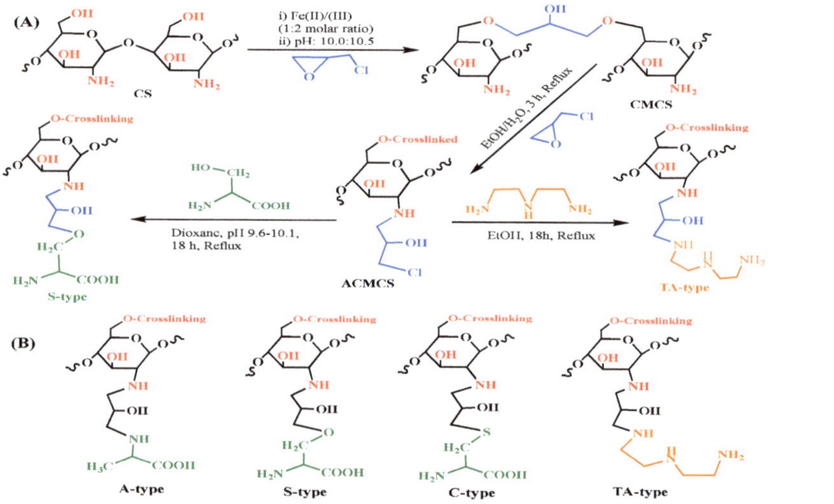

Scheme 1. S-type and TA-type nanohybrids synthesis (**A**) and structure of several adsorbent types (**B**).

2.2.1. Preparation of Activated Crosslinked Chitosan–Magnetite Nanohybrid

A precipitation method combined with hydrothermal treatment was applied to produce nanohybrid magnetic chitosan particles through a one-step process, which involved in situ co-precipitation of ferrous and ferric ions mixture simultaneously with the dissolved chitosan particles under alkaline conditions [35]. Chitosan solution was firstly prepared by dissolving 5 g of chitosan powder in 600 mL of acetic acid solution (5% w/w). Then 6.22 g of $FeCl_2$ and 9.60 g of $FeCl_3$ (with a molar ratio of 1:2, respectively) were added and homogeneously dispersed within the chitosan solution. The resulting solution mixture was subjected to chemical precipitation by the dropwise addition of NaOH (2 M), under steady stirring, at 40–45 °C, and the pH was controlled to 10.0–10.5. The suspension was kept under heating and constant stirring for 1 h at 90 °C before the nanohybrid particles were separated by decantation and magnetic attraction. Secondly, an alkaline solution of 0.01 M epichlorohydrin with pH 10.0 was added to the freshly obtained nanohybrid particles with a mass ratio of 1:1, and the reaction mixture was maintained under heating and continued stirring for 2 h at 50 °C. After that, the resulting material was collected by magnetic separation and washed extensively with ethanol and distilled water to remove any unreacted epichlorohydrin. Finally, the previously obtained crosslinked magnetic chitosan nanohybrid particles were suspended in 200 mL of ethanol/water mixture (1:1 v/v), followed by the addition of 10 mL of epichlorohydrin, and the mixture was heated and refluxed for 3 h at 70 °C. Thereafter, the activated nanohybrid particles were obtained, filtered, and washed extensively with ethanol and distilled water to remove any residual reagent [32,38].

2.2.2. Nanohybrid Functionalization

The sequential stages for the functionalization of the activated crosslinked magnetic chitosan nanohybrid with diethylenetriamine, as well as amino-acid functionalities, were presented in Scheme 1A. Briefly, the diethylenetriamine moiety was grafted as follows: the epichlorohydrin-activated nanohybrid (2.5 g) was suspended in 50 mL of ethanol, and 25 mL of diethylenetriamine was added. Then the reaction mixture was maintained under reflux for the next 18 h, at 75–80 °C [32,39].

On the other hand, alanine, cysteine, and serine moieties were grafted according to the following procedure: The epichlorohydrin-activated nanohybrid particles (2.5) were suspended in 150 mL of dioxin, and then 6.0 g of alanine/serine/cysteine was added and the pH of the mixture was adjusted to 9.5–10 by using 1 M NaOH solution. After that, the mixture was kept under reflux for the next 18 h at 95 °C [25]. After the reaction, all end-products were collected by magnetic separation and extensively washed with ethanol and double-distilled water. Finally, the obtained adsorbent materials were freeze-dried for at least 24 h.

2.3. Characterization Techniques

The obtained superparamagnetic multifunctional chitosan adsorbents were characterized by HR-TEM, CHNS/O, XRD, BET, pH_{ZPC} titration, FTIR, TGA/DTA, VSM, EDX, and XPS techniques. More detailed information for the different characterizations is offered in Supplementary Materials Section S1. To examine the material's chemical stability, a simple technique for testing and evaluating the chemical resistivity and durability of the nanohybrids was achieved by leaching under acidic and alkaline media [25]. So, various pHs (1.0–10.0) were used to submerge and then leach these materials (mass 10.0 mg/10 mL solution, at 25 °C for 48 h). After that, the treated samples were filtered and thoroughly washed (using deionized water), and both treated and untreated samples were dried for 8 h at 75 °C. The mass loss efficiency was calculated as follows: ($\Delta wt\% = ((m_0 - m_{eq}) \times 100)/m_0$), where m_0 is the untreated dry mass, and m_{eq} is the treated dry mass. After several adsorption/desorption cycles, the nanohybrid materials were further examined for functional stability, using FTIR spectroscopy.

2.4. Adsorption Experiments

Batch adsorption tests have been performed at a fixed adsorbent dose (0.5 g.L^{-1}) to investigate the pH influence (the pH_0 ranging from 2.0 to 7.0), agitation time (time was varied from 0 to 360 min at pH_0 5.0), sorption isotherms (at initial Cu(II) concentration from C_0: 0.128 to 1.243 mmol Cu.L^{-1} at temperatures ranging from 298 to 328 ± 2 K, and pH$_0$ 5.0), and adsorbent reusability at 298 ± 1 K and 150 rpm. The adsorption capacity (q_{eq}, in mmol Cu.g^{-1}) and the distribution coefficient (D, in L.g^{-1}) were evaluated using the following equations (Equations (1) and (2), respectively):

$$q_{eq} = \frac{(C_o - C_{eq}) \times V}{m} \quad (1)$$

$$D = \frac{q_{eq}}{C_{eq}} = \frac{(C_o - C_{eq})}{C_o} \times \frac{V}{m} \quad (2)$$

where C_0 and C_{eq} (in mmol Cu.L^{-1}) refer to the initial and remaining Cu(II) concentration after equilibrium, m (in g) points to is the adsorbent amount, and V (in L) corresponds to the volume.

Investigating the selectivity concerns in complicated solutions is required for assessing the potential of novel materials. A standardized study was carried out using the most common ions found in industrial wastewater, e.g., those from battery plants (such as Cu, Co, Zn, and Cd). For this test, an equimolar mixed solution was prepared (using standard atomic nitrate solutions). The adsorption test was conducted at pH$_0$ 5.0, for 4 h, at room temperature (298 ± 1 K) and 150 rpm, with an adsorbent dose of 0.5 g.L^{-1}. The supernatant concentrations were determined by AAS after centrifugation and separation. The adsorption characteristics could also be modulated by utilizing some chosen models (as shown in Supplementary Table S1) to match the adsorption kinetics and isotherms.

The adsorbent reusability was investigated after six adsorption–desorption cycles utilizing acidified thiourea (0.25 mol.L^{-1} at pH: 2) as eluent and regeneration agent (SD: 1 g.L^{-1}, 150 rpm and 298 ± 1 K for 60 min).

The desorption efficiency (D_E,) and the regeneration efficiency (RE) were described using the equations below (Equations (3) and (4)):

$$D_E = \frac{C_D \times V_L \times 100}{m_d \times q_d} \quad (3)$$

$$RE = \frac{q_d \times 100}{q_e} \quad (4)$$

where q_d (in mmol Cu.g^{-1}) points to the first adsorption capacity, C_D (in mmol Cu.L^{-1}) refers to eluate Cu(II) concentration, V_L relates to the eluent volume, and m_d (in g) represents the adsorbent weight used in desorption tests. All data are means of duplicates with a standard deviation of ±4–6%.

3. Results and Discussion

3.1. Fabrication Mechanism

Scheme 1A depicts the synthesis steps of multifunctionalized magnetic chitosan nanohybrid, whereas Scheme 1B depicts the chemical structure of the various derivatives. The nanohybrid was formed via heterogeneous nucleation through an in situ hydrothermal co-precipitation of Fe^{2+} and Fe^{3+} in chitosan solution at a pH of ~10.5 with NaOH solution. Firstly, Fe$_3$O$_4$ nanoparticles were utilized as a nucleation site for chitosan thin-layer formation [35]. The fundamental general process for producing mixed iron oxide (i.e., magnetite (Fe$_3$O$_4$)) growth of nanoparticles begins after NaOH is introduced to iron solutions, as shown in the following procedures in Equations (5)–(10) [25,40].

Deprotonation step:

$$Fe^{3+} + xH_2O \rightarrow 2Fe(OH)_x^{3-x} + xH^+ \quad (5)$$

$$Fe^{2+} + yH_2O \rightarrow Fe(OH)_y^{2-x} + yH^+ \quad (6)$$

Ferrihydrite intermediate production:

$$2Fe(OH)_2^+ + Fe(OH)^+ + 3OH^- \rightarrow \left(Fe^{3+}\right)_2 \left(Fe^{2+}\right)(OH^-)_8 \qquad (7)$$

Oxidation and dehydration step:

$$2Fe(OH)_x^{3-x} + Fe(OH)_y^{2-y} \rightarrow Fe_3O_4 + \frac{2x+y}{2}H_2O \qquad (8)$$

When ferrihydrite dehydrates, magnetite forms:

$$\left(Fe^{3+}\right)_2 \left(Fe^{2+}\right)(OH^-)_8 \rightarrow Fe_3O_4 + 4H_2O \qquad (9)$$

Overall reaction:

$$2FeCl_3 + FeCl_2 + 8NaOH \rightarrow Fe_3O_4 + 8NaCl + 4H_2O \qquad (10)$$

This reaction occurs in the presence of $N_2(g)$ to prevent Fe^{2+} from converting to Fe^{3+}; otherwise, different ferric hydroxide forms, such as maghemite, hematite, and goethite, can be formed by oxidizing magnetite [40].

Chitosan particles were crosslinked using epichlorohydrin to prevent them from dissolving under acidic environments and from losing adsorption ability when amine groups participated in the case of aldehyde linkage [41]. Indeed, the unfunctional crosslinking agent was employed to give covalent connections with the C_6–OH in chitosan moiety, resulting from the epoxide ring breakage and the release of a chlorine atom [25].

The plausible mechanism for the formation of the crosslinked chitosan (as depicted in Scheme 2A) involves the nucleophilic ring opening of the epoxy ring of the epichlorohydrin under basic conditions by primary OH of the less steric hindrance (C_6–OH). The second step involves the nucleophilic displacement of the chloride by the nucleophilic primary hydroxyl group of the second chitosan molecule from another chain, followed by the elimination of the HCl under basic conditions after forming the crosslinked chitosan. After that, the activation phase provides active spots on the surface of chitosan, allowing for chemical modification with amine and amino acid moieties.

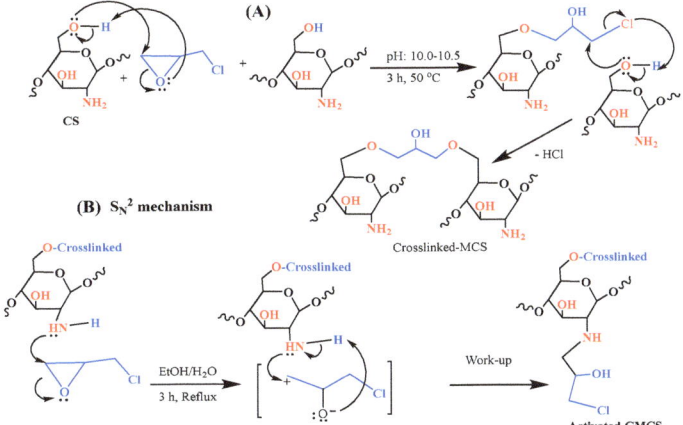

Scheme 2. Revised process for the epichlorohydrin crosslinking chitosan formation (**A**) and epoxy ring opening with crosslinked MCS (**B**).

Scheme 2B shows the S_N^2 pathway (S_N^2: substitution nucleophilic bimolecular), which might have a role in the nucleophilic assaults of the epoxy ring on nitrogen (CS–NH_2).

A carbonium ion is formed when an N-atom attacks the external carbon (which is less sterically inhibited).

Scheme 3 shows two pathways (A (S_N^1) or B (S_N^2)) which could be involved in the nucleophilic displacement of –Cl with electron-rich terminal hetero atoms (e.g., nitrogen (present in –NH_2), oxygen (as –OH in serine), and/or sulfur (as –SH in cysteine). For diethylenetriamine grafting, the postulated mechanism of the amination (Scheme 3A) is a typical substitution nucleophilic unimolecular (S_N^1) reaction that is preferred in protic solvents such as ethanol [39]. Chloride nucleophilic displacement by electron-rich nitrogen of the amine occurs via the formation of carbocation intermediates, and it is maintained and supported by ethanol solvation and hence favors product formation.

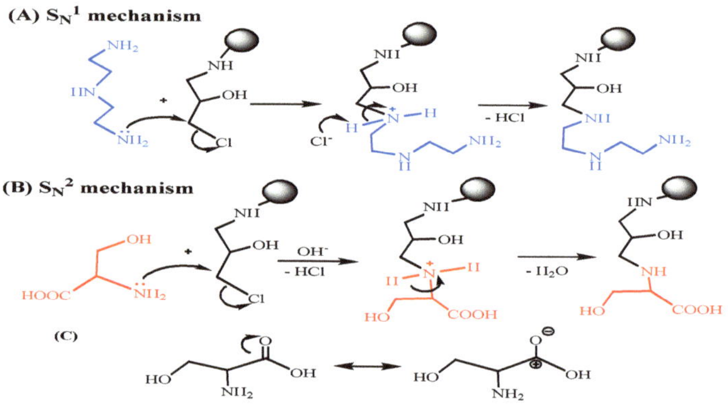

Scheme 3. The revised procedure for amine immobilization via S_N^1 mechanism (**A**) and for serine (as reprehensive for amino acids) grafting via S_N^2 mechanism (**B,C**).

Meanwhile, for amino-acid grafting (Scheme 3B), under alkaline conditions and heating, the usual S_N^2 reaction takes place between nucleophilic sites available at the terminal –NH_2 and/or –SH/–OH on alanine/cysteine/serine–amino acid and the C-atom connected to the Cl atom that is considered to become the promoter for the chemical reaction [25,38]. This reaction is enhanced by an alkaline aprotic solvent such as dioxane. For cysteine/serine grafting, the carboxylic group is an electron-withdrawing site that contributes to the reduction of the N-lone atom's electron pair, trying to make them lesser nucleophilic (Scheme 3C); this may explain why other –SH/–OH groups are favored versus –NH_2 where possible. Meanwhile, the –NH_2 group seems to be more active and preferred in alanine, despite the fact that certain amino groups may be implicated in the reaction. The majority of the grafting seems to occur using an alternate substituent via –SH/–OH active sites.

3.2. Materials Characterization

3.2.1. Physical Characterization

Nano-Structure Characterization—HR-TEM Analysis

The TEM micrographs of four adsorbents (A-type, S-type, C-type, and TA-type) were shown in Figure 1a–d. The Fe_3O_4 nanoparticles formed as compact spherical spots encircled by bright regions represented the polymeric shells [42] (due to the differences in electron-absorbing capacities of organic and inorganic components [34]); they have a uniform spherical morphology, are finely shaped, and have a mono-dispersed appearance. As a result of dipole/dipole magnetic interactions, the Fe_3O_4 particles pose an affinity to coalesce and agglomerate [41,43]. This exhibits the incorporation of Fe_3O_4 NPs into the polymer matrix, which was successfully prepared. The clumping validates the great magnetism, just like it was evaluated via magnetism tests. TEM images were used to determine and identify the exact the Fe_3O_4 nanoparticle size. The mean crystalline size for A-type, S-type, C-type, and TA-type was estimated to be 10.43 ± 1.9, 14.79 ± 3.2, 14.01 ± 3.7, and 8.45 ± 1.4 nm, utilizing histograms image software to count particles with a limited size distribution (Figure 1a–d), in line with XRD findings. In general, the diameter of Fe_3O_4 particles is smaller than 20 nm, suggesting that the adsorbents are nanocomposites.

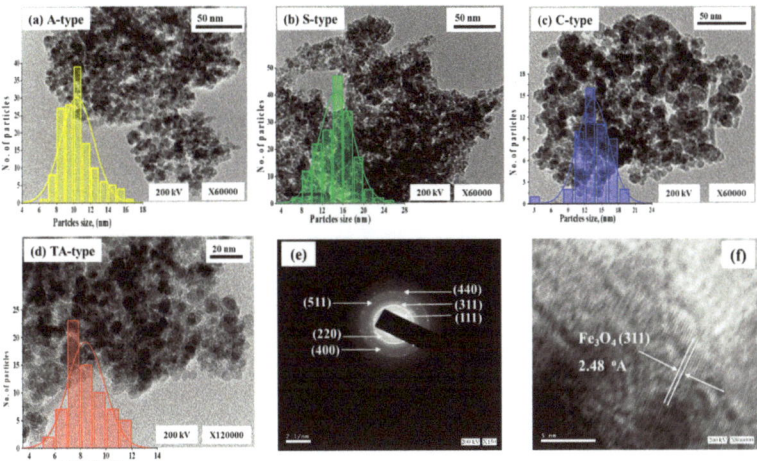

Figure 1. Particle sizes distribution using TEM micrographs and histograms for A-type, S-type, and C-type (**a–c**), and for TA-type (**d**), SAED analysis (**e**), and crystal lattice HRTEM picture (**f**).

The HR-TEM picture (Figure 1e,f) for TA-type, according to the SAED (selected area electron diffraction) pattern, exhibits that particles majorities visible and found in the bright field image with an ordered spherical shape (Figure 1e). This pattern has an inverted spinel structure (magnetite/hematite), which supports the two phases deduced from the XRD [39]. It is clear by precisely examining the SAED pattern that the spinel structure peak (311) related to the first intense magnetite broad ring of magnetite (or hematite) displays an outstanding efficient epitaxial formation of Fe_3O_4 single crystals [39,44]. Furthermore, the (511) and (220) planes of the Fe_3O_4 spinel structure were discernible. According to the HRTEM image, the nanoparticle ensembles are composed of a number of single crystals with polycrystalline particles (Figure 1e). Interestingly, the inter-planar distance (311) of the Fe_3O_4 planes is consistent with the crystalline fringe spacing of 0.248 nm (~2.48 Å) [24,44].

Crystalline Structure—XRD Analysis

The XRD spectra of the as-prepared nanohybrids are shown in Figure 2: magnetite (Fe_3O_4) has a crystalline cubic spinel shape [43]. Based on the JCPDS (PDF No. 65-3107) database file, all nanohybrids contain eight typical diffraction peaks for Fe_3O_4 at 2θ around 19°, 30°, 35°, 43°, 54°, 57°, 62°, and 74° [39,45]. Furthermore, the possibility of hematite (Fe_2O_3) coexisting with ferromagnetic properties cannot be left out., according to the findings of the XPS study, which is discussed later. The Fe2p band in the adsorbents (TA-type and C-type) spectrum may be deconvoluted by XPS into three signals (each signal having a unique doublet band) associated to Fe^{3+} in both octahedral (at BEs ~710.27–710.68 eV, ~712.81–713.23 eV, and ~732.42–732.39 eV for octahedral geometry) and tetrahedral (at BEs: 716.45–716.96 eV and ~727.12–727.69 eV), and Fe^{2+} (at BEs ~723.80–724.31 eV for tetrahedral geometry). The relevant satellite bands have a low resolution. Moreover, an $Fe2p_{3/2}$ signal (Fe^{3+}- octahedral form) at roughly 711.0–711.6 eV might be assigned to the most prevalent kinds of iron oxides, such as hematite (Fe_2O_3) or maghemite (g-Fe_2O_3) [46].

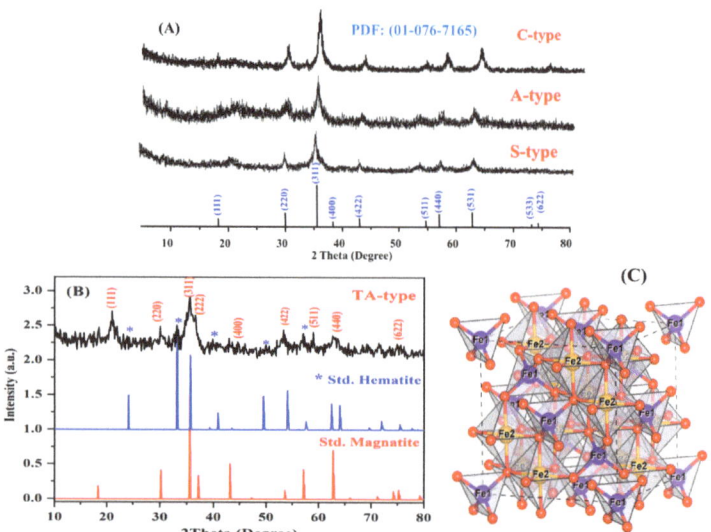

Figure 2. XRD analysis of the four different nanohybrids (**A**,**B**). The bar-type spectrum depicts the standard magnetite and hematite pattern (**B**). The spinel structure's unit cell crystal structures (**C**).

The SAED pattern indicated the nanohybrids' polycrystalline structure, which is characterized by the presence of sharp and continuous loops (Figure 1e). Furthermore, XRD spectra were used to evaluate the diameters of the crystalline Fe_3O_4 nanoparticles by using the Debye–Scherrer Equation: ($D = k\lambda/\beta Cos\theta$) [36,47], where D is the mean particle diameter (in nm), k is the shape factor (0.9), λ is X-ray radiation wavelength (~1.5418 Å), θ is the diffraction angle, and β is the FWHM (full width at half maximum) of the selected X-ray diffraction peaks. The size was estimated to be around 9.43, 10.83, 9.78, and 9.01 nm for A-type, S-type, C-type, and TA-type, respectively, for the most intense (311 index) peak (validated by SAED picture) [38].

From the aforementioned findings, Figure 2C depicts the structural Fe_3O_4 characteristics. Magnetite is composed of $Fe^{2+}Fe_2^{3+}O_4$ (generally so-called mixed oxide "Fe_3O_4"), where Fe1 and Fe2 are associated with Fe^{2+} and Fe^{3+}, respectively, showing cation positions inside the octahedron and tetrahedron, respectively, and O denotes the oxygen anion position. The unit cell of spinel magnetite is a cube composed of 32 O^{2-} anions that are formed by 8 $Fe^{2+}Fe_2^{3+}O_4$ molecules [39]. The packed face-centered cubic of Fe_3O_4 (FCC) is composed of oxygen anions that comprise 64 Fe1 (tetrahedral) and 32 Fe2 (octahedral) vacant spaces that are partially occupied by Fe^{3+} and Fe^{2+} cations [48].

Textural Properties—BET Surface Analysis

The textural features of the as-prepared nanohybrids were assessed using the BET method: $N_2(g)$ sorption (i.e., via the adsorption–desorption) and isotherms. Nitrogen isotherms of nanohybrid derivatives mirror Type-IV isotherms associated with mesoporous adsorbents (Supplementary Figure S1) [38,49]. The specific surface areas calculated using the BET method (S_{BET} in $m^2.g^{-1}$) for all nanohybrid derivatives were found and follow the following sequence: TA-type (75.27) > S-type (61.05) > A-type (~60.38) > C-type (~42.56). Moreover, the textural qualities of all nanohybrids are essentially similar: likely to be mesoporous in structure. The typical pore diameters are ~27.5, 26.9, 22.5, and 3.9 nm for TA-type, S-type, A-type, and C-type. However, the significant whole pore volume (WPV in $cm^3.g^{-1}$) was recorded: TA-type (0.57) > S-type (0.41) > A-type (0.36) > C-type (0.12).

Thermogravimetric Analysis (TGA)

The thermogravimetric properties of S-type are depicted in Figure 2A. Significant variations could be found in terms of weight loss and the numbers of degradation stages, as well as in the DTG diagrams (Supplementary Figure S2a,b). The deterioration profile is distinguished by four transitions: (a) The first phase (until 202–206 °C) is concerned with the liberation of absorbed water molecules at the adsorbent surface [47], and the consequent structural changes resulted in a 12% weight decrease. (b) The second phase (within 206–350 °C) results in a weight loss of 25–27% (cumulative wt. loss: 37–39%) [17], which is linked to the decomposition and destruction of amine moieties in chitosan (itself) and its amine and amino acid derivatives [35]. (c) The third phase (from 350 and 439 °C) resulted in a weight loss of around 4–7% (max. wt. loss: 41–44%) and includes depolymerization and pyranose rings disintegration [47]. (d) The final phase (up to 650 °C) corresponds to the char entire disintegration, which accounts for about 56.2%. This fraction corresponds to the magnetic core amount in the composites. The tracker of magnetite with char remaining at higher temperatures may throw this judgement off [35].

Furthermore, the DTG graph (Supplementary Figure S2b) indicates three significant valleys, namely at ~234.0 °C (strong and abrupt), at ~345.1 °C, and at ~415.2 °C, as well as a mild valley at 502.5 °C, related to the variations in the TGA profile slope and the release of adsorbed and structured water (where the samples were freeze-dried gave the strongest valleys), the breakdown of amine and amino acid moieties in adsorbent (i.e., chitosan and the derivatives), the depolymerization and breakdown in pyranose ring accompanying char formation, and, eventually, the char's thermal deterioration (final portion of deterioration) [35,47].

Magnetic Properties—VSM Analysis

The magnetic profiles of MCS and their derivatives were evaluated and analyzed using VSM (Supplementary Figure S2c). The hysteresis curves of all nanohybrids materials exhibit extremely small remanence magnetization (M_r), coercivity field (H_c), and no hysteresis loop (Supplementary Table S2: the basic physicochemical features of these nanohybrid materials; that is, the magnetization of the residue was almost zero, indicating superparamagnetic behavior. The magnetic saturation (Ms) is 42.97 emu.g^{-1} for MCS; nevertheless, after chemical modification via amine and amino acids immobilization, the Ms values (in emu.g^{-1}) were reduced and arranged as follows: r-MCS (42.97) > TA-type (31.13) > A-type (26.94) > S-type (23.21) > C-type (22.84). Because Ms values are connected to the amount of Fe_3O_4, the organic coating thickness increases with further chemical functionalization, resulting in a reduction of the iron oxide quantity and, hence, a decrease in Ms values [25,39]. Meanwhile, these Ms values are sufficient for simple and efficient magnetic separation from aqueous solutions [38,39].

3.2.2. Chemical Characterization

XPS Spectroscopy

The XPS survey signalizes the elemental composition of all nanocomposite derivatives (Supplementary Figure S3): C1s, N1s, O1s, Fe2p, and S2p signals at the binding energy (BE) of 286.97–288.29 eV, 400.17–401.37 eV, 532.74–533.88 eV, 711.96–713.41 eV, and 165.5 eV, respectively. Furthermore, two peaks at 710.3–711.4 eV (Fe2$p_{3/2}$) and 725–726 eV (Fe2$p_{1/2}$) distinguished the magnetite. The core levels' profiles for essential components are depicted in Figure 3 (and Supplementary Table S3). The deconvoluted spectra were determined by the assignments and average atomic fractions [50]. The C1s curve figures out the peaks at 285.6 (corresponding to C–H, C–C, C$_{advent}$), 286.3 (related to C–OH, C–N, C–O–C, C–S), and 287.5–288.8 (attributed to C=O (Amide), and –COOH). The N1s spectrum represents the N– containing functional groups since the related peaks to (N–H, C–N (amide) and (–NH$_2$, >NH (amine)) appeared at BE of ~399.3 and ~400.3 eV, respectively. A slight shift for amino acid derivatives was observed that could be due to the carboxylate environment. The O1s curve depicts the belonging peaks at BEs: 529.61 eV (corresponding to Lattice O

(Fe–O, Fe$_3$O$_4$)), 530.87 eV (refers to C=O, –O–C=O), and 532.51 eV (related to C–OH, O–H in H$_2$O, and C–O–C). The S2p graph depicts the S–containing functional groups at BEs: 164.17 eV (C-S (2p$_{3/2}$)) and at 168.05 eV C-S-C (S2p$_{1/2}$).

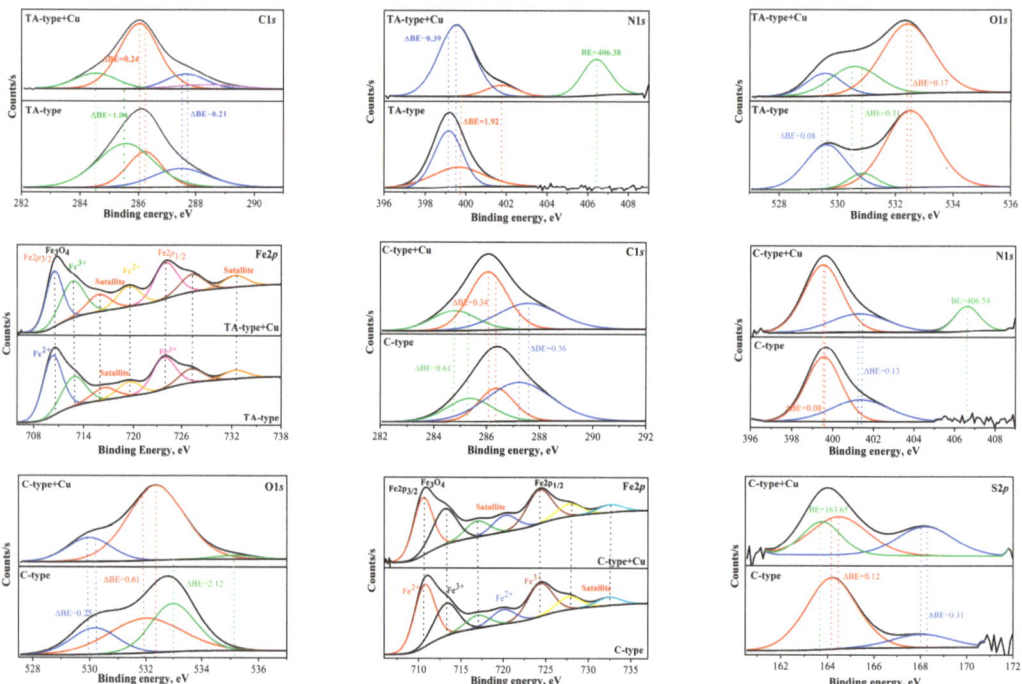

Figure 3. HR-XPS of TA-type and C-type for (C1s, N1s, O1s, Fe2p, and S2p spectra) before and after Cu(II) adsorption.

The primary peaks of Fe^{3+} 2$p_{3/2}$ and Fe^{3+} 2$p_{1/2}$ that are considered to be distinct Fe$_3$O$_4$ are located at ~710.27–711.67 and ~723.80–724.60 eV, respectively, in the Fe2p spectrum (Figure 3) [37,51]. After fitting the Fe2p dual peaks, we noted that Fe^{2+}2$p_{3/2}$ peaks emerge at ~712.16–713.02 eV and Fe^{3+} 2$p_{1/2}$ peaks appear at ~719.43–720.21 eV, confirming the existence of Fe^{2+} and Fe^{3+}, respectively, corresponding to Fe$_3$O$_4$. The Fe^{2+} 2$p_{3/2}$ peaks appear at ~712.16–713.02 eV, while the Fe^{3+} 2$p_{1/2}$ peaks arise at ~719.43–720.21 eV, proving the coexistence of Fe^{2+} and Fe^{3+}, respectively, which corresponds to Fe$_3$O$_4$ [37], whereas the Fe^{2+} 2$p_{1/2}$ satellite peak occurs at ~723.81–725.01 eV [37,52].

Element Analysis-CHNS/O

The effective chemical modification of chitosan backbone was confirmed using CHNS. Supplementary Table S2 summarizes and compares the elemental analysis of the various samples. The organic fraction (associated to C and N, wt. %) of raw chitosan (rCS) and magnetic chitosan (MCS) was examined and found to be substantially lower (52.7–53.9%) for rCS because of the magnetite content that represents ~53.5% of overall mass. After crosslinking process, the N-content (in mmol%) of crosslinked magnetic chitosan (CMCS) reduced from 2.56 to 2.17, owing to increased epichlorohydrin binding, which is consistent with the small decreases in the C and H contents. In the meantime, after activation via terminal –Cl atom insertion, the activated matrix (ACMCS) molar mass increased steadily; thus, the wt.% of C, H, and N was diluted and reduced [25]. Subsequently, immobilization of multifunctional moieties: the rise in C, H, N, and S content (in wt/wt) indicated the successful functionalization and grafting onto ACMCS. More particularly, the N content virtually increased from 1.41 mmol N.g^{-1} to 2.69, 2.58, 2.35, and 3.72 mmol N.g^{-1} for A-type, S-type, C-type, and TA-type, respectively [25,39]. Assuming C-type, the S content (in wt,%) of 3.01% supports effective cysteine grafting [38].

To estimate magnetite proportion of nanohybrid materials, the thermal degradation characteristic of the S-type was used (see Section Thermogravimetric Analysis–TGA). The S-type was a stand-in for distinct derivatives, since all nanohybrids have the same matrix and the only change is in the grafted moieties. The magnetic core portion accounted for 48–52 $\pm$0.5% of total weight. This fact is critical when considering the adsorption performance, as the inorganic fraction contributes half the total weight of the adsorbents and has a lower affinity for Cu(II) ions than the organic component of chitosan derivatives.

FTIR Spectroscopy

Synthesis stage: to characterize the functional groups of the nanohybrid adsorbents, infrared spectroscopy was utilized and is presented in Figure 4a. The spectra were essentially all fairly similar; the primary distinctions were in the relative strength. The band at 616–627 cm^{-1} corresponds to the υFe–O stretching vibration in Fe$_3$O$_4$ [25,36]. The peaks at ~533–558 cm^{-1} and ~456–480 cm^{-1} result from the split of the υ_1(Fe$_{tetra}$–O band) and υ_2(Fe$_{octa}$–O band) which is the blue-shift of bulk Fe$_3$O$_4$, respectively [25,31]. Generally, the strong and broad absorption peaks at ~3468–3400 cm^{-1} are relevant to the stretching vibration for –NH$_2$/>NH and –OH stretching vibration (and their overlapping) and hydrogen bonds in polysaccharides [31]. The peaks at 2923–2936 cm^{-1}, 2852–2869 cm^{-1}, and 1387–1392 cm^{-1} are assigned to C–H symmetric and asymmetric stretching vibration, respectively (in –CH$_2$ (methylene groups)), and C–H bending, respectively [26,35,53]. The peak around 2432–2334 cm^{-1} is caused by the CO$_2$ vibration. The substantial peak at 1602–1610 cm^{-1} could be ascribed to >NH bending vibration (–NH$_2$ and >NH amine bend, in plane deformation), and C=O (Amide I) and/or (COO$^-$) carboxylate group symmetric stretching vibration [23,30]. The bands at 1330–1256 cm^{-1} (referring to C–H (in CH$_2$) symmetric deformation and stretching vibration of C–N) and the ranges between 1112 and 11161 cm^{-1} and between 936 928 cm^{-1} (corresponding to stretching vibration of secondary –OH and C–O–C (in β–glucosidic bridge) [31], and β-D-glucose unit of carbohydrate ring) [54]. The peaks at 854–798 cm^{-1} and 722–711 cm^{-1} (corresponding to stretching vibration of secondary –OH and C–O–C bridge [54],) are caused by C–H out-of-plane bend bending vibration of C–H and –NH twist, respectively [26,35,39]. The C–C–N out-of-plane bending mode was assigned to the weak band at 464–480 cm^{-1} [23,55]. The absence of the C–S functional group's peak at 742 cm^{-1} could be due to the overlapping with other groups [41].

FTIR for characterization of Cu(II) interactions with TA-type and C-type nanohybrids: The FTIR spectra for both TA-type and C-type adsorbents are affected by Cu(II) adsorption (Figure 4b,c and Supplementary Table S4) as follows:

(a) For TA-type: The bands' intensities tend to strongly decline and red-shift at 3468, 3441, and 3398 cm^{-1} (assigned to –OH and –NH functionality); 2936 and 2857 cm^{-1} (–CH$_2$ groups); 1605 cm^{-1} (υsC=O (Amide I), (1°/2°) amine bend and their overlapping); and 850 cm^{-1} (C–H out-of-plane bend), shifting to 3467, 3436, 3391, 2923, 2858, 1602, and 818 cm^{-1}, respectively. Meanwhile the blue-shift at 1326–1260 cm^{-1} (υC–N stretching), 1113 cm^{-1} (C–O–C bridge υs), 928 cm^{-1} (υsC–O, 1° OH group, carbohydrate ring, and υsC–N), 710 cm^{-1} (–NH twist), 533 cm^{-1} (υ_1Fe$_{tetra}$–O), and 464 cm^{-1} (υ_2Fe$_{octa}$–O) moved to (1330–1261 cm^{-1}), 1117 cm^{-1}, 932 and 714 cm^{-1}, 541 and 472 cm^{-1}, respectively.

(b) For C-type: The bands' intensity alteration and blue-shift at 3436 cm^{-1}, (υs (–OH and –NH), 1404–1392 cm^{-1} (υsCH$_2$ def., (1°/2°) –OH bend, and COO$^-$ salt), 1330-1265 cm^{-1} (υC–N and –OH bending, –C–O str. (primary υs(–OH)), 1113 cm^{-1}, 855 cm^{-1}, 1113 cm^{-1}, 722(–NH twist), 627 (υsFe–O–Fe and C–H bending), and 591 and 480 cm^{-1}, (υ_1Fe$_{tetra}$–O and υ_2 Fe$_{octa}$–O, respectively) shifted to 3416 cm^{-1}, 1400–1387 cm^{-1}, 1329–1260 cm^{-1}, 1112 cm^{-1}, 853 cm^{-1}, 714 cm^{-1}, 624 cm^{-1}, 541 cm^{-1}, and 456 cm^{-1}. Meanwhile, the red-shift was at 2927 and 2857 cm^{-1} (–CH$_2$ groups), 1609 (υsC=O (Amide I), COO$^-$ salt, (1°/2°)

amine bend and their overlapping), and 932 cm^{-1} (υsC–O, 1° OH group, and υsC-N) to 2931 and 2858 cm^{-1}, 1610 cm^{-1}, and 936 cm^{-1}. The reactive groups involved in metal interaction were identified and discussed with other XPS and adsorption results and are presented in Scheme 4 (and Section 3.3.7).

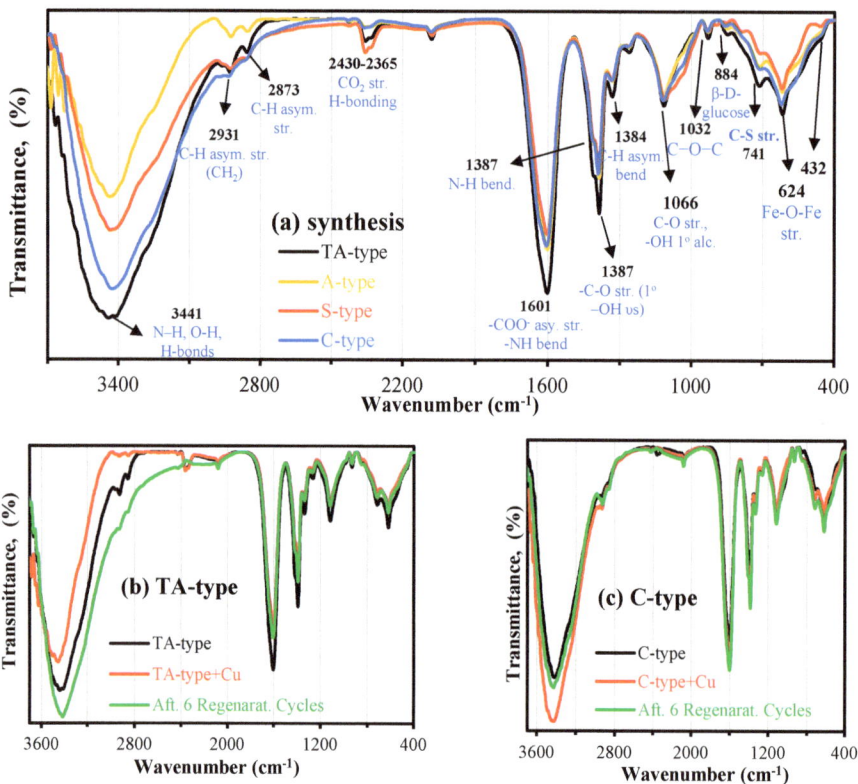

Figure 4. FTIR spectra (**a**) for the synthesized nanohybrids; (**b**) TA-type; and (**c**) C-type, pristine, after Cu(II) adsorption, and after 6th regeneration cycle.

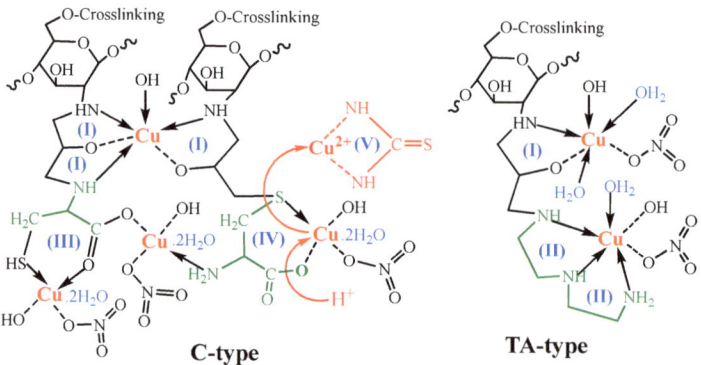

Scheme 4. Interaction and desorption mechanism of TA-type and C-type with Cu(II) complexes.

In both adsorbents (TA-type and C-type), numerous bands related to the stretching vibrations of M−OH and O−M−O were seen in the FTIR spectra at the 400–800 cm^{-1} range [23,35]. These changes were caused by Cu(II) binding through complexation with various reactive groups that alters the environment.

FTIR spectra after Cu(II) desorption: Figure 4b,c (and Supplementary Table S4) also display the TA-type and C-type nanohybrids' spectra after six regeneration cycles (adsorption/desorption of copper). The adsorbents' FTIR spectra were little impacted. Although the copper desorption was effective (Section 3.3.5), several tracer bands associated with some attached Cu(II) ions remained on each regenerated adsorbent's FTIR spectrum, whose spectrum between copper loaded and raw adsorbents can be thought of as "intermediary.

pH$_{ZPC}$—Drift Titration

Supplementary Figure S2d displays the rCS, MCS, and multifunctionalized nanohybrid titration patterns. The consecutive chemical changes have a significant impact on the acid–base properties [35]. The pH$_{ZPC}$ (is equivalent to ΔpH = 0) for rCS is 8.57 ± 0.02, and it correlates to the pKa of 6.3–6.6 for chitosan's –NH$_2$ groups [26]. After the incorporation of the Fe$_3$O$_4$ core in MCS, the pH$_{ZPC}$ dropped to 7.26 ± 0.03. The triamine/alanine/cysteine/serine immobilized onto MCS nanohybrid hardly influences the pH$_{ZPC}$: from 7.26 ± 0.03 for MCS to 7.88 ± 0.01, 7.66 ± 0.02, 7.78 ± 0.03, and 8.48 ± 0.02 for A-type, S-type, C-type, and TA-type, respectively, consistently with charge density, screening impact, and Fe$_3$O$_4$ amount [30]. The maximal obvious rise in ΔpH is at pH$_0$ 4.0, ascribed to the maximum protonation degree, which declines as the pH$_0$ increases.

Surely, after triamine/alanine/serine/cysteine grafting, the pH$_{ZPC}$ increased, most likely as a result of the extra functionalities grafting (e.g., amine and carboxylate moieties). Amino acids contain carboxylic acid groups having pKa values ranging from 2.11 to 2.35, whereas their amine groups have pKa values ranging from 9.15 to 9.69 [25]. The overall pH$_{ZPC}$ of amino acid nanohybrids were closest to those of the amine group of alanine/cysteine/serine in comparison to the carboxylate group end. After diethylenetriamine, immobilization was proportionate to the pKa values 3.58, 8.86, and 9.65 of the amine active sites [56]. Fraga et al. measured the pH$_{ZPC}$ for DETA grafted on graphene oxide, and it was found to be 8.21 [57], wherein the ΔpH$_{ZPC}$ was 0.29 pH unit (the change achieved 8.48), as a result of the support impact switched from graphene oxide to chitosan.

Chemical Analysis—Semi-Quantitative EDX

The EDX graphs of nanohybrid materials (Figure 5 and Supplementary Figure S4) exhibited the appearance of C, N, O, and Fe signals, with an additional S signal for the C-type. The EDX analysis indicates the existence of the characteristic peak for K$_\alpha$ Fe (at 6.403 keV) and L$_\alpha$ (poor peak) at 0.705 keV, as well as the emergence of a minor peak (having a low sensitivity signal at 0.393 keV for the K$_\alpha$ N), thus validating the hybrid formation. The O k$_{\alpha 1}$ signal (at 0.525 keV) and C k$_\alpha$ signal (at 0.277 keV) were observed. Moreover, the S element in the C-type is identified by S k$_\alpha$ and S L$_\alpha$ signals at 2.307 and 0.149 keV, respectively [58]. It should be noted that the semi-quantitative evaluation is restricted to a depth of penetration equivalent to the incident electron's wavelength, as well as a relative fraction with inadequate detection sensitivities [13]. The Cu(II) adsorption clearly shows a definite peaks of K$_\alpha$ Cu signal (at ~8.040 keV) and faint peak of L$_\alpha$ at ~0.930 KeV, respectively, indicating the existence of Cu(II) with an estimated weight percent reflecting efficient adsorption.

From the obtained results, it is clear that Cl atoms at the terminal end of activated magnetic nanohybrid (which consider the only main site for further functionalization) completely disappeared after functionalization with polyamine and amino acid, indicating that the grafting efficiency was more than 99%, and consequently the purity of the material should be like that seen in the XPS and EDX analysis of the final products adsorbent (see Figures 5, S3 and S4). Moreover, the high purity of all used reagents was high purity, and all products were extensively washed to remove unreacted species.

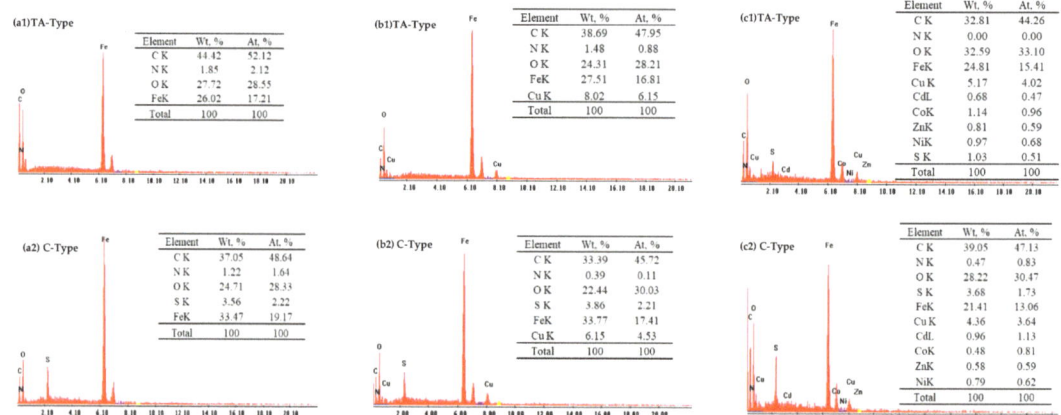

Figure 5. EDX measurements of TA-type and C-type adsorbents before (**a**) and after Cu(II) adsorption from synthetic solution (**b**) and complexed system (**c**).

3.3. Adsorption Investigations

3.3.1. Effect of Initial pH

The solution pH, as a significant physiochemical property, is crucial and has implications in the metal sorption. This parameter can alter the adsorbent's surface charge, as well as the metal's speciation [4,6,25]. Cu(II) adsorption was investigated throughout a pH_0 range of 2–7 to determine an appropriate pH range for adsorption. The pH value had a substantial influence on the adsorption performance of the r-MCS, TA-type, A-type, S-type, and C-type (Figure 6a), and the capacities trend versus pH (as a function of equilibrium pH (pH_{eq})) was also quite comparable. Figure 6a shows that the adsorption capacity rose initially and subsequently nearly fixed as the initial (pH_0: from 2.0 to 7.0) and equilibrium pH increased or climbed. The optimal pH_0 for the r-MCS, TA-type, A-type, S-type, and C-type was 5.0–7.0, and their adsorption capacities were 0.41, 1.56, 0.71, 0.83, and 0.93 mmol Cu.g^{-1}, respectively. The adsorption capabilities of the multifunctionalized materials were all greater than adsorption capacity of r-MCS (q_{eq}: 0.41 mmol Cu.g^{-1}) at optimal pH_0 = 5.5–7.0. Adsorption capabilities are quite low at $pH_0 < 5.0$, and rising pH equals greater capacities. In a low-pH environment, H$^+$ would biasedly bind to the functional groups located on the nanohybrids surface, making it harder for Cu(II) ions to bind. When the pH steadily rises at $pH_0 > 5.0$, the deprotonation process may cause Cu(II) to interact with functional groups, enhancing the nanohybrids' adsorption ability [4,6].

All in all, as the pH_0 went from 2.0 to 5.0, the adsorption increased, and then the pH rose over 5.0 and the Cu(II) adsorption remained steady. This process could be related to the precipitation since it impacts metal ion adsorption [4]. To prevent interference with the precipitation mechanism, an appropriate pH_0 value ~5.0 was chosen for further studies. Indeed, the initial solution pH is pushed toward lower pH values: the pH_{eq} is closely similar with somewhat higher at initial pH values (pH_0: 5.02–7.01) (Supplementary Figure S5a). However, the slight lowering is recorded as follows: for TA-type (at pH_{eq} of 6.16–6.35) > C-type (at pH_{eq} of 5.71–5.87) > A-type (at pH_{eq} of 5.46–5.58) > S-type (at pH_{eq} of 5.02–5.28); the solution appears to be buffered [38]. Supplementary Figure S5b depicts the relation between the $\log_{10} D$ plot vs. pH_{eq} as being linear. The line slope for r-MCS, TA-type, A-type, S-type, and C-type was 0.1796, 3.228, 0.2894, 0.3652, and 0.437, respectively, and it is related to the stoichiometric exchange in conjunction with the attached metal ions in the Cu(II) ion-exchange mechanism [32,59].

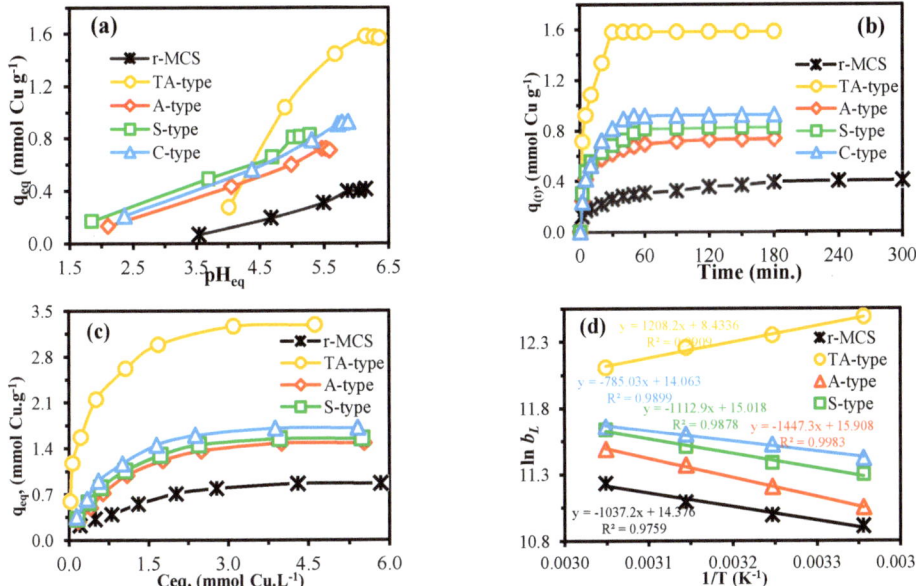

Figure 6. The pH effect (**a**), adsorption kinetics (**b**), isotherms (**c**), and thermodynamics (**d**).

3.3.2. Adsorption Time Effect and Kinetic Studies

The implications of adsorption period on the adsorption capacities were investigated further to investigate the kinetic parameters (Figure 6b). Adsorption reaction may be separated into two stages: the initial stage observed substantial increases in adsorption capacity, followed by a progressive slowing until adsorbents achieved saturation after a time duration. The kinetic curve in Figure 6b shows that, during the quick initial stage (within the first 10 min), the adsorbents remove approximately 68.6%, 72.3%, 68.2%, 57.4%, and 55.1% for the TA-type, A-type, S-type, C-type, and r-MCS, respectively, of the total sorption, representing a physisorption mechanism. Furthermore, the initial slope is greatly enhanced, and this may be connected to the adsorbent sensitivity for Cu(II) ions (particularly for the TA-type, S-type, C-type, and A-type over r-MCS, constantly, with a specific surface area for the different nanohybrids). The TA-type adsorbent exhibited the fastest adsorption, followed by amino acid derivatives, when compared to the raw chitosan. It should be noted that the adsorption equilibrium times were ordered as follows: TA-type (30 min) > C-type (50 min) > S-type (60 min) > A-type (90 min) > r-MCS (240 min).

As a result of complementary adsorption, the second slower step (within 10 and 40 min) follows, exemplifying a chemisorption mechanism (through electrostatic interaction and/or chelation). Finally, equilibrium between adsorption and desorption is reached. According to textural analysis, all nanohybrid adsorbents had mesopores that were wide enough (the pore size range ~3.4–8.3 nm) to enable Cu(II) to flow through (Shannon radius for $[Cu(H_2O)_6]^{2+}$ hydration is 0.62–0.94 Å [60]). (Corrected and done). This implies that Cu(II) ions can diffuse rather freely within the adsorbents' mesoporous network, thus showing that the delayed sorption step contributes to global sorption, which is most likely related to resistance to intraparticle diffusion.

To properly understand the adsorption kinetics, time-dependent data were simulated. The kinetic behavior was modeled using PFORE and PSORE (Supplementary Table S1) [61–63]. The experimental adsorption kinetics were exhibited after modeling and linearization with several models, as illustrated in Supplementary Figures S6 and S7 and Table S5. The PSORE kinetic model's correlation coefficient is closest to 1. As a result, the PSORE kinetic model may be used to explain Cu(II) adsorption. The PSORE model was founded on the concept that the adsorbent and adsorbate, i.e., between functionalized chitosan derivatives and Cu(II), are mostly chemisorption mechanisms [25,63]. Moreover, the goodness of the PSORE model was proven by approximately equal adsorption capacities ($q_{m,cal.}$ and $q_{m,exp.}$; Supplementary Table S5): the PSORE model's overestimated values ranged from 1.7% to 4.7%, whereas the PFORE model's underestimated values differ by 8.0–48.6%.

Supplementary Figure S7 depicts a multilinear plot (and Supplementary Table S5) that gives the values of the critical characteristics for intraparticle diffusion rate constants (K_{id}). Supplementary Figure S7 shows three-stage multilinear plot. The first two main phase with greater $K_{id,1}$ and $K_{id,2}$ values were sorted as follows: $K_{id,1} > K_{id,2}$, implying that the initial stage involves fast sorption at the most immediately accessible binding sites, as well as interior meso- and macropores [64]. The concentration tendency from the solution to the adsorbent therefore decreased, delaying the mass transfer of copper ions to inner reactive groups in mesopores. Finally, the $K_{id,3}$ was almost zero in the last part, owing to adsorption–desorption equilibria, assuming that the adsorbents are almost saturated [25,33].

3.3.3. Initial Cu(II) Concentration Effect

The adsorption equilibrium data as a function of $q_{eq} = f(C_{eq})$ indicate how the Cu(II) ions can be divided between both the aqueous phase (C_{eq}) and adsorbent surface (q_{eq}) (under various initial concentrations and selected experimental conditions) [10,35,61]. The initial metal-ion concentration propels the adsorption process by eliminating the mass transfer resistance between the liquid solution and the adsorbent surface phase [61]. The Cu(II) adsorption was examined by varying the initial Cu(II) concentration and the temperature. According to Figure 6c, all adsorbents follow the same pattern: at low concentrations, the adsorption capacity increased as the initial and equilibrium Cu(II) concentrations increased. The isotherm diagram compares Cu(II) adsorption capacities at pH_0 5.0 and 298 K and demonstrates that the maximal real adsorption capabilities (q_{eq}, in $mmol.Cu.g^{-1}$) are arranged as follows: TA-type (3.29) > C-type (1.71) > S-type (1.55) > A-type (1.48) > r-MCS (0.86). The benefit of r-MCS functionalization is clear; the peak adsorption capacity is about 3.33 times greater for TA-type and nearly doubled for amino-acid derivatives (1.72–1.95 times). Furthermore, this reveals the vast superiority of the TA-type's adsorption efficiency over that of the amino acid derivatives, which is compatible with the findings of pH influence and equilibration time.

The variation in the adsorption efficiency of the nanohybrids was assumed to be attributable to the differences in the functional and structural features of four amine and amino acid derivatives. As the polyamine–magnetic-chitosan derivative with diethylenetriamine, which is known as 2,2'-iminodiethylenetriamine [39]; meanwhile, for amino acid derivatives [36,65]: with alanine (a "simple" amino acid contains solely amino and carboxyl groups), Serine (a hydroxyl-containing amino acid) contains amino, carboxyl and hydroxyl groups [25] and cysteine (is a thiol-containing amino acid) possesses sulfhydryl, amino and carboxyl groups [38,42]: the major Cu(II) adsorption site is the amine/carboxyl group in addition to –SH/–S– or –OH/–O– groups. Moreover, this may explain why amine groups preferentially interact with alternative groups such as –SH or –OH (for cysteine and serine, respectively) and behave as coordination sites in adsorption reaction, as demonstrated by XPS analysis in the next section.

Table 1 shows a comparison with various adsorbents; however, a real comparison is impossible, owing to variances in experimental circumstances. The comparison shows that the TA-type and C-type adsorbents are the best, although the S-type and A-type adsorbents have an equivalent adsorption performance to conventional and traditional adsorbents. Despite this fact, some of the selected adsorbents had faster kinetics, such as Fe_3O_4/Cel-DETA [3], and N-aminorhodanine-modified chitosan hydrogel [17], within ~20 min, and greater adsorption capacities (e.g., urea calcium alginate beads functionalized with CR dye ~6.941 $mmol.g^{-1}$ [6], Cd–metal–organic framework MOF ~6.842 $mmol.g^{-1}$ [19], and Meso/micropore-controlled hierarchical porous carbon

~4.134 mmol.g^{-1} [21] and ethylenediamine-tetraacetic acid (EDTA) into layered double hydroxides (LDH) ~1.902 mmol.g^{-1} [66]). As a result, the TA-type and C-type appear to be more competitive adsorbents for Cu(II) recovery than the S-type and A-type. Adsorbents can be rated in terms of kinetics and adsorption capacity as follows: TA-type (30 min) >>> C-type (50 min) > S-type (60 min) > A-type (90 min) > r-MCS (240 min), which corresponds to the overall form of the fractional strategy to equilibria. Thus, all adsorbents have reasonable and good adsorption capacities and short equilibration times. Furthermore, with a pH range of 5.0–6.5, this implies that these nanohybrids adsorbents are suitable candidates for copper adsorption.

Table 1. Copper adsorption performances using different adsorbents.

Adsorbent	Operating Condition (pH$_0$, Time, Temp. (in °C), SD (g.L^{-1}))	q$_m$, mmol Cu.g^{-1}	Ref.
Pristine natural zeolite	5.5, 24 h, 26 °C, 1.0	0.235	[28]
PDA treated zeolite powders		0.45	
Zr-based MOFs (MOF-801)	5.5, 180 min, 27 °C, 0.4	0.278	[16]
DTPA modified sludge	3.0, 60 min, 25 °C, 0.4	0.557	[67]
Foamed geopolymer sphere	5.0, 48 h, 27 °C, 1.0	0.596	[68]
CS-GLA	5.0, 360 min, 26 °C, 5.0	0.601	[17]
Aminorhodanin@CS	5.0, 20 min, 26 °C, 5.0	0.984	
Melamine–HCHO–DTPA	4.5, 60 min, 20 °C, 10.0	0.729	[7]
N-Ch-Sal	5.0, 120 min, 26 °C, 1.0	1.331	[10]
Mag. Cel-DETA (MCGT)	5.4, 5 min, 25 °C, 1.0	1.449	[18]
Silkworm excrement biochar	5.0, 24 h, 25 °C, 1.0	1.471	[4]
MgAl-EDTA-LDH	5.0, 60 min, 25 °C, 1.25	1.901	[66]
Meso/microporous Carbon	5.0, 120 min, 25 °C, 0.35	4.134	[21]
Cd-terephthalate-MOF-2	6.0, 60 min, 27 °C, 0.5	6.842	[19]
CR dye@Urea Ca-alginate	6.5, 90 min, 55 °C, 1.5	6.94	[6]
r-MCS	5.0, 240 min, 55 °C, 0.5	0.987	
A-type	5.0, 90 min, 55 °C, 0.5	1.696	
S-typr	5.0, 60 min, 55 °C, 0.5	1.745	Here
C-type	5.0, 50 min, 55 °C, 0.5	1.921	
TA-type	5.0, 30 min, 25 °C, 0.5	3.287	

The Temkin, Langmuir, and Freundlich adsorption isotherms (Supplementary Table S1) were utilized and tested to simulate adsorption process (Supplementary Figure S8). As demonstrated in Supplementary Table S6, the correlation terms show that the Langmuir model had a greater association with the adsorption process than the Freundlich model, since the latter has a rather high correlation value (R^2). Furthermore, the adsorption capabilities were regularly overestimated for the Langmuir model (Δq_{eq}: 4.7–6.5%, 10.5–19.1%, and 10.8–20.1% for the TA-type amine derivative, amino acid derivatives, and r-MCS, respectively), whereas they were underestimated for the Freundlich model (Δq_{eq}: 41.6–46.6%, 26.1–30.6%, and 36.5–45.2%, respectively). Consequently, a limited number of identical active sites are scattered throughout the adsorbent's surface, resulting in monolayer uniform sorption. The resulting curve is close to the experimental locations, thus illustrating the Langmuir model's ability to suit and fit the adsorption isotherms [13,23,69].

The Temkin model also provides a fair match to experimental profiles (in some situations). This concept is often related to a linear lowering in adsorption heat with growing surface saturation (in contrast to the Freundlich model, assuming logarithmic variation in adsorption heat) [23,69]. Supplementary Table S6 and Figure S8c show the Temkin model variables. The B_T coefficient is the Temkin isotherm constant, which is proportional to the adsorption heat. The B_T values for the r-MCS, A-type, S-type, and C-type increased with the temperature (contrary to the TA-type). Additionally, the energetic parameter (A_T) yielded the same findings [23]; as previously stated as an example, consider the comparison between b_L and q_m in the Langmuir adsorption isotherm (Supplementary Table S6). Moreover, these findings agree with the thermodynamics results.

3.3.4. Temperature Effect and Thermodynamic Parameters

The effects of temperature on adsorption ability were studied and compared at four varying temperatures (e.g., 298, 308, 318, and 328 K). Supplementary Figure S9 demonstrates that the adsorption capacity and affinity that are associated with the initial slope for both the r-MCS and amino acid derivatives increase with increasing temperature: endothermic copper adsorption. Although the TA-type shows a reciprocal pattern, the adsorption capacity marginally reduces as the temperature rises; that is, TA-type adsorption is exothermic (Supplementary Figure S9). Furthermore, the Langmuir characteristic (b_L) is related to the adsorption energy: the greater the b_L-value is, the greater the adsorption energy and interaction of adsorbent–adsorbate [70]. Supplementary Table S7 demonstrates that, for amino-acid derivatives, the b_L values (in L mmol^{-1}) rise with temperature from 298 to 328 ±2 K and are ordered as follows: C-type (1.642–2.089) > S-type (1.454–2.042) > A-type (1.134–1.759) > r-MCS (0.988–1.360); thus, the copper adsorption is endothermic. Meanwhile, the TA-type exhibits a completely distinct reciprocal behavior: with the increasing temperature, the b_L values fall from 4.770 to 3.257 L mmol^{-1}; hence, the process is exothermic.

Supplementary Table S6 reports the modeling of the adsorption isotherms; the Langmuir models match the experimental profiles only slightly better. The thermodynamics might be investigated according to the Langmuir isotherm. Lima et al. state that the Langmuir's affinity coefficient (b_L) characteristics are utilized after converting (to L mol^{-1} from L mg^{-1}) and correcting for water molality [71]. The Van't Hoff equation and Gibbs function equation (Equations (11) and (12)) were utilized to compute thermodynamic parameters (ΔH°, (the enthalpy change, in kJ mol^{-1}), ΔS° (the entropy change, in J mol^{-1} K^{-1}), and ΔG° (the Gibbs free energy change, in kJ mol^{-1})) [35,38]:

$$\ln b_L = -\frac{\Delta H^0}{R} \times \frac{1}{T} + \frac{\Delta S^0}{R} \quad (11)$$

$$\Delta G^0 = \Delta H^0 - T\Delta S^0 \quad (12)$$

where the global gas constant is R = 8.314 J/mol.K, and the absolute temperature in Kelvin is T, in K. The slope and intercept of the relation diagram of ln b_L vs. 1/T plot (Figure 6d) are used to obtain the ΔH°, and ΔS° values. Figure 6d depicts two opposite trends obtained with the amino acid and amine derivatives with increasing the temperature.

Supplementary Table S7 shows the different three thermodynamic parameters derived using the Van't Hoff formula and the Gibbs function equation. For the amino acid derivatives, the positive ΔH° indicates that the Cu(II) adsorption is endothermic in nature [72], whilst the TA-type adsorbent has an entirely different behavior: the negative value indicates the exothermic nature [59,73]. Moreover, the ΔH° values < 40 kJ/mol demonstrate the presence of physical forces in the Cu(II) adsorption reaction. The net ΔH° is the summation of the dehydration enthalpy (ΔH_{dehydr}, that is thought to be positive because of the energy necessary to break the hydrated metal ions' ion–water and water–water bonds) and the enthalpy of complexation ($\Delta H_{complex}$, also negative) [47]. Furthermore, the endothermic process includes several phases, namely Cu(II) transfer to the solid surface, dehydration, and complex formation [74]. All of these reactions require heat to induce the adsorption (reciprocal to the TA-type).

The positive ΔS° revealed that adsorption improves the randomness of the state environment and global system and that the solution grew more disordered [23,37]. Moreover, the negative ΔG° values within a comparable range show that the sorption process is spontaneous. According to Zai [75], sorption is both physical and chemical in nature whenever the ΔG° values ranges from −20 to −80 kJ mol^{-1}. Moreover, Supplementary Table S7 indicates the absolute values ($|\Delta H^\circ| < |T\Delta S^\circ|$); the copper adsorption is regulated by entropic rather than enthalpic changes [26,35].

3.3.5. Adsorbent Reusability

Adsorption/desorption cycles up to six runs were carried out. The elution of Cu(II) was performed using thiourea (0.25 mol.L^{-1}) acidified with H_2SO_4 (at pH 2). The experimental adsorption and desorption conditions are systematically listed under Supplementary Table S8. Six sequential sorption/desorption cycles are shown in Supplementary Table S8, with a little decline in both adsorption and desorption efficiency for each sorption stage. After the sixth cycle, the efficiency losses in adsorption and desorption were about 5.3–9.1% and 4.5–5.3%, respectively. This confirmed that all adsorbents exhibited high levels of stability, reusability, and durability for repeated use.

3.3.6. Chemical Stability Examination

The nanohybrid samples were soaked and agitated for one day in various pH solutions, ranging from pH_0 1 to 10, in the pH_{ZPC} experiment. After collection and magnetic separation, the nanohybrids were constant and very stable. The dry samples' mass-loss efficiencies after acidic and alkaline treatment against the mass of the untreated samples (under identical conditions) were ~1.9–4.7%. The maximal mass loss values were recorded and observed at pH_0: 1.0 > pH_0: 2.0 because of the surficial Fe_3O_4 particles' dissolving [25]. Moreover, after the sixth adsorption/desorption cycle, C-type and TA-type FTIR spectra (as examples of nanohybrids) confirmed the functional stability and chemical resistivity (see Supplementary Table S2). Moreover, the little drop in adsorption/desorption characteristics supports and verifies the durability of all the nanohybrid materials [26]. These findings showed how extremely robust and durable these materials are.

3.3.7. Metal Adsorption Interaction and Complexation

XPS spectroscopy. XPS was employed to investigate the chemical environments of surface components, as well as the composition features of the survey XPS spectrum of TA-type and C-type (as a representative for amino acids-derivatives) before and after Cu(II) sorption (Figure 7). The core level signals of Cu2p are quite intense for TA-type and C-type adsorbents (see Figure 4). Cu(II) binding is underlined by the formation of a doublet at ~932.98–943.08 eV (Cu2$p_{3/2}$) and ~952.18–953.58 eV (Cu 2$p_{1/2}$).

Supplementary Table S3 outlines the band-fitting results from HRES spectra for the bare TA-type and C-type monohybrids that were considerably compared after Cu(II) binding for C1s, O2s, N1s, S2p, Fe2p, and Cu2p signals (Figure 7a). Notably, the most substantial changes are detected for the remaining chosen signals of the main elements, where Fe2p was hardly affected after Cu(II) adsorption (see Figure 3). Variations in the proportional contributions and change in BEs of the various bands provide the active sites engaged in the adsorption process that are mostly preceded by Cu(II) interacting with N–, S–, and O-containing groups such as (amine >NH/–NH$_2$), (–SH/thiolate), and (–OH/>C=O/–COO$^-$), respectively. The new significant Cu2p double peaks in the XPS spectrum clearly verified the Cu(II) adsorption by TA-type and C-type. The Cu2p excitations can be deconvoluted into some multiple-Cu2p peaks into two subpeaks' spin–orbit (L–S) split assigned to Cu2$p_{3/2}$ (~953 eV) and Cu2$p_{1/2}$ (~932 eV), accompanied by the corresponding ~8–9 eV satellite peaks from the Cu2$p_{3/2}$ (Figure 7b,c) [8]. These satellite peaks can be assigned as 3d→4s shake-up transitions [76].

The deconvolution of the Cu2$p_{3/2}$ displays two different BEs states that can be ascribed to CuO and cupric ions that reside in octahedral positions and interact strongly with both adsorbents [12], respectively. The two main large asymmetric peaks were fitted into two subpeaks for Cu$^+$ 2$p_{3/2}$ (at 932.70–933.38 eV), Cu^{2+} 2$p_{3/2}$ (at 934.73–935.52 eV), Cu$^+$ 2$p_{1/2}$ (at 945.26–952.94 eV), and Cu^{2+} 2$p_{1/2}$ (at 952.56–955.89 eV) which were situated (Supplementary Table S3) [8]. The Cu$^+$ 2$p_{3/2}$ at 932.70–933.38 eV, and Cu$^+$ 2$p_{1/2}$ at 945.26–952.94 eV were observed, owing to low values of the shake-up satellite/main peak ratio, likely related to a reduction in Cu$^+$ [12].

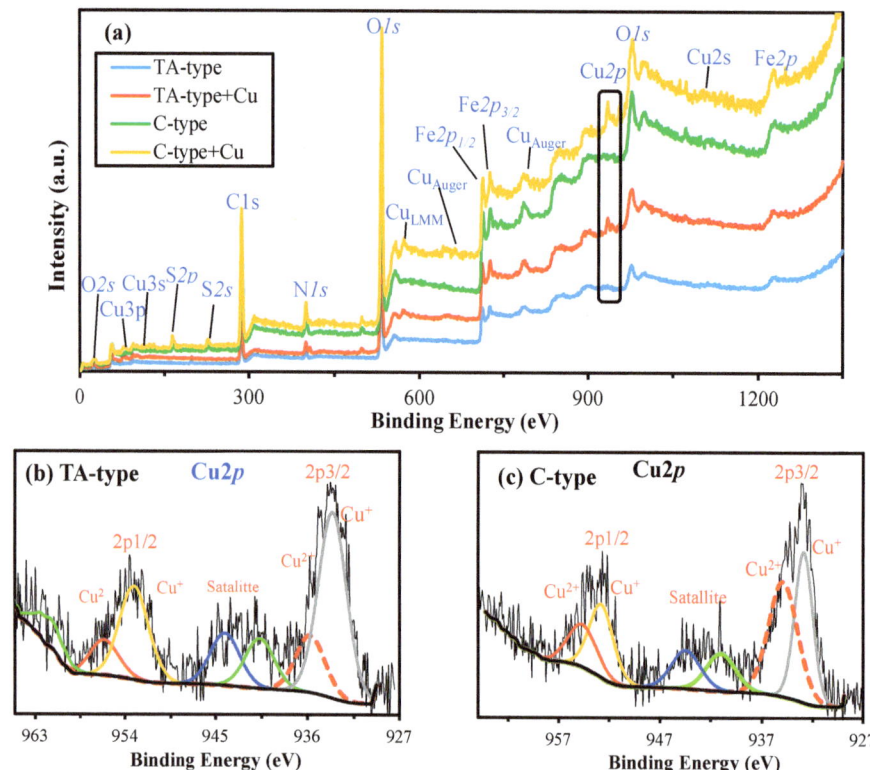

Figure 7. XPS survey spectra of raw TA-type and C-type adsorbent before and after Cu(II) adsorption (**a**) and core level for Cu2p (**b,c**).

In light of the mechanism discussion, various hypotheses for the coordination mechanism during adsorption reactions have been offered. Interaction mechanisms of Cu(II) with TA-type and C-type adsorbents were suggested from the FTIR and XPS findings. The coordination number for Cu(II) is 4–6. Cu(II) with six coordination number having d^9 configuration favors a distorted octahedral structure that is governed by a direct result of Jahn–Teller instabilities [77].

Based on XPS analyses, Cu(II) may exist in three different forms (Cu(II), [Cu(NO$_3$)]$^+$, and [Cu(OH)]$^+$), where NO$_3$$^-$ group from standard Cu(NO$_3$)$_2$ solution. Scheme 4 depicts the intuitive graph. Thus, two contribution styles (I and II) form chelating rings with five members; more chelate rings enhance the stability of the formed complex and subsequently improve the metal adsorption capability. Furthermore, Scheme 4 displays the hypothesized desorption mechanism (via employing acidified thiourea), since the solution's acidity promotes and breaks the chelation modes (model(V)). Moreover, thiourea is structurally identical to urea, with the exception that the oxygen atom is substituted with a sulfur atom. Thus, the CS(NH$_2$)$_2$ molecule may bind and coordinate with both terminal N atoms of –NH$_2$.

3.3.8. Adsorption Selectivity

To assess the efficacy of such nanohybrids adsorbents for Cu(II) removal and adsorption, the adsorption behavior of Cu(II) from equimolar solutions of the multicomponent system (C_0: 1 mmol.L^{-1} of divalent metal ions, e.g., Cd(II), Zn(II), Co(II), and Ni(II)), with a total concentration that reaches up to 5.01 mmol metal.L^{-1} at pH$_0$ 5.0 and room temperature 25 °C. For r-MCS and nanohybrid derivatives, the maximum adsorption capacity (q_m, mmol.metal.g^{-1}), distribution ratios (D, L.g^{-1}), and selectivity coefficients ($SC_{metal/Ni} = D_{metal}/D_{Ni}$, where Ni(II) was chosen as the standard because of its lower adsorption capacity), are shown in Figure 8. The cumulative adsorption capacities are ordered as follows (q_m, in mmol.g^{-1}): TA-type ($\sum q_m$: 3.33) > S-type ($\sum q_m$: 2.12) > C-type ($\sum q_m$: 2.07) > A-type ($\sum q_m$: 1.95) > r-MCS ($\sum q_m$: 1.16). These q_m-values were higher than those for Cu(II) in single solutions by 1.09%, 35.80%, 20.47%, 31.85%, and 30.79% for TA-type, S-type, C-type, A-type, and r-MCS, respectively. The enhanced accessibility of multi-central amine, hydroxyl, and thiol groups for metal ion adsorption and binding might explain the lack of selectivity for Cu(II). Copper adsorption capacities for TA-type, C-type, S-type, A-type, and r-MCS were 1.73, 1.29, 1.10, 1.07, and 0.58 mmol Cu.g^{-1}, respectively, and were lower than those of synthetic solutions. This suggests that all nanohybrid adsorbents retained a strong affinity, adsorption capacities, and selectivity for Cu(II) over other co-elements, despite a complex composition containing high salty and competing metals. This might be due to a differential in reactivity to other co-ions. This effect is most likely explained by the ionic radius differences and changes around 0.69–0.95 Å for Cd(II), Co(II), Zn(II), Ni(II), and ~0.75 Å for Cu(II). This difference in size provides more adaptability to reactive group accessibility, which could be impacted by steric hindrance.

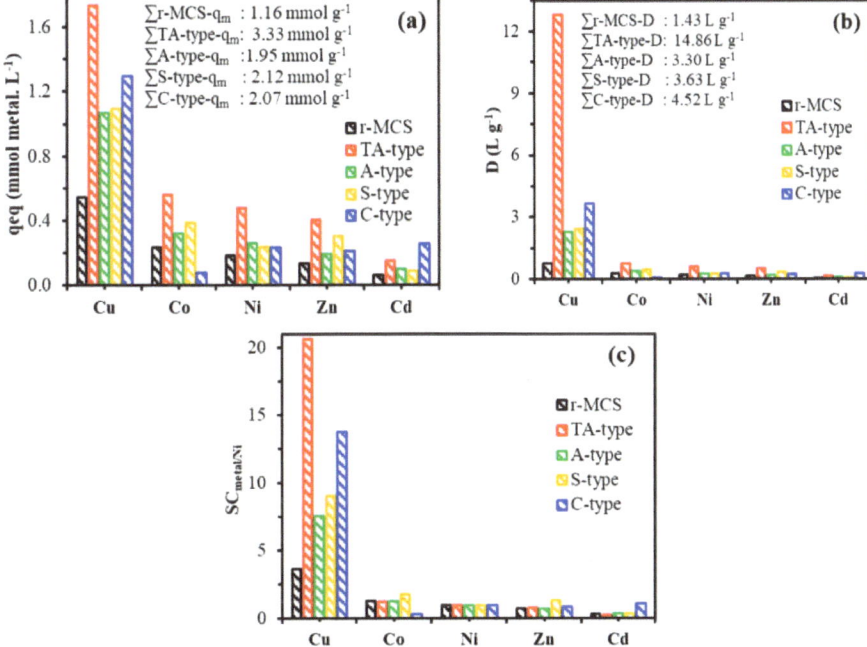

Figure 8. Selectivity experiments through sorption capacities (**a**), distribution coefficients (**b**), and selectivity coefficient ($SC_{metal/Ni}$) (**c**): (pH$_0$, 5.0; SD, 0.5 g.L^{-1}; T, 298 K; time, 120 min; 200 rpm).

Moreover, the C-type adsorbent is considerably more efficient and consistently superior for Cu(II) adsorption than other adsorbents. The second affinity chosen is Cd(II); this is most probably due to sulfur functional groups (i.e., regarded as a soft base), which are more reactive and sensitive to soft acids, such as Cd(II) [78], lowering the total affinity and selectivity of C-type adsorbent for Cu(II). Generally, no obvious association exists between HSAB (hard and soft acid-and-base theory) rank and metal adsorption affinity; Cd(II) is regarded as a soft acid. Other co-metals, on the other hand, are borderline metals (e.g., Ni(II), Zn(II), Co(II), and Cu(II)) [78].

The EDX measurements after multicomponent treatment reveal high selective Cu(II) adsorption for all materials (Figure 5 and Supplementary Figure S4). The other co-ions of multi-elements were identified by two characteristic peaks of Kα and poor Lα signals as follow: at 8.637 keV and 1.012 keV, respectively, for Zn signals; at 7.477 keV and 0.851 keV, respectively, for Ni signals; and at 6.929 keV and 0.776 keV, respectively, for Co signals. Meanwhile, there was the appearance of Cd signals of Kα at 23.175 keV and Lα at 3.133 keV [58].

These selectivity patterns were reflected in the D (distribution ratio) plots, which emphasize the increased affinity for Cu(II) in particular (Figure 8b). Therefore, Cu(II) showed highest D-values in $L.g^{-1}$ with the following orders: TA-type (D: 12.79) > C-type (D:3.65) >S-type (D: 2.43) > A-type (D: 2.29) > r-MCS (D: 0.81). Moreover, the graph of selectivity coefficients versus Ni(II) (i.e., $SC_{metal/Ni}$) presented in Figure 8c can be used to sort the metal ions' selectivity for each adsorbent:

TA-type: Cu(II) (20.58) >>> Co(II) (1.25) > Ni(II) (1.00) > Zn(II) (0.82) > Cd(II) (0.26).

C-type: Cu(II) (13.77) >>> Cd(II) (1.13) > Ni(II) (1.00) > Zn(II) (0.88) > Co(II) (0.31).

S-type: Cu(II) (9.04) >> Co(II) (1.78) > Zn(II) (1.39) > Ni(II) (1.00) > Cd(II) (0.36).

A-type: Cu(II) (7.59) >> Co(II) (1.28) > Ni(II) (1.00) >Zn(II) (0.72) > Cd(II) (0.35).

r-MCS: Cu(II) (4.12)> Co(II) (1.30) > Ni(II) (1.00) >Zn(II) (0.71) > Cd(II) (0.29).

The results show that the adsorbents have a high and strong affinity and selectivity for copper ions over other co-metals. In a trial to interpret these adsorption patterns via the physicochemical parameters of different co-metal cations, the concepts of QSAR were applied [79]; some key characteristics include hydrated ionic radius (in A°), electronegativity (Pauling units), and effective ionic charge (Supplementary Table S9 and Figure S10). The statistical data analysis (through R^2 values) showed skewing with the grouped co-cations (Cd(II), Zn(II), Co(II), and Ni(II)) versus Cu(II). The highest correlation coefficients were achieved when a direct relationship between adsorption capacity versus effective ionic charge (i.e., the most effective factor) and followed by atomic number (i.e., the second one) (Figure S10), except for the C-type, which may be controlled by HSAB rather than QSAR.

3.3.9. Direction of a Relationship between Variables

Meanwhile, there is not a strong relation between the QSAR tools and/or HSAB ranking and the obtained metal adsorption order.

3.4. Simulation and Graphical Mathematical Modeling

Recent multimedia advancements have expanded in the 3D (three-dimensional) information available. Therefore, effective data management involves appropriate approaches for its representation and processing. Nonlinear models have lately grown in popularity, not just for accuracy criteria but also for broadening the model's applicability [59,80]. The system requirements are effectively realized due to the focus on correctness throughout the simulation study. The latter is performed by the use of MATLAB software, a nonlinear regression, and a graphical technique. To measure the fitting and goodness of the suggested technique, the description evaluates the broad distribution of interactions between particular points (pH_0 and C_0 versus their response q_{eq}). The general Langmuir equation in nonlinear form ($q = q_m.KC^n/(1 + KC)^m$) was utilized to illustrate multiple 3D models [61,80] because the Langmuir isotherm matches the data better.

Both initial pH and Cu(II) concentration parameters versus their response Cu(II) adsorption can be connected by Equation (13) [80]. By adapting the Langmuir equation to integrate the initial pH and Cu(II) concentrations, we obtained the general formula and shown in Equation (13). It was adapted using ΔG (Gibbs energy change), as shown below in Equation (14):

$$\text{For } q_{(pH_0, C_0)} = \frac{q_m \times k \times pH^m \times C^n}{\left(1 + k \times pH^f \times C^e\right)^g} \quad (13)$$

$$\Delta G = -R \times T \times \log_e^k = \frac{-8.314 \times 298 \times \log_e^{(4.166 \times 63.546 \times 1000)}}{1000} = -30.9361 \quad (14)$$

where $q_{e(pH0,C0)}$ is the resultant adsorption capacity for a given pH_0, C_0, and q_m, the maximum adsorption capacity (for TA-type and A-type was 3.29 and 1.48 mmol Cu.g^{-1}, respectively). C_0 and pH_0 refer to initial Cu(II) concentration and initial pH, respectively [80]. K is the adjusted Langmuir constant. Letter symbols (e.g., m, n, f, e, and g) correspond to constant factors of the valuables.

Depending on the adsorption results at the various pH_0s and C_0s, in line with the methodical process in Equation (13), the experimental results were visually displayed in a 3D graph (Figure 9 and Supplementary Figure S11). This equation can be used for the pH_0 (range from 2.0 to 7.0) and C_0 (range from 0.32 to 6.27 mmol Cu.L^{-1}); these ranges suit the equation well. This mathematical formula often correlates better with experimental profiles: SSE, R^2, and Adj R^2 components support this fitting [80]. Furthermore, the calculated free energy (ΔG: −30.94 kJ.mol^{-1} at 298 K) is generally comparable with practical thermodynamic findings (see Supplementary Table S10).

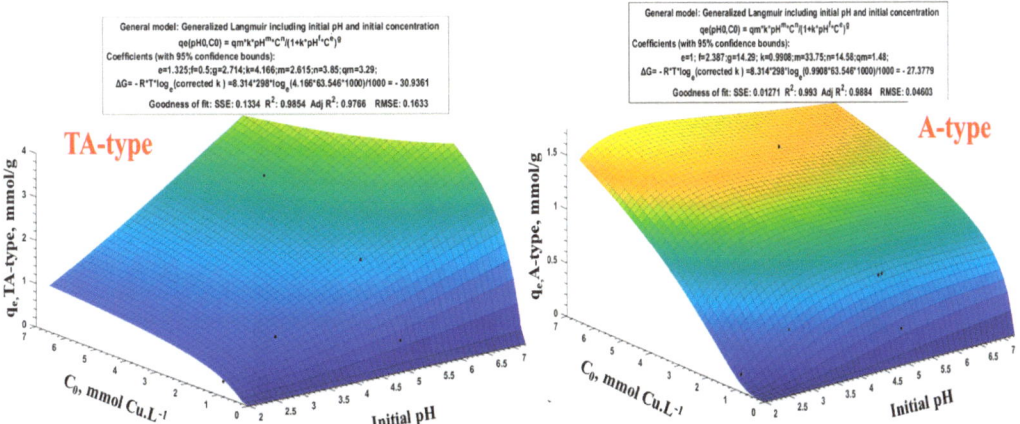

Figure 9. Generalized Langmuir model, including Cu(II) adsorption (q_e) vs. initial pH and C_0.

Figure 9 and Supplementary Figure S11 (and Equation (13)) illustrate how the function power is and present all parameters. The greater the order of exponential magnitude, the greater the effect on the influence and the response of $q_e(pH_0, C_0)$ [59]. According to the results of Equation (13)'s application, the numerical values of the exponential for all adsorbents are presented in Supplementary Table S10. For TA-type, the exponentials are ordered and arranged as follows: n = 3.85 > m = 2.615 > g = 2.714 > e = 1.325 > f = 0.50. Based on n and m values of C^n and pH^m, respectively, the most crucial element is the initial concentration (C_0), followed by the second component (pH_0). Moreover, this relates to the exponent values of the denominator's coefficients, i.e., e and f values for C_0 and pH_0, respectively, whilst the product of both ($pH_0 \times C_0$) is more effective than C_0 and pH_0 individually. This is congruent with the experimental findings.

The extent to which the theoretical findings coincide with the experimental data via the error percentage is used (Error % = ($q_{e,exp.} - q_{e,Math.}$) × 100)/$q_{e,exp.}$) to validate the mathematical methodology. Supplementary Table S11 illustrates the validation and comparative findings of the mathematical modeling for the pH_0 range (2.0 to 7.0) and initial concentration ranges (C_0: 0.32–6.27 mmol $Cu.L^{-1}$). The mathematical method's accuracy (Δ) is assessed and found to be overstated by ($\Delta < 7.78\%$) and underestimated by ($\Delta < 3.51\%$). The presence of a blank gap on the figure's surface in Figure 9 and Supplementary Figure S11 indicates that the results, in this case, cannot be applicable to the TA-type. Moreover, Supplementary Table S11 shows that the high error percent (Error %) for TA-type and other adsorbents was observed at pH_0 2.0 (except for A-type), and also at pH_0 7.0 for TA-type only.

4. Conclusions

Four eco-friendly superparamagnetic multifunctionalized chitosan nanohybrids were prepared suitably to be used as efficient adsorbents for Cu(II) ions. A facile method for synthesizing superparamagnetic Fe_3O_4/chitosan nanohybrids was reported via a heterogeneous nucleating of $Fe^{(II)/(III)}$ in a chitosan solution. The grafting of diethylenetriamine, alanine, serine, and cysteine moieties (via an intermediary chlorinated step of crosslinked epichlorohydrin–chitosan/Fe_3O_4 nanohybrid), giving TA-type, A-type, S-type, and C-type, respectively, improves the adsorption capacities (vs. non-functionalized chitosan nanohybrid) due to the specific reactivity of amine, carboxylate, hydroxyl, and sulfhydryl (thiol and/or thioether group) groups. The chemical changes are validated by CHNS, TEM, pH-titration, XPS, VSM, XRD, and FTIR analysis. The grafting of additional active sites considerably improves Cu(II) adsorption properties: the maximal adsorption capacity is more than triplicated (~3.33 times) for polyamine and nearly doubled for amino acid derivatives (~1.73–1.96 times) when compared to raw chitosan nanohybrid at pH_0 ~5.0. The equilibrium is achieved after 30–90 min of contact and matches the pseudo-second-order rate equation well. The Langmuir equation efficiently fits the sorption isotherms, and the Cu(II) adsorption process is endothermic for the amino acid derivatives (opposite to the amine derivative), spontaneous, and followed by increasing the randomness in the system. Copper desorption is exceedingly efficient (using 0.25 mol.L^{-1} thiourea solutions at pH 2.0), and the nanohybrid could be recycled at least six times. All nanohybrids materials display outstanding functionality, durability, performance, strength, and stability. In the selectivity test, copper adsorption from equimolar multicomponent systems demonstrates that all adsorbents have a surprising affinity toward Cu(II). The affinities are compatible with the metals' HSAB properties and the adsorbents softer/harder ordering. Furthermore, QSAR methods were employed to correlate intrinsic co-ion characteristics with their comparative affinity. Moreover, a mathematical simulator is being developed and built based on experimental data. The mathematical formula was validated and confirmed with measured values and the output findings.

The future perspectives of the presented research are as follows: this work can be tested for a wide range of heavy metal ions and cationic and anionic dye removal due to the wide effective pH range, high stability, and multifunctionalization with different active sites; thus, the developed chitosan hybrid in this study is a promising material in the field of wastewater treatment.

Supplementary Materials: The following supporting information can be downloaded at: https://www.mdpi.com/article/10.3390/polym15051157/s1, Table S1. Selected kinetics and adsorption isotherms models. Table S2. Effect of functionality on the adsorbent feature and adsorption characteristics. Table S3. Analysis of XPS spectra: assignment of the core-level signals (BEs and AF). Table S4. FTIR chemical assignments. Table S5. adsorption kinetics parameters of Cu(II) adsorption. Table S6. Adsorption isotherms parameters of Cu(II) adsorption. Table S7. Thermodynamic parameters for Cu(II) adsorption. Table S8. Metal desorption and adsorption cycles. Table S9. Chemical properties of selected metal ions. Table S10. The parameters and constants of Generalized Langmuir equation and Goodness of fit for all adsorbents. Table S11. Comparison for experimental adsorption capacity ($q_{e,exp}$) and mathematically calculated ($q_{e,cal.}$) for r-MCS, TA-type, A-type, S-type, and C-type adsorbents. Figure S1. Textural characterization of TA-type (a), S-type, A-type (b), and C-type (c): BET surface area analysis and pore size analysis (insert). Figure S2. Thermogravimetric analysis of S-type nanohybrid: TGA (a); DTG (b); magnetization curves (c); and the pH$_{ZPC}$ of r-CS, MCS, and all nanohybrids (d). Figure S3. XPS survey spectra of TA-type, A-type, S-type, and C-type nanocomposites. Figure S4. EDX measurements of raw r-MCS, A-type, and S-type materials before (a) and after Cu(II) adsorption from synthetic solution (b) and mixed system(c). Figure S5. Effect of initial pH versus equilibrium pH (a) and plot of distribution ratio (D, L g^{-1}) vs. pHeq (b). Figure S6. Kinetics models PFORE and PSORE for Cu(II) adsorption. Figure S7. Kinetics models of interparticle diffusion for Cu(II) sorption. Figure S8. (a) Langmuir isotherm models, (b) Freundlich isotherm models, and (c) Temkin isotherm models for Cu(II) adsorption. Figure S9. Adsorption isotherms for all adsorbent types at different temperatures (i.e., T: 298, 308, 318, and 328 K); time of 120 min for TA-, A-, S-, and C-type; and time of 240 min for r-MCS). Figure S10. Correlation between intrinsic physicochemical characteristics (atomic number (a), hydrated ionic radius (b), electronegativity (c), and effective ionic charge (d)) and their adsorption capacities (q_m). Figure S11. Generalized Langmuir including initial pH and initial Cu(II) concentration. References [23,25,30–32,34,38,39,43,53,54,61] are cited in Supplementary Materials.

Author Contributions: Conceptualization, A.A.A.-G., A.A.G. and S.W.; methodology, A.A.G. and S.W.; formal analysis, A.A.A.-G., A.A.G., A.A., Y.A.A.-T. and A.M.A.-A.; investigation, A.A.G., A.A. and S.W.; resources, A.A.A.-G., Y.A.A.-T. and A.M.A.-A.; data curation, A.A.G. and S.W.; writing—original draft preparation, A.A.G.; writing—review and editing, A.A.G. and S.W.; visualization, A.A.A.-G., A.A.G., A.A. and S.W.; supervision, A.A.A.-G. and S.W.; project administration, A.A.G., Y.A.A.-T., A.M.A.-A. and S.W.; funding acquisition, S.W. and A.A.A.-G. All authors have read and agreed to the published version of the manuscript.

Funding: This research was funded by Deputyship for Research & Innovation, Ministry of Education in Saudi Arabia, grant number (IFPNC-004- 130-2020)" and "The APC was funded by Deputyship for Research & Innovation, Ministry of Education in Saudi Arabia project number (IFPNC-004- 130-2020)".

Institutional Review Board Statement: Not applicable.

Informed Consent Statement: Not applicable.

Data Availability Statement: All data generated or analyzed during this study are included in this article.

Acknowledgments: The authors extend their appreciation to the Deputyship for Research & Innovation, Ministry of Education in Saudi Arabia for funding this research work through the project number (IFPNC-004- 130-2020) and King Abdulaziz University, DSR, Jeddah, Saudi Arabia.

Conflicts of Interest: The authors declare no conflict of interest.

References

1. Al-Saydeh, S.; El-Naas, M.; Zaidi, S. Copper removal from industrial wastewater: A comprehensive review. *J. Ind. Eng. Chem.* **2017**, *56*, 35–44. [CrossRef]
2. Zhang, L.; Zeng, Y.; Cheng, Z. Removal of heavy metal ions using chitosan and modified chitosan: A review. *J. Mol. Liq.* **2016**, *214*, 175–191. [CrossRef]
3. Donia, A.; Atia, A.; Abouzayed, F. Preparation and characterization of nano-magnetic cellulose with fast kinetic properties towards the adsorption of some metal ions. *Chem. Eng. J.* **2012**, *191*, 22–30. [CrossRef]
4. Bian, P.; Liu, Y.; Zheng, X.; Shen, W. Removal and mechanism of cadmium, lead and copper in water by functional modification of Silkworm excrement biochar. *Polymers* **2022**, *14*, 2889. [CrossRef] [PubMed]
5. Mondal, S.D.R. Removal of copper(II) from aqueous solution using zinc oxide nanoparticle impregnated mixed matrix hollow fiber membrane. *Environ. Technol. Innov.* **2022**, *26*, 102300. [CrossRef]
6. Elgarahy, A.; Elwakeel, K.; Mohammad, S.; Elshoubaky, G. Multifunctional eco-friendly sorbent based on marine brown algae and bivalve shells for subsequent uptake of Congo red dye and copper(II) ions. *J. Environ. Chem. Eng.* **2020**, *8*, 103915. [CrossRef]
7. Baraka, A.; Hall, P.; Heslop, M. Preparation and characterization of melamine–formaldehyde–DTPA chelating resin and its use as an adsorbent for heavy metals removal from wastewater. *React. Funct. Polym.* **2007**, *67*, 585–600. [CrossRef]
8. Zhang, Y.; Fan, B.; Jia, L.; Qiao, X.; Li, Z. Study on adsorption mechanism of mercury on Ce-Cu modified iron-based biochar. *Chem. Eng. J. Adv.* **2022**, *10*, 100259. [CrossRef]
9. Taylor, A.; Tsuji, J.; Garry, M.; McArdle, M.; Goodfellow, W.; Adams, W.; Menzie, C. Critical review of exposure and effects: Implications for setting regulatory health criteria for ingested copper. *Environ. Manag.* **2020**, *65*, 131–159. [CrossRef]
10. Hussain, M.; Musharraf, S.; Bhanger, M.; Malik, M. Salicylaldehyde derivative of nano-chitosan as an efficient adsorbent for lead(II), copper(II), and cadmium(II) ions. *Int. J. Biol. Macromol.* **2020**, *147*, 643–652. [CrossRef]

11. Crini, G.; Lichtfouse, E. Advantages and disadvantages of techniques used for wastewater treatment. *Environ. Chem. Lett.* **2019**, *17*, 145–155. [CrossRef]
12. Vieira, R.; Oliveira, M.; Guibal, E.; Rodríguez-Castellón, E.; Beppu, M. Copper, mercury and chromium adsorption on natural and crosslinked chitosan films: An XPS investigation of mechanism. *Colloids Surf. A Physicochem. Eng. Asp.* **2011**, *374*, 108–114. [CrossRef]
13. Neiber, R.; Galhoum, A.; El Sayed, I.E.-T.; Guibal, E.; Xin, J.; Lu, X. Selective lead (II) sorption using aminophosphonate-based sorbents: Effect of amine linker, characterization and sorption performance. *Chem. Eng. J.* **2022**, *442*, 136300. [CrossRef]
14. Jjagwe, J.; Olupot, P.; Menya, E.; Kalibbala, H. Synthesis and Application of Granular Activated Carbon from Biomass Waste Materials for Water Treatment: A Review. *J. Bioresour. Bioprod.* **2021**, *6*, 292–322. [CrossRef]
15. Kayalvizhi, K.; Alhaji, N.; Saravanakkumar, D.; Mohamed, S.; Kaviyarasud, K.; Ayeshamariam, A.; Al-Mohaimeed, A.; Abdel-Gawwad, M.; Elshikh, M. Adsorption of copper and nickel by using sawdust chitosan nanocomposite beads—A kinetic and thermodynamic study. *Enviro. Res.* **2022**, *203*, 111814. [CrossRef]
16. Tan, T.; Somat, H.; Latif, M.; Rashid, S. One-pot solvothermal synthesis of Zr-based MOFs with enhanced adsorption capacity for Cu^{2+} ions removal. *J. Solid State Chem.* **2022**, *15*, 123429. [CrossRef]
17. Zidan, T.; Abdelhamid, A.; Zaki, E. N-Aminorhodanine modified chitosan hydrogel for antibacterial and copper ions removal from aqueous solutions. *Int. J. Biol. Macromol.* **2020**, *158*, 32–42. [CrossRef]
18. Qi, L.; Xu, Z.; Jiang, X.; Hu, C.; Zou, X. Preparation and antibacterial activity of chitosan nanoparticles. *Carbohydr. Res.* **2004**, *339*, 2693–2700. [CrossRef]
19. Ghaedi, A.; Panahimehr, M.; Nejad, A.; Hosseini, S.; Vafaei, A.; Baneshi, M. Factorial experimental design for the optimization of highly selective adsorption removal of lead and copper ions using metal organic framework MOF-2 (Cd). *J. Mol. Liq.* **2018**, *272*, 15–26. [CrossRef]
20. Ma, X.; Zhao, S.; Tian, Z.; Duan, G.; Pan, H.; Yue, Y.; Li, S.; Jian, S.; Yang, W.; Liu, K.; et al. MOFs meet wood: Reusable magnetic hydrophilic composites toward efficient water treatment with super-high dye adsorption capacity at high dye concentration. *Chem. Eng. J.* **2022**, *446*, 136851. [CrossRef]
21. Cuong, D.; Liu, N.-L.; Nguyen, V.; Hou, C.-H. Meso/micropore-controlled hierarchical porous carbon derived from activated biochar as a high-performance adsorbent for copper removal. *Sci. Total Environ.* **2019**, *692*, 844–853. [CrossRef] [PubMed]
22. Obey, G.; Adelaide, M.; Ramaraj, R. Biochar derived from non-customized matamba fruit shell as an adsorbent for wastewater treatment. *J. Bioresour. Bioprod.* **2022**, *7*, 109–115. [CrossRef]
23. Galhoum, A.; Akashi, T.; Linnolahti, M.; Hirvi, J.; Al-Sehemid, A.; Kalam, A.; Guibal, E. Functionalization of poly(glycidylmethacrylate) with iminodiacetate and imino phosphonate groups for enhanced sorption of neodymium—Sorption performance and molecular modeling. *React. Funct. Polym.* **2022**, *180*, 105389. [CrossRef]
24. Galhoum, A.; Eisa, W.; El-Sayed, I.E.-T.; Tolba, A.; Shalaby, Z.; Mohamady, S.; Muhammad, S.; Hussien, S.; Akashi, T.; Guibal, E. A new route for manufacturing poly(aminophosphonic)-functionalized poly(glycidyl methacrylate)-magnetic nanocomposite-Application to uranium sorption from ore leachate. *Environ. Pollut.* **2020**, *264*, 114797. [CrossRef] [PubMed]
25. Al-Ghamdi, A.; Galhoum, A.; Alshahrie, A.; Al-Turki, Y.; Al-Amri, A.; Wageh, S. Mechanistic studies of uranyl interaction with functionalized mesoporous chitosan-superparamagnetic nanocomposites for selective sorption: Characterization and sorption performance. *Mater. Today Commun.* **2022**, *33*, 104536. [CrossRef]
26. Imam, E.; El-Sayed, I.E.-T.; Mahfouz, M.; Tolba, A.; Akashi, T.; Galhoum, A.; Guibal, E. Synthesis of α-aminophosphonate functionalized chitosan sorbents: Effect of methyl vs phenyl group on uranium sorption. *Chem. Eng. J.* **2018**, *352*, 1022–1034. [CrossRef]
27. Kravanja, G.; Primožič, M.; Knez, Ž.; Leitgeb, M. Chitosan-based (nano)materials for novel biomedical applications. *Molecules* **2019**, *24*, 1960. [CrossRef]
28. Yu, Y.; Shapter, J.; Popelka-Filcoff, R.; Bennett, J.; Ellis, A. Copper removal using bio-inspired polydopamine coated natural zeolites. *J. Hazard. Mat.* **2014**, *273*, 174–182. [CrossRef]
29. Crini, G. Recent developments in polysaccharide-based materials used as adsorbents in wastewater treatment. *Prog. Polym. Sci.* **2005**, *30*, 38–70. [CrossRef]
30. Benettayeb, A.; Morsli, A.; Elwakeel, K.; Hamza, M.; Guibal, E. Recovery of heavy metal ions using magnetic glycine-modified chitosan—Application to aqueous solutions and tailing leachate. *Appl. Sci.* **2021**, *11*, 8377. [CrossRef]
31. Yamaura, M.; Camilo, R.; Sampaio, L.; Macêdo, M.; Nakamura, M.; Toma, H. Preparation and characterization of (3-aminopropyl)triethoxysilane-coated magnetite nanoparticles. *J. Magn. Magn. Mater.* **2004**, *279*, 210–217. [CrossRef]
32. Galhoum, A.; Mahfouz, M.; Gomaa, N.; Vincent, T.; Guibal, E. Chemical modifications of chitosan nano-based magnetic particles for enhanced uranyl sorption. *Hydrometallurgy* **2017**, *168*, 127–134. [CrossRef]
33. Jiang, W.; Wang, W.; Pan, B.; Zhang, Q.; Zhang, W.; Lv, L. Facile Fabrication of Magnetic Chitosan Beads of Fast Kinetics and High Capacity for Copper Removal. *ACS Appl. Mater. Interfaces* **2014**, *6*, 3421–3426. [CrossRef]
34. Pourmortazavi, S.; Sahebi, H.; Zandavar, H.; Mirsadeghi, S. Fabrication of Fe_3O_4 nanoparticles coated by extracted shrimp peels chitosan as sustainable adsorbents for removal of chromium contaminates from wastewater: The design of experiment. *Compos. B Eng.* **2019**, *175*, 107130. [CrossRef]
35. Morshedy, A.; Galhoum, A.; Aleem, A.A.H.A.; El-din, M.S.; Okaba, D.; Mostafa, M.; Mira, H.; Yang, Z.; El-Sayed, I.E. Functionalized aminophosphonate chitosan-magnetic nanocomposites for Cd(II) removal from aqueous solutions: Performance and mechanisms of sorption. *Appl. Surf. Sci.* **2021**, *561*, 150069. [CrossRef]

36. Galhoum, A.; Atia, A.; Mahfouz, M.; Abdel-Rehem, S.; Gomaa, N.; Vincent, T.; Guibal, E. Dy(III) recovery from dilute solutions using magnetic-chitosan nano-based particles grafted with amino acids. *J. Mater. Sci.* **2015**, *50*, 2832–2848. [CrossRef]
37. Esfandiari, N.; Kashefi, M.; Mirjalili, M.; Afsharnezhad, S. Role of silica mid-layer in thermal and chemical stability of hierarchical Fe_3O_4-SiO_2-TiO_2 nanoparticles for improvement of lead adsorption: Kinetics, thermodynamic and deep XPS investigation. *Mater. Sci. Eng. B.* **2020**, *262*, 114690. [CrossRef]
38. Al-Ghamdi, A.; Galhoum, A.; Alshahrie, A.; Al-Turki, Y.; Al-Amri, A.; Wageh, S. Mesoporous magnetic cysteine functionalized chitosan nanocomposite for selective uranyl ions sorption: Experimental, structural characterization, and mechanistic studies. *Polymers* **2022**, *14*, 2568. [CrossRef]
39. Alghamdi, N.A. Mesoporous magnetic-polyaminated-chitosan nanocomposite for selective uranium removal: Performance and mechanistic studies. *Int. J. Environ. Sci. Technol.* **2022**, *2022*, 1–22. [CrossRef]
40. Mahdavi, M.; Ahmad, M.; Haron, M.; Namvar, F.; Nadi, B.; Rahman, M.; Amin, J. Synthesis, Surface Modification and Characterisation of Biocompatible Magnetic Iron Oxide Nanoparticles for Biomedical Applications. *Molecules* **2013**, *18*, 7533–7548. [CrossRef]
41. Galhoum, A.; Mafhouz, M.; Abdel-Rehem, S.; Gomaa, N.; Atia, A.; Vincent, T.; Guibal, E. Cysteine-Functionalized Chitosan Magnetic Nano-Based Particles for the Recovery of Light and Heavy Rare Earth Metals: Uptake Kinetics and Sorption Isotherms. *Nanomaterials* **2015**, *5*, 154–179. [CrossRef] [PubMed]
42. Yang, W.; Wang, Y.; Wang, Q.; Wu, J.; Duan, G.; Xu, W.; Jian, S. Magnetically separable and recyclable Fe_3O_4@PDA covalent grafted by l-cysteine core-shell nanoparticles toward efficient removal of Pb^{2+}. *Vacuum* **2021**, *189*, 110229. [CrossRef]
43. Stoia, M.; Istratie, R.; Păcurariu, C. Investigation of magnetite nanoparticles stability in air by thermal analysis and FTIR spectroscopy. *J. Therm. Anal. Calorim.* **2016**, *125*, 1185–1198. [CrossRef]
44. Shahrashoub, M.; Bakhtiari, S.; Afroosheh, F.; Googheri, M. Recovery of iron from direct reduction iron sludge and biosynthesis of magnetite nanoparticles using green tea extract. *Colloids Surf. A Physicochem. Eng. Asp.* **2021**, *622*, 126675. [CrossRef]
45. Zhang, X.; Jiao, C.; Wang, J.; Liu, Q.; Li, R.; Yang, P.; Zhang, M. Removal of uranium(VI) from aqueous solutions by magnetic Schiff base: Kinetic and thermodynamic investigation. *Chem. Eng. J.* **2012**, *198–199*, 412–419. [CrossRef]
46. Kong, H.; Song, J.; Jang, J. One-step fabrication of magnetic γ-Fe_2O_3/polyrhodanine nanoparticles using in situ chemical oxidation polymerization and their antibacterial properties. *Chem. Comm.* **2010**, *46*, 6735–6737. [CrossRef]
47. El-Magied, M.A.; Galhoum, A.; Atia, A.; Tolba, A.; Maize, M.; Vincent, T.; Guibal, E. Cellulose and chitosan derivatives for enhanced sorption of erbium(III). *Colloids Surf. A Physicochem. Eng. Asp.* **2017**, *529*, 580–593. [CrossRef]
48. Sharma, J.; Srivastava, P.; Singh, G.; Virk, H. Nanoferrites of Transition Metals and Their Catalytic Activity. *Diffus. Defect Data Pt. B Solid State Phenom.* **2016**, *241*, 126–138. [CrossRef]
49. Sing, K.S.W. Reporting physisorption data for gas/solid system with special reference to the determination of surface area and porosity (Recommendations 1984). *Pure Appl. Chem.* **1985**, *57*, 603–619. [CrossRef]
50. The International XPS Database of XPS Reference Spectra, Peak-Fits & Six (6) BE Tables. Available online: https://xpsdatabase.com/ (accessed on 16 October 2020).
51. Grosvenor, A.; Kobe, B.; Biesinger, M.; McIntyre, N. Investigation of multiplet splitting of Fe 2p XPS spectra and bonding in iron compounds. *Surf. Interface Anal.* **2004**, *36*, 1564–1574. [CrossRef]
52. Wang, J.; Ma, X.; Qu, F.; Asiri, A.; Sun, X. Fe-Doped Ni2P Nanosheet Array for High-Efficiency Electrochemical Water Oxidation. *Inorg. Chem.* **2017**, *56*, 1041–1044. [CrossRef]
53. Oh, S.; Yoo, D.; Shin, Y.; Kim, H.; Kim, H.; Chung, Y.; Park, W.; Youk, J. Crystalline structure analysis of cellulose treated with sodium hydroxide and carbon dioxide by means of X-ray diffraction and FTIR spectroscopy. *Carbohydr. Res.* **2005**, *340*, 2376–2391. [CrossRef]
54. Coates, J. Interpretation of infrared spectra: Apractical Approach. In *Encyclopedia of Analytical Chemistry*; John Wiley & Sons Ltd: Chochester, UK, 2000.
55. Muthuselvi, C.; Pandiarajan, S.; Ravikumar, B.; Athimoolam, S.; Srinivasan, N.; Krishnakumar, R. FT-IR and FT-Raman spectroscopic analyzes of indeno quinoxaline derivative crystal. *Asian J. Appl. Sci.* **2018**, *11*, 83–91. [CrossRef]
56. Willianms, R. Pka Data Compiled by R. Willianms. 2017, pp. 1–33. Available online: https://organicchemistrydata.org/hansreich/resources/pka/pka_data/pka-compilation-williams.pdf\{pKa_compilation-1-Williams.pdf\} (accessed on 25 August 2020).
57. Fraga, T.; de Lima, L.; de Souza, Z.; Carvalho, M.; Freire, E.; Ghislandi, M.; da Motta, M. Amino-Fe_3O_4-functionalized graphene oxide as a novel adsorbent of Methylene Blue: Kinetics, equilibrium, and recyclability aspects. *Environ. Sci. Pollut. Res.* **2019**, *26*, 28593–28602. [CrossRef]
58. Thompson, A.; Attwood, D.; Gullikson, E.; Howells, M.; Kim, K.-J.; Kirz, J.; Kortright, J.; Lindau, I.; Liu, Y.; Pianetta, P.; et al. *X-Ray Data Booklet*; Lawrence Berkeley National Laboratory—University of California: Berkeley, CA, USA, 2009.
59. Fouda, S.; El-Sayed, I.; Attia, N.; Abdeen, M.; Aleem, A.A.; Nassar, I.; Mira, H.; Gawad, E.; Kalam, A.; Al-Ghamdi, A.; et al. Mechanistic study of Hg(II) interaction with three different α-aminophosphonate adsorbents: Insights from batch experiments and theoretical calculations. *Chemosphere* **2022**, *304*, 135253. [CrossRef]
60. Persson, I. Hydrated metal ions in aqueous solution: How regular are their structures? *Pure Appl. Chem.* **2010**, *82*, 1901–1917. [CrossRef]
61. Tien, C. *Adsorption Calculations and Modeling*; Butterworth-Heinemann: Boston, MA, USA, 1994.
62. Hu, X.; Chen, C.; Zhang, D.; Xue, Y. Kinetics, isotherm and chemical speciation analysis of Hg(II) adsorption over oxygen-containing MXene adsorbent. *Chemosphere* **2021**, *278*, 130206. [CrossRef]

63. Hubbe, M.; Azizian, S.; Douven, S. Implications of apparent pseudo-second-order adsorption kinetics onto cellulosic materials: A review. *BioResources* **2019**, *14*, 7582–7626. [CrossRef]
64. Doğan, M.; Özdemir, Y.; Alkan, M. Adsorption kinetics and mechanism of cationic methyl violet and methylene blue dyes onto sepiolite. *Dyes Pigm* **2007**, *75*, 701–713. [CrossRef]
65. Zhang, G.; Fang, Y.; Wang, Y.; Liu, L.; Mei, D.; Ma, F.; Meng, Y.; Dong, H.; Zhang, C. Synthesis of amino acid modified MIL-101 and efficient uranium adsorption from water. *J. Mol. Liq.* **2022**, *349*, 118095. [CrossRef]
66. Chen, H.; Lin, J.; Zhang, N.; Chen, L.; Zhong, S.; Wang, Y.; Zhang, W.; Ling, Q. Preparation of MgAl-EDTA-LDH based electrospun nanofiber membrane and its adsorption properties of copper(II) from wastewater. *J. Hazard. Mat.* **2018**, *345*, 1–9. [CrossRef] [PubMed]
67. Saleem, A.; Wang, J.; Sun, T.; Sharaf, F.; Haris, M.; Lei, S. Enhanced and selective adsorption of Copper ions from acidic conditions by diethylenetriaminepentaacetic acid-chitosan sewage sludge composite. *J. Environ. Chem. Eng.* **2020**, *8*, 104430. [CrossRef]
68. Tan, T.; Mo, K.; Lai, S.; Ling, T.-C. Investigation on the copper ion removal potential of a facile-fabricated foamed geopolymer sphere for wastewater remediation. *Clean. Mater.* **2022**, *4*, 100088. [CrossRef]
69. Al-Ghouti, M.; Da, D. Guidelines for the use and interpretation of adsorption isotherm models: A review. *J. Hazard. Mater.* **2020**, *393*, 122383. [CrossRef] [PubMed]
70. Rangabhashiyam, S.; Anu, N.; Nandagopal, M.G.; Selvaraju, N. Relevance of isotherm models in biosorption of pollutants by agricultural byproducts. *J. Environ. Chem. Eng.* **2014**, *2*, 398–414. [CrossRef]
71. Lima, E.; Hosseini-Bandegharaei, A.; Moreno-Piraján, J.; Anastopoulos, I. A critical review of the estimation of the thermodynamic parameters on adsorption equilibria. Wrong use of equilibrium constant in the Van't Hoof equation for calculation of thermodynamic parameters of adsorption. *J. Mol. Liq.* **2019**, *273*, 425–434. [CrossRef]
72. Xia, D.; Liu, Y.; Cheng, X.; Gu, P.; Chen, Q.; Zhang, Z. Temperature-tuned fish-scale biochar with two-dimensional homogeneous porous structure: A promising uranium extractant. *Appl. Surf. Sci.* **2022**, *591*, 153136. [CrossRef]
73. Aslani, C.; Amik, O. Active Carbon/PAN composite adsorbent for uranium removal: Modeling adsorption isotherm data, thermodynamic and kinetic studies. *Appl. Radiat. Isot.* **2021**, *168*, 109474. [CrossRef]
74. Liu, Y.; Zhao, Z.; Yuan, D.; Wang, Y.; Dai, Y.; Zhu, Y.; Chew, J. Introduction of amino groups into polyphosphazene framework supported on CNT and coated Fe_3O_4 nanoparticles for enhanced selective U(VI) adsorption. *Appl. Surf. Sci.* **2019**, *466*, 893–902. [CrossRef]
75. Zhai, Q.-Z. Use of SBA-15 ordered nano mesoporous silica for removal of copper(II) from aqueous media: Studies on equilibrium, isotherm, kinetics and thermodynamics. *J. Environ. Chem. Eng.* **2019**, *7*, 103069. [CrossRef]
76. Ertl, G.; Hierl, R.; Knözinger, H.; Thiele, N.; Urbach, H. XPS study of copper aluminate catalysts. *Appl. Surf. Sci.* **1980**, *5*, 49–64. [CrossRef]
77. Borsari, M. Cadmium: Coordination Chemistry. In *Encyclopedia of Inorganic and Bioinorganic Chemistry*; Wiley Online Library: Hoboken, NJ, USA, 2011; pp. 1–16.
78. Pearson, R.G. Acids and bases. *Science* **1966**, *151*, 172–177. [CrossRef]
79. Chen, C.; Wang, J. Correlating metal ionic characteristics with biosorption capacity using QSAR model. *Chemosphere* **2007**, *69*, 1610–1616.
80. Boyadjiev, C. *Theoretical Chemical Engineering: Modeling and Simulation*; Springer: Berlin/Heidelberg, Germany, 2010.

Disclaimer/Publisher's Note: The statements, opinions and data contained in all publications are solely those of the individual author(s) and contributor(s) and not of MDPI and/or the editor(s). MDPI and/or the editor(s) disclaim responsibility for any injury to people or property resulting from any ideas, methods, instructions or products referred to in the content.

Article

Study on the Mechanism and Experiment of Styrene Butadiene Rubber Reinforcement by Spent Fluid Catalytic Cracking Catalyst

Tilun Shan [1,2,3], Huiguang Bian [3], Donglin Zhu [1], Kongshuo Wang [1], Chuansheng Wang [1,2,3] and Xiaolong Tian [1,2,3,*]

[1] National Engineering Laboratory of Advanced Tire Equipment and Key Materials, Qingdao University of Science and Technology, Qingdao 266061, China
[2] Shandong Key Laboratory of Advanced Manufacturing of Polymer Materials, Qingdao 266061, China
[3] College of Electromechanical Engineering, Qingdao University of Science and Technology, Qingdao 266061, China
* Correspondence: tianxiaolong@qust.edu.cn

Abstract: Spent Fluid Catalytic Cracking (FCC) Catalyst is a major waste in the field of the petroleum processing field, with a large output and serious pollution. The treatment cost of these waste catalysts is high, and how to achieve their efficient reuse has become a key topic of research at home and abroad. To this end, this paper conducted a mechanistic and experimental study on the replacement of some carbon blacks by spent FCC catalysts for the preparation of rubber products and explored the synergistic reinforcing effect of spent catalysts and carbon blacks, in order to extend the reuse methods of spent catalysts and reduce the pollution caused by them to the environment. The experimental results demonstrated that the filler dispersion and distribution in the compound are more uniform after replacing the carbon black with modified spent FCC catalysts. The crosslinking density of rubber increases, the Payne effect is decreased, and the dynamic mechanical properties and aging resistance are improved. When the number of replacement parts reached 15, the comprehensive performance of the rubber composites remained the same as that of the control group. In this paper, the spent FCC catalysts modified by the physical method instead of the carbon-black-filled SBR can not only improve the performance of rubber products, but also can provide basic technical and theoretical support to realize the recycling of spent FCC catalysts and reduce the environmental pressure. The feasibility of preparing rubber composites by spent catalysts is also verified.

Keywords: spent FCC catalyst; physical modification; SBR composites; reinforcement mechanism

Citation: Shan, T.; Bian, H.; Zhu, D.; Wang, K.; Wang, C.; Tian, X. Study on the Mechanism and Experiment of Styrene Butadiene Rubber Reinforcement by Spent Fluid Catalytic Cracking Catalyst. *Polymers* 2023, 15, 1000. https://doi.org/10.3390/polym15041000

Academic Editors: Giorgio Luciano and Maurizio Vignolo

Received: 7 January 2023
Revised: 9 February 2023
Accepted: 13 February 2023
Published: 17 February 2023

Copyright: © 2023 by the authors. Licensee MDPI, Basel, Switzerland. This article is an open access article distributed under the terms and conditions of the Creative Commons Attribution (CC BY) license (https:// creativecommons.org/licenses/by/ 4.0/).

1. Introduction

With the increase in car ownership, the demand for tires is growing rapidly every year [1,2]. Rubber composite is widely used in the tire industry for its excellent comprehensive performance and processing performance [3,4]. In order to meet the requirements of tire use, rubber compounds need to be prepared with a variety of reinforcing fillers to improve their performance [5,6]. In the process of tire preparation, the amount of filler accounts for more than 1/3 of the amount of tire rubber formula, and is becoming one of the most important materials in the tire rubber composition [7,8]. Compared with the high economic cost and serious environmental pollution caused by traditional fillers (carbon black), the "green tire" has become the main direction of automotive tire design and development [9–12].

Fluid Catalytic Cracking (FCC) is one of the important refinery processing technologies and has a pivotal position in the petroleum refining industry [13–15]. In the process of the petroleum catalytic cracking, the amount of the FCC catalyst is very large, and a large number of spent catalysts are also produced, which is called spent FCC catalysts in the industry [16]. This part of the spent FCC catalysts cannot meet the catalytic cracking requirements of petroleum; however, these catalysts still have high activity, and the residual activity of spent FCC catalysts can be used again in other fields. After our extensive preliminary research, we found that waste FCC catalysts applied to the waste tire pyrolysis process can significantly improve the quality of pyrolysis oil and meet the requirements of waste tire pyrolysis [17–19]. Therefore, the regeneration, recycling, and harmless application of spent catalysts is a key research direction in the future [20,21].

The FCC catalyst is a kind of porous, microsphere granular solid acid catalyst, which is made of active components (Y and ZSM-5 molecular sieve), matrix (kaolin), and binder (silica, alumina, etc.) by spray drying [22,23]. Kaolin, silica, and other substances are the main additives in rubber. Agustini S et al. [24] used kaolin as a filler for natural rubber to prepare solid tires for scooters. The vulcanization properties, mechanical properties, and thermal properties of the rubber compound were investigated. The experimental results demonstrated that the amount of kaolin had a great influence on the maximum torque, scorch time, optimum curing time, and mechanical properties of the vulcanized rubber. The thermogravimetric analysis demonstrated that the thermal stability of the rubber was influenced by the amount of the kaolin-filling fraction. Tan J et al. [25] ground anthracite coal to replace carbon-black-filled styrene–butadiene rubber (SBR) to prepare composites of styrene–butadiene rubber (SBR) and modified anthracite coal (MA). The experimental results demonstrate that the anthracite flakes can be well dispersed in the rubber matrix, providing good reinforcement properties. In addition, the low content of carbon black or silica composite fillers further promoted the dispersion of coal particles in the rubber, which effectively enhanced the mechanical reinforcing properties of coal particles and the thermal stability of rubber composites. Wang Z et al. [26] investigated the possibility of illite as an alternative natural rubber (NR) filler. The experimental results demonstrated that illite treated with cetyltrimethylammonium bromide (CTAB) could enhance the crosslinking density and dispersion of illite-NR, and the mechanical properties and wear resistance of illite/NR composites could be improved. Phuhiangpa N et al. [27] investigated the effect of the nano calcium carbonate (NCC) and micron calcium carbonate (MCC) on natural rubber composites. The experimental results demonstrate that two kinds of fillers (MCC and NCC) and filled rubber composites showed the same trend, but the effect of the small particle size in NCC on the composite properties was more pronounced and could be better used to adjust the rubber product characteristics and processing properties.

To extend the reuse methods of spent catalysts and reduce the pollution caused by them to the environment, this paper conducted a mechanistic and experimental study on the replacement of some carbon blacks by spent FCC catalysts for the preparation of rubber products, and the feasibility of the application of the spent catalysts to prepare rubber composite was also explored. The use of the spent FCC catalyst as a filler to replace part of the carbon black for rubber composite preparation can not only realize the recycling of waste rubber products, but also reduce the environmental pollution caused by the improper treatment of the spent catalyst. The feasibility of preparing rubber composites by spent catalysts is also verified.

2. Experiment

2.1. Experimental Scheme

To explore the influence of the number of replacement parts of spent FCC catalysts and the vulcanization system on the properties of rubber composites, the experimental formulations used in this study are shown in Table 1.

The purpose of the $1^{\#}$–$4^{\#}$ in Table 1 is to study the influence of replacing different ratios of carbon black on the mechanical properties of the rubber composites before the modification of the spent FCC catalyst. The purpose of the $5^{\#}$–$7^{\#}$ is to study the influence of particle size and pore size variation on the mechanical properties of rubber composites after the modification of spent FCC catalysts. The purpose of the $8^{\#}$–$10^{\#}$ is to study the influence of adding the S, accelerator NS, and silane coupling agent Si69 on the reinforcement performance of the filler with the increase in the replacement amount of the modified spent FCC catalyst.

Table 1. Experimental formulation.

Samples	1#	2#	3#	4#	5#	6#	7#	8#	9#	10#	Manufacturer
SBR1500	100	100	100	100	100	100	100	100	100	100	PetroChina Dushanzi Petrochemical Company, Karamay, China
Zinc oxide	3	3	3	3	3	3	3	3	3	3	Shijiazhuang Yunpo Chemical Technology Co., Ltd., Shijiazhuang, China
Accelerator NS	1	1	1	1	1	1	1	1.07	1.13	1.2	Shandong Shangshun Chemical Co., Ltd., Weifang, China
Sulphur (S)	1.75	1.75	1.75	1.75	1.75	1.75	1.75	1.87	1.98	2.1	Chaoyang Tianming Industry & Trade Co., Ltd., Beijing, China
Silane coupling agent Si69	\	\	\	\	\	\	\	0.25	0.5	0.75	Shandong Xiya Chemical Co., Ltd., Linyi, China
Carbon black N660	50	45	40	35	45	40	35	45	40	35	Shanghai Cabot Chemical Co., Ltd., Shanghai, China
Spent fcc catalyst	\	5	10	15	\	\	\	\	\	\	Sinopec Jinan Oil Refinery, Jinan, China
Modified spent FCC catalyst	\	\	\	\	5	10	15	5	10	15	Sinopec Jinan Oil Refinery, Jinan, China

Note: The unit of component dosage in the table is g.

2.2. Experimental Process

2.2.1. Spent FCC Catalyst Modification

First, the spent FCC Catalyst was ball milled (all-round planetary ball mill, QM-QX4, Nanjing Nanda Instruments Co., Ltd., Nanjing, China) at 180 r/min for 2 h. Then, the spent FCC catalyst was calcined (tube furnace, MFLGKD 405-12, Shanghai Muffle Furnace Technology Instruments Co., Ltd., Shanghai, China) at 600 °C for 3 h. Subsequently, the spent FCC catalyst and deionized water were mixed 1:5 for ultrasonic (ultrasonic disperser, VCY-1500, Shanghai Yanyong Ultrasonic Equipment Co., Ltd., Shanghai, China) treatment for 1 h, with continuous stirring during the ultrasonic process. Finally, the spent FCC catalyst was filtered and dried (electric blast drying oven, DHG-9240A, Shanghai Yiheng Scientific Instrument Co., Ltd., Shanghai, China) to complete the preparation of the spent FCC catalyst. The modification process of the spent FCC catalyst was roughly shown in Figure 1.

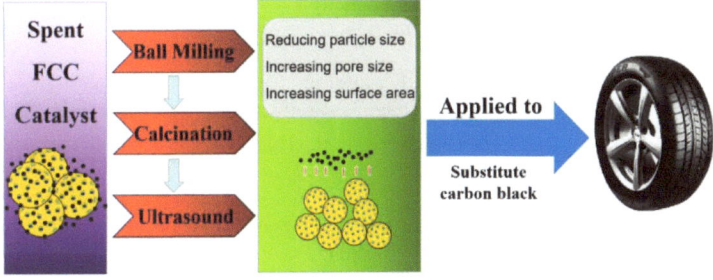

Figure 1. Flow chart of spent FCC catalyst modification.

2.2.2. Preparation of Rubber Composites

First, the butadiene rubber was plasticized into thin pieces of about 4mm in the open mixing machine (open mixing machine, X (S) K-160, Shanghai Rubber Machinery Factory, Shanghai, China) for later cutting and compacting. Subsequently, rubber, fillers (carbon black and spent FCC catalyst), and small materials were added to the mixer (laboratory small mixer, 0.3 L, Harbin Harper Electric Technology Co., Ltd., Harbin, China) to complete the preliminary preparation of rubber products. Then, the S and accelerator NS were added to the open mixing machine to complete the preparation of the compounded rubber. Finally, the mixed rubber pieces after standing for 12 h were used for the initial testing and vulcanization.

The vulcanization time T90 of each group of blends is measured by a rotorless sulfurometer (rotorless vulcanization instrument, M-2000-AN, high-speed rail detection instrument (Dongguan) Co., Ltd., Dongguan, China). Using a flat vulcanizing machine (flat vulcanizing machine, XLD-400X400X2, Qingdao Yilang Rubber Equipment Co., Ltd., Qingdao, China) for vulcanization, the vulcanization conditions are 150 °C × 1.3 T90.

2.3. Characterization

The Payne effect of the compound was tested by a rubber processing analyzer (Rubber Processing Analyzer, RPA2000, Alpha Technologies, Inc., Akron, OH, USA). The frequency was 1Hz, the strain range was between 0.28% and 40%, and the temperature was 60 °C.

The Mooney viscosity of the compound was tested by the Mooney viscometer (Mooney viscometer, PremierMV, Alpha Technology Co., Ltd., Akron, OH, USA) according to the standard ISO 289-1: 1994.

Vulcanization properties of the compound were tested by a rotorless rheometer according to the standard ISO 6502:1991, and the test temperature was 150 °C.

The hardness test of rubber was carried out using a Shore hardness tester (Shore hardness tester, LX-A, Shanghai Liuling Instrument Factory, Shanghai, China) according to the standard ISO 7619-1:2004. The rubber rebound rate was tested using a rubber rebound tester (rubber rebound tester, DIN-53512, Dongguan Songjiao Testing Instruments Co., Ltd., Dongguan, China) according to the standard ISO 4662-1986. The abrasion was tested by the DIN roller abrasion tester (DIN roller abrasion tester, GT-2012-D, Taichung Gao Tai Testing Machine Co., Ltd., Taiwan, China) according to the standard ISO 4649-2002. Tensile tearing properties were tested using the tensile testing machine (Tensile Testing Machine, AI-7000-MGD, Gautech Testing Instruments (Dongguan) Co., Ltd.) according to the standard of ISO 37-2005 and ISO 34-1:2004.

The dynamic mechanical properties of vulcanized rubber were tested by a dynamic thermo-mechanical analyzer (dynamic thermo-mechanical analyzer, EPLEXOR 150N, GABO, Ahlden, Germany). The dynamic strain was 0.25%, the static strain was 7%, the heating rate was 2 °C/min, the temperature range was −65~65 °C, and the frequency was 10 Hz.

The aging experiment of rubber was tested by using a hot air aging box (hot air aging box, GT-7017-M, High-Tech Testing Instruments Co., Ltd., Taiwan, China) according to the standard ISO 188-1998.

3. Results and Discussion

3.1. Analysis of Physical–Chemical Characteristics of Spent FCC Catalysts

The characterization results of XRD spectra of the spent FCC catalyst before and after modification are shown in Figure 2.

The crystal plane spacing of the spent FCC catalyst before and after the modification has not changed much in Figure 2. There was no new diffraction peak, indicating that the spent FCC catalyst did not change its phase composition after the modification [28,29].

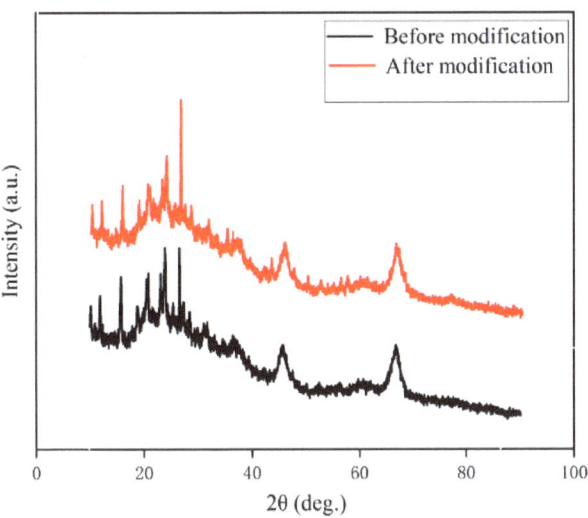

Figure 2. XRD spectrum of spent FCC catalyst.

The volume change and distribution of the particle size of the spent FCC catalyst before and after the modification are shown in Table 2 and Figure 3.

Table 2. Changes in particle size and volume of spent FCC catalyst before and after modification.

	Before Modification	After Modification	Reduce Proportion (%)
Dv (10)/μm	0.0306	0.0259	15.36
Dv (50)/μm	12.8	0.163	98.73
Dv (90)/μm	55.0	18.7	66.0

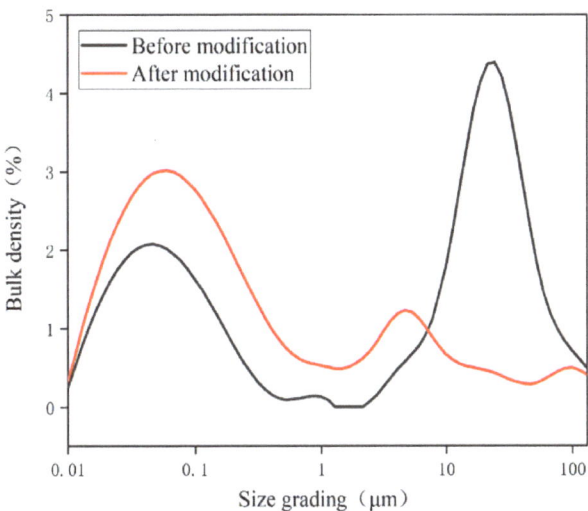

Figure 3. Particle size distribution of spent FCC catalyst.

The particle size of the spent FCC catalyst decreased significantly after a modification from Table 2 and Figure 3. The average particle size of the spent FCC catalyst decreased from 12.8 μm to 0.163 μm, which may be due to the serious agglomeration phenomenon before the modification of the spent FCC catalyst. During the ball milling process, the collision between the ball, the cylinder wall, and the spent FCC catalyst or the spent FCC catalyst self-grinded with each other, which contributed to the particle size reduction.

The changes of the specific surface area of the spent FCC catalyst before and after the modification are shown in Table 3.

Table 3. Brunauer–Emmett–Teller (BET) characterization results.

	Specific Surface Area (m^2/g)	Substrate Surface (m^2/g)	Micropore Surface Area (m^2/g)	Total Pore Volume (mL/g)	Micropore Volume (mL/g)
Before modification	75	17	58	0.1564	0.0104
After modification	89	29	60	0.1884	0.0152

The changes of the pore size morphology before and after the modification of spent FCC catalysts are shown in Figure 4.

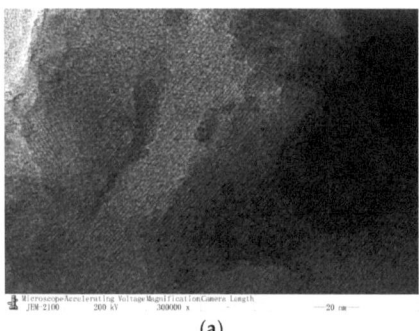
(a)

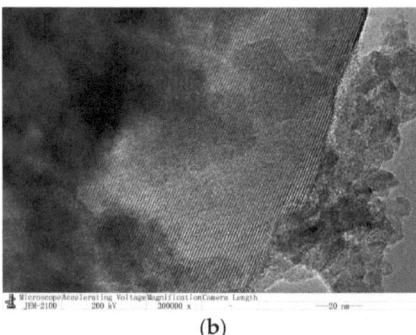
(b)

Figure 4. Changes in pore morphology of spent FCC catalysts before and after modification. (a) Before modification; (b) after modification.

From Table 3 and Figure 4, it can be observed that the specific surface area and pore volume of the spent FCC catalyst increase after the modification. Because the ball milling reduces the catalyst particle size; the combination of calcination and sonication reduces the adsorbed material on the catalyst (on the surface and inside the pores), thus increasing the specific surface area and pore volume [30].

The relative percentages of major elements before and after the modification of the spent FCC catalysts are shown in Figure 5.

It can be observed from Figure 5 that the elements contained in the spent FCC catalyst before the modification mainly include aluminum (Al), silicon (Si), lanthanum (La), etc. The relative content of Si and Al elements increased after the modification of the spent FCC catalyst, while the content of the cerium (Ce) remained basically unchanged and the relative content of the other elements decreased. This may be due to the volatilization of coke and other substances adsorbed from crude oil during the high-temperature calcination process, which caused the elemental content of the spent FCC catalyst to change.

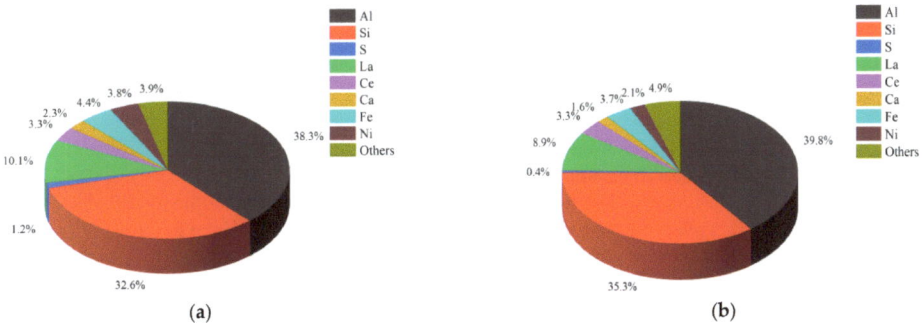

Figure 5. Relative proportions of main elements. (**a**) Before modification; (**b**) after modification.

3.2. Processing Performance Analysis of Rubber Composites

The rubber vulcanization characteristics of the different proportions of carbon black replaced by the spent FCC catalyst before the modification are shown in Table 4.

Table 4. Vulcanization characteristics of rubber composites with spent FCC catalyst before modification.

	1[#]	2[#]	3[#]	4[#]
Mooney viscosity/MU	70.90	70.14	69.27	68.77
ML/(N·m)	2.13	2.06	1.87	2.19
MH/(N·m)	20.27	12.75	9.96	9.85
MH-ML/(N·m)	18.14	10.69	8.09	7.66
T10/min	11.92	10.07	10.2	8.87
T90/min	29.53	36.83	44.11	46.00
T100/min	56.33	59.59	59.72	59.93

Note: 1[#], 2[#], 3[#], respectively corresponding to the experimental group in Table 1.

The rubber vulcanization characteristics of modified spent FCC catalyst replacing carbon black with different proportions are shown in Table 5.

Table 5. Vulcanization characteristics of rubber composites after modification with spent FCC catalyst.

	5[#]	6[#]	7[#]	8[#]	9[#]	10[#]
Mooney viscosity/MU	70.68	69.13	68.95	70.31	69.27	68.46
ML/(N·m)	1.94	1.92	1.84	2.15	2.08	1.99
MH/(N·m)	18.43	18.2	16.12	22.08	21.70	21.29
MH-ML/(N·m)	16.53	16.28	14.28	19.93	19.62	19.30
T10/min	10.8	12.05	12.22	10.22	11.06	11.27
T90/min	31.47	36.03	38.50	26.69	27.70	28.33
T100/min	59.18	59.87	59.95	48.75	52.91	53.00

Note: 5[#], 6[#], 7[#], 8[#], 9[#], 10[#] respectively corresponding to the experimental group in Table 1.

From Table 4, it can be observed that with the increase in the number of parts of carbon black replaced by the spent FCC catalyst, the Mooney viscosity of the rubber compound demonstrated a decreasing trend, indicating that the addition of the spent FCC catalyst improved the plasticity and processability of the compound [31]. The value of MH-ML is considered to be positively related to the crosslinking density of the rubber [32,33]. The higher the value, the higher the crosslinking density. According to the data in the table, the MH-ML value decreases as the number of parts increases, which is related to the weak reinforcement of the spent FCC catalyst and is consistent with the trend of the Menny viscosity. T10 is generally considered to be related to the rubber scorch time; the larger the value, the higher the processing safety, and vice versa [34]. It can be observed from the data that the processing safety decreases as the number of replacement parts of the spent FCC catalyst increases. The elemental analysis demonstrates that the spent FCC catalyst contains elements such as S, which leads to early local cross-linking during the mixing process, therefore resulting in scorching. T90 is the positive vulcanization time. From the data, it can be observed that as the number of replacement parts increases, the positive vulcanization time becomes longer, which leads to an increase in vulcanization time and a decrease in economic efficiency [35].

As can be observed from Table 5, the reinforcement of the modified spent FCC catalyst is greatly improved, and it can be used as a reinforcing filler to replace a part of the carbon black for the rubber filler. From the MH-ML difference, T10, and T90 in the data of columns 5#–7#, it can be observed that the crosslink density is significantly increased compared with that before the modification. There is a slight increase in T10, and the processing safety is improved. There is a significant decrease in the positive vulcanization time of T90. This is due to the decrease in the particle size, the increase in the specific surface area and pore volume, and the decrease in impurities of the modified spent FCC catalyst, which leads to the enhancement of interfacial bonding with the polymer and the improvement of performance [36]. From the data in columns 8#–10#, it can be observed that S, accelerator NS, and the silane coupling agent Si69 increased equiproportionally with the increase in the spent FCC catalyst replacement. The MH-ML difference and T10 and T90 parameters are greatly improved; thus, the performance is even more excellent.

3.3. Rubber Composite Payne Effect Analysis

The change of the storage modulus of rubber composites with different parts of the carbon black replaced by the spent FCC catalyst is shown in Figure 6.

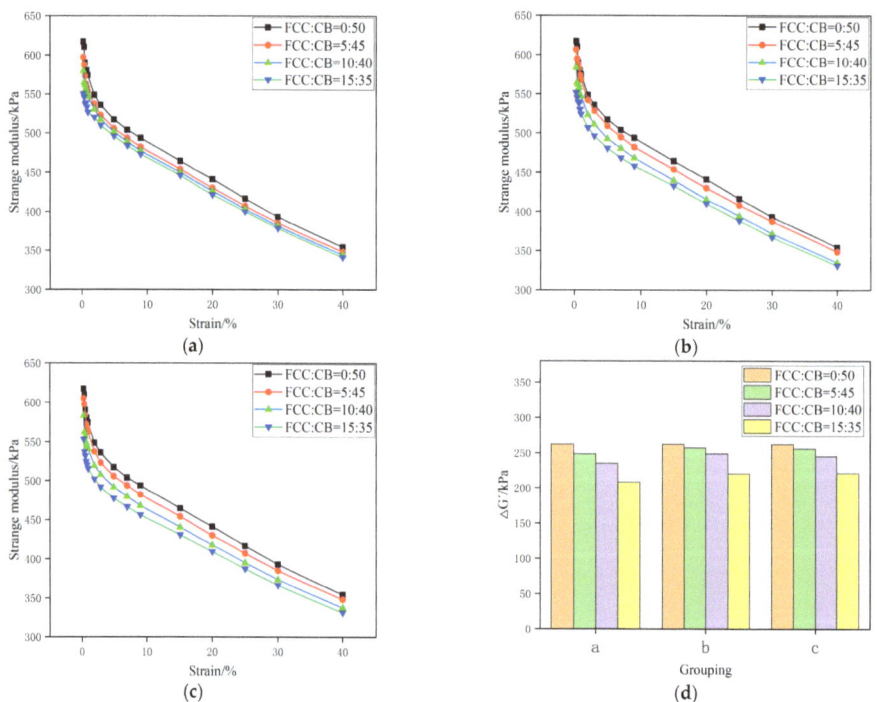

Figure 6. Storage modulus change diagram of rubber composite (abscissa a, b, and c in (**d**) correspond to three storage modulus change diagrams: a, b, and c, respectively). (**a**) Formulation 1#–4#; (**b**) formulation 1#, 5#–7#; (**c**) formulation 1#, 8#-1; (**d**) change of storage modulus.

Figure 6 is the storage modulus and strain curve of rubber composites under the spent FCC catalyst/carbon black. With the increase in strain, the storage modulus of filler–filled rubber composites decreases, which is called the Payne effect [37,38]. $\Delta G'$ represents the degree of the network structure of the filler, and $\Delta G'$ is the difference between the storage modulus at a 40% strain and the storage modulus at a 0.28% strain. The smaller $\Delta G'$ indicates that the Payne effect is weak and the filler has a better dispersion in the rubber matrix. From Figure 6d, it can be observed that with the increase in the number of carbon black parts replaced by the spent FCC catalyst, the $\Delta G'$ of the compounded rubber shows a decreasing trend. This is mainly because the spent FCC catalyst can form more filler–rubber network structures with rubber, with less filler agglomeration, and exhibit a low modulus in torsion tests; thus, the Payne effect is weakened.

3.4. Analysis of Physical and Mechanical Properties of Rubber Composites

The function of fillers in rubber products is mostly to improve the mechanical properties such as hardness, tensile strength, and elongation at the break of rubber composites. The mechanical properties corresponding to different experimental formulations in Table 1 are shown in Table 6 (Table 1, formulation 1[#]–4[#]) and Table 7 (Table 1, formulation 5[#]–10[#]).

Table 6. Mechanical properties of vulcanized rubber.

	1[#]	2[#]	3[#]	4[#]
Hardness/Shore A	60.0	48.5	45.5	45.0
10% tensile stress/MPa	0.54	0.56	0.55	0.54
100% tensile stress/MPa	2.74	1.67	1.36	1.18
300% tensile stress/MPa	12.82	6.57	4.25	2.55
Tensile strength/MPa	19.61	18.22	13.62	8.48
Elongation at break/%	454.85	657.80	693.20	752.91
Tensile product	8919.61	11,985.12	9441.38	6384.68
Tearing strength/N	77.78	54.39	50.41	39.54
Specific gravity/g·cm^{-3}	1.142	1.149	1.156	1.163
DIN abrasion/cm^3	0.108	0.137	0.149	0.166
Rebound rate/%	60.3	60.9	61.4	61.7

Note: 1[#], 2[#], 3[#], respectively corresponding to the experimental group in Table 1.

Table 7. Mechanical properties of vulcanized rubber.

	5[#]	6[#]	7[#]	8[#]	9[#]	10[#]
Hardness/Shore A	58.0	57.5	55.0	62.0	61.5	61.0
10% tensile stress/MPa	0.60	0.56	0.53	0.62	0.57	0.56
100% tensile stress/MPa	2.54	2.37	2.04	3.16	3.01	3.00
300% tensile stress/MPa	11.03	9.67	7.50	13.12	13.10	12.88
Tensile strength/MPa	19.46	19.35	19.01	20.04	19.95	19.79
Elongation at break/%	478.22	505.41	580.31	441.34	443.60	443.87
Tensile product	9306.16	9830.22	11,031.69	8846.31	8848.27	8784.10
Tearing strength/N	73.68	71.44	68.37	87.13	83.25	81.46
Specific gravity/g·cm^{-3}	1.146	1.152	1.162	1.147	1.153	1.161
DIN abrasion/cm^3	0.128	0.129	0.131	0.109	0.119	0.124
Rebound rate/%	62.0	62.5	63.1	64.00	65.50	66.70

Note: 5[#], 6[#], 7[#], 8[#], 9[#], 10[#] respectively corresponding to the experimental group in Table 1.

As shown in Table 6, the mechanical properties of the rubber material filled with the spent FCC catalyst directly instead of the carbon black rubber were poor. The tensile strength, tear strength, and abrasion properties decreased significantly as the number of replacement parts increased.

From Table 7 (formulation 5[#]–7[#]), it can be observed that the hardness and tensile stress of the rubber composites demonstrated a decreasing trend as the number of replacement parts of the spent FCC catalyst increased. This may be due to the combination of the spent

FCC catalyst with the rubber molecular chain being weaker compared to the carbon black, resulting in the decrease in the hardness and tensile stress of the rubber composites.

From Table 7 (formulation 8#–10#), it can be observed that with the increase in the number of replacement carbon black parts after the modification of the spent FCC catalyst, the fixed tensile strength, tensile strength, and tear strength of the rubber composites were enhanced after the addition of S, accelerator NS, and the silane coupling agent Si69 in an equal proportion, so that the properties of the three groups of the rubber composites remained basically the same as the control group (formulation 1#). This is mainly because with the increase in the accelerator NS, S, and silane coupling agent Si69, the interfacial bonding between silica and rubber molecular chains in the spent FCC catalyst is enhanced. At the same time, the increase in the accelerator NS and S improves the degree of vulcanization and increases the crosslinking density; this shows the characteristics of a high modulus and high elongation.

The wear resistance of rubber composites after replacing the carbon black with the spent FCC catalyst was slightly lower than that of the control group. This may be due to the particle size of the spent FCC catalyst being larger than that of the carbon black (N660 particle size 49–60 nm), and the combination of the spent FCC catalyst and rubber molecular chains is weaker compared to the carbon black, which leads to the easy fall off of the spent FCC catalyst during the process of abrasion. After the spent FCC catalyst falls off, the surface of the rubber composite is defective, resulting in more rubber particles being worn off and therefore with increased wear. The spent FCC catalysts contain reinforcing substances such as SiO_2. With the increase in the amount of the accelerator NS, S and silane coupling agent Si69, the interfacial combination of silica in the spent FCC catalyst and rubber molecular chain is enhanced, and the spent FCC catalyst does not easily fall off during the wear process; thus, the wear consumption is reduced.

3.5. Microscopic Morphology

The microscopic morphologies corresponding to the rubber composites prepared by formulations 1#, 5#, 6#, and 7# are shown in Figure 7.

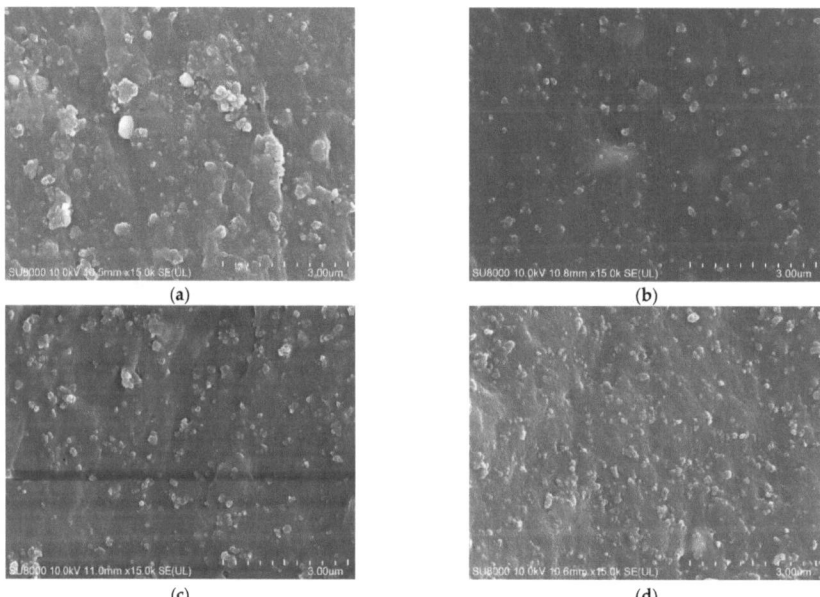

Figure 7. Scanning electron micrograph of rubber composites prepared by replacing part of carbon black with spent FCC catalyst. (**a**) 1#; (**b**) 5#; (**c**) 6#; (**d**) 7#.

As can be observed from Figure 7, the modified spent FCC catalyst, instead of the carbon black filler in the rubber, has a better dispersion and no more obvious agglomeration phenomenon. When 15 phr of the modified spent FCC catalyst was used to replace the equal mass of the carbon black, the dispersion and distribution of the filler in the rubber was optimal, which was consistent with the Payne effect results shown in Figure 8.

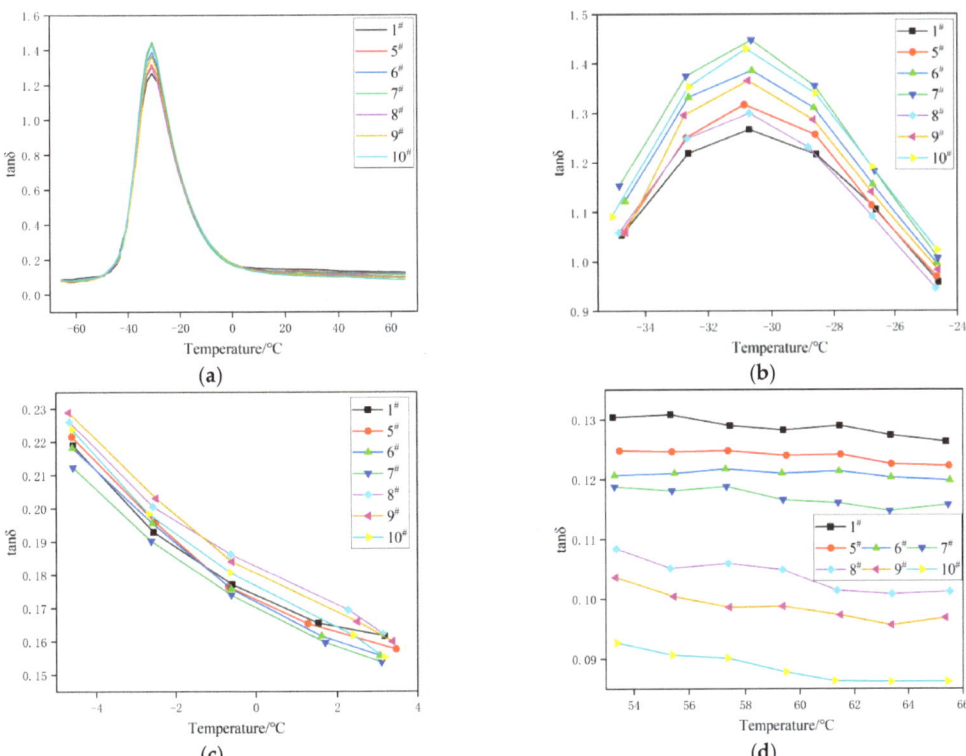

Figure 8. Dynamic mechanical properties of rubber composites.(**a**) tanδ-T Curve of the composites; (**b**) is magnified graphs of curves around −30 °C; (**c**) is magnified graphs of curves around 0 °C; (**d**) is magnified graphs of curves around 60 °C.

3.6. Dynamic Properties Analysis of Rubber Composites

Figure 8 shows the change curve of the loss factor (tanδ) with a temperature for rubber composites at −65~65 °C [39,40]. Usually, the higher the peak value of tanδ, the better the dispersion of the filler. The tanδ at 0 °C represents the wet skid resistance of the tire, and the larger the value, the better the wet skid resistance. The tanδ at 60 °C represents the rolling resistance of the tire, and the smaller the value, the lower the rolling resistance. The peak value of tanδ from high to low is 7[#] ≈ 10[#] > 6[#] ≈ 9[#] > 5[#] ≈ 8[#] > 1[#], which is basically consistent with the characterization results in Figure 6, indicating that the replacement of the carbon black by the spent FCC catalyst for rubber filling can improve the dispersion of the filler in the rubber matrix. At 0 °C, the curves of formulation 5[#]–7[#] are lower than those of the control group (formulation 1[#]), indicating that the wet skid resistance of the tires decreases after replacing the carbon black with the spent FCC catalyst. The corresponding curves of formulation 8[#]–10[#] are higher than those of the control group (formulation 1[#]). With the increase in the number of replacement parts of spent FCC catalysts, the equal proportional addition of S, accelerator NS and the silane coupling agent Si69 improved the anti-slip properties of the tires. Mao C et al. [41] believed that there is an inevitable relationship between the surface roughness of the sample and the wet skid resistance. The rougher the surface, the higher the coefficient of the wet friction and the better the wet skid resistance. Therefore, the wet skid resistance of the rubber polymer in this experiment may be related to the particle size of the spent FCC catalyst and the interfacial bonding strength of the filler and the rubber. At 60 °C, the corresponding curves of formulations $5^{\#}$–$10^{\#}$ are lower than those of formulation $1^{\#}$, while the corresponding curves of formulations $8^{\#}$–$10^{\#}$ continue to show a downward trend and are lower than those of formulations $5^{\#}$–$7^{\#}$, indicating that with the increase in replacement parts of the spent FCC catalyst, the addition of S, accelerator NS and the silane coupling agent Si69 in equal proportion could reduce the rolling resistance of the tire.

3.7. Aging Properties' Analysis

The aging properties of the vulcanized rubber composites (formulations $1^{\#}$–$10^{\#}$) were tested after aging at 100 °C for 24 h. The aging tensile properties are shown in Table 8.

Table 8. Aging tensile properties of rubber composites.

Test Items		$1^{\#}$	$5^{\#}$	$6^{\#}$	$7^{\#}$	$8^{\#}$	$9^{\#}$	$10^{\#}$
Elongation at break/%	Before aging	454.85	478.22	505.41	580.31	441.34	443.60	443.87
	After aging	270.49	311.26	331.82	385.11	295.08	297.68	298.86
Aging property change rate/%		40.53	34.91	34.35	33.64	33.14	32.89	32.67
Tensile strength/MPa	Before aging	19.61	19.46	19.35	19.01	20.04	19.85	19.79
	After aging	16.82	18.24	17.43	16.82	19.06	18.49	18.28
Aging property change rate/%		14.23	6.27	9.92	11.52	4.89	6.85	8.59
Tensile product	Before aging	8919.61	9306.16	9779.68	11,031.69	8844.45	8805.46	8784.19
	After aging	4550.26	5677.38	5783.62	6477.55	5624.22	5504.10	5463.16
Aging property change rate/%		48.99	38.99	40.86	41.28	36.41	37.49	37.81
Aging coefficient		0.51	0.61	0.59	0.59	0.64	0.63	0.62

As shown in Table 8, the aging property change rates of the elongation at the break, and the tensile strength and tensile product of the rubber composites prepared by replacing the carbon black with spent FCC catalysts (formulations $5^{\#}$–$7^{\#}$), are lower than those of the control group (formulation $1^{\#}$), indicating that the aging resistance is better than that of the control group. With the increase in the number of replacement parts of the spent FCC catalyst, the aging property change rate of the composites was reduced again by adding S, accelerator NS and the silane coupling agent Si69 (formulation $8^{\#}$–$10^{\#}$) in equal proportion; the anti-aging property is improved again and is significantly better than that of the control group. This is mainly because the spent FCC catalyst is porous material and can adsorb some of the vulcanization system and other rubber additives to become the carrier of the slow-release agent. During the aging process, the S absorbed in the pore size is released and re-involved in the vulcanization reaction. At the same time, the silane coupling agent Si69 contains four elemental sulfur, which will also be re-involved in the vulcanization reaction during the aging process to enhance the anti-aging properties of rubber.

3.8. Synergistic Reinforcement Mechanism of Spent FCC Catalyst and Carbon Black for SBR

The spent FCC catalyst is uniformly and disorderly dispersed in the SBR matrix. During the mixing process of the filler (spent FCC catalyst) and rubber, rubber molecular chains are adsorbed on the surface of the filler to form bound rubber; meanwhile, some rubber molecular chains enter the voids and pores of the filler to form a small amount of inclusion rubber (retention rubber). The interaction between the spent FCC catalyst and rubber mainly includes the physical adsorption and chemical combination, which can form a close filler–polymer network structure with rubber, so as to improve the basic physical properties of rubber products. When the rubber products are stretched by an external force, the existence of the bound rubber is conducive to prolonging the crack expansion path so that the rubber can withstand a greater external force, and the existence of the inclusion rubber is conducive to limiting the displacement of the molecular chains around the filler, thereby improving the mechanical properties of the rubber.

4. Conclusions

In this paper, we explore the spent FCC catalysts to replace different parts of the carbon black for the rubber product preparation to realize the resource utilization and diversified applications of spent FCC catalysts. The use of the spent FCC catalyst as a filler to replace part of the carbon black for the rubber composite preparation can not only realize the recycling of waste rubber products, but also reduce the environmental pollution caused by the improper treatment of the spent catalyst.

The spent FCC catalyst as a filler to replace carbon black in the rubber has good dispersibility, which is conducive to the improvement of the comprehensive performance of rubber products (stretching, stretching, rebound, tearing, etc.). The Payne effect is decreased by about 3–16%, and the rolling resistance is obviously reduced. The spent FCC catalyst is a porous material, which can be used as a carrier of the slow release agent. In the aging process, S in the pore size of the spent FCC catalyst will be released again to participate in the vulcanization reaction and improve the anti-aging properties of rubber products.

Author Contributions: Conceptualization, X.T. and T.S.; methodology, X.T. and C.W.; software, K.W.; validation, D.Z.; data curation, H.B.; writing—original draft preparation, X.T. and T.S.; writing—review and editing, X.T.; visualization, K.W.; supervision, H.B. All authors have read and agreed to the published version of the manuscript.

Funding: This work was financially supported by the National Natural Science Foundation of China (52103117), Natural Science Foundation of Shandong Province Key Projects (ZR2020KE037), Qingdao Science and Technology Benefit People Demonstration Guide Special Project (22-3-7-cspz-18-nsh), Shandong postdoctoral innovation project (202103024), and Qingdao Postdoctoral Applied Research Project.

Data Availability Statement: The data supporting the findings of this study are available within the article.

Conflicts of Interest: The authors declare no conflict of interest.

References

1. Formela, K. Sustainable development of waste tires recycling technologies–recent advances, challenges and future trends. *Adv. Ind. Eng. Polym. Res.* **2021**, *4*, 209–222. [CrossRef]
2. Bockstal, L.; Berchem, T.; Schmetz, Q.; Richel, A. Devulcanisation and reclaiming of tires and rubber by physical and chemical processes: A review. *J. Clean. Prod.* **2019**, *236*, 117574. [CrossRef]
3. Kunanusont, N.; Samthong, C.; Bowen, F.; Yamaguchi, M.; Somwangthanaroj, A. Effect of mixing method on properties of ethylene vinyl acetate copolymer/natural rubber thermoplastic vulcanizates. *Polymers* **2020**, *12*, 1739. [CrossRef]
4. Shoul, B.; Marfavi, Y.; Sadeghi, B.; Kowsari, E.; Sadeghi, P.; Ramakrishna, S. Investigating the potential of sustainable use of green silica in the green tire industry: A review. *Environ. Sci. Pollut. R.* **2022**, *29*, 51298–51317. [CrossRef]
5. Sökmen, S.; Oßwald, K.; Reincke, K.; Ilisch, S. Influence of Treated Distillate Aromatic Extract (TDAE) Content and Addition Time on Rubber-Filler Interactions in Silica Filled SBR/BR Blends. *Polymers* **2021**, *13*, 698. [CrossRef]
6. Jong, L. Improved mechanical properties of silica reinforced rubber with natural polymer. *Polym. Test.* **2019**, *79*, 106009. [CrossRef]
7. Song, S.H. Study on silica-based rubber composites with epoxidized natural rubber and solution styrene butadiene rubber. *Polym. Polym. Compos.* **2021**, *29*, 1422–1429. [CrossRef]
8. Zhang, X.; Cai, L.; He, A.; Ma, H.; Li, Y.; Hu, Y.; Zhang, X.; Liu, L. Technology. Facile strategies for green tire tread with enhanced filler-matrix interfacial interactions and dynamic mechanical properties. *Compos. Sci. Technol.* **2021**, *203*, 108601. [CrossRef]
9. Zhang, C.; Tang, Y.; Tian, Q.; Xie, X.; Xu, L.; Li, X.; Ding, T. Preparation of dispersible nanosilica surface-capped by hexamethyl disilazane via an in situ surface-modification method and investigation of its effects on the mechanical properties of styrene–butadiene/butadiene rubber. *J. Appl. Polym. Sci.* **2019**, *136*, 47763. [CrossRef]
10. Lee, C.K.; Seo, J.G.; Kim, H.J.; Song, S.H. Novel green composites from styrene butadiene rubber and palm oil derivatives for high performance tires. *J. Appl. Polym. Sci.* **2019**, *136*, 47672. [CrossRef]
11. Seo, J.G.; Lee, C.K.; Lee, D.; Song, S.H. High-performance tires based on graphene coated with Zn-free coupling agents. *J. Ind. Eng. Chem.* **2018**, *66*, 78–85. [CrossRef]

12. Yang, C.; Gao, D.; Zhang, R.; Deng, W.; Coltd, X.J. Attempt to New Processing Technology of Special Natural Rubber for Tire with High Performance. *Trop. Agric. Sci. Technol.* **2019**, *42*.
13. Ni, Y.; Liu, Z.; Tian, P.; Chen, Z.; Fu, Y.; Zhu, W.; Liu, Z. A dual-bed catalyst for producing ethylene and propylene from syngas. *J. Energy. Chem.* **2022**, *66*, 190–194. [CrossRef]
14. Suganuma, S.; Katada, N. Innovation of catalytic technology for upgrading of crude oil in petroleum refinery. *Fuel Process. Technol.* **2020**, *208*, 106518. [CrossRef]
15. Zhao, R.; Heng, M.; Chen, C.; Li, T.; Shi, Y.; Wang, J. Catalytic effects of Al_2O_3 nano-particles on thermal cracking of heavy oil during in-situ combustion process. *J. Petrol. Sci. Eng.* **2021**, *205*, 108978. [CrossRef]
16. Luan, H.; Lin, J.; Xiu, G.; Ju, F.; Ling, H. Study on compositions of FCC flue gas and pollutant precursors from FCC catalysts. *Chemosphere* **2020**, *245*, 125528. [CrossRef]
17. Wang, C.; Tian, X.; Zhao, B.; Zhu, L.; Li, S. Experimental study on spent FCC catalysts for the catalytic cracking process of waste tires. *Processes* **2019**, *7*, 335. [CrossRef]
18. Tian, X.; Wang, K.; Shan, T.; Li, Z.; Wang, C.; Zong, D.; Jiao, D. Study of waste rubber catalytic pyrolysis in a rotary kiln reactor with spent fluid-catalytic-cracking catalysts. *J. Anal. Appl. Pyrolysis* **2022**, *167*, 105686. [CrossRef]
19. Tian, X.; Han, S.; Wang, K.; Shan, T.; Li, Z.; Li, S.; Wang, C. Waste resource utilization: Spent FCC catalyst-based composite catalyst for waste tire pyrolysis. *Fuel* **2022**, *328*, 125236. [CrossRef]
20. Abd Rahman, N.A.; Fermoso, J.; Sanna, A. Stability of Li-LSX Zeolite in the Catalytic Pyrolysis of Non-Treated and Acid Pre-Treated Isochrysis sp. Microalgae. *Energies* **2020**, *13*, 959. [CrossRef]
21. Trinh, H.B.; Lee, J.-C.; Suh, Y.-J.; Lee, J. A review on the recycling processes of spent auto-catalysts: Towards the development of sustainable metallurgy. *Waste Manag.* **2020**, *114*, 148–165. [CrossRef]
22. Uzcátegui, G.; de Klerk, A. Causes of deactivation of an amorphous silica-alumina catalyst used for processing of thermally cracked naphtha in a bitumen partial upgrading process. *Fuel* **2021**, *293*, 120479. [CrossRef]
23. Salam, M.A.; Cheah, Y.W.; Ho, P.H.; Olsson, L.; Creaser, D. Hydrotreatment of lignin dimers over NiMoS-USY: Effect of silica/alumina ratio. *Sustain. Energ. Fuels* **2021**, *5*, 3445–3457. [CrossRef]
24. Agustini, S.; Sholeh, M. Utilization of Kaolin as a Filling Material for Rubber Solid Tire Compounds for Two-wheeled Electric Scooters. In *IOP Conference Series: Materials Science and Engineering*; IOP Publishing: Bristol, UK, 2021; p. 012010.
25. Tan, J.; Cheng, H.; Wei, L.; Gui, X. Thermal and mechanical enhancement of styrene–butadiene rubber by filling with modified anthracite coal. *J. Appl. Polym. Sci.* **2019**, *136*, 48203. [CrossRef]
26. Wang, Z.; Wang, S.; Yu, X.; Zhang, H.; Yan, S. Study on the use of CTAB-treated illite as an alternative filler for natural rubber. *Acs Omega* **2021**, *6*, 19017–19025. [CrossRef]
27. Phuhiangpa, N.; Ponloa, W.; Phongphanphanee, S.; Smitthipong, W. Performance of nano-and microcalcium carbonate in uncrosslinked natural rubber composites: New results of structure–properties relationship. *Polymers* **2020**, *12*, 2002. [CrossRef]
28. Xue, B.; Wang, X.; Sui, J.; Xu, D.; Zhu, Y.; Liu, X. A facile ball milling method to produce sustainable pyrolytic rice husk bio-filler for reinforcement of rubber mechanical property. *Ind. Crops Prod.* **2019**, *141*, 111791. [CrossRef]
29. Bu, X.; Chen, Y.; Ma, G.; Sun, Y.; Ni, C.; Xie, G. Wet and dry grinding of coal in a laboratory-scale ball mill: Particle-size distributions. *Powder Technol.* **2020**, *359*, 305–313. [CrossRef]
30. Wang, T.; Li, G.; Yang, K.; Zhang, X.; Wang, K.; Cai, J.; Zheng, J. Enhanced ammonium removal on biochar from a new forestry waste by ultrasonic activation: Characteristics, mechanisms and evaluation. *Sci. Total. Environ.* **2021**, *778*, 146295. [CrossRef]
31. Zong, X.; Wang, S.; Li, N.; Li, H.; Zhang, X.; He, A. Regulation effects of trans-1, 4-poly (isoprene-co-butadiene) copolymer on the processability, aggregation structure and properties of chloroprene rubber. *Polymer* **2021**, *213*, 123325. [CrossRef]
32. Hassanabadi, M.; Najafi, M.; Motlagh, G.H.; Garakani, S. Synthesis and characterization of end-functionalized solution polymerized styrene-butadiene rubber and study the impact of silica dispersion improvement on the wear behavior of the composite. *Polym. Test.* **2020**, *85*, 106431. [CrossRef]
33. Zachariah, A.K.; Chandra, A.K.; Mohammed, P.; Thomas, S. Vulcanization kinetics and mechanical properties of organically modified nanoclay incorporated natural and chlorobutyl rubber nanocomposites. *Polym. Test.* **2019**, *76*, 154–165. [CrossRef]
34. Barghamadi, M.; Karrabi, M.; Ghoreishy, M.H.R.; Mohammadian-Gezaz, S. Effects of two types of nanoparticles on the cure, rheological, and mechanical properties of rubber nanocomposites based on the NBR/PVC blends. *J. Appl. Polym. Sci.* **2019**, *136*, 47550. [CrossRef]
35. Wei, Y.-C.; Liu, G.-X.; Zhang, H.-F.; Zhao, F.; Luo, M.-C.; Liao, S. Non-rubber components tuning mechanical properties of natural rubber from vulcanization kinetics. *Polymer* **2019**, *183*, 121911. [CrossRef]
36. Fan, Y.; Fowler, G.D.; Zhao, M. The past, present and future of carbon black as a rubber reinforcing filler—A review. *J. Clean. Prod.* **2020**, *247*, 119115. [CrossRef]
37. Li, X.; Tian, C.; Li, H.; Liu, X.; Zhang, L.; Hong, S.; Ning, N.; Tian, M. Combined effect of volume fractions of nanofillers and filler-polymer interactions on 3D multiscale dispersion of nanofiller and Payne effect. *Compos. Part A-Appl. Sci. Manuf.* **2022**, *152*, 106722. [CrossRef]
38. Sattayanurak, S.; Sahakaro, K.; Kaewsakul, W.; Dierkes, W.K.; Reuvekamp, L.A.; Blume, A.; Noordermeer, J.W. Synergistic effect by high specific surface area carbon black as secondary filler in silica reinforced natural rubber tire tread compounds. *Polym. Test.* **2020**, *81*, 106173. [CrossRef]

39. Greiner, M.; Unrau, H.-J.; Gauterin, F. A model for prediction of the transient rolling resistance of tyres based on inner-liner temperatures. *Vehicle Syst. Dyn.* **2018**, *56*, 78–94. [CrossRef]
40. Choi, S.S.; Kwon, H.M.; Kim, Y.; Ko, E.; Lee, K.S. Hybrid factors influencing wet grip and rolling resistance properties of solution styrene-butadiene rubber composites. *Fortschr. Phys.* **2018**, *67*, 340–346. [CrossRef]
41. Mao, C.; Li, X.; Liu, S. The Effect of Roughness on the Wet Skid Resistance of Tire Tread Compounds. *J. Phys. Conf. Ser.* **2020**, *1649*, 012032. [CrossRef]

Disclaimer/Publisher's Note: The statements, opinions and data contained in all publications are solely those of the individual author(s) and contributor(s) and not of MDPI and/or the editor(s). MDPI and/or the editor(s) disclaim responsibility for any injury to people or property resulting from any ideas, methods, instructions or products referred to in the content.

Article

Effect of Cementitious Capillary Crystalline Waterproofing Materials on the Mechanical and Impermeability Properties of Engineered Cementitious Composites with Microscopic Analysis

Yan Tan [1,*], Ben Zhao [1], Jiangtao Yu [2], Henglin Xiao [1], Xiong Long [1] and Jian Meng [1]

[1] College of Civil Engineering, Architecture and Environment, Hubei University of Technology, Wuhan 430068, China
[2] School of Civil Engineering, Tongji University, Shanghai 200092, China
* Correspondence: tanyan@hbut.edu.cn; Tel.: +86-185-7150-2382

Citation: Tan, Y.; Zhao, B.; Yu, J.; Xiao, H.; Long, X.; Meng, J. Effect of Cementitious Capillary Crystalline Waterproofing Materials on the Mechanical and Impermeability Properties of Engineered Cementitious Composites with Microscopic Analysis. *Polymers* 2023, *15*, 1013. https://doi.org/10.3390/polym15041013

Academic Editors: Giorgio Luciano and Maurizio Vignolo

Received: 25 January 2023
Revised: 13 February 2023
Accepted: 15 February 2023
Published: 17 February 2023

Copyright: © 2023 by the authors. Licensee MDPI, Basel, Switzerland. This article is an open access article distributed under the terms and conditions of the Creative Commons Attribution (CC BY) license (https://creativecommons.org/licenses/by/4.0/).

Abstract: Building structures are prone to cracking, leakage, and corrosion under complex loads and harsh marine environments, which seriously affect their durability performance. To design cementitious composites with excellent mechanical and impermeability properties, Engineered Cementitious Composites (ECCs) doped with ultrahigh molecular weight polyethylene short-cut fibers (PE-ECCs) were used as the reference group. Different types (XYPEX-type from Canada, SY1000-type from China) and doses (0%, 0.5%, 1.0%, 1.5%, 2.0%) of Cementitious Capillary Crystalline Waterproofing materials (CCCWs) were incorporated. The effect of CCCWs on the mechanical and impermeability properties of PE-ECCs, and the microscopic changes, were investigated to determine the best type of CCCW to use and the best amount of doping. The results showed that with increasing the CCCW dosage, the effects of both CCCWs on the mechanical and impermeability properties of PE-ECC increased and then decreased, and that the best mechanical and impermeability properties of PE-ECC were achieved when the CCCW dosing was 1.0%. The mechanical properties of the PE-ECC were more obviously improved by XYPEX-type CCCW, with a compressive strength of 53.8 MPa, flexural strength of 11.8 MPa, an ultimate tensile stress of 5.56 MPa, and an ultimate tensile strain of 7.53 MPa, which were 37.95%, 53.25%, 14.17%, and 21.65% higher than those of the reference group, respectively. The effects of the two CCCWs on impermeability were comparable. CCCW-PE-ECC(X1.0%) and CCCW-PE-ECC(S1.0%) showed the smallest permeation heights, 2.6 mm and 2.8 mm, respectively. The chloride ion diffusion coefficients of CCCW-PE-ECC(X1.0%) and CCCW-PE-ECC(S1.0%) exhibited the smallest values, 0.15×10^{-12} m^2/s and 0.10×10^{-12} m^2/s, respectively. Micromorphological tests showed that the particle size of the XYPEX-type CCCW was finer, and the intensity of the diffraction peaks of C-S-H and CaCO$_3$ of PE-ECC increased after doping with two suitable doping amounts of CCCW. The pore structure was improved, the surface of the matrix was smoother, and the degree of erosion of hydration products on the fiber surface was reduced after chloride ion penetration. XYPEX-type CCCW demonstrated a more obvious improvement in the PE-ECC pore structure.

Keywords: cementitious capillary crystalline waterproofing material; engineered cementitious composites; mechanical properties; impermeability properties; chloride ion diffusion coefficient

1. Introduction

Concrete is one of the main construction materials in modern civil engineering [1]. It is widely utilized in building coastal and marine infrastructures because of its good mechanical properties, simple construction, and low cost [2]. However, concrete, as a typical inorganic composite material, has hydrophilic and porous characteristics [3]. Additionally, concrete without reinforcement is prone to fracture under tensile and bending loads, thus providing an influx channel to aggressive ions under the combined actions of water, air, and temperature variations [4–7]. This greatly accelerates the decline in the mechanical properties and durability of concrete structures [8,9], affecting the safety and service lives of the structures [10].

Engineered Cementitious Composites (ECCs) are green, high-performance construction materials, designed based on micromechanics and fracture mechanics, that exhibit excellent tensile ductility at relatively low fiber content (typically ≤2% by volume fraction) [11,12]. The tensile strain of ECCs reaches at least 2%, which is more than 200 times that of ordinary concrete [13,14]. ECCs exhibit strain hardening and saturation cracking under increasing tensile forces, with crack widths typically ranging from 0.05 mm~0.1 mm [15–18]. Studies have shown that the permeability stabilizes within 3–4 days when the ECC fracture width is <60 μm, and takes 7–10 days, or even longer, to stabilize when the fracture width is >100 μm. A tightly cracked ECC can complete self-healing in a short period of time and can effectively prevent water molecules and corrosive ions from attacking the material [19]. The chloride ion diffusion coefficient of ECCs is only approximately 1/2 that of mortar and 10–35% that of concrete [20–22], and the chloride ion diffusion coefficient of ordinary cement mortar increases exponentially with preload deformation, while ECCs show a linear trend. This indicates that the chloride ion diffusion coefficient of ECCs is significantly lower than that of ordinary mortar under any preloading deformation [23]. It is generally believed that multiple fine cracks produced by an ECC under load can effectively limit the chloride ion penetration, but studies have shown that the resistance of an ECC to chloride ion intrusion depends mainly on the accumulated crack width rather than the maximum width [24,25]. Under prolonged exposure to chloride ions, ECCs still exhibited high durability, ductility, and multiple cracks, with a strain capacity of more than 2% [26–28].

Most ECCs are often prepared using polyvinyl alcohol (PVA) and polypropylene (PP) fibers, which typically achieve a 3% to 5% strain capacity [29]. Due to the presence of hydroxyl groups in the molecular chains, the chemical bonding of PVA fibers to the matrix is very strong, resulting in a large number of fibers that cannot be pulled out of the matrix but break directly. Oiling the surface of PVA fibers can significantly reduce the interfacial bond strength between the fibers and the matrix, to achieve the desired multiple-cracking behavior [30]. Compared with PVA fibers, PE fibers have a higher strength and elastic modulus. More importantly, unlike the hydrophilic nature of PVA, PE fibers are hydrophobic, which reduces the chemical bonding between the fibers and the substrate and makes PE fibers less likely to break during the pull-out process. These properties of PE fibers favor defect size tolerance and fiber bridging complementary energy [31,32], with great potential for the preparation of stronger ECCs [33–35]. With increasing PE fiber content, the tensile strain capacity, bending deformation capacity, and fiber bridging capacity increased significantly, while the first cracking stress, peak stress, flexural strength, and fiber bridging strength increased and then decreased, and the composites containing a 1.5% volume fraction of PE fibers had the highest peak stress, compressive strength, and excellent workability [32]. However, under the influence of extreme environments such as high corrosion, the fiber-matrix interface of PE-ECCs may be altered, thus affecting the durability performance of PE-ECCs. First, the PE-ECC is doped with a large number of polymer fibers, and as the PE fiber content increases, the interface area between the fibers and the matrix increases accordingly, providing more channels for ion transport. Second, during the hydration reaction of cement, a water film is easily formed around the polymer fibers, resulting in a high water-cement ratio and increased porosity in the area around the fibers, and the fine cracks of ECC still cannot effectively block the rapid invasion of water molecules and chloride ions under long-term immersion in seawater or extreme environments [36]. Therefore, enhancing the permeation resistance of PE-ECCs can help to improve their durability performance in extremely corrosive environments.

Cementitious Capillary Crystalline Waterproofing materials (CCCWs) are rigid waterproofing materials that offer the advantages of environmental protection, waterproofing, corrosion resistance, and self-healing [37,38]. K. Zheng's studies have shown that the addition of CCCW promotes the hydration of cement and increases the condensation rate within a certain period of time, and that an appropriate amount of CCCW can greatly improve the mechanical and durability properties of cement-based materials [39]. When a CCCW is used as a surface coating, it has little effect on the compressive strength of the concrete, but when a CCCW is used as an admixture, it can greatly improve the compressive strength of the concrete [40]. This is because CCCWs can be uniformly dispersed in the structure during the mixing process, filling the pores through capillary crystallization reactions [41], which makes the structure more dense. Escoffres showed that under constant load, the incorporation of calcium is beneficial for restoring the mechanical properties of high-performance fiber-reinforced concrete, and the synergistic effect of fibers and calcium enables self-healing products to better bind to fibers [42]. Huang argues that when a structure cracks due to external conditions, internal fibers can play a crack resistance role, which helps CCCWs to better exert their performance [43]. When cracks appear in the concrete structure, the active chemicals in the CCCW penetrate into the concrete through the water, which acts as a carrier, which promotes a chemical reaction between the CCCW and cement hydration products and forms water-insoluble crystals, thereby blocking the pores and microcracks, effectively alleviating the penetration of chloride ions, prolonging the service life of the structure, and greatly reducing the maintenance costs caused by corrosion [44,45].

The improvement in structural impermeability imparted by CCCWs is related to the improvement in the internal pore structure, which depends on many factors, such as the CCCW content, composition, and environmental conditions [37]. Zheng et al. [39] showed that the total porosity of the structure decreased with increasing CCCW content. Azarsa et al. [41] demonstrated that CCCW increased the surface resistivity of concrete and reduced the chloride ion flux, thus improving the resistance of concrete to chloride ion penetration. CCCWs were more effective at improving the permeability resistance of the structural matrix as an admixture, compared to CCCW-coated specimens [37]. In general, the addition of a CCCW to concrete can improve the strain capacity of the material [46]. The CCCW has little effect on the material's compatibility, and when the amount of the CCCW is appropriate, it can greatly improve the mechanical properties and durability of cementitious materials [37], while there are differences in the effects of different types of CCCWs on material properties. Addition of a CCCW may increase the types of self-healing products, change their morphology and quantity, and improve the properties of the structure [47]. The types and proportions of self-healing products vary depending on the location of the cracks, and the self-healing products of cracks on the surfaces of structures were mainly $CaCO_3$ [48].

At present, there have been more studies on the role of CCCWs in ordinary concrete, but less research has been conducted on the impact of PE-ECC performance, especially the degree of impact on tensile and impermeability properties, which remains to be studied. Therefore, in this study, ECCs doped with ultrahigh molecular weight polyethylene short-cut fibers (PE-ECCs) were used as a benchmark group, and were mixed with different types and amounts of cementitious capillary crystalline waterproofing materials (CCCWs). Compressive tests, flexural tests, and tensile tests were used to test their mechanical properties, and the permeability height method and rapid chloride migration coefficient method were used to analyze their impermeabilities. Laser particle size distribution, XRD, MIP, and SEM measurements were used to examine the particle size distribution, phase composition, pore structure, and micromorphology of the materials, respectively, providing implications for the protection of structures and buildings under complex loading conditions and in extremely harsh marine environments.

2. Experimental Procedure

2.1. Materials and Mix Ratio

The raw materials prepared for casting specimens included PII 52.5 Portland cement, Class fly ash, fine sand, and polycarboxylic high-range water reducer (HRWR). The chemical compositions of the cement and fly ash, measured by XRF (X-ray fluorescence spectroscopy), are shown in Table 1. The fine sand was ultra-fine sand produced by Shanghai Fengxian Sand Factory, with specifications of 70~110 mesh and a maximum particle size of 0.21 mm. The water reducing agent was the high efficiency powder water reducing agent of polycarboxylic acid, produced by Shanghai Sanrui Company. Ultrahigh molecular weight polyethylene cropped fibers (PE) was selected as the reinforcement. The appearance and microscopic morphology of the PE fibers are shown in Figure 1a,b, the fibers have a disordered distribution. The fiber length was 12 mm, fiber diameter was 25 μm, and the density and elastic modulus of the fiber were 0.97 g/cm^3 and 116 GPa, respectively. The volume fraction of PE fibers in the mix was 1.5%.

Table 1. Chemical compositions of the raw materials.

Ingredients (%)	SiO$_2$	K$_2$O	TiO$_2$	Fe$_2$O$_3$	CaO	Al$_2$O$_3$	SO$_3$
Cement	19.90	0.79	0.21	3.00	64.90	4.42	2.67
Fly ash	51.70	1.40	1.19	5.22	7.65	23.90	0.91

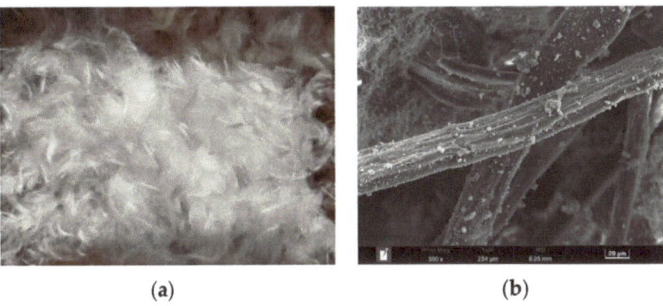

Figure 1. PE fiber: (a) Appearance; (b) SEM microtopography.

XYPEX-type CCCW from Canada, and SY1000-type CCCW developed by Yu Jianying's team from the Wuhan University of Technology in China, were selected. The particle size distributions of the XYPEX-type CCCW and SY1000-type CCCW are presented in Figure 2. For the XYPEX-type CCCW, most of the particle sizes were less than 102.2 μm. Fifty percent and 90% of the particles had sizes of less than 14.82 μm and 44.68 μm, respectively. The most probable distribution of the particle size was 23.96 μm. For SY1000, most of the particle sizes were less than 114.6 μm. Fifty percent and 90% of the particles had sizes less than 38.86 μm and 70.83 μm, respectively. The most probable distribution of the particle size was 46.86 μm. Clearly, the XYPEX-type CCCW is more fine than the SY1000-type.

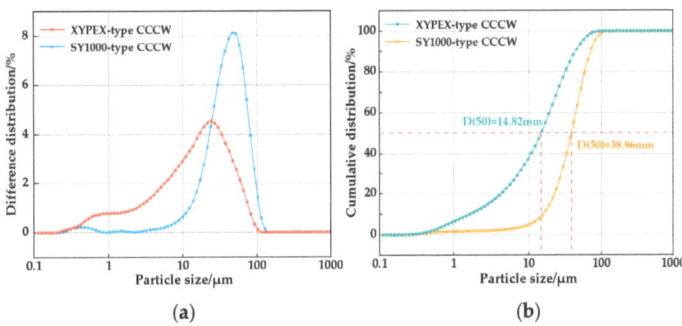

Figure 2. Particle size distribution of the XYPEX-type and SY1000-type CCCWs: (a) Difference particle size distributions; (b) Cumulative particle size distributions.

After 28 days of curing, X-ray diffraction analysis (D8, Advance, Bruker) was performed to identify the chemical compositions of the two different kinds of CCCWs, the phase composition (Figure 3).

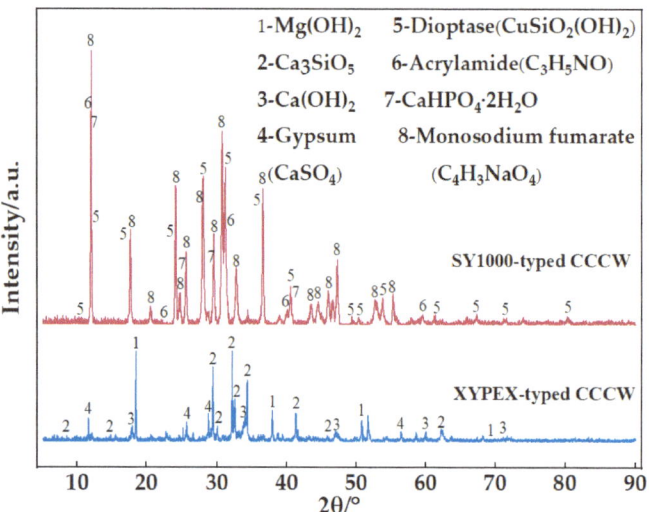

Figure 3. XRD patterns of PE-ECC, CCCW-PE-ECC (X1.0%), and CCCW-PE-ECC (S1.0%).

The main components of the XYPEX-type CCCW included $Ca(OH)_2$, $Mg(OH)_2$, Ca_3SiO_5, and $CaSO_4$. Ca_3SiO_5 reacted with H_2O to produce calcium silicate hydrate (C-S-H) gel, which can effectively fill the cracks and pores in cement-based materials. As the compensators of Ca^{2+}, $Ca(OH)_2$ and $CaSO_4$ provide a large amount of Ca^{2+} as Ca^{2+} complexants, and Ca^{2+} complexants can reduce the activation energy of hydration products reacting with calcium ions. When reaching the area where the cement gel is enriched, due to the different solubilities and stabilities of the products, the anions in the complexing agent will be replaced by silicate and aluminate ions. This can generate a large number of stable insoluble Ca^{2+} complexes in the water environment, promote the healing of cracks in the matrix material, and effectively resist the erosion of chloride ions.

The SY1000-type CCCW mainly contained clairvorite ($CuSiO_2(OH)_2$), acrylamide (C_3H_5NO), and dicalcium hydrogen phosphate ($CaHPO_4 \cdot 2H_2O$), and sodium fumarate ($C_4H_3NaO_4$). $CaHPO_4 \cdot 2H_2O$ was deemed an enhancer of Ca^{2+} as it can provide large quantities of Ca^{2+} in the water. Therefore, more Ca^{2+} complexes tended to be generated, which was conducive to the filling of cracks and pores. C_3H_5NO can not only improve the physical properties of synthetic fibers but also has the ability to prevent corrosion. In combination with $C_4H_3NaO_4$, a new type of antiseptic, it obviously promoted the anticorrosion of the materials. The two advantages, i.e., the favorable formation of Ca^{2+} complexes and anticorrosion effect, demonstrated the better chloride ion resistance of the SY1000-type CCCW.

The mass fractions of CCCW used were set as 0%, 0.5%, 1.0%, 1.5%, and 2.0% of the cementitious material. The mix proportions are listed in Table 2.

Table 2. Mix proportion (kg/m^3) of PE-ECC CCCW-PE-ECC(X) and CCCW-PE-ECC(S).

	Sand	Cement	Fly Ash	Water	HRWR	Fiber	CCCW
PE-ECC	474.4	593.0	711.6	387.1	4.0	14.7	0
CCCW-PE-ECC(X0.5%)	474.4	593.0	711.6	387.1	4.0	14.7	6.52
CCCW-PE-ECC(X1.0%)	474.4	593.0	711.6	387.1	4.0	14.7	13.05
CCCW-PE-ECC(X1.5%)	474.4	593.0	711.6	387.1	4.0	14.7	19.57
CCCW-PE-ECC(X2.0%)	474.4	593.0	711.6	387.1	4.0	14.7	26.09
CCCW-PE-ECC(S0.5%)	474.4	593.0	711.6	387.1	4.0	14.7	6.52
CCCW-PE-ECC(S1.0%)	474.4	593.0	711.6	387.1	4.0	14.7	13.05
CCCW-PE-ECC(S1.5%)	474.4	593.0	711.6	387.1	4.0	14.7	19.57
CCCW-PE-ECC(S2.0%)	474.4	593.0	711.6	387.1	4.0	14.7	26.09

Note: PE-ECC is the reference group, CCCW-PE-ECC(X) indicates that the XYPEX-type CCCW is added. CCCW-PE-ECC(S) indicates that the SY1000-type CCCW is added. CCCW-PE-ECC(X0.5%) indicates that the XYPEX-type CCCW is added and its mass fraction is 0.5% of the cementitious material, CCCW-PE-ECC(S0.5%) indicates that the SY1000-type CCCW is added and its mass fraction is 0.5% of the cementitious material.

2.2. Experimental Programs

An LJ-XLG50E blender was used to mix the raw materials. First, 80% of the water was added to the mixer, and then all the cement, sand, fly ash, and HRWR were added and mixed for 1 min. The remaining 20% of the water was added during the mixing process. After mixing for 3–5 min, PE fibers were gradually added and stirred for 5 min until the fibers were well dispersed. Finally, the fresh PE-ECC mortar was placed into the mold. All the specimens were demolded after one day and cured in a constant temperature and humidity container for 28 days.

The compressive strength and flexural strength were tested using a DYE-300S hydraulic servo loading system. The dimensions of the specimens used for testing were 70.7 mm × 70.7 mm × 70.7 mm and 40 mm × 40 mm × 160 mm, respectively. The tensile tests were conducted using a WDW-100C electronic servo loading system, and the loading rate was set to 1.5 mm/min. As shown in Figure 4a, electronic extensometers were used to measure the elongation.

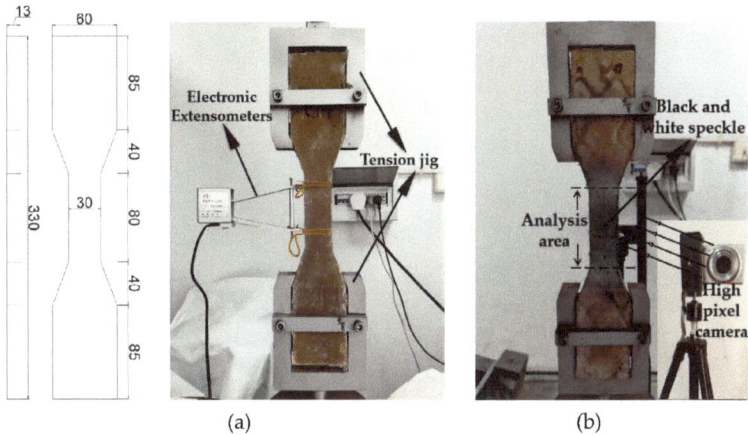

Figure 4. (**a**) Dog-bone specimen dimensions and test device; (**b**) DIC test.

The crack morphology and distribution were tested by digital image correlation (DIC). First, black and white paint was sprayed on the specimen to form a random speckle pattern, as shown in Figure 4b. Then, high-resolution photographs of the target area on the specimen surface were taken every 10 s during uniaxial tensile loading, and finally, the photographs were imported into the MATLAB software to calculate the local strain by tracking the movement of pixels in small areas before and after deformation, to generate strain maps at different loading stages.

The permeability test was performed using the permeability height method, via an SS-15 mortar permeability meter, test equipment, and the specific specimen size, as shown in Figure 5a,b. After the test began, the water pressure was maintained at a constant (1.2 ± 0.05) MPa and maintained for 24 h. At the end of the test the specimens were split into two halves, using a waterproof pen to trace the watermarks and a steel ruler along the watermarks at equal intervals to measure 10 measuring points. Finally, the average value was used as the water penetration height of the group of specimens.

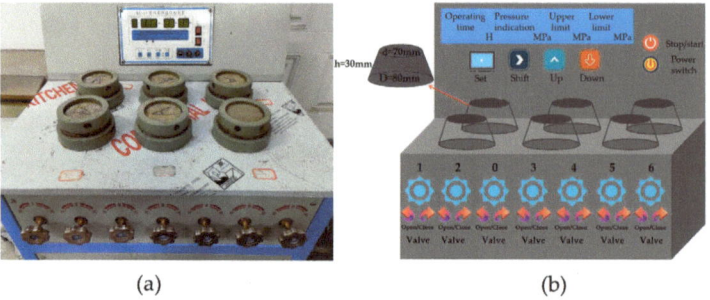

Figure 5. (**a**) SS-15 Mortar penetrometer; (**b**) Schematic diagram.

The chloride ion penetration test was performed based on the rapid chloride-ion migration coefficient method (RCM) stipulated in GB/T50082-2009 [49]. The diameter and height of the cylinder specimens were 100 mm and 50 mm, respectively. Before the test, a water saturation machine was used to vacuum water the specimens for (18 ± 2) h. Approximately 300 mL of NaOH solution, with a concentration of 0.3 mol/L, was injected into the rubber sleeve, and 12 L of NaCl solution, with a mass concentration of 10%, was injected into the cathode test tank and made flush with the liquid level of the NaOH solution in the rubber sleeve, as shown in the test diagram and schematic diagram in Figure 6a,b. After the experiment, the cylinders were cut into halves along the diameter and then sprayed with 0.1 mol/l AgNO$_3$ solutions onto the cut sections. After that, the penetration depth of chloride ions was measured by using a KS-105 wireless electronic meter, as shown in Figure 7a–c.

$$D_{\text{RCM}} = \frac{0.0239 \times (273 + T)L}{(U - 2)t}\left(X_d - 0.0238\sqrt{\frac{(273 + T)LX_d}{U - 2}}\right) \quad (1)$$

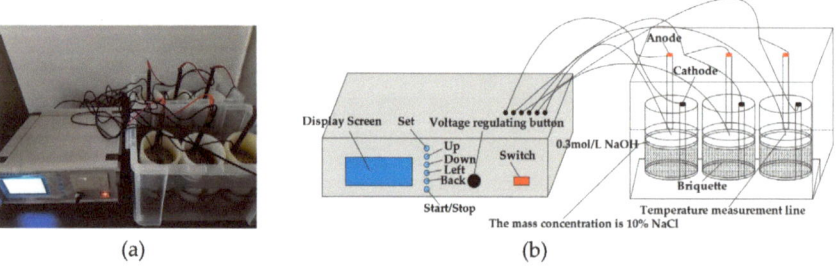

Figure 6. (**a**) Anti-chloride ion penetration test; (**b**) Schematic diagram.

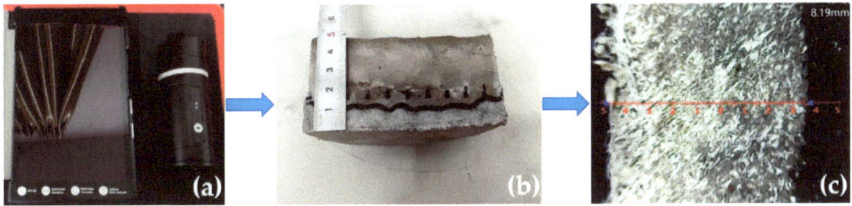

Figure 7. Chloride ion penetration depth measurement: (**a**) KS-105 wireless electronic meter; (**b**) Penetration depth of chloride ions; (**c**) Penetration depth test.

D_{RCM}—Nonstationary chloride ion mobility coefficient; U—absolute value of the voltage used (V); T—average of the initial and ending temperatures of the anode solution (°C); L—specimen thickness (mm); X_d—average value of chloride ion penetration depth (mm); and t—duration of the test (h).

Furthermore, a BT-9300S laser particle size distribution instrument was used to detect the particle size of the two types of CCCWs, and X-ray diffractometry (XRD; Brooke D8 advanced X-ray diffractometer) was used to test the phase composition of the two types of CCCWs, as well as the hydration products of PE-ECC and CCCW-PE-ECC. The PE-ECC and CCCW-PE-ECC pore distributions were determined using an AutoPore Iv 9520 mercury pressure meter. Scanning electron microscopy (SEM) was used to observe the micromorphologies of the PE-ECC and CCCW-PE-ECC specimens before and after chloride ion penetration.

3. Results and Discussion

3.1. Mechanical Properties

The compressive strength and flexural strength of CCCW-PE-ECC(X) and CCCW-PE-ECC(S) are shown in Figure 8a,b. The compressive strength and flexural strength of PE-ECC were increased to a certain extent after mixing two kinds of CCCW separately, because the CCCW contains not only active substances but also a large amount of Ca^{2+} and SiO_3^{2-}, which can generate more calcium carbonate ($CaCO_3$) and calcium silicate hydrate (C-S-H) and other gel products while catalyzing the hydration reaction of cement, filling the pores and cracks. The density of the matrix was improved. The compressive and flexural strengths of CCCW-PE-ECC(X) and CCCW-PE-ECC(S) were enhanced and then weakened with increased CCCW doping. The compressive strengths of CCCW-PE-ECC(X1.0%) and CCCW-PE-ECC(S1.0%) reached maximum values of 53.8 Mpa and 51.3 Mpa, respectively, which are 37.95% and 31.54% higher than those of the PE-ECC. The flexural strengths of CCCW-PE-ECC(X1.0%) and CCCW-PE-ECC(S1.0%) were 11.8 Mpa and 9.5 Mpa, respectively, which are 53.25% and 23.38% higher than those of the PE-ECC. Moreover, the effect of the XYPEX-type CCCW on the compressive and flexural strengths is more obvious. This may be because the particle size of the XYPEX-type is finer than that of the SY1000-type, and the hydration reaction of cement is more efficient, which results in a denser microstructure and matrix-fiber interface.

As shown in Figure 9a, under uniaxial tension, the PE-ECC showed excellent crack control ability. Different from the brittle failure that occurred in ordinary concrete, the failure modes of PE-ECC were ductile failures, extending from one single crack to multiple fine cracks. All the PE-ECCs demonstrated tensile strain-hardening behaviors under tension, indicating that PE-ECC can still bear a higher load with a continuous increase in strain. Additionally, the cracking at the first crack point gradually developed into multiple cracking in the whole range. The microscopic morphology of the matrix material and PE fibers following tensile damage is shown in Figure 9b, where the PE fibers are pulled out instead of being pulled off. These properties of PE fibers favor defect size tolerance and fiber bridging complementary energy. This indicates that the PE fibers can withstand greater deformation and that the material ductility was significantly enhanced.

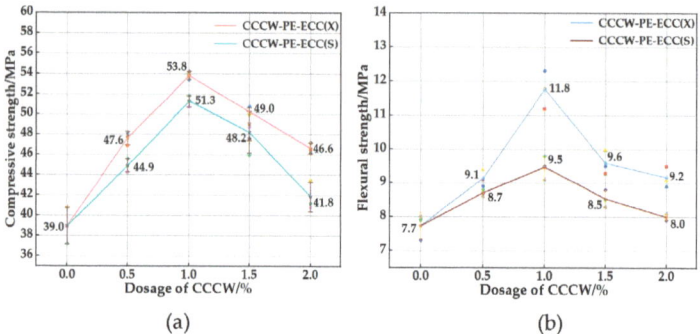

Figure 8. (**a**) Compressive strength; (**b**) Flexural strength.

Figure 9. (**a**) Tensile crack distribution; (**b**) Microscopic morphology at the fracture.

The stress-strain curves of CCCW-PE-ECC(X) and CCCW-PE-ECC(S) are shown in Figure 10a,b. The stress-strain curves can be roughly divided into three stages: elastic stage, crack stable development stage, and local crack expansion fiber pull-out stage. From the beginning of loading to the appearance of the first crack, the stress and strain increase linearly, and in the crack stable development stage the strain hardening phenomenon is presented. Finally, in the local crack expansion fiber pull-out stage, no new cracks are generated, and the local cracks keep expanding until the maximum stress appears and the specimen is pulled out. The ultimate tensile stress and ultimate tensile strain are shown in Figure 11a,b. With increasing CCCW doping, the ultimate tensile stress and ultimate tensile strain of CCCW-PE-ECC(X) and CCCW-PE-ECC(S) show an increasing and then decreasing trend. The ultimate tensile stress and ultimate tensile strain of CCCW-PE-ECC(X1.0%) and CCCW-PE-ECC(S1.0%) reach ultimate tensile stress values of 5.56 N/mm^2 and 5.28 N/mm^2, which are 14.17% and 8.42% increases, respectively, compared with the reference group. The ultimate tensile strains were 7.53% and 7.11%, which are 21.65% and 14.86% increases, respectively, compared with the reference group. The ultimate tensile stress and ultimate tensile strain of CCCW-PE-ECC(X2.0%) and CCCW-PE-ECC(S2.0%) exhibit the minimum values. The ultimate tensile stress values are 4.15 N/mm^2 and 4.27 N/mm^2, which are 14.78% and 12.32% lower than the benchmark group, and the ultimate tensile strains are 5.43% and 5.33%, which are 12.28% and 13.89% lower than the benchmark group, respectively. This shows that the tensile properties of the PE-ECC are improved to a certain extent when the right amount of CCCW is incorporated, while an excessive amount of CCCW reduces the tensile properties of the PE-ECC. When an appropriate amount of CCCW was added, the generated gel products effectively filled the internal pores and cracks of the matrix, enhanced its compactness, promoted the bridging properties between the matrix and fibers, and thus enhanced its tensile properties. In contrast, excessive CCCW led to the generation of excessive hydration products, which caused the internal volume of the matrix to expand to a certain extent and destroyed the balance of mechanical interaction among the fibers, matrix, and composite interface in PE-ECC. The balance of the mechanical interaction among the three in the PE-ECC led to the weakening of its tensile properties.

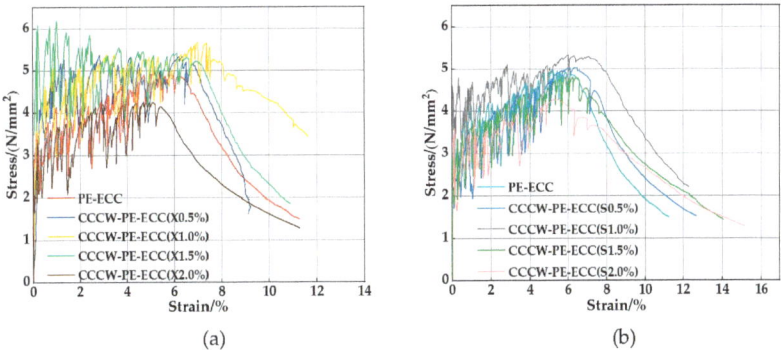

Figure 10. Stress-strain curves of (**a**) CCCW-PE-ECC(X) and (**b**) CCCW-PE-ECC(S).

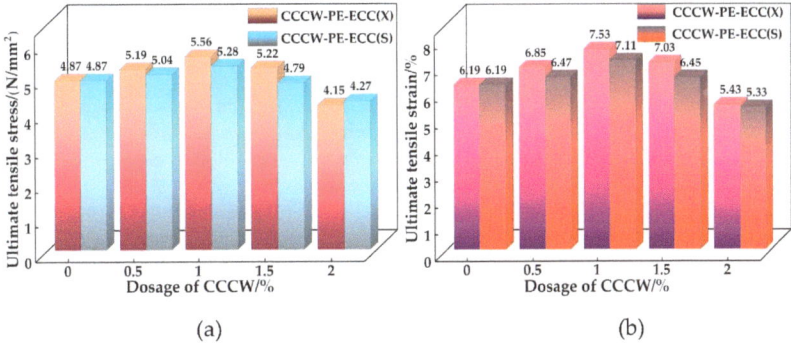

Figure 11. Ultimate tensile stress and ultimate tensile strain of CCCW-PE-ECC(X) and CCCW-PE-ECC(S): (**a**) Ultimate tensile stress; (**b**) Ultimate tensile strain.

The regulation of crack width has a significant influence on improving the mechanical properties and durability of cementitious materials. The standard requires that for structural components exposed to an adverse environment, or specially designed for anti-seepage, the crack width should be less than 200 μm [50]. As shown in Figure 12a–c, uniformly distributed cracks were narrowed when PE-ECC, CCCW-PE-ECC(X), and CCCW-PE-ECC(S) were subjected to tension, and the corresponding average crack widths were 60 μm, 70 μm, 65 μm, respectively. Evidently, all the crack widths were far less than the specification requirements [50], indicating that the aforementioned ECCs had acceptable durability. With the addition of CCCW, the crack spacing of PE-ECC became wider. Additionally, the number of cracks within the 80 mm gauge length of dog-bone-shaped specimens cast with PE-ECC, CCCW-PE-ECC(X), and CCCW-PE-ECC(S) was 42, 63, and 56, respectively. This demonstrates that the appropriate addition of CCCW caused an increase in the crack width and the number of cracks, thus enhancing the tensile deformation capacity.

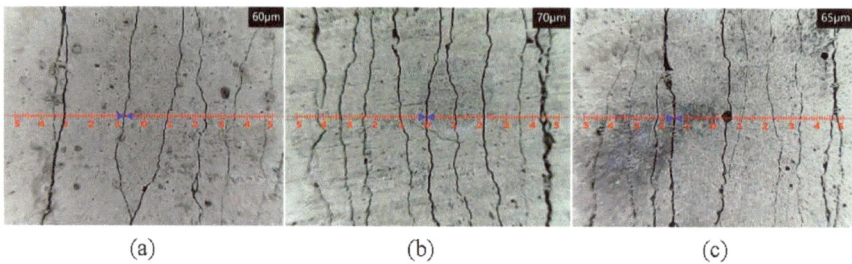

Figure 12. Crack distribution after stretching of (**a**) PE-ECC, (**b**) CCCW-PE-ECC (X1.0%), and (**c**) CCCW-PE-ECC (S1.0%).

The crack morphology and distribution were tested by digital image correlation (DIC). As shown in Figure 13a–c, as the load increases from zero, the tensile stress inside the cement matrix composite increases linearly until it reaches the elastic limit state. When the stress intensity factor is equal to the local matrix fracture toughness, the maximum crack internal defect, or weak zone, causes microcrack extension. When the tensile stress reaches the initial cracking strength, flattened matrix microcracks expand almost instantaneously on the specimen surface. After satisfying the multiple crack energy criterion, the cracks on the specimen surface expand in a steady-state flat crack expansion mode. A slight decrease in tensile stress occurs due to the sudden loss of the load transfer capability of the matrix. However, the fiber bridging stress does not exceed the matrix cracking stress, and as the tensile strain increases, the tensile stress not only recovers but exceeds the initial crack strength. The tensile stress continues to increase until another microcrack appears at the next largest defect. New flattened cracks continue to form and expand until the tensile stress at the weakest cross section of all cracked sections of the specimen exceeds the fiber bridging capacity. Eventually, the damage is concentrated at the weakest part of the bridging action. Compared with PE-ECC, CCCW-PE-ECC(X1.0%) and CCCW-PE-ECC(S1.0%) have a more uniform distribution of the main cracks throughout the middle region. Numerous fine saturated microcracks can be formed between the main cracks, the total deformation in the intermediate region increases, and the ductility is enhanced. Thus, the tensile deformation capacity of the material has been improved.

3.2. Water Seepage Resistance

The water seepage heights of CCCW-PE-ECC(X) and CCCW-PE-ECC(S) are shown in Figure 14. With increasing CCCW doping, the water seepage heights of both CCCW-PE-ECC(X) and CCCW-PE-ECC(S) show a trend of first decreasing and then increasing. CCCW-PE-ECC(X1.0%) and CCCW-PE-ECC(S1.0%) show the smallest permeation heights, 2.6 mm and 2.8 mm, respectively, which are 69.77% and 68.18% lower than those of the baseline group. Both CCCWs improve the impermeability of the PE-ECC to a similar extent. A CCCW can act synergistically with the PE fibers to promote the compactness of the matrix material. The principle is that when the cementitious material is in a dry environment, the active substance in CCCW is in a dormant state inside it, and when it is in a water environment or a wet state, the water pressure prompts the water molecules to penetrate into the pores and cracks inside the cementitious material with the active substance, and the active substance and Ca^{2+} in the material undergo a precipitation reaction to produce insoluble Ca^{2+} precipitate. At this time, part of the precipitate attaches to the surface of the PE fibers, which strengthens the bond between the fibers and the matrix, and the other part fills the pores and cracks, blocks the transmission of water molecules inside the matrix, and improves the water seepage resistance of the cementitious materials. However, excessive CCCW leads to an increase in hydration products and a certain expansion of the matrix volume, resulting in an increase in pores and cracks and a slight decrease in the water seepage resistance.

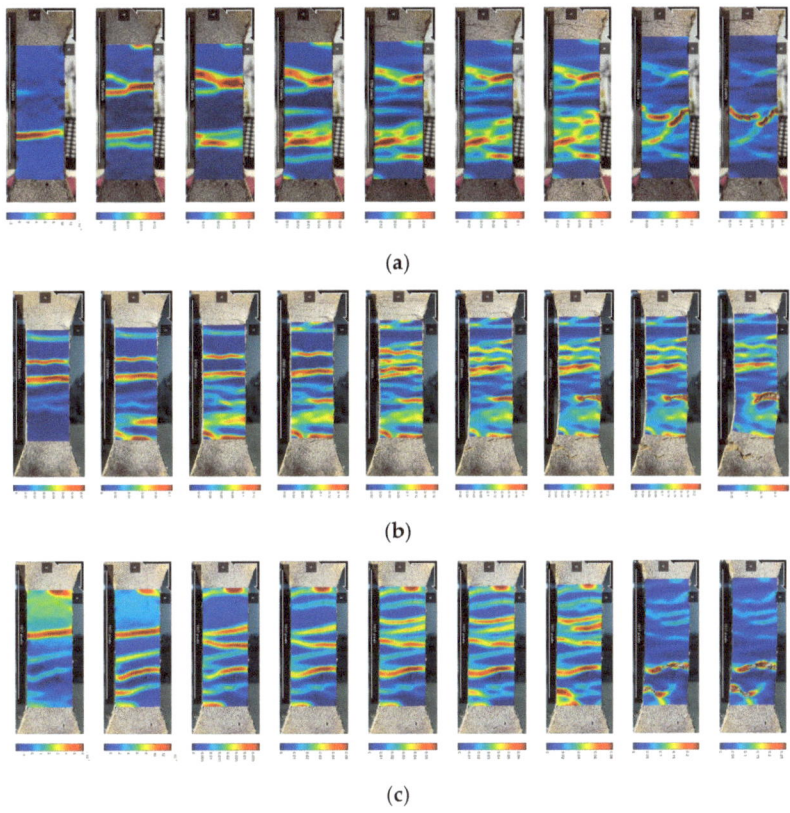

Figure 13. Strain clouds of (**a**) PE-ECC, (**b**) CCCW-PE-ECC(X1.0%) and (**c**) CCCW-PE-ECC(S1.0%).

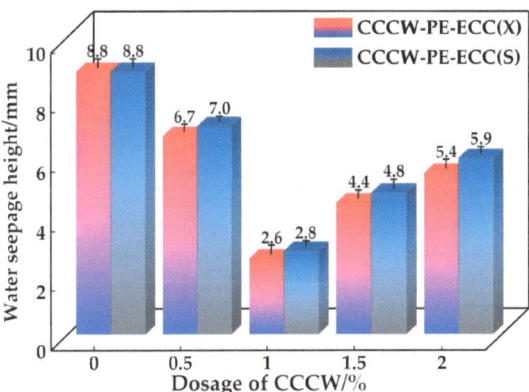

Figure 14. Water seepage height of CCCW-PE-ECC(X) and CCCW-PE-ECC(S).

3.3. Anti-Chloride Ion Penetration Performance

The chloride ion permeation heights and chloride ion diffusion coefficients of CCCW-PE-ECC(X) and CCCW-PE-ECC(S) are shown in Figure 15a,b, respectively. With increasing CCCW doping, the chloride ion permeation heights of CCCW-PE-ECC(X) and CCCW-PE-ECC(S) show a decreasing and then increasing trend, and the chloride ion permeation heights of CCCW-PE-ECC(X1.0%) and CCCW-PE-ECC(S1.0%) show minimum chloride ion permeation heights of 3.13 mm and 2.18 mm, respectively, which are 64.39% and 75.20% lower than the baseline group. The chloride ion diffusion coefficients shows a decreasing and then increasing trend, and CCCW-PE-ECC(X1.0%) and CCCW-PE-ECC(S1.0%) exhibit the smallest values, 0.15×10^{-12} m^2/s and 0.10×10^{-12} m^2/s, respectively, which are 68.75% and 79.17% lower than those of the reference group.

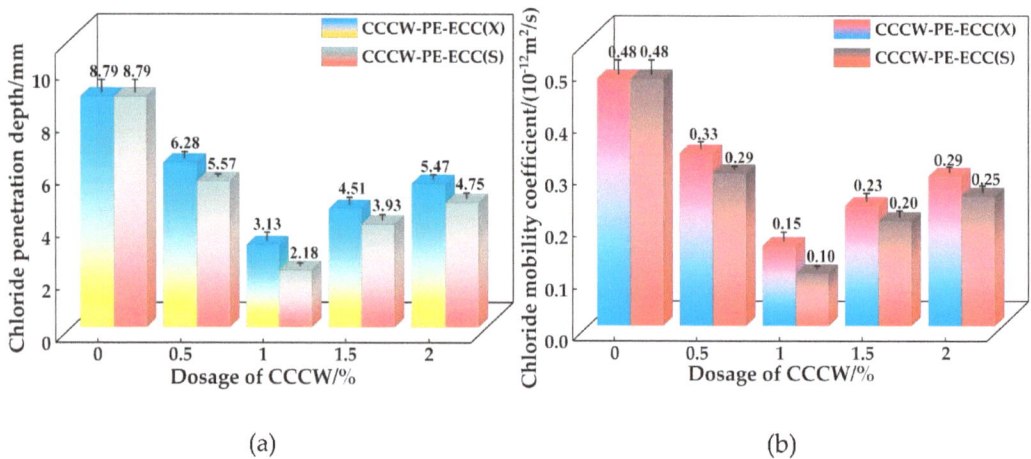

Figure 15. (a) Chloride ion permeation height; (b) Chloride ion mobility coefficient.

The anti-chloride ion permeation performance of PE-ECC is enhanced to a certain extent by incorporating an appropriate amount of CCCW, while the incorporation of excessive CCCW leads to a specific decrease in the anti-chloride ion permeation performance. This is because the incorporation of an appropriate amount of CCCW can effectively promote the hydration reaction of cement, generating a more suitable content of hydration products to fill the pores, and blocking the transmission of chloride ions inside the matrix, effectively inhibiting the erosion of chloride ions and enhancing the resistance to chloride ion penetration. However, an excessive amount of the CCCW will lead to the generation of excessive hydration products, resulting in matrix cracking damage, some pores, and cracks, and if there is an excessive amount of calcium alumina (AFT) in the hydration products, then excessive volume expansion occurs, which can have a negative impact on the cement matrix, thus leading to a decline in the chloride ion penetration resistance. The combined results of the chloride ion permeation heights and chloride ion diffusion coefficients show that the SY1000-type CCCW had better impermeability resistance than the XYPEX-type CCCW. This may be because the SY1000-type CCCW contains some anti-corrosive substances, such as acrylamide (C_3H_5NO) and sodium fumarate ($C_4H_3NaO_4$), which can effectively neutralize Cl^-, reduce the erosion of Cl^- on the matrix material, and effectively improve the material's resistance to chloride ion penetration.

4. Microscopic Analysis

4.1. Phase Composition Testing and Analysis

After 28 days of curing, X-ray diffraction analysis (D8, Advance, Bruker) was performed to identify the chemical compositions of the hydration products with PE-ECC, CCCW-PE-ECC(X1.0%), and CCCW-PE-ECC(S1.0%).

According to Figure 16, the same hydration products, including SiO_2, C-S-H, $CaCO_3$, and ettringite, exist in PE-ECC, CCCW-PE-ECC(X1.0%), and CCCW-PE-ECC(S1.0%), in which SiO_2 accounted for the largest proportions and the intensity diffraction peaks of C-S-H, $CaCO_3$, and ettringite in the CCCW-PE-ECC(X1.0%) and CCCW-PE-ECC(S1.0%) are higher than those in the reference PE-ECC. $CaCO_3$ serves as a nucleation matrix, decreasing the potential barrier of nucleation and thus accelerating cement hydration. C-S-H gel is prone to forming in pores, which can refine the microstructures and then increase the compactness. Despite the existence of the volume expansion effect, the formation of ettringite is not necessarily regarded negatively. The expansion can somehow be helpful in promoting strength development at an early stage, compensating for shrinkage. It is safe to say that the addition of CCCW could promote the hydration reaction of the matrix, due to differences in concentration and pressure, the active substances in the CCCW can penetrate into a matrix with water through microcracks and then react with free lime and oxides in the pores to form an insoluble crystalline substance. In a dry environment, the active chemical precipitates out into its solid form and remains dormant, and when microcracks appear in the matrix, the water re-excites the activity of the active chemical, causing it to continue to diffuse and react until the crack is filled and compacted.

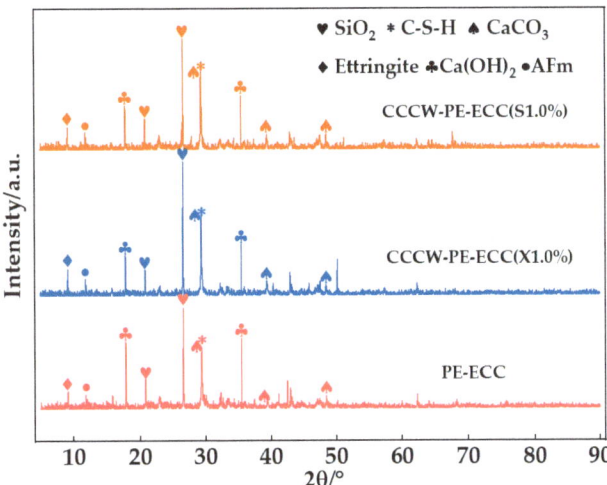

Figure 16. XRD patterns of PE-ECC, CCCW-PE-ECC(X1.0%) and CCCW-PE-ECC(S1.0%).

The peak intensity diffraction of $Ca(OH)_2$ in CCCW-PE-ECC(X1.0%) and CCCW-PE-ECC(S1.0%) was reduced compared to PE-ECC because the active substance in CCCW was able to consume the $Ca(OH)_2$ present in the cement base. This plays an important role in terms of the durability. The chemical reaction to reduce the Ca^{2+} content in substances is beneficial for the strength and compactness of the material, thus improving the overall strength and impermeability of the material.

4.2. Pore Structure Testing and Analysis

CCCWs can optimize the pore structure in EECs by improving their structural compactness [43]. The data obtained for PE-ECC, CCCW-PE-ECC(X1.0%), and CCCW-PE-ECC(S1.0%) are shown in Table 3, where the total pore volume, total pore area, and average pore size of CCCW-PE-ECC(X1.0%) and CCCW-PE-ECC(S1.0%) were decreased, while the bulk density and apparent density were increased, leading to a decrease in permeability. It is implied that the structure of the PE-ECC was more compact after incorporating the CCCW.

Table 3. MIP results of PE-ECC, CCCW-PE-ECC(X1.0%) and CCCW-PE-ECC(S1.0%) for 28 days.

	Total Intrusion Volume (mL/g)	Total Pore Area (m²/g)	Average Pore Diameter (nm)	Bulk Density (g/mL)	Apparent Density (g/mL)	Porosity (%)	Permeability (md)
PE-ECC	0.21	11.90	62.32	1.24	1.54	24.74	476.58
CCCW-PE-ECC(X1.0%)	0.14	9.43	47.54	1.49	2.16	19.32	122.34
CCCW-PE-ECC(S1.0%)	0.16	9.98	53.21	1.40	1.75	20.09	197.61

According to pore size, pores in concrete can be divided into air pores (d > 1 μm), capillaries (10 nm~1 μm), and gel pores (d < 10 nm) [51,52]. As shown in Figure 17a, the total pore volume of CCCW-PE-ECC(X1.0%) was 0.14 mL/g, and the total pore volume of CCCW-PE-ECC(S1.0%) was 0.16 mL/g. While the total pore volume of PE-ECC was 0.21 mL/g, compared with PE-ECC, the total pore volumes of CCCW-PE-ECC(X1.0%) and CCCW-PE-ECC(S1.0%) were reduced by 33.33% and 23.81%, respectively, and the capillary pores and gel pores were greatly reduced, indicating that the internal pore structure of PE-ECC was effectively improved after incorporation of the CCCW, the pores were reduced and the structure was more compact. The pore volume varies with pore size, as shown in Figure 17b. The PE-ECC pore volume varies greatly in the range of stomatal pores, the pore volume of CCCW-PE-ECC(S1.0%) changes greatly in the pore range, and CCCW-PE-ECC(X1.0%) varies greatly in the gel pore range, indicating that the CCCW-PE-ECC(X1.0%) pores were more uniformly distributed in the range of stomatal pores and pores, and the overall pore volume was small. The trend of total pore area with pore size is shown in Figure 17c. When the diameter was greater than 1000 nm, the pore area was almost zero, and the proportion of large pores was very small, indicating that the pores were mainly composed of pores and gel pores. The total pore area of PE-ECC was 11.90 m²/g, and the total pore areas of CCCW-PE-ECC(X1.0%) and CCCW-PE-ECC(S1.0%) were 9.43 m²/g and 9.98 m²/g, respectively, which were 20.76% and 16.13% lower than that of PE-ECC. This indicates that the number of PE-ECC pores decreased after incorporation of the CCCW. The pore density distribution function is shown in Figure 17d. The physical significance of the pore density distribution function is to divide the entire aperture distribution range into several pores of 1 nm. If there is a hole with a certain nanometer size, then the pore capacity value of this hole is expressed in ordinate coordinates. When the pore sizes of PE-ECC, CCCW-PE-ECC(X1.0%), and CCCW-PE-ECC(S1.0%) were 9.06 nm, 9.07 nm, and 5.49 nm, respectively, the pore volume was the largest, and most of the pore structure was capillary and gel pores. When d > 1000 nm, the pore volume was almost zero, indicating that although there are some large pores, the pore depth was small. The pore structure distribution is shown in Figure 17e. The peak of each stage is the most permeable pore size in the aperture range; that is, the pore volume here was the largest, and the peak of CCCW-PE-ECC(X1.0%) in the range of pores and gel pores is less than that of PE-ECC and CCCW-PE-ECC(S1.0%), indicating that its maximum pore volume in each pore size range was small. The measured and predicted values of the total pore volume with pressure are shown in Figure 17f. When the pressure was close to 20,000 Pa, the pores in the material were almost filled with mercury. When the pressure exceeded 20,000 Pa, the total pore volume did not change much.

In summary, after having an appropriate amount of CCCW incorporated into it, the pore structure of PE-ECC was improved. Both the total pore volume and total pore area, or the porosity and permeability, demonstrated decreases. This is because, on the one hand, the CCCW can promote the cement hydration reaction, and the generated $CaCO_3$, C-S-H gel, and calcium alumina (AFT) are beneficial for filling pores and cracks. On the other hand, the active substances in the CCCW can replace Ca^{2+} in unhydrated $Ca(OH)_2$ in the cement and generate more stable C-S-H gel and $CaCO_3$ crystals to fill capillaries and gel pores, improving the pore structure. The XYPEX-type CCCW more obviously improved the pore structure of the PE-ECC.

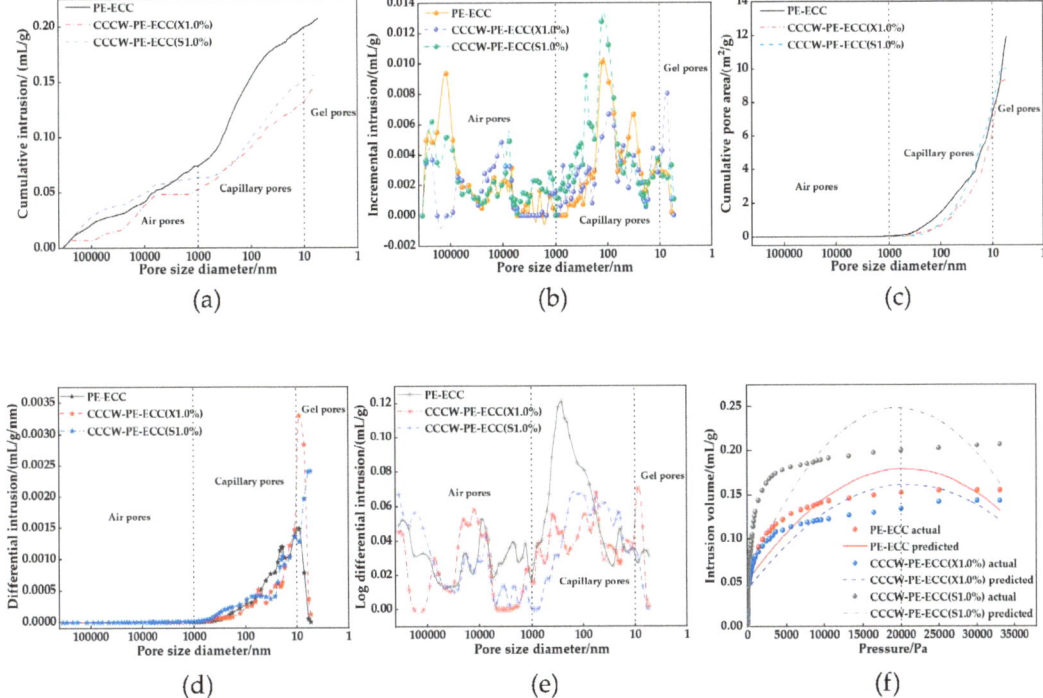

Figure 17. (**a**) Total pore volume vs. pore size; (**b**) Amount of variation in the pore; (**c**) The total pore area varies with pore size; (**d**) Pore size density distribution function; (**e**) Distribution map of the pore structure; (**f**) The total pore volume changes with pressure.

4.3. Micromorphology Test Analysis

The microscopic topographies of PE-ECC, CCCW-PE-ECC(X1.0%) and CCCW-PE-ECC(S1.0%) after chloride ion penetration are presented in Figure 18a–c, respectively. THe XRD results showed that the hydration products may be $CaCO_3$, C-S-H, and AFT on the fiber surface. The amount and volume of hydration products on the surface of the PE-ECC fibers were decreased compared with CCCW-PE-ECC(X1.0%) and CCCW-PE-ECC(S1.0%). Hydration products on a fiber's surface not only promote the connection between the fiber and the matrix but also fill the pores. Due to the reduction of hydration products, the pores originally filled by hydration products will reappear, resulting in the decrease of matrix compactness and the degradation of PE-ECC performance to a certain extent. The hydration products of CCCW-PE-ECC(X1.0%) and CCCW-PE-ECC(S1.0%) are greater in number and more dense than those of the PE-ECC after chloride ion permeation, indicating that they are more resistant to chloride ion attack than PE-ECC, which further indicates that the incorporation of a CCCW can effectively improve the chloride ion permeation resistance of PE-ECC.

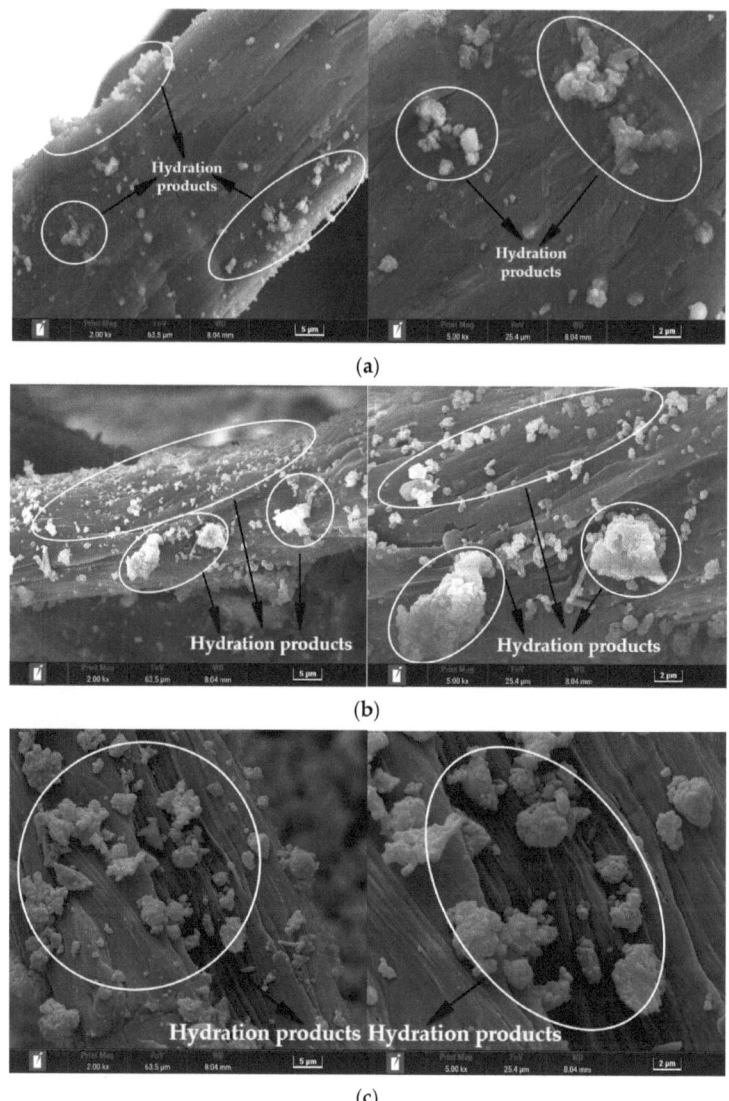

Figure 18. After chloride ion penetration: (**a**) PE-ECC; (**b**) CCCW-PE-ECC(X1.0%); (**c**) CCCW-PE-ECC(S1.0%).

5. Conclusions

To design cementitious composites with excellent mechanical and impermeability properties, the excellent tensile properties of PE-ECC and the unique impermeability properties of CCCW were combined. Different types and doses of CCCW were incorporated into the PE-ECC to study the effect of the CCCW on the mechanical and impermeability properties of the PE-ECC, and to analyze them at the microscopic level. The main conclusions are described as follows:

(1) With increasing CCCW doping, the mechanical properties of the PE-ECC tended to increase first and then decrease, and the mechanical properties were best when the doping amount was 1%. The mechanical properties of the PE-ECC were more

obviously improved by the XYPEX-type CCCW, with a compressive strength of 53.8 MPa, flexural strength of 11.8 MPa, an ultimate tensile stress of 5.56 MPa, and an ultimate tensile strain of 7.53 MPa, which were 37.95%, 53.25%, 14.17%, and 21.65% higher than those of the reference, respectively.

(2) According to crack width meter and DIC analyses, the number of cracks in the middle region of the dog-bone specimens increased, the crack tolerance increased, the distribution was more uniform, and the crack control ability and tensile ductility were enhanced, after incorporating a suitable amount of the CCCW.

(3) With increased CCCW dosing, the seepage resistance of PE-ECC tended to increase and then decrease, and the best performance of PE-ECC was achieved when the dosing was 1%. CCCW-PE-ECC(X1.0%) and CCCW-PE-ECC(S1.0%) showed the smallest permeation heights, 2.6 mm and 2.8 mm, respectively, which are 69.77% and 68.18% lower than that of the baseline. The chloride ion diffusion coefficients of CCCW-PE-ECC(X1.0%) and CCCW-PE-ECC(S1.0%) exhibit the smallest values, 0.15×10^{-12} m^2/s and 0.10×10^{-12} m^2/s, respectively, which are 68.75% and 79.17% lower than that of the reference.

(4) Laser particle size distribution meter analysis showed that the XYPEX-type CCCW particle size was finer than that of SY1000. The XRD analysis showed that both CCCWs, with suitable doping, can enhance the C-S-H gel and CaCO$_3$ intensity diffraction peaks of the PE-ECC and that the enhancement of the XYPEX-type CCCW was more obvious. The enhancement of the XYPEX-type CCCW on the PE-ECC against chloride ion permeation is mainly due to the generation of more hydration products to fill the pores and improve the structural compactness, while the SY1000-type CCCW mainly improves the performance against chloride ion permeation because it contains acrylamide and sodium fumarate, which perform an anti-corrosion function.

(5) MIP and SEM showed that the total pore volume, total pore area, permeability, and porosity of the PE-ECC decreased, and that the structure was more compact, after doping with two suitable doses of CCCW. The improvement in the pore structure of the PE-ECC was more obvious after doping with XYPEX-type CCCW. After doping with CCCW, the surface of the PE-ECC matrix was flatter, and the degree of erosion of hydration products on the PE fiber surface was reduced after chloride ion penetration.

Author Contributions: Conceptualization and methodology, Y.T. and B.Z.; software and validation, B.Z. and J.Y.; formal analysis and investigation, Y.T. and B.Z.; resources and data curation, B.Z. and H.X.; writing—original draft preparation, B.Z.; writing—review and editing, X.L. and J.M.; visualization and supervision, J.Y. and H.X.; project administration, Y.T.; funding acquisition, Y.T. and H.X. All authors have read and agreed to the published version of the manuscript.

Funding: This research was funded by the Innovation Group Project of Hubei Science and Technology Department (No. 2020CFA046), Hubei Provincial Teaching Research Project (2021302) and State Key Laboratory of Bridge Structural Health and Safety (BHSKL19-04-KF).

Institutional Review Board Statement: Not applicable.

Data Availability Statement: The data presented in this study are available on request from the corresponding author upon reasonable request.

Acknowledgments: This research was supported by the Hubei University of Technology.

Conflicts of Interest: The authors declare no conflict of interest.

References

1. Beauchemin, S.; Fournier, B.; Duchesne, J. Evaluation of the concrete prisms test method for assessing the potential alkali-aggregate reactivity of recycled concrete aggregates. *Cem. Concr. Res.* **2018**, *104*, 25–36. [CrossRef]
2. Weinberg, K.; Khosravani, M.R. On the tensile resistance of UHPC at impact. *Eur. Phys. J. Spec. Top.* **2018**, *227*, 167–177. [CrossRef]
3. Liu, P.; Feng, C.; Wang, F.; Gao, Y.; Yang, J.; Zhang, W.; Yang, L. Hydrophobic and water-resisting behavior of Portland cement incorporated by oleic acid modified fly ash. *Mater. Struct.* **2018**, *51*, 38. [CrossRef]

4. Song, J.; Li, Y.; Xu, W.; Liu, H.; Lu, Y. Inexpensive and non-fluorinated superhydrophobic concrete coating for anti-icing and anti-corrosion. *J. Colloid Interface Sci.* **2019**, *541*, 86–92. [CrossRef] [PubMed]
5. Song, J.; Zhao, D.; Han, Z.; Xu, W.; Lu, Y.; Liu, X.; Liu, B.; Carmalt, C.J.; Deng, X.; Parkin, I.P. Super-robust superhydrophobic concrete. *J. Mater. Chem. A* **2017**, *5*, 14542–14550. [CrossRef]
6. Yin, B.; Xu, T.; Hou, D.; Zhao, E.; Hua, X.; Han, K.; Zhang, Y.; Zhang, J. Superhydrophobic anticorrosive coating for concrete through in-situ bionic induction and gradient mineralization. *Constr. Build. Mater.* **2020**, *257*, 119510. [CrossRef]
7. Yeganeh, M.; Omidi, M.; Mortazavi, H.; Etemad, A.; Rostami, M.R.; Shafiei, M.E. Enhancement routes of corrosion resistance in the steel reinforced concrete by using nanomaterials. In *Smart Nanoconcretes and Cement-Based Materials*; Elsevier: Amsterdam, The Netherlands, 2020; pp. 583–599.
8. Qiao, C.; Suraneni, P.; Weiss, J. Flexural strength reduction of cement pastes exposed to $CaCl_2$ solutions. *Cem. Concr. Compos.* **2018**, *86*, 297–305. [CrossRef]
9. Yang, Y.; Zhang, Y.; She, W.; Liu, N.; Liu, Z. In situ observing the erosion process of cement pastes exposed to different sulfate solutions with X-ray computed tomography. *Construct. Build. Mater.* **2018**, *176*, 556–565. [CrossRef]
10. Kayondo, M.; Combrinck, R.; Boshoff, W.P. State-of-the-art review on plastic cracking of concrete. *Constr. Build. Mater.* **2019**, *225*, 886–899. [CrossRef]
11. Li, V.C. On engineered cementitious composites (ECC). *J. Adv. Concr. Technol.* **2003**, *1*, 215–230. [CrossRef]
12. Li, V.C. Tailoring ECC for special attributes: A review. *Int. J. Concr. Struct. Mater.* **2012**, *6*, 135–144. [CrossRef]
13. Li, V.C.; Wu, H.C. Conditions for pseudo-strain-hardening in fiber reinforced brittle matrix composites. *Appl. Mech. Rev.* **1992**, *45*, 390–394. [CrossRef]
14. Li, V.C. *Engineered Cementitious Composites (ECC)—Bendable Concrete for Sustainable and Resilient Infrastructure*; Springer: Berlin/Heidelberg, Germany, 2019.
15. Zhu, H.; Yu, K.Q.; Li, V.C. Sprayable engineered cementitious composites (ECC) using calcined clay-limestone cement (LC3) and PP fiber. *Cem. Concr. Compos.* **2021**, *115*, 103868. [CrossRef]
16. Figueiredo, T.C.S.P.; Curosu, I.; Gonzales, G.L.G.; Hering, M.; Silva, F.D.; Curbach, M.; Mechtcherine, V. Mechanical behavior of strain-hardening cement-based composites (SHCC) subjected to torsional loading and to combined torsional and axial loading. *Mater. Des.* **2021**, *198*, 109371. [CrossRef]
17. Arce, G.A.; Noorvand, H.; Hassan, M.M.; Rupnow, T.; Dhakal, N. Feasibility of low fiber content PVA-ECC for jointless pavement application. *Construct. Build. Mater.* **2021**, *268*, 121131. [CrossRef]
18. Lin, J.X.; Song, Y.; Xie, Z.H.; Guo, Y.C.; Yuan, B.; Zeng, J.J.; Wei, X. Static and dynamic mechanical behavior of engineered cementitious composites with PP and PVA fibers. *J. Build. Eng.* **2020**, *29*, 101097. [CrossRef]
19. Reinhardt, H.W.; Jooss, M. Permeability and self-healing of cracked concrete as a function of temperature and crack width. *Cem. Concr. Res.* **2003**, *33*, 981–985. [CrossRef]
20. Sahmaran, M.; Li, M.; Li, V.C. Transport properties of engineered cementitious composites under chloride exposure. *ACI Mater. J.* **2007**, *104*, 604–611.
21. Miyazato, S.; Hiraishi, Y. Transport properties and steel corrosion in ductile fiber reinforced cement composites. In Proceedings of the Eleventh International Conference on Fracture, Turin, Italy, 20–25 March 2005; pp. 20–25.
22. Zhuang, S.; Wang, Q. Inhibition mechanisms of steel slag on the early-age hydration of cement. *Cem. Concr. Res.* **2021**, *140*, 106283. [CrossRef]
23. Ma, H.; Yi, C.; Wu, C. Review and outlook on durability of engineered cementitious composite (ECC), Review and outlook on durability of engineered cementitious composite (ECC). *Constr. Build. Mater.* **2021**, *287*, 122719. [CrossRef]
24. Kobayashi, K.; Le Ahn, D.; Rokugo, K. Effects of crack properties and water-cement ratio on the chloride proofing performance of cracked SHCC suffering from chloride attack. *Cem. Concr. Compos.* **2016**, *69*, 18–27. [CrossRef]
25. Kobayashi, K.; Kojima, Y. Effect of fine crack width and water cement ratio of SHCC on chloride ingress and rebar corrosion. *Cem. Concr. Compos.* **2017**, *80*, 235–244. [CrossRef]
26. Li, M.; Li, V.C. Cracking and healing of engineered cementitious composites under chloride environment. *ACI Mater. J.* **2011**, *108*, 333–340.
27. Liu, H.; Zhang, Q.; Li, V.; Su, H.; Gu, C. Durability study on engineered cementitious composites (ECC) under sulfate and chloride environment. *Constr. Build. Mater.* **2017**, *133*, 171–181. [CrossRef]
28. Liu, H.; Zhang, Q.; Gu, C.; Su, H.; Li, V. Self-healing of microcracks in engineered cementitious Composites under sulfate and chloride environment. *Constr. Build. Mater.* **2017**, *153*, 948–956. [CrossRef]
29. Pakravan, H.R.; Ozbakkaloglu, T. Synthetic fibers for cementitious composites: A critical and in-depth review of recent advances. *Constr. Build. Mater.* **2019**, *207*, 491–518. [CrossRef]
30. Arain, M.F.; Wang, M.X.; Chen, J.Y.; Zhang, H.P. Study on PVA fiber surface modification for strain-hardening cementitious composites (PVA-SHCC). *Constr. Build. Mater.* **2019**, *197*, 107–116. [CrossRef]
31. Yun, H.D.; Kim, S.W.; Lee, Y.O.; Rokugo, K. Tensile behavior of synthetic fiber-reinforced strain-hardening cement-based composite (SHCC) after freezing and thawing exposure. *Cold Reg. Sci. Technol.* **2011**, *67*, 49–57. [CrossRef]
32. Wang, Y.C.; Liu, F.C.; Yu, J.T. Effect of polyethylene fiber content on physical and mechanical properties of engineered cementitious composites. *Constr. Build. Mater.* **2020**, *251*, 118917. [CrossRef]

33. Yu, J.T.; Ye, J.H.; Zhao, B.; Xu, S.L.; Wang, B.; Yu, K.Q. Dynamic response of concrete frames including plain ductile cementitious composites. *J. Struct. Eng.* **2019**, *6*, 145. [CrossRef]
34. Yu, K.Q.; Wang, Y.C.; Yu, J.T.; Xu, S.L. A strain-hardening cementitious composites with the tensile capacity up to 8%. *Construct. Build. Mater.* **2017**, *137*, 410–419. [CrossRef]
35. Ye, J.H.; Yu, J.T.; Cui, C.; Wang, Y.C. Flexural size effect of ultra-high ductile concrete under different damage and ductility levels. *Cem. Concr. Compos.* **2021**, *115*, 103852. [CrossRef]
36. Yu, J.T.; Dong, F.Y.; Ye, J.H. Experimental study on the size effect of ultra-high ductile cementitious composites. *Construct. Build. Mater.* **2020**, *240*, 117963. [CrossRef]
37. Hu, X.Y.; Xiao, J.; Zhang, Z.D.; Wang, C.H.; Long, C.Y.; Dai, L. Effects of CCCW on properties of cement-based materials: A review. *J. Build. Eng.* **2022**, *50*, 104184. [CrossRef]
38. Zha, Y.; Yu, J.; Wang, R.; He, P.; Cao, Z. Effect of ion chelating agent on self-healing performance of Cement-based materials. *Construct. Build. Mater.* **2018**, *190*, 308–316. [CrossRef]
39. Zheng, K.; Yang, X.; Chen, R.; Xu, L. Application of a capillary crystalline material to enhance cement grout for sealing tunnel leakage. *Construct. Build. Mater.* **2019**, *214*, 497–505. [CrossRef]
40. Cappellesso, V.G.; Dos Santos Petry, N.; Dal Molin, D.C.C.; Masuero, A.B. Use of crystalline waterproofing to reduce capillary porosity in concrete. *J. Build. Pathol. Rehabil.* **2016**, *1*, 9. [CrossRef]
41. Azarsa, P.; Gupta, R.; Biparva, A. Assessment of self-healing and durability parameters of concretes incorporating crystalline admixtures and Portland Limestone Cement. *Cem. Concr. Compos.* **2019**, *99*, 17–31. [CrossRef]
42. Escoffres, P.; Desmettre, C.; Charron, J.P. Effect of a crystalline admixture on the self-healing capability of high-performance fiber reinforced concrete in service conditions. *Constr. Build. Mater.* **2018**, *173*, 763–774. [CrossRef]
43. Huang, W.; Wang, P.; Yin, W.; Zhang, L.; Shu, W. The crack resistance and impermeability properties of mortar with PP fiber and capillary crystalline material and their mechanism. *Adv. Mater. Res.* **2010**, *168–170*, 1381–1387. [CrossRef]
44. Yu, Y.; Sun, T. Polymer-modified cement waterproofing coating and cementitious capillary crystalline waterproofing materials: Mechanism and applications. *Key Eng. Mater.* **2017**, *726*, 527–531. [CrossRef]
45. Li, G.Y.; Huang, X.F.; Lin, J.S.; Jiang, X.; Zhang, X.Y. Activated chemicals of cementitious capillary crystalline waterproofing materials and their self-healing behavior. *Construct. Build. Mater.* **2019**, *200*, 36–45. [CrossRef]
46. Jo, B.W.; Sikandar, M.A.; Baloch, Z.; Khan, R.M.A. Effect of incorporation of self healing admixture (SHA) on physical and mechanical properties of mortars. *J. Ceram. Process. Res.* **2015**, *16*, s138–s143.
47. Zhang, Y.; Yang, J.; Zhang, W.; Shen, C.; Cai, X.; Zuo, L. Microstructure and mechanism analysis of cement-based capillary crystalline waterproofing material. *New Build. Mater.* **2017**, *44*, 68–70. [CrossRef]
48. Sisomphon, K.; Copuroglu, O.; Koenders, E.A.B. Self-healing of surface cracks in mortars with expansive additive and crystalline additive. *Cem. Concr. Compos.* **2012**, *4*, 566–574. [CrossRef]
49. *GB/T50082-2009*; Standard for Long-Term Performance and Durability of Ordinary Concrete. China Building Industry Press: Beijing, China, 2009. (In Chinese)
50. *GB/T 50476-2008*; Design Specification for Durability of Concrete Structures. China Construction Industry Press: Beijing, China, 2008; p. 162. (In Chinese)
51. Chousidis, N.; Ioannou, I.; Rakanta, E. Effect of fly ash chemical composition on the reinforcement corrosion, thermal diffusion and strength of blended cement concretes. *Construct. Build. Mater.* **2016**, *126*, 86–97. [CrossRef]
52. Liu, J.; Ou, G.F.; Qiu, Q.W. Chloride transport and microstructure of concrete with/without fly ash under atmospheric chloride condition. *Construct. Build. Mater.* **2017**, *146*, 493–501. [CrossRef]

Disclaimer/Publisher's Note: The statements, opinions and data contained in all publications are solely those of the individual author(s) and contributor(s) and not of MDPI and/or the editor(s). MDPI and/or the editor(s) disclaim responsibility for any injury to people or property resulting from any ideas, methods, instructions or products referred to in the content.

Article

Towards Highly Efficient Nitrogen Dioxide Gas Sensors in Humid and Wet Environments Using Triggerable-Polymer Metasurfaces

Octavian Danila [1,*] and Barry M. Gross [2,3]

1. Physics Department, University Politehnica of Bucharest, 060042 Bucharest, Romania
2. Optical Remote Sensing Laboratory, The City College of New York, New York, NY 10031, USA
3. NOAA—Cooperative Science Center for Earth System Sciences and Remote Sensing Technologies, New York, NY 10031, USA
* Correspondence: octavian.danila@upb.ro

Citation: Danila, O.; Gross, B.M. Towards Highly Efficient Nitrogen Dioxide Gas Sensors in Humid and Wet Environments Using Triggerable-Polymer Metasurfaces. *Polymers* **2023**, *15*, 545. https://doi.org/10.3390/polym15030545

Academic Editor: Hyeonseok Yoon

Received: 4 January 2023
Revised: 18 January 2023
Accepted: 18 January 2023
Published: 20 January 2023

Copyright: © 2023 by the authors. Licensee MDPI, Basel, Switzerland. This article is an open access article distributed under the terms and conditions of the Creative Commons Attribution (CC BY) license (https:// creativecommons.org/licenses/by/ 4.0/).

Abstract: We report simulations on a highly-sensitive class of metasurface-based nitrogen dioxide (NO_2) gas sensors, operating in the telecom C band around the 1550 nm line and exhibiting strong variations in terms of the reflection coefficient after assimilation of NO_2 molecules. The unit architecture employs a polymer-based (polyvinylidene fluoride—PVDF or polyimide—PI) motif of either half-rings, rods, or disks having selected sizes and orientations, deposited on a gold substrate. On top of this, we add a layer of hydrophyllic polymer (POEGMA) functionalized with a NO_2-responsive monomer (PAPUEMA), which is able to adsorb water molecules only in the presence of NO_2 molecules. In this process, the POEGMA raises its hidrophyllicity, while not triggering a phase change in the bulk material, which, in turn, modifies its electrical properties. Contrary to absorption-based gas detection and electrical signal-based sensors, which experience considerable limitations in humid or wet environments, our method stands out by simple exploitation of the basic material properties of the functionalized polymer. The results show that NO_2-triggered water molecule adsorption from humid and wet environments can be used in conjunction with our metasurface architecture in order to provide a highly-sensitive response in the desired spectral window. Additionally, instead of measuring the absorption spectrum of the NO_2 gas, in which humidity counts as a parasitic effect due to spectral overlap, this method allows tuning to a desired wavelength at which the water molecules are transparent, by scaling the geometry and thicknesses of the layers to respond to a desired wavelength. All these advantages make our proposed sensor architecture an extremely-viable candidate for both biological and atmospheric NO_2 gas-sensing applications.

Keywords: metasurfaces; gas sensors; frequency-selective surface; optical sensing

1. Introduction

Metasurfaces are artificial materials that have the ability to influence and control the electromagnetic field by means of almost any degree of freedom in their construction [1,2]. As a result, a significant number of effects have been observed, some of them recreating the ones in bulk materials and some completely new. The effects that can also be observed in conventional bulk materials are spectral filtering by absorption [3–5], and wavefront and polarization control [6–10]. New, 'exotic' effects that were reported include electromagnetic cloaking [11–13], generalized reflection/refraction [14–16], perfect absorption via anisotropy [17], and enhanced emission by means of epsilon-near-zero materials [18]. Initially, metasurface architectures consisted of a unit cell with a certain pattern: metallic rods, rings, crosses, C-shapes, V-shapes, Y-shapes, or circular and elliptical disks, made of either gold, silver, copper, beryllium, and aluminum were deposited on either a silicon or a silicon oxide (SiO_2) substrate and illuminated with a certain electromagnetic field in the spectral window of interest. The metallic nanostructures favored the creation of surface plasmons which oscillated at a resonance frequency, providing on-demand absorption and spectral filtering. However, for some optical and terahertz applications, such architectures exhibited significant losses in the metallic layers, which, together with heat dissipation and subsequent dilation, offered a relatively unstable resonant behavior. This issue was addressed by the creation of hybrid (metallic-semiconductor) [19] and all-dielectric metasurface architectures [20,21]. In the case of some all-dielectric metasurfaces, the materials and geometries chosen generate a dielectric permittivity that follows a hyperbolic dispersion rather than an elliptical one [22,23], which leads to new ways of manipulating the wavefronts of the reflected and transmitted waves. More recently, adaptive metasurfaces using externally-addressable materials have opened the way for active tuning and increased versatility towards the spectral domain required by the application [24–26]. Metasurfaces can also exhibit sensitivity to certain molecules deposited on them, which can induce a variation in their electric or magnetic properties, leading to a perturbed response with respect to an ideal case. This arrangement generates the possibility to realize both biological [27,28] and atmospheric gas sensors. Of the latter, the broadest attention has been dedicated to the development of CO_2 gas sensors, which can be obtained by layering a thin, CO_2 sensitive polymer on top of the metasurface structure [29] or by mounting the whole metasurface on a resistive heater and observing the absorbed thermal emission [30]. By comparison, much less attention has been dedicated to the high-sensitivity detection of nitrogen dioxide (NO_2), which is a direct result of fuel combustion, and considered a pollutant gas in higher concentrations. In nature, NO_2 composes the ozone layer in the Earth's troposhpere and creates low-concentration nitrate aerosols. While not as efficient as CO_2 or other nitrous oxides (such as N_2O), NO_2 can also have positive radiative forcing effects, making it a greenhouse gas [31]. Recently, a class of NO_2-responsive polymers, such as N-(2-aminophenyl) methacrylamide hydrochloride (NAPMA) [32,33], as well as poly(oligo(ethylene glycol) methyl ether methacrylate) (POEGMA), functionalized with endogenous NO-monomers, such as poly(2-(3-(2-aminophenyl) ureido)ethyl methacrylate) (PAPUEMA) [34], have been reported. Under the presence of NO_2 molecules, the two polymers have been shown to exhibit strong hydrophylic behaviors which, in turn, in the presence of water molecules, induces high gradients in their electric properties such as dielectric permittivity and conductivity [35,36].

In this paper, we report simulations on a metasurface-based nitrogen dioxide gas detector based on a metasurface architecture layered with a PAPUEMA-functionalized POEGMA polymer, tailored to increase their sensitivity to NO_2 molecules by means of the change in electric permittivity. We model the electric properties of the polymer layers when used in conjunction with the metasurface architecture and highlight the frequency response around the central wavelength $\lambda_0 = 1.55$ μm, in the presence of NO_2 and as a function of humidity in the proximity of the sensor. The selection of this spectral window was made for two reasons: firstly, the laser beam is not attenuated by the water droplets in high-humidity conditions, and, secondly, the telecom C band around 1550 nm has a multitude of off-the-shelf components which can be used to create low-cost gas sensors. Recent reports [37], in which InP nanowire arrays are used to monitor NO_2 at room temperature by measuring an electrical signal have provided a significant level of insight in the realization of NO_2 gas sensors in typical ambient environments. Our method differs from the former due to the fact that our sensor is designed for use in humid and wet environments, where the recording of small-level electrical signals can become challenging. The simulation results show that, when appropriately tuned, the architecture shows enhanced detection sensitivity for NO_2 molecules in high-humidity conditions. This makes our proposed architecture suitable for a series of biological and atmospheric NO_2 sensors which use humidity to their advantage, instead of considering it a parasitic effect.

2. Design Considerations

All our envisioned architectures follow the following layering scheme: a polymer-based metasurface having thickness h_{ms} similar to the penetration depth of the incoming radiation is deposited on a gold substrate having thickness h_{sb} much larger than the penetration depth of the incoming radiation. It should be noted that the penetration depth for mid-infrared radiation is $h_{pd} \simeq 0.2$ µm. On top of this, another NO_2-sensitive polymer layer having h_{ply} is deposited. For our study, we chose the POEGMA polymer functionalized with a NO_2-reactive monomer known as PAPUEMA [34]. The thickness of the polymer represents a degree of freedom which is subject to optimization: as the thickness increases, so does the sensitivity to NO_2 molecules, however, as the thickness reaches several times the penetration depth, the interaction with the metasurface pattern layered underneath it is no longer possible, and the whole architecture attains the behavior of the upper surface of the NO_2-sensitive polymer layer [38]. The layering scheme is presented in Figure 1a. Assuming near-optimal polymer layer thickness, the operating principle of the NO_2-based sensor is the following: in the reference configuration, the whole architecture is placed in a humid environment which lacks any NO_2 molecules. This allows the upper layer to absorb trace amounts of water molecules, which contribute to the baseline levels of its dielectric permittivity and electric conductivity. The architecture is illuminated with mid-infrared radiation, which for our case was the doubled CO_2 laser line operating at $\lambda_0 = 5.8$ µm, and the absorption and reflection spectra are recorded. Due to the large thickness of the gold layer, the transmission through the sample is considered negligible. When performing measurements in an environment with NO_2 molecules, their adsorption by the PAPUEMA monomer will trigger a change in the hydrophyllic nature of the POEGMA component layer, allowing it to absorb a considerably-larger amount of water molecules, and, therefore, change its electric parameters over a large scale of values. In the case of POEGMA, the reported change in the values of the relative permittivity was from 8.04 in a 'quasi-dry' state to almost 63 in a 'wet' state [36], accompanied by an increase in its electric conductivity by at least one order of magnitude. The absorption spectra obtained for the cell under test is always compared to the one of the reference cell. The operating principle is presented in Figure 1b. The metasurface architecture extends the range of basic resonator shapes previously reported in [39,40] to incomplete rings, rods, and cylinders fabricated out of polyvinylidene fluoride (PVDF) or polyimide (PI). The geometry comprises of a super-cell of unit geometries (either rings, rods, or cylinders) with different sizes spread symmetrically across the square surface of the super-cell having linear size a. The layouts are presented in Figure 1c for the incomplete rings, Figure 1d for the rods and Figure 1e for the cylinders. The geometric sizes are specific to each architecture so that they produce a significant response in the desired spectral window, and their coordinates (x, y) across the cell are set in such a way that the centers of the shapes are located in points $(a/4, 3a/4)$, $(3a/4, 3a/4)$, $(3a/4, a/4)$, and $(a/4, a/4)$, respectively, to preserve a degree of symmetry to the design.

In terms of material properties, we considered the architecture non-responsive to magnetic fields (i.e., $\mu_r = 1$). Due to the fact that the architecture is subjected to high-frequency radiation, the DC values of the dielectric permittivity and electric conductivity have to be adjusted for the gold substrate. In the study, we have considered a Drude conduction and relaxation model, with DC conductivity $\sigma_{Au,DC} = 4.517 \times 10^7$ S/m. The Drude model then yields an effective conductivity for the gold substrate [41]:

$$\sigma_{Au,eff} = \frac{\sigma_0}{1 + j\omega\tau} \quad (1)$$

where $j = \sqrt{-1}$ is the imaginary unit, $\omega = 2\pi f$ is the angular frequency associated to the radiation frequency f, and $\tau \simeq 27$ fs is the average relaxation time of the gold atoms. From this, assuming the effective relaxation-effect model, the relative permittivity can be calculated as [41]:

$$\epsilon_{r,Au,eff} = \frac{1}{j\omega\epsilon_0}\left(\sigma' + j(\omega\epsilon_0 - \sigma'')\right) \quad (2)$$

where $\sigma' = Re\{\sigma\}$ and $\sigma'' = Im\{\sigma\}$ are the complex components of the electric conductivity, and ϵ_0 is the vacuum permittivity. For dielectrics, the relaxation time is increased from femtoseconds to nanoseconds, and, therefore, the classical "skin-effect" model applies.

In this assumption, the conductivity and permittivity values at THz frequencies can be approximated by the DC or low frequency (kHz to MHz) values. The DC/MHz values for the PVDF and PI polymer elements are recreated from previous works [26,42], and the baseline properties of the POEGMA layer are extracted from [36]. The electrical properties are centralized in Table 1.

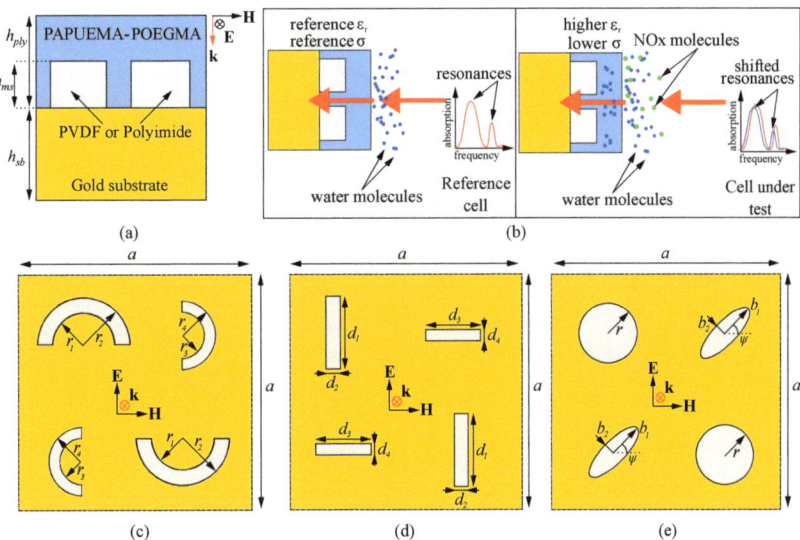

Figure 1. The structure and operating principles of the metasurface-based NO$_2$ gas sensors: (**a**) the layers comprising the architecture with their respective thicknesses; (**b**) example of operating the reference and test metasurfaces: the reference cell (left side) is operated in the absence of NO$_2$ molecules, and the PAPUEMA-POEGMA layer exhibits reference values of the relative permittivity ϵ_r and electric conductivity σ. The test cell (right side) is operated in the presence of NO$_2$ molecules, which increase the hydrophyllic state of the PAPUEMA-POEGMA and considerably increase both ϵ_r and σ. The absorption readout of the cell under test is modified considerably with respect to the reference cell, usually in the form of prominent absorption peaks, if the reference is set to a fully-reflective behavior; (**c**) the incomplete ring metasurface layout; (**d**) the rod metasurface layout; and (**e**) the cylinder metasurface layout.

Table 1. Electrical properties of the polymer elements in the metasurface architecture.

$\epsilon_{r,PVDF}$	σ_{PVDF} (S/m)	$\epsilon_{r,PI}$	σ_{PI} (S/m)	$\epsilon_{r,POEGMA}$	σ_{POEGMA} (S/m)
$9.55 - j \times 0.05$	1.01×10^{-3}	$3.13 - j \times 0.02$	$(2 - j \times 0.01) \times 10^{-9}$	$8.04 - j \times 0.02$	1.5×10^{-10}

To express the variation of POEGMA's dielectric constant and electric conductivity as a function of the trigger rate resulting from the action of the NO$_2$ molecules, we assume a linear dependence of the form:

$$\epsilon_r = \epsilon_{r,dry} + \beta_\epsilon \gamma w \quad \sigma = \sigma_{dry} - \beta_\sigma \gamma w \qquad (3)$$

where $\gamma = N_x/N_w$ is the molar ratio between the NO$_2$ and water molecules in the interaction volume close to the PAPUEMA-POEGMA layer, and w is the air humidity expressed in parts per million (ppm). In the 'wet' state, where we assume $w = 1$, the reported dielectric constant is 68 [36], the calculated permittivity wetting factor β_ϵ is calculated at 59.96 ppm^{-1}. Upon wetting, the conductivity of the POEGMA layer decreases, which is consistent with the increase in permittivity. We also assume a linear dependence $\sigma = \sigma_{dry} - \beta_\sigma w$. In the 'wet' state, the conductivity decreases by a factor of almost 20, which leads to a a conductivity wetting factor $\beta_\sigma = 1.9 \times 10^{-8}$ S·m^{-1}·ppm^{-1}. To model the 'triggering' effect of the wetting in the presence of NO$_2$, we assume that the chance for a NO$_2$-mediated water molecule absorption is proportional to the instant number of NO$_2$ molecules around the POEGMA layer, modified by the triggering efficiency α of a NO$_2$ molecule with the layer. This leads to an exponential rate of triggering, similar to any population rate equation:

$$P(n) = C \exp(\alpha n) \qquad (4)$$

where n is the instant number of NO$_2$ molecules in close proximity of the POEGMA layer. Assuming that the probability of triggering an absorption is close to unity at some large number N of NO$_2$ molecules, the normalizing constant C becomes:

$$C \simeq \left(\int_0^N \exp(\alpha n) dn \right)^{-1} = \frac{\alpha}{\exp(\alpha N) - 1} \qquad (5)$$

Following the two relations above, for $n \simeq N$ sufficiently large, the trigger rate then becomes $P(n) \simeq \alpha$. This is indeed the case for regular atmospheric concentrations of NO$_2$ in urban areas, where the typical values of 50 ppb (parts per billion) is approximately $N_x = 3 \times 10^{16}$ NO$_2$ molecules per mole of air. In our simulations, this value corresponds to 100% humidity. To model imperfect NO$_2$ triggering, we chose conservative values of α at 25% as the NO$_2$ trigger rate. In dry air, the water vapor concentration is approximately zero, whereas in humid air, the concentration can reach $w = 15,740$ ppm [43], corresponding to $N_w = 9.48 \times 10^{21}$ water molecules per mole of air. Combining all the considerations performed under all the assumptions above, the net variation in the values of the electrical parameters in the presence of NO$_2$ is:

$$\epsilon_r = \epsilon_{r,dry} + (\beta_\epsilon \gamma w)\alpha; \quad \sigma = \sigma_{dry} - (\beta_\sigma \gamma w)\alpha \qquad (6)$$

Within this framework, the variation of humidity from zero to one produces variations of the electric permittivity function at the second decimal, however, due to the subwavelength nature of the architecture, this variation is enough to produce significant variations in the response. This increased sensitivity is key in the design and potential applications of our proposed considerations. It is worth mentioning, however, that real implementations of such architectures suffer both from imperfections in the geometrical sizes of the elements, thickness of the layers, and induced trigger rates from the nitrogen dioxide molecules. Moreover, the duty-cycle of any potential sensors relying on this architecture will have to take into account imperfect drying cycles of the PAPUEMA-POEGMA polymer layer, which may lead to long-term deviations from the reference response. These real-life situations, however, can be accounted for with initial and periodic calibration of the sensor, in conjunction with the recording of the device's behavior of the device across multiple wet-dry cycles. Variations in the geometric response affect the response by shifting the resonance peaks from that particular wavelength. Small variations (5–10%) from the reference configuration provide an almost linear shift in the resonance peaks, as indicated by previous studies in polymeric metasurfaces [26]. Depending on the sizes of the metasurface, the shift can be compensated by using any readily available tunable laser source operating in the telecom C-band. Most state-of-the-art nano-imprint and deposition techniques are able to operate within this tolerance. For larger variation, a calibration of the sensor using a wideband spectrometer setup is needed. To account for wetting–drying cycles, we use the results obtained in Ref. [36], in which the PAPUEMA-POEGMA retained water droplets equivalent to 5% humidity over multiple wet–dry cycles. The results obtained indicate that the variation in the resonant response at 5% humidity are negligible with respect to the reference configuration.

In terms of simulation conditions, we used a finite element method (FEM)-based commercial solution, namely COMSOL Multiphysics, RF Module. The solution allows for accurate simulation with iterative error estimation. Meshing of the complete simulation environment was set to a size of $\lambda_0/20$, where $\lambda_0 = 1.55$ μm is the central working wavelength, with an increase in the spatial resolution of narrow regions to nanometer size. The accuracy of the iterative solver was increased in such a way that any solution was considered accurate if the estimated error was below 10^{-6}. To simulate the periodicity of the unit cell, all side walls of the simulation cell were set with Floquet periodicity, and a Floquet wave-vector that was directly taken from the simulation port. Modeling the skin depth of the PVDF and PI elements was performed by attributing a transition boundary condition on the interface between the elements and the gold substrate. Additionally, to save computing time, the back plate of the gold substrate was set to an impedance boundary condition, since no transmission occurs through the architecture.

3. Results and Discussion

Regardless of the configuration used, the first step of our investigation was tuning the geometric sizes and layer thicknesses in such a way as to achieve a significant response close to λ_0 as possible, when the metasurface is in its reference configuration (i.e., dry environment). The spectral behavior was quantified by means of measuring the logarithmic S_{11} parameter (negative return loss factor), defined as:

$$S_{11} = 10 \log \left(\frac{P_{refl}}{P_{in}} \right) \quad (7)$$

where P_{in} and P_{refl} are the input and reflected power levels, respectively. We remind that, typically, the S_{11} parameter is negatively-valued. A 0 dB value of the S_{11} parameter is equivalent to a fully-reflective interface (no return loss), while a negative value represents partial absorption of the input field (positive return loss value). The representation of S_{11} allows more sensitivity in the determination of absorption peaks when compared to standard optical visualizations. Finally, for each situation, we envision an application scenario, where the metasurface is used as a NO_2 gas detector. In these scenarios, the change in response can be quantified via two methods: the first method, known as the wavelength interrogation method, assesses the shift of a certain resonance peak as a function of the increase in humidity levels. The second, known as the signal level interrogation method, assesses the change in the level of the S_{11} parameter at a single frequency as a function of increasing humidity levels. Given a reference state p and a measured state q, each method comes with its own average figure of merit, defined as the rate of change of the interrogated physical quantity (i.e., wavelength or reflection coefficient) as a function of the applied quantity (i.e., humidity level or polarization state) between the two states. Representing as a function of humidity factor w, the figures of merit characterizing the deviation from reference are defined as:

$$FOM_\lambda^{(pq)}(w) = \frac{\lambda_q - \lambda_p}{w_q - w_p}; \quad FOM_S^{(pq)}(w) = \left. \frac{S_q - S_p}{w_q - w_p} \right|_{\lambda_0} \quad (8)$$

with the same formulas being used for the interrogation methods as a function of the change in state of polarization for the input field.

3.1. Half-Ring Architecture

The results for the half-ring structure are presented in Figure 2, in which a strongly resonant response in the desired window is evident, followed by a significant modification in the metasurface behavior as a function of external humidity factor and input field polarization.

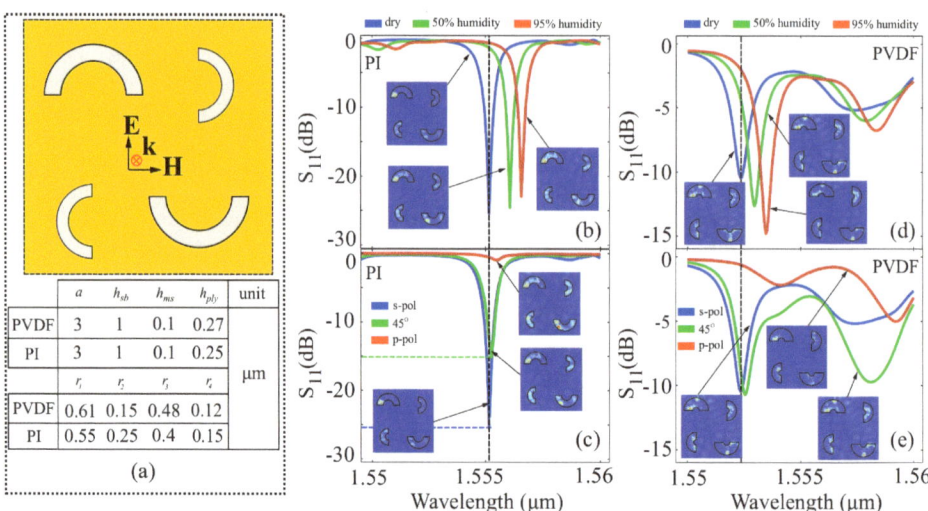

Figure 2. Results obtained for the half-ring structure: (**a**) layout of the ring elements and element sizes chosen for the PVDF and PI in order to have a resonant response around the working wavelength; (**b**) spectral behavior of the PI half-ring element structure at different humidity levels and (**c**) at different input polarization states; and (**d**) spectral behavior of the PVDF half-ring element structure at different humidity levels and (**e**) at different input polarization states. For subfigures (**b**–**e**), the insets show the local field enhancement in the interface plane taken at maximum resonance.

Based on the results obtained above, in the reference configuration shown in Figure 2b, the PI-element metasurface exhibits a strong resonance at 1555 nm, with a 26 dB return loss at the input port, which is well below the 1% reflection threshold. In the presence of NO_2 molecules and 50% external humidity, the resonance peak is shifted towards 1557 nm, and the return loss moves to 23 dB, whereas in the case of 95% external humidity, the resonance peak is shifted at 1558 nm, with a 22 dB return loss. The bandwidth of the resonances is roughly 0.8 nm, which implies that in applications, the spectrum can be readily sampled with off-the-shelf narrowband filters, some of which are exhibiting less than 50 picometer resolution. Assuming level detection is performed via wavelength interrogation, the figures of merit for the wavelength interrogation method are $FOM_\lambda^{(21)}(w) = 1.016$ nm·ppm^{-1} for the dry-to-50% humidity increase scenario and $FOM_\lambda^{(31)}(w) = 1.021$ nm·ppm^{-1} for the dry-to-95% humidity increase scenario. For the signal level interrogation method, the figures of are $FOM_S^{(21)}(w) = 3.2 \times 10^{-3}$ dB·ppm^{-1} for the zero-to-50% humidity increase scenario, and $FOM_S^{(31)}(w) = 3.6 \times 10^{-3}$ dB·ppm^{-1}. When changing the input field polarization, the PI-element metasurface exhibits high sensitivity and low morphing behavior, preserving its absorption peak at 1555.2 nm. Assuming the same behavior in terms of humidity increase, the preservation of the wavelength corresponding to the resonance implies that the wavelength interrogation method can be reliably used with unpolarized radiation. In terms of signal level interrogation, the polarization figures of merit are $FOM_S^{(21)}(SOP) = 15.27$ dB·rad^{-1} for a 45° polarization state rotation, and $FOM_S^{(31)}(SOP) = 15.91$ dB·rad^{-1} for the 90° rotation. For the PVDF element, the chosen geometry provides a resonance response at 1552.5 nm in the reference configuration. As depicted in Figure 2d, the increase in humidity shifts the resonance peak to 1553.5 nm in the case of 50% humidity and to 1554.3 nm for 100% humidity. The corresponding figures of merit are $FOM_\lambda^{(21)}(w) = 1.27 \times 10^{-4}$ nm·ppm^{-1} for the zero-to-50% humidity scenario, and $FOM_\lambda^{(31)}(w) = 1.2 \times 10^{-4}$ nm·ppm^{-1} for the zero-to-100% humidity scenario. Regarding signal-level interrogation taken at reference configuration resonance frequency, the signal increases from −11 dB to −4 dB and −2.5 dB for the 50% and 100% humidity scenarios, respectively. The corresponding figures of merit are $FOM_S^{(21)}(w) = 8.9 \times 10^{-4}$ dB·ppm^{-1} for the dry-to 50% humidity shift, and $FOM_S^{(31)}(w) = 6.35 \times 10^{-4}$ dB·ppm^{-1} for the dry-to-95% humidity. In terms of polarization behavior, shown in Figure 2e, the PVDF-element metasurface is sensitive to input polarization in the reference configuration. The resonance peak at 1552.5 nm does not shift, however it is reduced drastically when the input polarization state is close to p-polarization. The structure also exhibits a second resonance for an input polarization of 45°, which reaches −10 dB at around 1557.5 nm. Since the morphing behavior of the spectrum is quite strong, a wavelength-based figure of merit does not provide any relevant information, since resonance peaks are not preserved for all polarization states. For a single wavelength, however, the signal level interrogation method can still be applied. Taking the reference resonance peak wavelengths, the figures of merit are $FOM_S^{(21)}(SOP) = -0.127$ dB·rad^{-1} for a 45° rotation, and $FOM_S^{(31)}(SOP) = 6.04$ dB·rad^{-1}. As a common property of all half-ring structures under study, it can be seen that regardless of the variational parameter (humidity or input field polarization) the local field enhancement maps in the insets show a localization of the oscillating plasmon in the ring structure, which confirm the strong resonances observed.

3.2. Rod Architecture

In the case of the rod architecture, the PVDF and PI elements were arranged in an equivalent manner to the half-rings. The structure and element sizes are presented in Figure 3a.

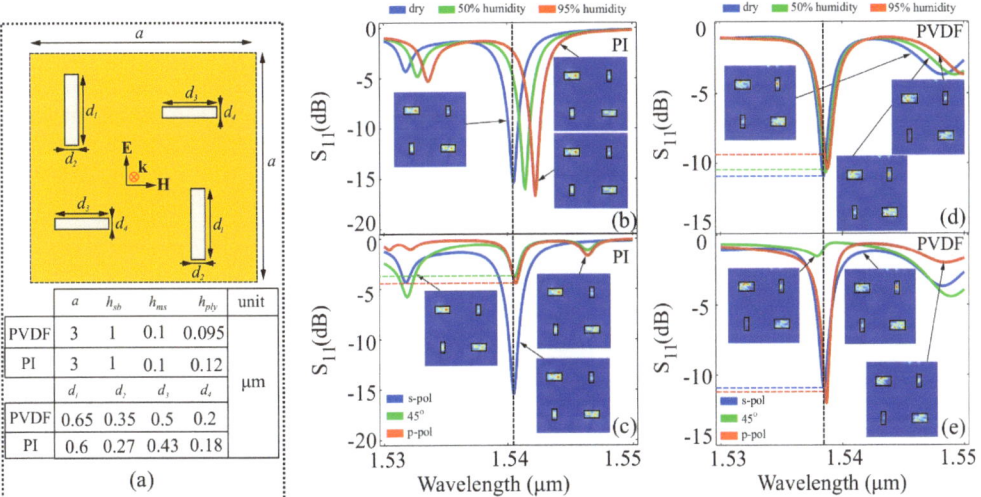

Figure 3. Results obtained for the rod-based structure. (**a**) Layout of the ring elements and element sizes chosen for the PVDF and PI in order to have a resonant response around the working wavelength; (**b**) spectral behavior of the PI rod-element structure at different humidity levels and (**c**) at different input polarization states; and (**d**) spectral behavior of the PVDF rod-element structure at different humidity levels and (**e**) at different input polarization states. For subfigures (**b**–**e**), the insets show the local field enhancement in the interface plane taken at the resonance peak.

Simulations performed on the PI element structure indicate that for an appropriate element size and analyte height configuration, a strong resonance, with a return loss of 15.3 dB is observed at 1540.5 nm, in the reference configuration. The resonance is reasonably narrow, with a FWHM bandwidth of approximately 1.5 nm. When increasing external humidity to 50%, the resonance peak shifts towards 1542 nm, while maintaining its FWHM, and attaining a return loss of 16 dB. For 95% humidity, the peak is shifted towards 1543 nm, with an associated return loss of 16.5 dB and negligible FWHM modification. The response is presented in Figure 3b. When applying the two interrogation methods, the figure merits are calculated as $FOM_\lambda^{21}(w) = 1.9 \times 10^{-4}$ nm·ppm^{-1} and $FOM_\lambda^{31}(w) = 1.67 \times 10^{-4}$ nm·ppm^{-1} for the resonance peak interrogation method, and $FOM_S^{(21)}(w) = 0.88 \times 10^{-4}$ dB·ppm^{-1} and $FOM_S^{(31)}(w) = 0.88 \times 10^{-4}$ dB·ppm^{-1}. For polarization sensitivity measurements of the reference configuration, presented in Figure 3c, the resonances have negligible shifts, however, the signal suffers strong return loss attenuation, from -15.3 dB to -4.6 dB in the case of a $45°$ rotation, and -4.1 dB in the case of a $90°$ rotation. The calculated figures of merit are: $FOM_\lambda^{(21)}(SOP) \simeq FOM_\lambda^{(31)}(SOP) < 10^{-6}$ nm·rad^{-1}. Signal interrogation figures of merit, however, are $FOM_S^{(21)}(SOP) = 13.62$ dB·rad^{-1} for a $45°$ rotation, and $FOM_S^{(31)}(SOP) = 14.26$ dB·rad^{-1} for a $90°$ rotation. For the PVDF-element architecture, the effect of humidity in shifting the resonance peaks is relatively negligible, as presented in Figure 3d. The resonance peak shifts from 1539.2 nm to 1539.25 nm for an increase in humidity from zero to 50%, and from 1539.2 nm to 1539.27 nm for an increase in humidity from zero to 95%. The associated figures of merit for the resonance wavelength interrogation method are $FOM_\lambda^{(21)}(w) \simeq FOM_\lambda^{(31)}(w) < 10^{-6}$ dB·ppm^{-1} for both humidity shift scenarios. In terms of signal interrogation, the return loss factor varies from -10.8 dB to -10.3 dB for the dry-to-50% humidity shift and to -9.3 dB for the dry-to-95% humidity shift, respectively. The associated figures of merit are $FOM^{(21)} = 6.35 \times 10^{-5}$ dB·ppm^{-1} for the dry-to-50% humidity shift, and $FOM^{(31)} = 1.002 \times 10^{-4}$ dB·ppm^{-1} for the dry-to-95% humidity shift. The polarization-dependent response of the PVDF presented in Figure 3e exhibits significant changes. Specifically, for a horizontal and vertical input SOP, the behavior of the metasurface remains unchanged, with a negligible shift in the resonance peak, and a slight signal level modification. However, for a $45°$ SOP, the response has a relatively-small resonance peak, which also exhibits Fano-like asymmetry. For the wavelength interrogation method, the figures of merit are virtually negligible, as the resonance peak moves only slightly. For the signal level interrogation method, due to the fact that the response of the metasurface is not changing monotonously across the polarization spectrum, the figures of merit do not offer relevant information. The insets of Figure 3b–e also show the local field enhancement in the interface plane taken at the resonance peak. It can be seen that for the majority of cases, the electric field enhancement is localized in the PI and PVDF elements, which is in total agreement with theoretical models regarding dielectric resonators.

3.3. Disk Architecture

Just as before, we have adjusted the dimensions of the disk elements so that resonances are produced in the desired spectral window. The structure and element sizes are presented in Figure 4a.

For the PI disk-element architecture, we observe a strong resonance peak at 1558.6 nm, with a return loss of 32.5 dB. As it can be inferred from Figure 4b. The spectral response retains its shape when varying the humidity factor, with an associated shift in the resonance peaks which does not exhibit monotony. When increasing humidity to 50%, the resonance peak does not shift, and only has its return loss attenuated to 12.5 dB. When increasing humidity to 95%, the resonance peak shifts to 1560 nm, and the shift is associated with a return loss factor of 28.5 dB. The polarization-dependent response, shown in Figure 4c, shows that the reference configuration suffers considerable modifications when the SOP is switched to $45°$ and $90°$, respectively. For both SOP rotations, the spectrum becomes flatter, with resonances up to -15 dB, as well as shifted, with the highest resonances being observed at 1557 nm and 1552 nm, respectively. Due to the fact that the modification of the spectral response is not monotonous, calculating the figures of merit does not offer relevant information on the behavior of the metasurface. The humidity-dependent behavior of the PVDF disk-element architecture, presented in Figure 4d, exhibits a resonance peak at 1562.5 nm, and a highly-monotonous morphing behavior as a function of humidity. The main resonance peaks reaching -20 dB are shifted to 1563.9 nm, at 50% humidity, and to 1564.4 nm at 95%

humidity. The response also exhibits another resonance peak reaching −10 dB at 1564 nm, which also has high morphing monotony as the humidity factor is increased. The figures of merit for the wavelength interrogation method are: $FOM_\lambda^{21}(w) = 1.28 \times 10^{-4}$ nm·ppm^{-1}, and $FOM_\lambda^{31}(w) = 1.06 \times 10^{-4}$ nm·ppm^{-1}. Assuming linearity in the shift, an average figure of merit associated to the resonance wavelength is $FOM_\lambda(w) = 1.17 \times 10^{-4}$ nm·ppm^{-1}. The same technique can be applied to determine the figures of merit corresponding to the wavelength shift of the smaller resonance peak obtained in at 1565.2 nm for the reference configuration, and at 1565.4 nm and 1565.7 nm for the 50% and 95% humidity configurations, respectively. For the signal interrogation method taken at reference resonance wavelength, the signal varies significantly, from −20 dB in the reference configuration, to −6 dB and −3 dB in the 50% and 95% humidity configurations, respectively. The associated figures of merit are $FOM_S^{21}(w) = 1.77 \times 10^{-3}$ dB·ppm^{-1} and $FOM_S^{31}(w) = 1.13 \times 10^{-3}$ dB·ppm^{-1}, with an average figure of merit $FOM_S(w) = 1.45$ dB·ppm^{-1}. When looking at the polarization picture, we observe a strong blueshift in the resonance when the reference metasurface is illuminated with a 45° polarization, and virtually no resonance behavior when a vertical input polarization is used. The same behavior is observed for the smaller resonances at 1565.5 nm, where the resonance is either shifted or disappears completely, depending on the input polarization. Due to the fact that the wavelength shift is not monotonous in terms of polarization, a linear figure of merit cannot be defined. In terms of signal interrogation taken at the main resonance wavelength of the reference configuration, the resonance decreases from −20 dB to −5 dB in the case of the 45° input polarization configuration, and to 0 dB in the case of the vertically polarized input field configuration. The associated figures of merit are: $FOM_S^{21}(SOP) = 19.1$ dB·rad^{-1}, and $FOM_{31} = 12.73$ dB· rad^{-1}, with an average figure of merit $FOM_S(SOP) = 15.9$ dB· rad^{-1}. The local field amplification taken at resonance frequencies and presented in the insets of Figure 4b–e show that for the majority of the cases, the local field is concentrated in the PI/PVDF elements, which supports theoretical assumptions regarding the appearance of resonances in such structures.

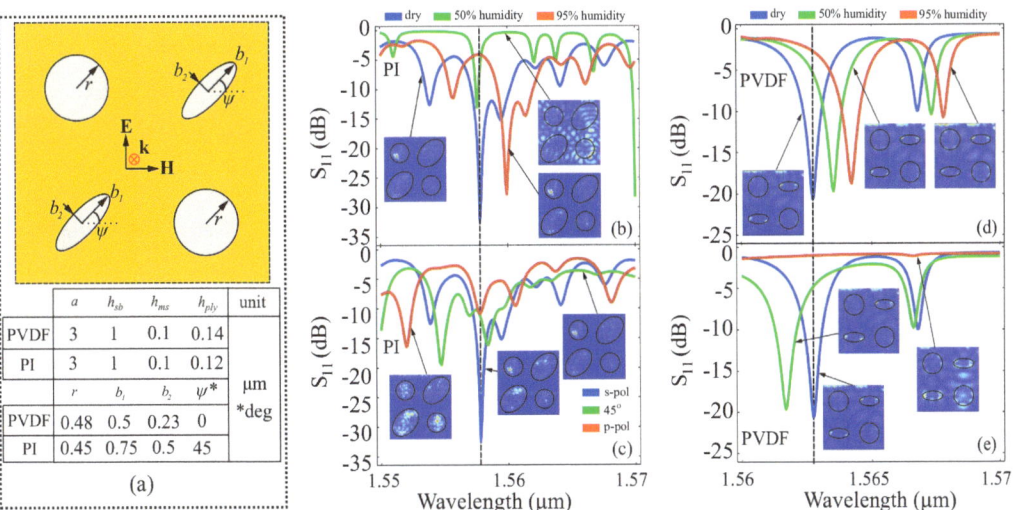

Figure 4. Results obtained for the disk-element architecture. (**a**) Layout of the ring elements and element sizes chosen for the PVDF and PI in order to have a resonant response around the working wavelength; (**b**) spectral behavior of the PI disk-element structure at different humidity levels and (**c**) at different input polarization states; and (**d**) spectral behavior of the PVDF disk-element structure at different humidity levels and (**e**) at different input polarization states. For subfigures (**b**–**e**), the insets show the local field enhancement in the interface plane taken at the resonance peak.

As a final note, in terms of angular and polarization stability, all envisioned architectures have been designed to operate under normal incidence of the input field, under linear polarization. A small variation in the incidence angle (less than 5°) produces negligible modification in the result. However, a small modification of the input polarization breaks the in-plane asymmetry, resulting in a high polarization sensitivity. This limitation can be overcome by providing a high-stability polarization laser via high-quality polarizers.

4. Conclusions and Outlook

In this paper, we have created and conducted simulations on a NO_2 molecule-triggered responsive metasurface architecture in the near-infrared spectral domain around 1550 nm, which have the ability to use external humidity to its advantage. The architecture combines a metasurface layer made from a gold substrate with either a polyvinylidene fluoride (PVDF) or polyimide (PI) subwavelength element configuration, in the shape of either half-rings, rods or disks, with a NO_2-responsive functionalized PAPUEMA-POEGMA layer. This layer has been proven to exhibit significant changes in its dielectric constant in the presence of NO_2 molecules and humidity, by changing its hydrophyllic state. Owing to its functionalization with the PAPUEMA monomer, the POEGMA polymer is able to change its hydrophillic state only in the presence of NO_2 molecules. The results obtained show that even in the conservative assumption of an imperfect trigger rate and small variation (at the second decimal) of the dielectric constant due to humidity increase, the subwavelength nature of the architecture is tuned to be extremely sensitive both to humidity increase and to polarization. To offer some rough quantitative characterization of the response modifications we introduced the wavelength and signal interrogation methods certain figures of merit, that, in the monotonous response variation assumption, offer information on the resonance shift and signal variation taken at the reference resonance. Real devices have to take into account, however any deviation from this ideal configuration, in terms of metasurface element and layer thickness size, relative displacement, as well as the imperfect restoration of the dielectric constant across multiple wetting and drying cycles. This impediment can be resolved, however, by conducting an initial and periodic calibration of the sensor thus designed. Overall, we believe the studies conducted here to be a promising stepping stone in the development of highly sensitive, low-cost NO_2 sensors in the near-infrared telecom C band, around the 1550 nm wavelength.

Author Contributions: Conceptualization, O.D. and B.M.G.; methodology, O.D.; software, O.D.; validation, O.D.; formal analysis, O.D.; investigation, O.D.; resources, B.M.G.; data curation, O.D.; writing—original draft preparation, O.D.; writing—review and editing, O.D. and B.M.G.; visualization, O.D.; supervision, B.M.G.; project administration, B.M.G.; funding acquisition, O.D. and B.M.G. All authors have read and agreed to the published version of the manuscript.

Funding: This research was funded by the Romanian-US Fulbright Commission, throught the Visiting Scholar Grant No. 783/2022. The APC was funded by the Polytechnic University of Bucharest, through the PubArt Programme.

Institutional Review Board Statement: Not applicable.

Informed Consent Statement: Not applicable.

Data Availability Statement: Not applicable.

Acknowledgments: O.D. acknowledges the Romanian—US Fulbright Commission Scholar Visiting grant no 783/2022.

Conflicts of Interest: The authors declare no conflict of interest.

References

1. Pendry, J.B.; Holden, A.J.; Stewart, A.J. Extremely low frequency plasmons in metallic mesostructures. *Phys. Rev. Lett.* **1997**, *78*, 4289–4292. [CrossRef]
2. Pendry, J.B.; Holden, A.J.; Robbins, D.J.; Stewart, W.J. Magnetism from conductors and enhanced nonlinear phenomena. *IEEE Trans. Microwave Theory Tech.* **1999**, *47*, 2075–2084. [CrossRef]
3. Munk, B.A. *Frequency Selective Surfaces: Theory and Design*; Wiley Online Library, Wiley Interscience: New York, NY, USA, 2000.

4. Chiu, C.N.; Chang, K.P. A novel miniaturized-element frequency selective surface having a stable resonance. *IEEE Antennas Wirel. Propag. Lett.* **2009**, *8*, 1175–1177. [CrossRef]
5. Li, D.; Li, T.W.; Li, E.P.; Zhang, Y.J.A. A 2.5-D angularly stable frequency selective surface using via-based structure for 5G EMI shielding. *IEEE Trans. Electromagn. Compat.* **2018**, *60*, 768–775. [CrossRef]
6. Pendry, J.B.; Ramakrishna, S.A. Near-field lenses in two dimensions. *J. Phys. Condens. Matter* **2002**, *14*, 7409–7416. [CrossRef]
7. Abdelrahman, A.H.; Elsherbeni, A.Z.; Yang, F. Transmission phase limit of multilayer frequency-selective surfaces for transmitarray designs. *IEEE Trans. Antennas Propag.* **2014**, *62*, 690–697. [CrossRef]
8. Li, B.; Shen, Z. Three-dimensional dual-polarized frequency selective structure with wide out-of-band rejection. *IEEE Trans. Antennas Propag.* **2014**, *62*, 130–137. [CrossRef]
9. Zhang, B.; Zhang, Y.; Duan, J.; Zhang, W.; Wang, W. An omnidirectional polarization detector based on a metamaterial absorber. *Sensors* **2016**, *16*, 1153. [CrossRef]
10. Dorrah, A.H.; Rubin, N.A.; Zaidi, A.; Tamagnone, M.; Capasso, F. Metasurface optics for on-demand polarization transformations along the optical path. *Nat. Photon.* **2021**, *15*, 287–296. [CrossRef]
11. Pendry, J.B.; Schurig, D.; Smith, D.R. Controlling electromagnetic fields. *Science* **2006**, *312*, 1780–1782. [CrossRef]
12. Schurig, D.; Mock, J.J.; Justice, B.J.; Cummer, S.A.; Pendry, J.B.; Starr, A.F.; Smith, D.R. Metamaterial electromagnetic cloak at microwave frequencies. *Science* **2006**, *314*, 977–980. [CrossRef]
13. Leonhardt, U. Optical conformal mapping. *Science* **2006**, *312*, 1777–1780. [CrossRef]
14. Pendry, J.B. Negative refraction makes a perfect lens. *Phys. Rev. Lett.* **2000**, *85*, 3966–3969. [CrossRef]
15. Yu, N.; Genevet, P.; Kats, M.A.; Aieta, F.; Tetienne, J.P.; Capasso, F.; Gaburro, Z. Light propagation with phase discontinuities: generalized laws of reflection and refraction. *Science* **2011**, *334*, 333–337. [CrossRef]
16. Danila, O.; Manaila-Maximean, D. Bifunctional metamaterials using spatial phase gradient architectures: Generalized reflection and refraction considerations. *Materials* **2021**, *14*, 2201. [CrossRef]
17. Liang, Y.; Koshelev, K.; Zhang, F.; Lin, H.; Lin, S.; Wu, J.; Jia, B.; Kishvar, Y. Bound states in the continuum in anisotropic plasmonic metasurfaces. *Nano Lett.* **2020**, *20*, 6351–6356. [CrossRef]
18. Minerbi, E.; Sideris, S.; Khurgin, J.B.; Ellenbogen, T. The role of epsilon near zero and hot electrons in enhanced dynamic THz emission from nonlinear metasurfaces. *Nano Lett.* **2022**, *22*, 6194–6199. [CrossRef]
19. Danila, O. Spectroscopic assessment of a simple hybrid Si-Au cell metasurface-based sensor in the mid-infrared domain. *J. Quant. Spectr. Rad. Tans.* **2020**, *254*, 107209. [CrossRef]
20. Zhan, A.; Colburn, S.; Trivedi, R.; Fryett, T.K.; Dodson, C.M.; Majumdar, A. Low-contrast dielectric metasurface optics. *ACS Photon.* **2016**, *3*, 209–214. [CrossRef]
21. Arbabi, A.; Horie, Y.; Bagheri, M.; Faraon, A. Dielectric metasurfaces for complete control of phase and polarization with subwavelength spatial resolution and high transmission. *Nat. Nanotechnol.* **2015**, *10*, 937–943. [CrossRef]
22. Gomez-Diaz, J.S.; Alù, A. Flatland optics with hyperbolic metasurfaces. *ACS Photon.* **2016**, *3*, 2211–2224. [CrossRef]
23. Mekawy, A.; Alù, A. Hyperbolic surface wave propagation in mid-infrared metasurfaces with extreme anisotropy. *J. Phys. Photonics* **2021**, *3*, 034006. [CrossRef]
24. Vassos, E.; Churm, J.; Feresidis, A. Ultra-low-loss tunable piezoelectric-actuated metasurfaces achieving 360 and 180 dynamic phase shift at millimeter-waves. *Sci. Rep.* **2020**, *10*, 15679. [CrossRef] [PubMed]
25. Zhu, H.; Patnaik, S.; Walsh, T.F.; Semperlotti, F. Nonlocal elastic metasurfaces: Enabling broadband wave control via intentional nonlocality. *Proc. Natl. Acad. Sci. USA* **2020**, *117*, 26099–26108. [CrossRef] [PubMed]
26. Danila, O. Polyvinylidene fluoride-based metasurface for high-quality active switching and spectrum shaping in the terahertz G-band. *Polymers* **2021**, *13*, 1860. [CrossRef] [PubMed]
27. Tseng, M.L.; Jahani, Y.; Leitis, A.; Altug, H. Dielectric metasurface enabling advanced optical biosensors. *ACS Photon.* **2021**, *8*, 47–60. [CrossRef]
28. Iwanaga, M. All-dielectric metasurface fluorescence biosensors for high-sensitivity antibody/antigen detection. *ACS Nano* **2020**, *14*, 17458–17467. [CrossRef]
29. Kazanskiy, N.L.; Butt, M.A.; Khonina, S.N. Carbon dioxide gas sensor based on polyhexamethylene biguanide polymer deposited on silicon nano-cylinders metasurface. *Sensors* **2021**, *21*, 378. [CrossRef]
30. Miyazaki, H.T.; Kasaya, T.; Iwanaga, M.; Choi, B.; Sugimoto, Y.; Sakoda, K. Dual-band infrared metasurface thermal emitter for CO_2 sensing. *Appl. Phys. Lett.* **2014**, *105*, 121107. [CrossRef]
31. Edenhofer, O.; Pichs-Madruga, R.; Sokona, Y.; Farahani, E.; Kadner, S.; Seyboth, K.; Adler, A.; Baum, I.; Brunner, S.; Eickeimeier, P.; et al. *Intergovernmental Panel on Climate Change AR-5 Synthesis Report 2014: Mitigation of Climate Change*; IPCC: Geneva, Switzerland, 2014.
32. Hu, J.; Whittaker, M.R.; Duong, H.; Li, Y.; Boyer, C.; Davis, T.P. Biomimetic polymers responsive to a biological signaling molecule: nitric oxide triggered reversible self-assembly of single macromolecular chains into nanoparticles. *Angew. Chem. Int. Ed.* **2014**, *53*, 7779–7784. [CrossRef]
33. Draghi, E.; Gupta, A.; Blayac, S.; Saunier, S.; Benaben, P. Characterization of sintered inkjet-printed silicon nanoparticle thin films for thermoelectric devices. *Phys. Status Solidi A* **2014**, *211*, 1301–1307. [CrossRef]
34. Hu, J.; Whittaker, M.R.; Duong, H.; Li, Y.; Boyer, C.; Davis, T.P. The use of endogenous gaseous molecules (NO and CO_2) to regulate the self-assembly of dual-responsive triblock copolymer. *Polym. Chem.* **2015**, *6*, 2407–2415. [CrossRef]

35. Gavrilova, N.D.; Vorob'ev, A.V.; Malyshkina, I.A.; Makhaeva, E.E.; Novik, V.K. Effect of change in the physical properties of water at its peculiar temperature points on the dielectric behavior of sodium polyacrylate. *Polym. Sci. Ser. A* **2016**, *58*, 33–41. [CrossRef]
36. Joh, D.Y.; McGuire, F.; Abedini-Nassab, R.; Andrews, J.B.; Achar, R.K.; Zimmers, Z.; Mozhdehi, D.; Blair, R.; Albarghouthi, F.; Oles, W.; et al. Poly(oligo(ethylene glycol) methyl ether methacrylate) brushes on high-k metal oxide dielectric surfaces for bioelectrical environments. *ACS Appl. Mater. Interfaces* **2017**, *9*, 5522–5529. [CrossRef]
37. Wei, S.; Li, Z.; Murugaappan, K.; Li, Z.; Zhang, F.; Saraswathyvilasam, A.G.; Lysevych, M.; Tan, H.H.; Jagadish, C.; Tricoli, A.; et al. A self-powered portable nanowire array gas sensor for dynamic NO_2 monitoring at room temperature. *Adv. Mater.* **2022**, *1*. [CrossRef]
38. Tabassum, S.; Nayemuzzaman, S.K.; Kala, M.; Mishra, A.K.; Satyendra, K.M. Metasurfaces for sensing applications: Gas, bio and chemical. *Sensors* **2022**, *22*, 6896. [CrossRef]
39. Hasan, D.; Lee, C. Hybrid metamaterial absorber platform for sensing of CO_2 gas at mid-IR. *Adv. Sci.* **2018**, *5*, 1700581. [CrossRef]
40. Tan, X.; Zhang, H.; Li, J.; Wan, H.; Guo, Q.; Zhu, H.; Liu, H.; Yi, F. Non-dispersive infrared multi-gas sensing via nanoantenna-integrated narrowband detectors. *Nat. Comm.* **2020**, *11*, 5245. [CrossRef]
41. Zhou, Y.; Lucyszyn, S. HFSS modelling with THz metal-pipe rectangular waveguide structures at room temperature. *PIERS Online* **2009**, *5*. [CrossRef]
42. Chisca, S.; Sava, I.; Musteata, V.E.; Bruma, M. Dielectric and conduction properties of polyimide films. In Proceedings of the 2011 International Semiconductor Conference, Sinaia, Romania, 17–19 October 2011.
43. Techniques HFT. Typical Composition of the Earth's Atmosphere in PPM. 2020. Available online: www.huntingdonfusion.com (accessed on 3 January 2023).

Disclaimer/Publisher's Note: The statements, opinions and data contained in all publications are solely those of the individual author(s) and contributor(s) and not of MDPI and/or the editor(s). MDPI and/or the editor(s) disclaim responsibility for any injury to people or property resulting from any ideas, methods, instructions or products referred to in the content.

Achieving High Thermal Conductivity and Satisfactory Insulating Properties of Elastomer Composites by Self-Assembling BN@GO Hybrids

Xing Xie [1,2] and Dan Yang [1,*]

1. College of Materials Science and Engineering, Beijing University of Chemical Technology, Beijing 100029, China
2. College of New Materials and Chemical Engineering, Beijing Institute of Petrochemical Technology, Beijing 102617, China
* Correspondence: danyang@buct.edu.cn

Abstract: With increasing heat accumulation in advanced modern electronic devices, dielectric materials with high thermal conductivity (λ) and excellent electrical insulation have attracted extensive attention in recent years. Inspired by mussel, hexagonal boron nitride (hBN) and graphene oxide (GO) are assembled to construct mhBN@GO hybrids with the assistance of poly(catechol-polyamine). Then, mhBN@GO hybrids are dispersed in carboxy nitrile rubber (XNBR) latex via emulsion coprecipitation to form elastomer composites with a high λ and satisfactory insulating properties. Thanks to the uniform dispersion of mhBN@GO hybrids, the continuous heat conduction pathways exert a significant effect on enhancing the λ and decreasing the interface thermal resistance of XNBR composites. In particular, the λ value of 30 vol% mhBN@GO/XNBR composite reaches 0.4348 W/(m·K), which is 2.7 times that of the neat XNBR (0.1623 W/(m·K)). Meanwhile, the insulating hBN platelets hinder the electron transfer between adjacent GO sheets, leading to satisfactory electrical insulation in XNBR composites, whose AC conductivity is as low as 10^{-10} S/cm below 100 Hz. This strategy opens up new prospects in the assembly of ceramic and carbonaceous fillers to prepare dielectric elastomer composites with high λ and satisfactory electrical insulation, making them promising for modern electrical systems.

Keywords: thermal conductivity; insulating properties; surface modification; self-assembling; dielectric

1. Introduction

With the high degree of integration and miniaturization of electronic devices, a large amount of heat is accumulated in a small volume of the electronic components in operation. In that regard, efficient heat dissipation has emerged as a critical issue that urgently needs to be addressed to ensure the service life, reliability, and stability of electronic devices [1–4]. Polymeric composites are widely utilized as thermal management materials in electronic packing and engineering due to their excellent characteristics, including high resistivity, low cost, lightweight, and easy processing [5,6]. Elastomers, in particular, are usually used as thermal pads or thermal greases [6–8]. Due to its large amounts of strong polar groups, the carboxy nitrile rubber (XNBR), which shows a relatively high dielectric constant, has been considered as a candidate thermal management material in electron components. Nevertheless, the intrinsic thermal conductivity (λ) of elastomers is usually low (0.16–0.22 W/(m·K)), which does not meet the requirements for heat transfer in advanced electronic components [9–11].

In order to increase the λ of polymers, an effective solution consists of incorporating high thermally conductive fillers into the polymeric matrix, such as ceramics, metals, and carbon materials [2,12,13]. Hexagonal boron nitride (hBN) has attracted considerable attention from researchers due to its high λ and excellent electrical insulation. Unsatisfactorily, the pristine hBN is poorly compatible with polymers and exhibits a nonuniform dispersion within the polymeric matrix, thereby restraining the growth of λ in the final composites [14,15]. In recent years, the carbonaceous fillers with high electrical conductivity, such as graphene and graphene oxide (GO), have been shown to be able to increase the λ of polymeric composites at a low loading [16]. Nevertheless, the application of such fillers is unavoidably limited in fields where electrical insulation is required. Therefore, designing and preparing dielectric polymeric composites with high λ and outstanding electrical insulation still remain a challenge [17].

Blocking the electron transmission between neighboring carbonaceous fillers by depositing insulating layers or particles is a reliable way to enhance the λ of the polymeric composites while maintaining their electrical insulating properties [15,18]. Shen et al. [15] utilized silica nanoparticles to coat graphene nanoplatelets (GNPs) to construct Silica@GNPs complexes. Then, the obtained Silica@GNPs complexes were incorporated into a polydimethylsiloxane (PDMS) matrix to fabricate Silica@GNP/PDMS composites whose λ value (0.497 W/(m·K)) was found to be 155% that of the neat PDMS (0.195 W/(m·K)). Nevertheless, the electrical resistivity of PDMS composites was still as high as 10^{13} Ω·cm, ensuring their excellent insulation. Guo et al. [2] first functionalized Al_2O_3 nanoparticles by γ-aminopropyltriethoxysilane (denoted as f-Al_2O_3) and then covered them with GO to form f-Al_2O_3@RGO hybrids, which were afterward used as thermally conductive fillers for improving the λ of the nanofibrillated cellulose (NFC) matrix. Given the synergistic effect of both Al_2O_3 and RGO, the in-plane λ of the as-prepared f-Al_2O_3@RGO/NFC composite could be increased to 8.3 W/(m·K) at the RGO and f-Al_2O_3 loadings of 30 wt% and 5.6 wt%, respectively. This result was 2075% of the neat NFC (0.4 W/(m·K)). Meanwhile, the measured electrical resistivity of f-Al_2O_3@RGO/NFC composites exceeded 10^9 Ω·cm, thus meeting the requirements for electrical insulation.

Inspired by marine mussels, Messersmith et al. [19] discovered that dopamine containing catechol and amine functional groups can self-polymerize to form strongly adhesive polydopamine (PDA) in the alkaline buffer aqueous solution. Nevertheless, dopamine is unsuitable for widespread use because of its high cost. Therefore, a cheaper substance as a replacement for the expensive dopamine must be found. Luckily, the inexpensive catechol can react with the economical polyamine to synthesize poly(catechol/polyamine) (PCPA) which exhibits adhesion properties similar to PDA [20].

In this work, hBN and GO are assembled to construct mhBN@GO hybrids with the assistance of PCPA to improve the λ of XNBR. Meanwhile, the insulating hBN effectively prevented the connection between adjacent GO sheets, leading to satisfactory electrical insulation for the XNBR composites. The as-prepared XNBR composites might be promising thermal management materials in the future electronic industry.

2. Experimental

2.1. Materials

The XNBR latex (Zeon International Trading Co., Ltd., Tokyo, Japan) was chosen as the polymeric matrix. The hBN platelets were obtained from Dandong Rijin Technology Co., Ltd., Dandong, China. The GO sheets were purchased from Chengdu Organic Chemicals Co., Ltd., Chengdu, China. Polyamine (Tetraethylenepentamine, TEPA) and catechol were provided by Bailingwei Technology Co., Ltd., Beijing, China. Dicumyl peroxide (DCP), tris-acid, and other reagents were produced by Maclean's Reagent Co., Ltd., Beijing, China.

2.2. Preparation Process of mhBN@GO/XNBR Composites

The fabrication method of mhBN@GO/XNBR composites is illustrated in Figure 1. At the beginning, 5 g of hBN were dispersed in 250 mL of deionized water, and the pH of the mixed solution was adjusted to 9.5 with tris-acid. Then, 0.75 g of catechol and 0.25 g of TEPA were added to the above solution and mechanically stirred at 40 °C for 3 h to obtain the modified hBN hybrids (denoted as mhBN). Subsequently, the GO sheets were dispersed into the blend, which was afterward stirred at 60 °C for 8 h. Finally, the as-obtained mixture was vacuum filtered, washed with deionized water, and vacuum-dried at 60 °C overnight to obtain mhBN@GO hybrids.

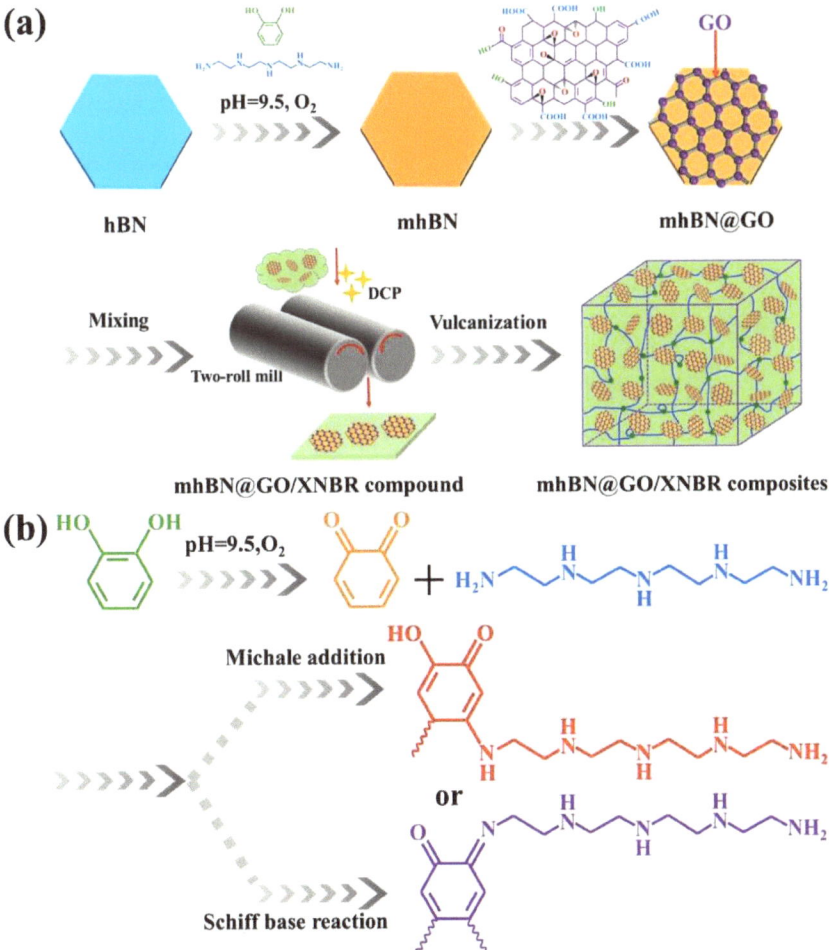

Figure 1. (**a**) Preparation of mhBN@GO hybrids and mhBN@GO/XNBR composites. (**b**) The reaction mechanism between catechol and tetraethylenepentamine.

After that, different contents of hBN or mhBN@GO (the volume fractions of 10, 20, and 30 vol%) were incorporated into the XNBR latex via the emulsion coprecipitation method. Initially, the XNBR latex and filler were stirred for 30 min to achieve a uniform dispersion. Then, 1 wt% $CaCl_2$ was added as a flocculant to the solution. Next, the mixture was water-washed and vacuum-dried at 60 °C to obtain dry compounds. Finally, the compounds and a certain amount of DCP were blended in a double roll open mixer. After 12 h, all as-prepared compounds were compressed by molding at 160 °C and 12 MPa for their vulcanizing time (T_{C90}) to obtain the cured XNBR composites. The T_{C90} values were based on the rheometer (M-3000AU) test data.

2.3. Characterization

The thermogravimetric curves were acquired using a thermogravimetric analysis instrument (TGA, TA SDT650, TA Instruments, New Castle, DE, USA). The surface chemical compositions and surface morphologies of hBN, mhBN, and mhBN@GO were analyzed by means of an X-ray photoelectron spectrometer (XPS, ESCALAB 250, Thermo Electron Corporation, Madison, WI, USA) and a high-resolution transmission electron microscope (HR-TEM, Hitachi H9000, Hitachi Instrument, Tokyo, Japan), respectively. The cross-section morphologies of the XNBR composites were observed by a JSM-5600LV FE-SEM (Thermo Fisher Scientific, Hillsboro, OR, USA). The mechanical properties of the XNBR composites were evaluated using a tensile machine (Instron-3366, Instron Corporation, Norwood, MA, USA). The dielectric behaviors of the composites were studied with a Concept 40 dielectric property tester at room temperature. The λ of the XNBR samples were measured using a flat-panel thermal conductivity meter (DRL-III, Xiangyi Instrument, Xiangtan, China).

3. Results and Discussion

Figure 1a shows the preparation process of the mhBN@GO/XNBR composites. The hBN was first modified by PCPA (denoted as mhBN), which was then assembled with GO to synthesize the mhBN@GO hybrids. Next, the as-prepared mhBN@GO was dispersed into the XNBR matrix to obtain mhBN@GO/XNBR composites via a latex mixing method. A possible reaction mechanism for the PCPA has been added to Figure 1b. First, the catechol was oxidized into quinoid structures in an alkalescent tris-acid buffer solution, and then the generated quinoid structures react with polyamine through Michael addition or Schiff base reactions to form an intermolecularly cross-linked PCPA network, which is finally deposited on the surfaces of the hBN platelets [21].

The TGA curves of hBN, mhBN, and mhBN@GO are shown in Figure 2. It can be seen that the hBN displayed a small weight loss in the range of 30–800 °C, which is ascribed to the excellent thermal stability of hBN. However, a slight weight loss of 2.71% was observed in the TGA curve of mhBN at 600 °C. The larger weight loss of mhBN compared with hBN might be attributed to the degradation of the PCPA layer at the high temperature. In addition, there was also a significant weight loss of 11.64% in the TGA curve of mhBN@GO within the range of 30–600 °C. The increased weight loss of mhBN@GO compared with mhBN was due to the pyrolysis of unstable oxygen-containing groups in GO [22].

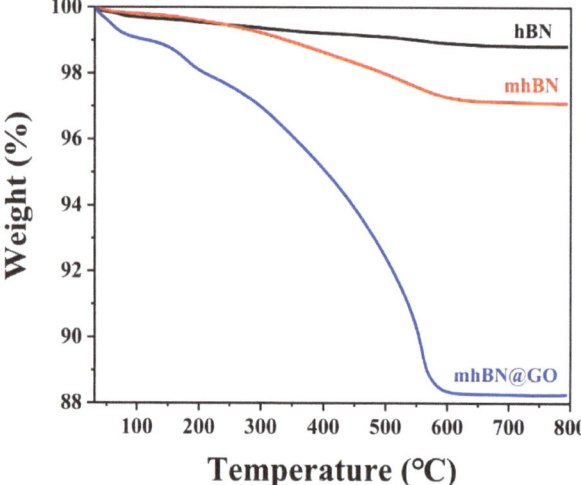

Figure 2. TGA curves of hBN, mhBN, and mhBN@GO.

The surface chemical elements of hBN, mhBN, and mhBN@GO are analyzed via XPS, and the results are shown in Figure 3 and Table 1. The clearly distinguishable peaks were observed in all cases at 191, 285, 398, and 532 eV, which corresponded to boron (B 1s), carbon (C 1s), nitrogen (N 1s), and oxygen (O 1s), respectively. Unexpectedly, the C 1s and O 1s peaks also appeared in the pristine hBN (Figure 3a), which might be due to contamination [23]. Meanwhile, the C 1s core-level spectrum of the hBN was decomposed into three peaks that were attributed to C-C (284.8 eV), C-N (285.5 eV), and C-O (286.6 eV) bonds. However, one new peak corresponding to C=O (288.3 eV) bonds emerged in the mhBN, which was derived from the PCPA coating. In comparison with Figure 3b, another additional peak ascribed to O-C=O (289.4 eV) bonds was observed in mhBN@GO (Figure 3c), which was caused by the -COOH groups in GO. This indicated that the GO could be attached to the mhBN surface through the hydrogen bonds between the -COOH groups in GO and the -OH groups in PCPA [24].

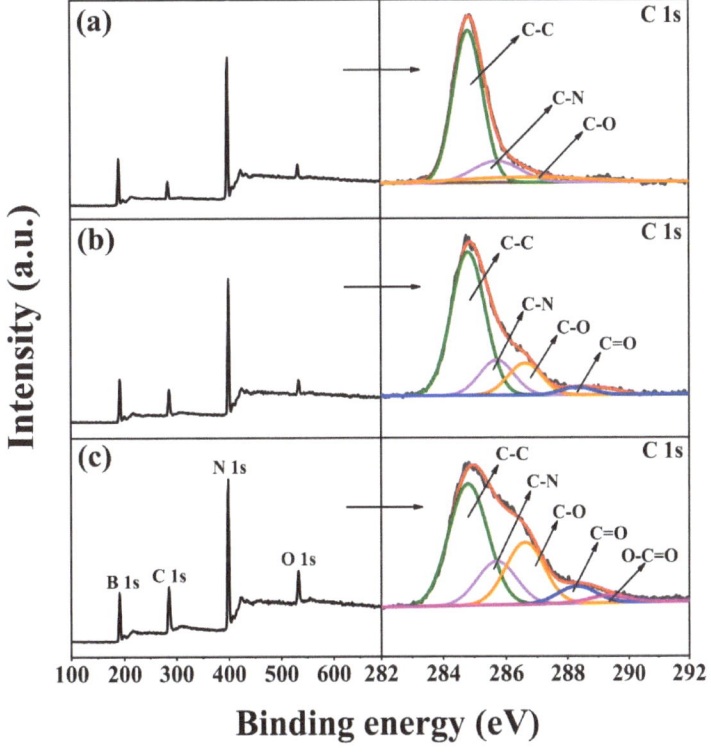

Figure 3. XPS spectra and decomposed C 1s spectra of (**a**) hBN, (**b**) mhBN, and (**c**) mhBN@GO.

Table 1. Chemical element compositions of hBN, mhBN, and mhBN@GO.

Samples	Elemental Analysis (wt%)			
	B	N	C	O
hBN	50.54	39.17	7.79	2.50
mhBN	46.07	38.07	13.03	2.83
mhBN@GO	40.67	34.48	19.31	5.54

Moreover, according to Table 1, the C content in the pristine hBN was only 7.79 wt%, whereas that in mhBN dramatically increased to 13.03 wt%, which was due to the deposition of PCPA onto the hBN surface [16]. In addition, the content of O increased from 2.50 wt% for hBN to 5.54 wt% for mhBN@GO, which was attributed to the oxygen-containing groups in GO. These results provided compelling evidence of the successful assembly of the PCPA layer and GO sheets on the hBN surface.

Figure 4 displays the HR-TEM images of the hBN, mhBN, and mhBN@GO. In Figure 4a, the pristine hBN exhibited a homogeneous flat surface without any impurities. Compared with Figure 4a, the surface of the mhBN was coated by a smooth layer with a thickness of approximately 5 nm (Figure 4b), which was due to the formation of a PCPA layer on the hBN surface. As shown in Figure 4c, after the introduction of GO, a large and plicate flake was successfully bonded to the surface of mhBN, indicating strong adhesion of PCPA.

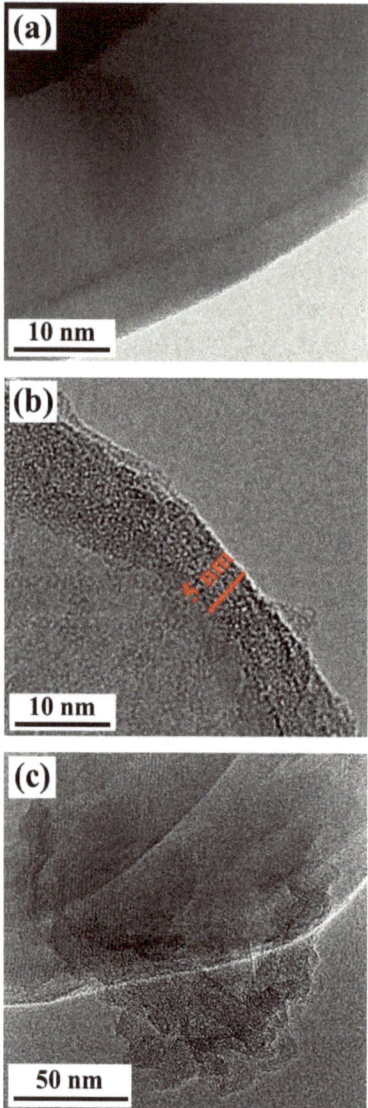

Figure 4. HR-TEM images of (**a**) hBN, (**b**) mhBN, and (**c**) mhBN@GO.

The cross-sectional SEM images of XNBR composites with different filler contents are shown in Figure 5. As seen from Figure 5a–c, multiple hBN platelets were exposed within the XNBR matrix, indicating their poor compatibility and weak interface interaction. Moreover, the number and size of aggregates between hBN platelets in the XNBR matrix increased with the increase in filler content. Compared to the hBN/XNBR composites, the mhBN@GO hybrids were uniformly dispersed in the XNBR matrix according to certain orientations (see the red arrows in Figure 5d–f). These orientations were beneficial for the formation of heat conduction pathways in the XNBR composites. Furthermore, good compatibility and strong interface bonding between the mhBN@GO hybrids and XNBR matrix would be critical in reducing the interface thermal resistance and phonon scattering of elastomer composites [3,25].

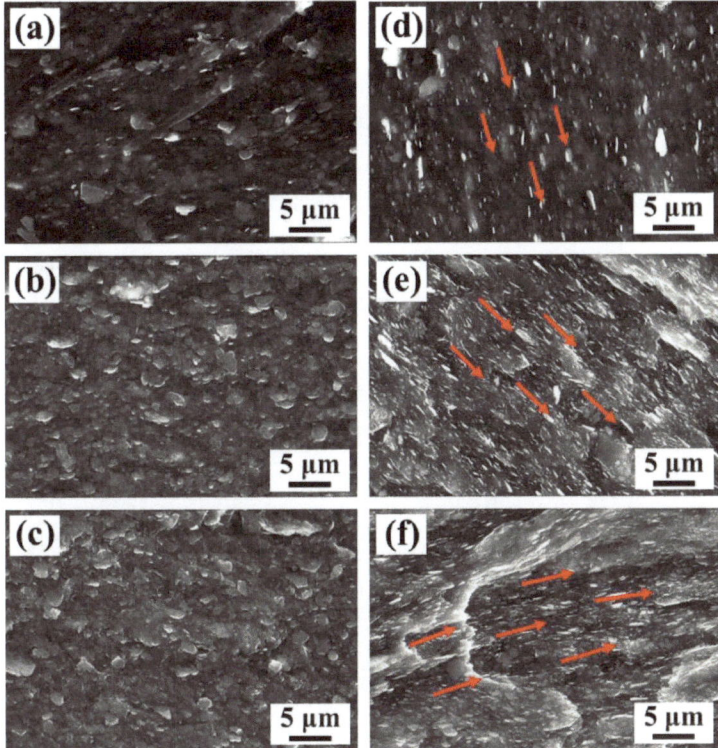

Figure 5. Cross-sectional SEM images of (**a**) 10 vol% hBN/XNBR, (**b**) 20 vol% hBN/XNBR, (**c**) 30 vol% hBN/XNBR, (**d**) 10 vol% mhBN@GO/XNBR, (**e**) 20 vol% mhBN@GO/XNBR, and (**f**) 30 vol% mhBN@GO/XNBR composites. The red arrows represent the orientated dispersion of mhBN@GO.

Figure 6 shows the mechanical properties of XNBR composites filled with various volume fractions of pristine hBN platelets and mhBN@GO hybrids. With the increase in filler content, the tensile strength of both hBN/XNBR and mhBN@GO/XNBR composites increased. However, the tensile strength of mhBN@GO/XNBR composites exceeded that of hBN/XNBR composites at the same filler content. This could be explained by the enforcement effect of mhBN@GO hybrids and the strong interfacial coupling between GO and XNBR [24]. As for the mhBN@GO/XNBR composite with 30 vol% hybrids, the tensile strength increased to 15.35 MPa, which was about 6.50 times that of the neat XNBR (2.36 MPa). Nevertheless, with increasing filler content, the elongation at break of both hBN/XNBR and mhBN@GO/XNBR composites decreased. The mhBN@GO/XNBR composites showed lower elongation at break than the hBN/XNBR composites with the same filler loading, which could be explained by the following two aspects: first, the strong interface interaction between the mhBN@GO hybrids and XNBR matrix might have restricted the slippage of XNBR chains [26]; secondly, owing to the good dispersion and compatibility of mhBN@GO hybrids in XNBR, the filler networks gradually formed in XNBR composites with the increase in mhBN@GO hybrids content [27,28]. The strong interface bonding between the filler and the matrix and the filler networks in polymeric composites usually leads to a lower elongation at break.

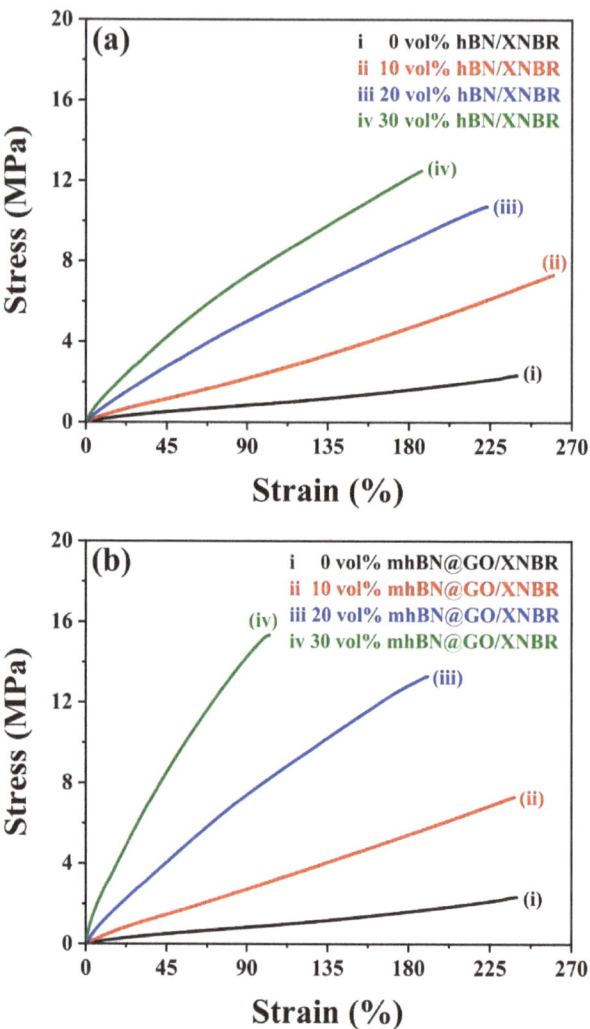

Figure 6. Mechanical properties of (**a**) hBN/XNBR and (**b**) mhBN@GO/XNBR composites.

Figure 7 displays the frequency dependence of the dielectric behaviors for neat XNBR and XNBR-based composites over the range of 10–10^7 Hz. In Figure 7a,b, the dielectric constant (ε_r) of the XNBR composites decreased gradually with increasing frequency. This suggested that the interface polarization between the filler and the XNBR matrix did not have enough relaxation time to catch up with the change in electric field frequency [29]. Besides, the ε_r of the XNBR composites decreased with the increase of filler content, which was mainly ascribed to the hBN platelets with a ε_r (about 4 at 1 kHz) smaller than that of the neat XNBR (11.30 at 1 kHz) [30]. However, at the same filler content, the mhBN@GO/XNBR composites displayed a higher ε_r than the hBN/XNBR composites (Figure 7a,b). For instance, the ε_r of 30 vol% hBN/XNBR composites decreased to 7.13 at 1 kHz, which was inferior to that of 30 vol% mhBN@GO/XNBR composites (8.64 at 1 kHz). This finding could be explained in terms of the following factors. The first was related to the so-called mini-capacitor principle. With the incorporation of GO, many mini-capacitors with GO as electrodes and XNBR as dielectrics are generated in the composites, leading to a high ε_r [31,32]. Another one was the enhanced interface polarization between the thermally conductive fillers and the XNBR matrix. The interface polarization in the mhBN@GO/XNBR composites occurred at the interfaces between hBN and PCPA, between PCPA and GO, and between GO and XNBR, whereas that in the hBN/XNBR composites was only between the hBN filler and the XNBR matrix [29].

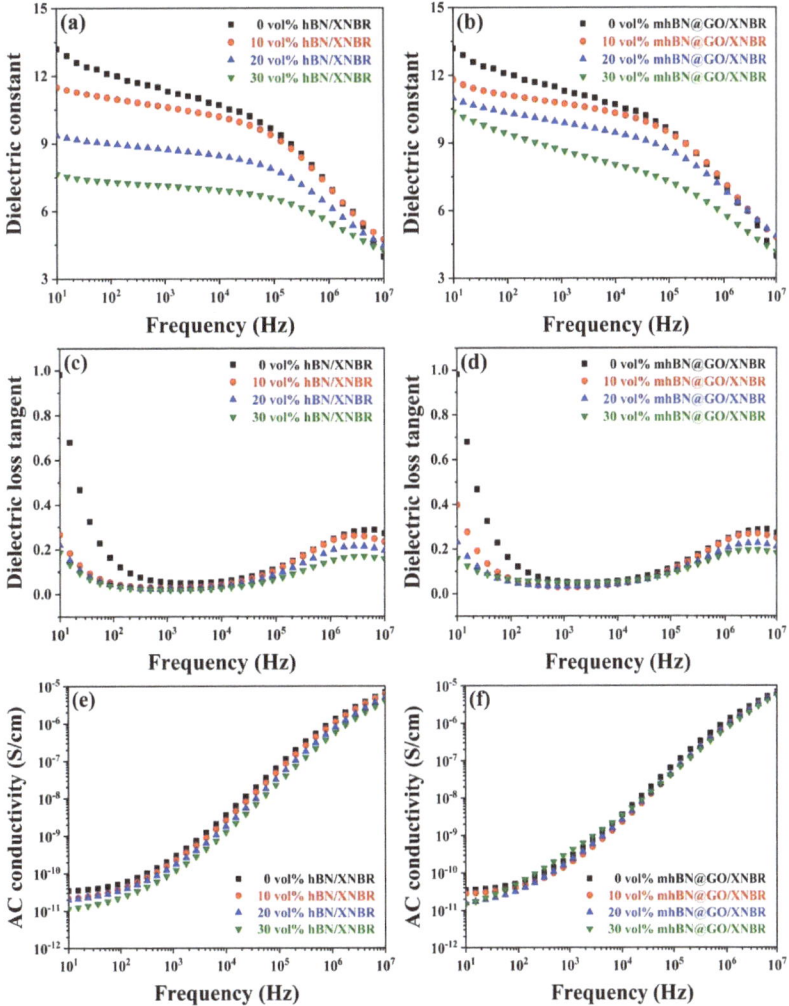

Figure 7. (**a**,**b**) Dielectric constant, (**c**,**d**) dielectric loss tangent, and (**e**,**f**) AC conductivity as functions of frequency for hBN/XNBR and mhBN@GO/XNBR composites, respectively.

The variation of the dielectric loss tangent (tanδ) with frequency in XNBR composites is shown in Figure 7c,d. The tanδ of neat XNBR and XNBR composites first decreases with increasing frequency in the low-frequency range (below 10^3 Hz) and then increases slightly in the high-frequency region (10^3–10^7 Hz). The decreased tanδ in the low-frequency range could be attributed to Maxwell-Wagner-Sillars (MWS) polarization. However, the slightly increased tanδ in the high-frequency region is mainly due to dipolar relaxation [29]. It is evident that the tanδ of the hBN/XNBR composites has decreased compared to that of the neat XNBR. This was because the hBN acted as an effective insulator to prevent the space charge and leakage current [33]. With increasing content of mhBN@GO hybrids, the tanδ of the mhBN@GO/XNBR composites increased from 0.030 to 0.049 at 1 kHz, which was due to the increased leakage current caused by GO [34]. However, the tanδ of hBN/XNBR and mhBN@GO/XNBR composites were still maintained at low levels (<0.1 at 1 kHz), indicating that the mhBN effectively hindered the electrical connection between GO [35].

Figure 7e,f depict the frequency-dependent AC conductivity of XNBR composites. According to Maxwell's equations, the current density j = σ^*E and the time derivative of the dielectric displacement dD/dt = $i\omega\varepsilon^*\varepsilon_0$E are equivalent, where $\sigma^*(\omega)$ is the complex conductivity. Therefore, the conductivity of samples can be calculated according to the following Equations (1) and (2):

$$\sigma*(\omega) = \sigma'(\omega) + i\sigma''(\omega) = i\omega\varepsilon 0 \varepsilon*(\omega) \quad (1)$$

$$\sigma'(\omega) = \omega\varepsilon 0 \varepsilon''(\omega). \ \sigma''(\omega) = \omega\varepsilon 0 \varepsilon'(\omega) \quad (2)$$

where σ^* is the complex conductivity, σ' and σ'' are the real and imaginary parts of the complex conductivity, respectively; ε^* is the complex dielectric function or permittivity, ε' and ε'' are the real and imaginary parts of the complex dielectric function, respectively; ε_0 is the dielectric permittivity of vacuum (ε_0 = 8.854 × 10^{-12} F/m), i is imaginary unit $i = \sqrt{-1}$, and ω is the radial frequency [36]. The AC conductivity of XNBR composites increases gradually with increasing frequency over the range of 10–10^7 Hz. Obviously, the AC conductivity of the XNBR composites is dependent on the frequency, and it might belong to non-Ohmic conduction [32]. After the addition of hBN or mhBN@GO hybrids, the AC conductivity of the XNBR composites decreased with increasing filler content. The reduction in AC conductivity was mainly due to the lower insulation of the thermally conductive fillers [33]. Notably, the AC conductivity of the XNBR composites was lower than 10^{-10} S/cm below 100 Hz, meaning that the as-prepared mhBN@GO/XNBR composites still maintained excellent electrical insulation [37].

The λ values of the hBN/XNBR and mhBN@GO/XNBR composites are plotted in Figure 8a. It is worth noting that the λ values of the XNBR composites have been gradually enhanced with increasing additions of filler. Moreover, at the same filler loading, the mhBN@GO/XNBR composites exhibited a higher λ compared to that of the hBN/XNBR composites. As the volume fraction of the mhBN@GO hybrids increased to 30%, the corresponding λ increased to 0.4348 W/(m·K), being about 2.7 times of the neat XNBR (0.1623 W/(m·K)). Figure 8b displays the enhancement in λ of XNBR composites relative to that of the neat XNBR. Compared to the hBN/XNBR composites, a stronger λ enhancement could be observed in the mhBN@GO/XNBR composites. The reasons for this difference were as follows: first, the uniform dispersion of the mhBN@GO hybrids and the construction of continuous heat conduction pathways played an essential role in the heat transfer process within the mhBN@GO/XNBR composites; secondly, due to the low interface thermal resistance, the mhBN@GO/XNBR composites exhibited higher efficiently phonon transmission than the hBN/XNBR composites.

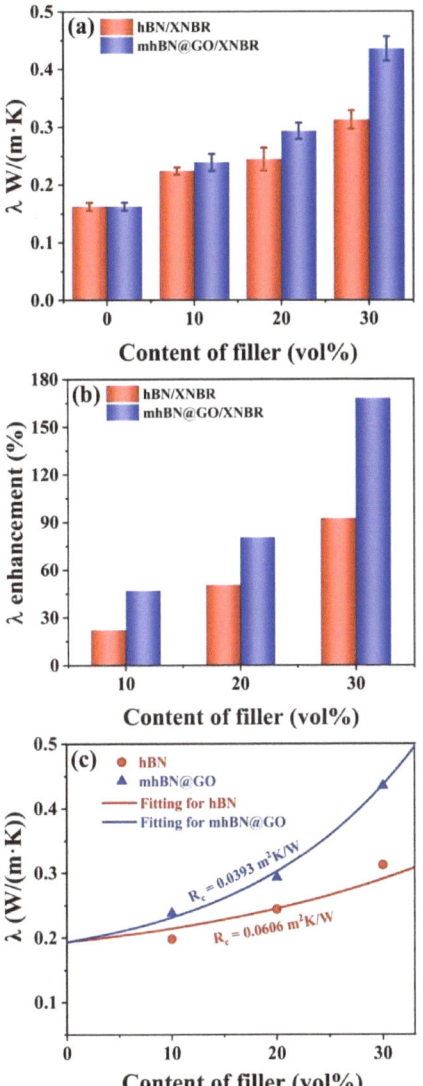

Figure 8. (a) The λ of hBN/XNBR and mhBN@GO/XNBR composites with different filler loading. (b) The enhancement in λ of hBN/XNBR and mhBN@GO/XNBR composites relative to that of the neat XNBR. (c) Fitting λ of hBN/XNBR and mhBN@GO/XNBR composites using the modified Hashin-Shtrikman model.

To expound the internal relationship between the filler and the XNBR matrix, the experimental λ values of XNBR composites (Figure 8c) were fitted using the modified Hashin-Shtrikman model (Equation (3)) [38,39].

$$\lambda_{eff} = \lambda_m \frac{(2KV_f + 1)\lambda_c - (2KV_f - 1)\lambda_m}{(KV_f + 1)\lambda_m - (KV_f - 1)\lambda_c} \quad (3)$$

where λ_m is the λ of the XNBR matrix, λ_c is the λ of the XNBR composites, V_f is the volume fraction of the filler, and K is a coefficient related to the interface thermal resistance of the XNBR composites and defined as Equation (4).

$$K = 13.3347 \exp(-13.2701 \frac{R_c \lambda_m}{L}) \quad (4)$$

where R_c stands for the interface thermal resistance and L represents the characteristic length of the unit model [3,40]. According to Figure 8c, the interface thermal resistance of the mhBN@GO/XNBR composites (R_c = 0.0393 m²K/W) was lower than that of the hBN/XNBR composites (R_c = 0.0606 m²K/W), demonstrating that PCPA acted as an interface modifier, effectively reducing the interface thermal resistance of XNBR composites.

Figure 9 displays the schematic models of phonon transmission in the hBN/XNBR and mhBN@GO/XNBR composites. In the hBN/XNBR composites, the hBN platelets are inevitably aggregated in the XNBR matrix, leading to severe scattering of phonons and high interface thermal resistance [41]. In addition, ineffective thermal transfer channels are rendered due to poor dispersion of the hBN in the XNBR matrix, limiting the improvement of λ. As a result, the heat transfer process of the hBN/XNBR composites is like a car moving on a mountain road. However, the heat transfer process of the mhBN@GO/XNBR composites is similar to a car speeding on the highway. Because of the reduced interface thermal resistance, the phonons travel along the continuous heat conduction channels with few barriers. In addition, the ordered heat conduction channels formed by mhBN@GO hybrids make the thermal transfer channels shorter [23].

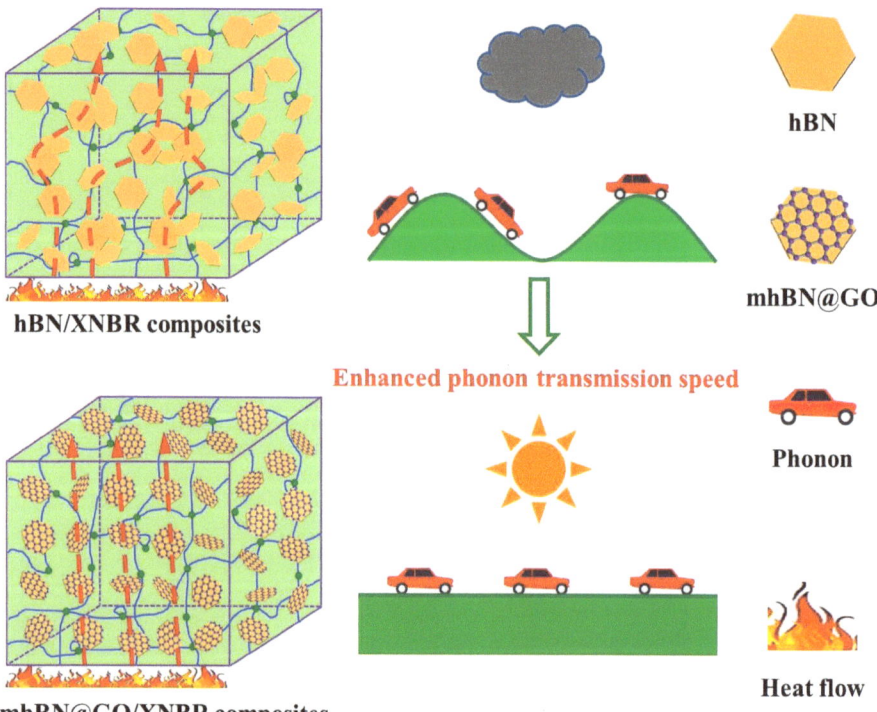

Figure 9. Schematic models of phonon transmission in hBN/XNBR and mhBN@GO/XNBR composites.

4. Conclusions

The thermally conductive mhBN@GO hybrids were successfully synthesized via a facile and green self-assembly method and then were embedded in XNBR latex to obtain high λ and satisfactory insulating properties of mhBN@GO/XNBR composites. Owing to the formation of hydrogen bonds between carboxyl groups in XNBR and hydroxyl groups in GO, the interfacial interaction between the XNBR matrix and mhBN@GO hybrids was obviously improved, leading to reduced interface thermal resistance and the formation of effective thermal transport pathways in the mhBN@GO/XNBR composites. Thus, the obtained mhBN@GO/XNBR composites exhibited a relatively high λ value of 0.4348 W/(m·K), which was 2.7 times that of the neat XNBR (0.1623 W/(m·K)). However, the insulating hBN effectively prevented the connection between adjacent GO sheets, leading to the mhBN@GO/XNBR composites showing a satisfactory electrical insulating property (less than 10^{-10} S/cm below 100 Hz). Therefore, the effective and eco-friendly method proposed in this work allows one to prepare thermal management materials with insulation properties conforming to the heat dissipation requirements for modern electronic devices.

Author Contributions: Date curation, validation, formal analysis, investigation, and writing—original draft, writing—review and editing, X.X. and D.Y.; methodology, funding acquisition, resources, supervision, D.Y. All authors have read and agreed to the published version of the manuscript.

Funding: This research was funded by the National Natural Science Foundation of China (No. 52273259), the Beijing Nova Program (Z201100006820036) from Beijing Municipal Science & Technology Commission, and the Fundamental Research Funds for the Central Universities (buctrc202217), China.

Institutional Review Board Statement: Not Applicable.

Informed Consent Statement: Not Applicable.

Data Availability Statement: Not Applicable.

Conflicts of Interest: The authors declare no conflict of interest.

References

1. Han, J.; Du, G.; Gao, W.; Bai, H. An anisotropically high thermal conductive boron nitride/epoxy composite based on nacre-mimetic 3D network. *Adv. Funct. Mater.* **2019**, *29*, 1900412. [CrossRef]
2. Guo, S.; Zheng, R.; Jiang, J.; Yu, J.; Dai, K.; Yan, C. Enhanced thermal conductivity and retained electrical insulation of heat spreader by incorporating alumina-deposited graphene filler in nano-fibrillated cellulose. *Compos. Part B Eng.* **2019**, *178*, 107489. [CrossRef]
3. Guo, Y.; Lyu, Z.; Yang, X.; Lu, Y.; Ruan, K.; Wu, Y.; Kong, J.; Gu, J. Enhanced thermal conductivities and decreased thermal resistances of functionalized boron nitride/polyimide composites. *Compos. Part B Eng.* **2019**, *164*, 732–739. [CrossRef]
4. Tang, L.; He, M.; Na, X.; Guan, X.; Zhang, R.; Zhang, J.; Gu, J. Functionalized glass fibers cloth/spherical BN filler/epoxy laminated composites with excellent thermal conductivities and electrical insulation properties. *Compos. Commun.* **2019**, *16*, 5–10. [CrossRef]
5. Xu, X.; Zhou, J.; Chen, J. Thermal transport in conductive polymer-based materials. *Adv. Funct. Mater.* **2020**, *30*, 1904704. [CrossRef]
6. Niu, H.; Ren, Y.; Guo, H.; Małycha, K.; Orzechowski, K.; Bai, S.-L. Recent progress on thermally conductive and electrical insulating rubber composites: Design, processing and applications. *Compos. Commun.* **2020**, *22*, 100430. [CrossRef]
7. Xue, Y.; Li, X.; Wang, H.; Zhao, F.; Zhang, D.; Chen, Y. Improvement in thermal conductivity of through-plane aligned boron nitride/silicone rubber composites. *Mater. Des.* **2019**, *165*, 107580. [CrossRef]
8. Gan, L.; Dong, M.; Han, Y.; Xiao, Y.; Yang, L.; Huang, J. Connection-improved conductive network of carbon nanotubes in a rubber cross-link network. *ACS Appl. Mater. Interfaces* **2018**, *10*, 18213–18219. [CrossRef]
9. Liu, C.; Wu, W.; Drummer, D.; Shen, W.; Wang, Y.; Schneider, K.; Tomiak, F. ZnO nanowire-decorated Al2O3 hybrids for improving the thermal conductivity of polymer composites. *J. Mater. Chem. C* **2020**, *8*, 5380–5388. [CrossRef]
10. Guo, Y.; Qiu, H.; Ruan, K.; Wang, S.; Zhang, Y.; Gu, J. Flexible and insulating silicone rubber composites with sandwich structure for thermal management and electromagnetic interference shielding. *Compos. Sci. Technol.* **2022**, *219*, 109253. [CrossRef]
11. Yang, X.; Guo, Y.; Han, Y.; Li, Y.; Ma, T.; Chen, M.; Kong, J.; Zhu, J.; Gu, J. Significant improvement of thermal conductivities for BNNS/PVA composite films via electrospinning followed by hot-pressing technology. *Compos. Part B Eng.* **2019**, *175*, 107070. [CrossRef]

12. Chen, Y.; Hou, X.; Liao, M.; Dai, W.; Wang, Z.; Yan, C.; Li, H.; Lin, C.-T.; Jiang, N.; Yu, J. Constructing a "pea-pod-like" alumina-graphene binary architecture for enhancing thermal conductivity of epoxy composite. *Chem. Eng. J.* **2020**, *381*, 122690. [CrossRef]
13. Yang, X.; Fan, S.; Li, Y.; Guo, Y.; Li, Y.; Ruan, K.; Zhang, S.; Zhang, J.; Kong, J.; Gu, J. Synchronously improved electromagnetic interference shielding and thermal conductivity for epoxy nanocomposites by constructing 3D copper nanowires/thermally annealed graphene aerogel framework. *Compos. Part A* **2020**, *128*, 105670. [CrossRef]
14. Zhang, J.; Du, Z.; Zou, W.; Li, H.; Zhang, C. MgO nanoparticles-decorated carbon fibers hybrid for improving thermal conductive and electrical insulating properties of Nylon 6 composite. *Compos. Sci. Technol.* **2017**, *148*, 1–8. [CrossRef]
15. Shen, C.; Wang, H.; Zhang, T.; Zeng, Y. Silica coating onto graphene for improving thermal conductivity and electrical insulation of graphene/polydimethylsiloxane nanocomposites. *J. Mater. Sci. Technol.* **2019**, *35*, 36–43. [CrossRef]
16. Yang, D.; Ni, Y.; Kong, X.; Gao, D.; Wang, Y.; Hu, T.; Zhang, L. Mussel-inspired modification of boron nitride for natural rubber composites with high thermal conductivity and low dielectric constant. *Compos. Sci. Technol.* **2019**, *177*, 18–25. [CrossRef]
17. Guo, Y.; Wang, S.; Ruan, K.; Zhang, H.; Gu, J. Highly thermally conductive carbon nanotubes pillared exfoliated graphite/polyimide composites. *NPJ Flex. Electron.* **2021**, *5*, 1–9. [CrossRef]
18. Jiang, F.; Cui, X.; Song, N.; Shi, L.; Ding, P. Synergistic effect of functionalized graphene/boron nitride on the thermal conductivity of polystyrene composites. *Compos. Commun.* **2020**, *20*, 100350. [CrossRef]
19. Oh, J.; Jo, H.; Lee, H.; Kim, H.-T.; Lee, Y.M.; Ryou, M.-H. Polydopamine-treated three-dimensional carbon fiber-coated separator for achieving high-performance lithium metal batteries. *J. Power Sources* **2019**, *430*, 130–136. [CrossRef]
20. Dong, X.; Ding, B.; Guo, H.; Dou, H.; Zhang, X. Superlithiated polydopamine derivative for high-capacity and high-rate anode for lithium-ion batteries. *ACS Appl. Mater. Interfaces* **2018**, *10*, 38101–38108. [CrossRef]
21. Zhu, T.; Chen, Q.; Xie, D.; Liu, J.; Chen, X.; Nan, J.; Zuo, X. Low-cost and heat-resistant poly(catechol/polyamine)-silica composite membrane for high-performance lithium-ion batteries. *ChemElectroChem* **2021**, *8*, 1369–1376. [CrossRef]
22. Guo, Y.; Yang, X.; Ruan, K.; Kong, J.; Dong, M.; Zhang, J.; Gu, J.; Guo, Z. Reduced graphene oxide heterostructured silver nanoparticles significantly enhanced thermal conductivities in hot-pressed electrospun polyimide nanocomposites. *ACS Appl. Mater. Interfaces* **2019**, *11*, 25465–25473. [CrossRef]
23. Xiong, S.-W.; Zhang, P.; Xia, Y.; Zou, Q.; Jiang, M.; Gai, J.-G. Unique antimicrobial/thermally conductive polymer composites for use in medical electronic devices. *J. Appl. Polym. Sci.* **2021**, *138*, 50113. [CrossRef]
24. Wei, Q.; Yang, D.; Yu, L.; Ni, Y.; Zhang, L. Fabrication of carboxyl nitrile butadiene rubber composites with high dielectric constant and thermal conductivity using Al2O3@PCPA@GO hybrids. *Compos. Sci. Technol.* **2020**, *199*, 108344. [CrossRef]
25. Zhang, Z.; Qu, J.; Feng, Y.; Feng, W. Assembly of graphene-aligned polymer composites for thermal conductive applications. *Compos. Commun.* **2018**, *9*, 33–41. [CrossRef]
26. Xiao, C.; Song, Q.; Shen, Q.; Wang, T.; Xie, W. Understanding on interlaminar nano-reinforcement induced mechanical performance improvement of carbon/carbon composites after silicon infiltration. *Compos. Part B Eng.* **2022**, *239*, 109946. [CrossRef]
27. Han, Y.; Ruan, K.; Gu, J. Janus (BNNS/ANF)-(AgNWs/ANF) thermal conductivity composite films with superior electromagnetic interference shielding and joule heating performances. *Nano Res.* **2022**, *15*, 4747–4755. [CrossRef]
28. Cetin, M.S.; Toprakci, H.A.K. Flexible electronics from hybrid nanocomposites and their application as piezoresistive strain sensors. *Compos. Part B Eng.* **2021**, *224*, 109199. [CrossRef]
29. Yu, L.; Yang, D.; Wei, Q.; Zhang, L. Constructing of strawberry-like core-shell structured Al2O3 nanoparticles for improving thermal conductivity of nitrile butadiene rubber composites. *Compos. Sci. Technol.* **2021**, *209*, 108786. [CrossRef]
30. Ji, S.-Y.; Jung, H.-B.; Kim, M.-K.; Lim, J.-H.; Kim, J.-Y.; Ryu, J.; Jeong, D.-Y. Enhanced energy storage performance of polymer/ceramic/metal composites by increase of thermal conductivity and coulomb-blockade effect. *ACS Appl. Mater. Interfaces* **2021**, *13*, 27343–27352. [CrossRef]
31. Tawade, B.V.; Apata, I.E.; Singh, M.; Das, P.; Pradhan, N.; Al-Enizi, A.M.; Karim, A.; Raghavan, D. Recent developments in the synthesis of chemically modified nanomaterials for use in dielectric and electronics applications. *Nanotechnology* **2021**, *32*, 142004. [CrossRef] [PubMed]
32. Wang, R.; Xie, C.; Luo, S.; Xu, H.; Gou, B.; Zeng, L. Preparation and properties of MWCNTs-BNNSs/epoxy composites with high thermal conductivity and low dielectric loss. *Mater. Today Commun.* **2020**, *24*, 100985. [CrossRef]
33. Hu, B.; Guo, H.; Wang, Q.; Zhang, W.; Song, S.; Li, X.; Li, Y.; Li, B. Enhanced thermal conductivity by constructing 3D-networks in poly(vinylidene fluoride) composites via positively charged hexagonal boron nitride and silica coated carbon nanotubes. *Compos. Part A: Appl. Sci. Manuf.* **2020**, *137*, 106038. [CrossRef]
34. Wang, B.; Yin, X.H.; Peng, D.; Lv, R.H.; Na, B.; Liu, H.S.; Gu, X.B.; Wu, W.; Zhou, J.L.; Zhang, Y. Achieving thermally conductive low loss PVDF-based dielectric composites via surface functionalization and orientation of SiC nanowires. *Express Polym. Lett.* **2020**, *14*, 2–11. [CrossRef]
35. Hao, Y.; Li, Q.; Pang, X.; Gong, B.; Wei, C.; Ren, J. Synergistic enhanced thermal conductivity and dielectric constant of epoxy composites with mesoporous silica coated carbon nanotube and boron nitride nanosheets. *Materials* **2021**, *14*, 5251. [CrossRef]
36. Schönhals, A.; Kremer, F. *Analysis of Dielectric Spectra*; Springer: Berlin/Heidelberg, Germany, 2003; pp. 59–98.
37. Shen, Z.; Feng, J. Achieving vertically aligned SiC microwires networks in a uniform cold environment for polymer composites with high through-plane thermal conductivity enhancement. *Compos. Sci. Technol.* **2019**, *170*, 135–140. [CrossRef]

38. Ngo, I.L.; Byon, C.; Lee, B.J. Analytical study on thermal conductivity enhancement of hybrid-filler polymer composites under high thermal contact resistance. *Int. J. Heat Mass Transf.* **2018**, *126*, 474–484. [CrossRef]
39. Ngo, I.L.; Vattikuti, P.S.V.; Byon, C. A modified Hashin-Shtrikman model for predicting the thermal conductivity of polymer composites reinforced with randomly distributed hybrid fillers. *Int. J. Heat Mass Tranf.* **2017**, *114*, 727–734. [CrossRef]
40. Liu, X.; Gao, Y.; Shang, Y.; Zhu, X.; Jiang, Z.; Zhou, C.; Han, J.; Zhang, H. Non-covalent modification of boron nitride nanoparticle-reinforced PEEK composite: Thermally conductive, interfacial, and mechanical properties. *Polymer* **2020**, *203*, 122763. [CrossRef]
41. Du, C.; Li, M.; Cao, M.; Song, S.; Feng, S.; Li, X.; Guo, H.; Li, B. Mussel-inspired and magnetic co-functionalization of hexagonal boron nitride in poly(vinylidene fluoride) composites toward enhanced thermal and mechanical performance for heat exchangers. *ACS Appl. Mater. Interfaces* **2018**, *10*, 34674–34682. [CrossRef]

Disclaimer/Publisher's Note: The statements, opinions and data contained in all publications are solely those of the individual author(s) and contributor(s) and not of MDPI and/or the editor(s). MDPI and/or the editor(s) disclaim responsibility for any injury to people or property resulting from any ideas, methods, instructions or products referred to in the content.

Article

Analysis of the Structure and the Thermal Conductivity of Semi-Crystalline Polyetheretherketone/Boron Nitride Sheet Composites Using All-Atom Molecular Dynamics Simulation

Yuna Oh [1,2], Kwak Jin Bae [1], Yonjig Kim [2] and Jaesang Yu [1,*]

[1] Composite Materials Application Research Center, Institute of Advanced Composite Materials, Korea Institute of Science and Technology (KIST), Chudong-ro 92, Bongdong-eup, Wanju-gun 55324, Republic of Korea
[2] Department of Mechanical Design Engineering, Jeonbuk National University, Baekje-daero 567, Deokjin-gu, Jeonju 54896, Republic of Korea
* Correspondence: jamesyu@kist.re.kr

Abstract: Thermal transport simulations were performed to investigate the important factors affecting the thermal conductivity based on the structure of semi-crystalline polyetheretherketone (PEEK), and the addition of boron nitride (BN) sheets. The molecular-level structural analysis facilitated the prediction of the thermal conductivity of the optimal structure of PEEK reflecting the best parameter value of the length of amorphous chains, and the ratio of linkage conformations, such as loops, tails, and bridges. It was found that the long heat transfer paths of polymer chains were induced by the addition of BN sheets, which led to the improvement of the thermal conductivities of the PEEK/BN composites. In addition, the convergence of the thermal conductivities of the PEEK/BN composites in relation to BN sheet size was verified by the disconnection of the heat transfer path due to aggregation of the BN sheets.

Keywords: semi-crystalline polyetheretherketone; boron nitride sheet; thermal conductivity; polymer composite

Citation: Oh, Y.; Bae, K.J.; Kim, Y.; Yu, J. Analysis of the Structure and the Thermal Conductivity of Semi-Crystalline Polyetheretherketone/ Boron Nitride Sheet Composites Using All-Atom Molecular Dynamics Simulation. *Polymers* **2023**, *15*, 450. https://doi.org/10.3390/polym15020450

Academic Editor: Mikhail G. Kiselev

Received: 26 December 2022
Revised: 9 January 2023
Accepted: 12 January 2023
Published: 14 January 2023

Copyright: © 2023 by the authors. Licensee MDPI, Basel, Switzerland. This article is an open access article distributed under the terms and conditions of the Creative Commons Attribution (CC BY) license (https://creativecommons.org/licenses/by/4.0/).

1. Introduction

Polymers are widely used in many industrial fields, including the aerospace, automobile, semiconductor, and biomedical fields because they are light weight, easy to process, have low production costs, good mechanical properties, and chemical resistance [1–4]. Depending on their degree of crystallinity, polymers are classified as an amorphous polymer, a semi-crystalline polymer, or a crystalline polymer. Analysis of linkage conformations such as loops, tails, and bridges affecting the material properties of semi-crystalline polymers is important because the structure, mechanical properties, and thermal properties of the crystalline structure and the amorphous structure are different [5–7]. All-atom molecular dynamics (MD) simulation, which deals with all atoms of molecules of a simulation model, can facilitate a detailed analysis of material properties, such as the behavior of polymer chains, the radius of gyration, and the interaction between materials at the atomic level [8–10]. Zhu et al. [11] investigated the tensile property of semi-crystalline polyurethane using an MD simulation. The tensile yield stress of the model with many bridge chains connecting the crystalline and amorphous domains was 19 times higher than that of a model with no bridge due to the straightening and stretching of the bridge chain. In addition, He et al. [12] reported that the thermal conductivity of the interphase region in the semi-crystalline polyethylene was determined by the stretching of the bridge chains. While the effects of bridge chains on material properties have been analyzed in many studies, the effect of the other linkage conformations of loops and tails on the material properties have not yet been concretely considered in MD simulations. We found that the thermal conductivity of a semi-crystalline polymer depends on the contents of the linkage conformations of the loops, tails, and bridges. Therefore, the effect of linkage conformation on thermal properties was investigated to determine the ideal structure of a semi-crystalline polymer with high thermal conductivity. Among semi-crystalline polymers, semi-crystalline polyetheretherketone (PEEK) is a high-performance engineering thermoplastic with excellent wear resistance and heat resistance. PEEK with a low thermal conductivity of 0.25 W/mK was used to investigate the ratio of the optimal combination of loops, tails, and bridges, which leads to the improvement of the thermal conductivity [13].

Adding fillers is a general approach for improving the thermal properties of polymers among many methods, such as the crystallite orientation of polymer and increasing crystallinity [14–16]. A boron nitride (BN) sheet has been used as a filler in polymer composites for various applications, including aerospace, electronic devices, and semiconductors, because it leads to excellent mechanical properties and high thermal transport performance [17,18]. In a report by Ghosh et al. [19], the thermal conductivity of PEEK increased after the addition of hexagonal boron nitride (h-BN). The thermal conductivities of polymer composites reinforced with BN sheets were higher than those of untreated polymers due to the effect of not only the high thermal conductivities of the fillers, but also the increase of phonon velocities between the polymer chains and fillers [20]. Accordingly, the factor responsible for increasing the phonon mobility should be investigated to improve the thermal conductivity of the composite. In addition, it is known that the thermal conductivity of a polymer composite depends on the arrangement of the polymer chains. Li et al. [21] reported that the thermal conductivity of the poly(dimethylsiloxane)/h-BN composite was improved because the arrangement of polymer chains completely covered the surface of the h-BN sheet, inducing strong phonon mobility. The factors affecting the distribution of polymer chains should be investigated because the arrangement of polymer chains on the surface of a BN sheet is related to the thermal conductivity of the composite. Although the thermal conductivity of the composite increased with the addition of fillers, it converged with the aggregation of the fillers [22,23]. The factor inducing the aggregation of fillers should be investigated to understand the reason for the convergence of the thermal conductivities of the composites.

In this study, a thermal transport simulation was performed to investigate thermal conductivity in relation to the ratio of the linkage conformations between the crystalline domain and amorphous domain linked by loops, tails, and bridges. The optimal chain length of the amorphous PEEK was selected with the length at the point of convergence of the thermal conductivities of the semi-crystalline PEEK models. The interphase thermal conductivity between the crystalline domain and the amorphous domain, and the heat transfer path of the semi-crystalline PEEK were investigated after the thermal transport simulation, in accordance with the various combinations of loops, tails, and bridges. The thermal transport simulations of the semi-crystalline PEEK/BN composites reinforced with BN sheets were performed to verify the effect of the addition of BN sheets and the mobility of phonon between the BN sheets and PEEK chains. In addition, the factor inducing the aggregation of fillers in the PEEK/BN composite was investigated by visualizing the equilibrated simulation models.

2. Computational Methods

2.1. Semi-Crystalline Polyetheretherketone Modeling

The semi-crystalline PEEK model consisted of a crystalline domain with aligned chains and an amorphous domain with randomly entangled chains in the periodic boundary system. The simulation model was constructed based on experimental results, as shown in Figure 1. The degree of crystallinity of the semi-crystalline PEEK was about 30%. It is identical to the experimental result obtained with a differential scanning calorimeter (DSC) [24]. The size of a crystalline PEEK domain is 124 Å (x) × 50 Å (y) × 50 Å (z). The atomic content of crystalline PEEK with 30,800 atoms, relative to the total number of atoms in the semi-crystalline PEEK, is 30%. The amorphous PEEK was set to 70% relative to the total number of atoms in the semi-crystalline PEEK. The total size of the semi-crystalline

PEEK model was 460 Å (x) × 50 Å (y) × 50 Å (z) in the periodic boundary condition. The density distribution of the amorphous domain well matched the density of the amorphous PEEK, which was of 1.26 g/cm^3 as obtained from the experiment [25]. The crystalline PEEK had a density of about 1.40 g/cm^3 [25]. This was well matched with the density of the crystalline domain in this work. A range of 10 to 100 monomer units was used to verify the change in thermal conductivity for various chain lengths of amorphous PEEK. An analysis of thermal transport according to the ratio of the linkage conformation between the crystalline domain and the amorphous domain was performed using the selected chain length of amorphous PEEK. The crystalline domain and the amorphous domain of semi-crystalline PEEK were linked by loops, tails, and bridges (Figure 1). Both ends of a loop chain are connected to the same side of the crystalline domain. A tail chain only connects to one end of the crystalline domain. The bridge chain is connected to the crystalline domains by one polymer chain. The simulation models were constructed using Materials Studio 2017 software with a Dreiding force field [26].

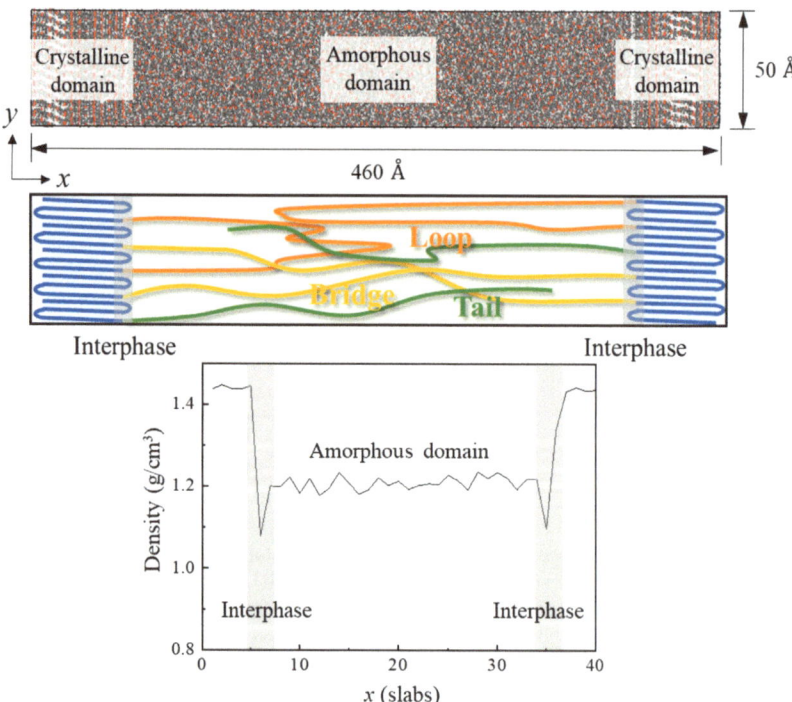

Figure 1. Structure and density of a semi-crystalline PEEK in the periodic boundary condition.

The effects of the chain length of the amorphous PEEK and the contents of linkage conformations on thermal conductivity were analyzed following the thermal transport simulations of the semi-crystalline PEEK models. The radius of gyration (R_g), which is based on the size of the molecule from the center of mass in a polymer chain, was calculated by

$$R_g^2 = \frac{1}{M} \sum m_i (r_i - r_{cm})^2 \quad (1)$$

where M is the total mass of the polymer chain, m_i is the mass of the atom, r_i is the coordinate of the atom, and r_{cm} is the coordinate of the center of mass on a polymer chain. The radius of gyration indicates the distribution size of a single polymer chain. The $R_{a,g}$ is the average value of the radius of gyration of the amorphous polymer chains. In the simulation model, a large $R_{a,g}$ means a long heat transfer path. The radius of gyration on the bulk polymer (R_{bulk}) in the amorphous domain was calculated by

$$R_{bulk}^2 = \frac{1}{M_b} \sum m_i (r_i - r_{cm,bulk})^2 \qquad (2)$$

where M_b is the total mass of the amorphous polymer chains and $r_{cm,bulk}$ is the coordinate of the center of mass on the bulk amorphous polymer. In the simulation model, R_{bulk} is the size of the bulk polymer from the center of mass on the bulk polymer to all amorphous polymer chains. R_{bulk} indicates the degree of entanglement of the polymer chains. A large R_{bulk} means that amorphous polymer chains are widely distributed with low entanglement. BN sheets were used to improve the thermal property of the semi-crystalline PEEK. In our previous study [27], the newly proposed Dreiding force field (N-DFF) was developed to calculate the potential energy of boron and nitride atoms using all-atom MD simulation. The force constants of the N-DFF were parameterized based on the potential energy obtained from the density functional theory (DFT). Consequently, the N-DFF can be used to help accurately predict the material properties of the BN sheet, such as its mechanical and thermal properties. In this work, the N-DFF was used to analyze the thermal transport of the PEEK/BN composite. The thermal properties of the semi-crystalline PEEK were calculated using the original Dreiding force field (DFF) because the DFF provides a more accurate prediction of material properties when the polymers consist of carbon, oxygen, and hydrogen atoms [8,28]. In addition, the parameters of the bond stretching terms of the B–H and N–H bonds were obtained using the linear least square fitting (LLSF) method for the accurate force constants of the bonded term. A detailed description of the calculation method can be found in the Supplementary Materials.

2.2. Thermal Transport Using Molecular Dynamics Simulation

All of the simulation models were optimized using the conjugate gradient method to minimize the total potential energy. The optimized models were equilibrated using an isothermal–isobaric ensemble (NPT) at room temperature for 3 ns with a time step of 1 fs. After the equilibration process, the thermal conductivity was calculated using a reverse non-equilibrium molecular dynamics (RNEMD) simulation. A swap of kinetic energy between the atoms in a heat sink and the atoms in a heat source was performed under the microcanonical ensemble (NVE) for 3 ns (Figure 2). A temperature gradient was generated during the heat transfer process with a swap time of 100 fs. The semi-crystalline PEEK models and the composite models composed of the semi-crystalline PEEK and BN sheets were divided into 40 slabs along the x direction. The thermal conductivity (K) was calculated by

$$J = \frac{1}{2tA} \sum_{N_{swap}} \frac{m_h v_h^2 - m_c v_c^2}{2} \qquad (3)$$

$$K = \frac{J}{\partial T / \partial x} \qquad (4)$$

where J is the total heat flux during the heat transfer time t, A is the cross-sectional area of the simulation model, m_h and m_c are the atomic masses of the heat source and the heat sink, respectively, and v is the velocity of the atom. The phonon behavior is closely related to the thermal transport in the materials. The phonon density of states (PDOS) of the simulation models was calculated using a Fourier transform of the velocity autocorrelation function (VACF). The VACF was averaged over the value of the BN atoms and PEEK atoms in the composite model obtained during the equilibration process. The PDOS was calculated by

$$PDOS(\omega) = \frac{1}{\sqrt{2\pi}} \int_0^\tau VACF(t) e^{-i\omega t} dt \qquad (5)$$

$$VACF(t) = \langle v(t) v(0) \rangle \qquad (6)$$

where ω is the frequency, τ is a total integration time, and $v(t)$ is the velocity of atoms at time t. The angle bracket denotes the ensemble average of all atoms in the simulation model. All equilibration and thermal transport simulations were performed using the Large-scale Atomic/Molecular Massively Parallel Simulator (LAMMPS) [29].

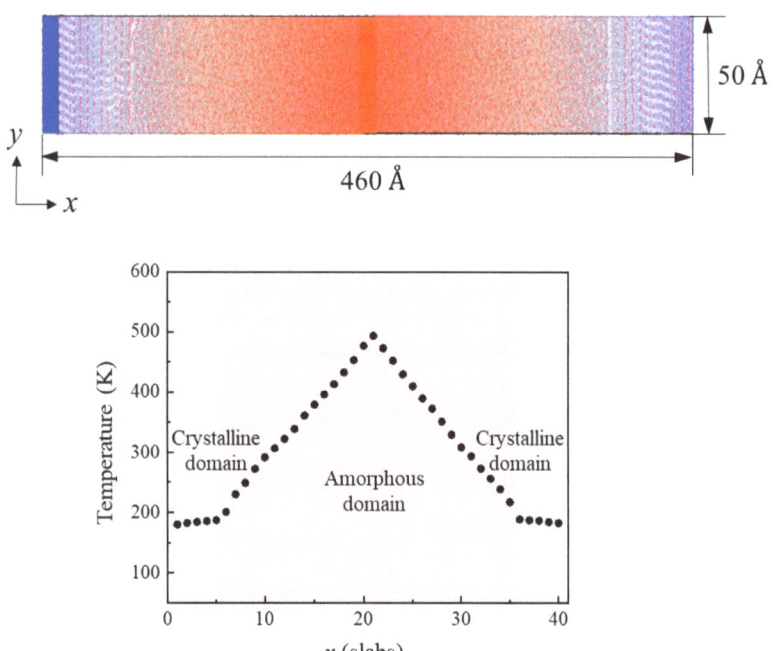

Figure 2. Temperature gradient of a semi-crystalline PEEK model during RNEMD simulation.

3. Results and Discussion

3.1. Effect of the Chain Length of Amorphous PEEK on the Thermal Conductivity

Thermal transport simulations of the semi-crystalline PEEK models for various chain lengths were performed to determine the optimal repeat unit of the amorphous chain. For example, PEEK10 was a semi-crystalline PEEK model consisting of a crystalline domain and the amorphous domain with an amorphous chain of 10 repeat units. Thermal conductivity was observed to increase from PEEK10 to PEEK70. It converged at about 0.24 W/mK when there were 70 chain repeat units (Figure 3a). The convergence of bending, torsion, and non-bonding potential energies on the polymer molecule according to the chain lengths led to the convergence of the thermal conductivities [10]. The thermal conductivity from the simulation matched the experimental value well. It was about 0.25 W/mK when the chain repeat units exceeded 70 [13]. The semi-crystalline PEEK model facilitated the reliable calculation of thermal conductivity. The thermal conductivity of PEEK100 with an $R_{a,g}$ of about 93 Å was higher than that of PEEK10 because PEEK100 had a longer heat transfer path than PEEK10 (Figure 3b). The thermal conductivity of a polymer is associated with phonon transfer in a polymer chain. The thermal conductivity of PEEK100 with its long polymer chain was higher than that of PEEK10 with the short polymer chain, due to an increase of phonon mobility caused by the increase of bond vibration [30]. In addition, thermal transport is generated by collisions among the chains. The thermal conductivity of PEEK10 was lower than those of other models because of its low phonon mobility, induced by the reduction in collision between shorter chains compared with the chains of other models [31]. In addition, the entanglement of polymer chains affected the thermal conductivities of the PEEK models in this study. PEEK100 with an R_{bulk} of about 143 Å had the highest thermal conductivity due to the increase of the heat transfer path induced by a reduction of entanglement of the polymer chains (Figure 3c). These results indicate that the heat transfer path on a polymer chain and the entanglement of polymer chains are both important factors affecting thermal conductivity. The chain with the 70 repeat units at the point of convergence was selected as an amorphous chain to investigate the effect of linkage conformation connecting the crystalline domain and the amorphous domain.

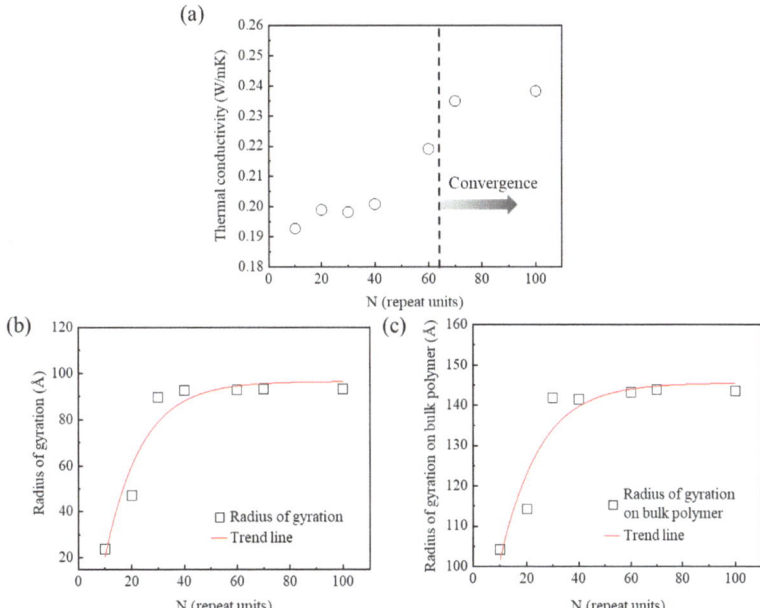

Figure 3. Thermal transport according to the length of the amorphous chains in semi-crystalline PEEK: (**a**) thermal conductivities, (**b**) radius of gyration, and (**c**) radius of gyration on the bulk polymer.

3.2. Effect of Loop and Tail Conformations Connecting the Crystalline Domain and the Amorphous Domain on Thermal Conductivity

The thermal conductivities of PEEK models were predicted according to the various ratios of loop and tail conformations. The PEEK models were labeled loop30tail70, loop50tail50, and loop70tail30 models, according to the ratio of the linkage conformation on the amorphous chain. For example, loop30tail70 model was set to 30% loop and 70% tail conformations, relative to the total number of amorphous chains. The thermal conductivity of the loop70tail30 model, which had a high content of loop conformation, was about 0.41 W/mK (Figure 4a). This was about 14% higher than those of loop30tail70 and loop50tail50. The loop70tail30 model had the longest heat transfer path, induced by the largest $R_{a,g}$ (Figure 4b). The R_{bulk} values of three models were the same, about 118 Å (Figure 4c). This result means that the radial distribution of the bulk polymer did not change the thermal conductivities of the PEEK models in accordance with the ratio of loop and tail conformations. The thermal conductivity in the model, according to the ratio of loop and tail conformations, was related to the interphase thermal conductivity in the interphase region between the crystalline domain and the amorphous domain. The thermal conductivities of the loop30tail70, loop50tail50, and loop70tail30 models in the interphase region were about 0.36, 0.36, and 0.52 W/mK, respectively (Figure 4d). The thermal conductivity of the interphase region of loop30tail70 was lower than that of loop70tail30 due to the low heat transfer path induced by the tail chains, which are connected to the crystalline domain by only one end of the chain. On the other hand, the thermal conductivity of the interphase of the loop70tail30 model was about 44% higher than those of the other models due to the increase of the heat transfer path, because both ends of the loop chain were connected with the crystalline domain. Consequently, the thermal conductivity of the loop70tail30 model, with the high content of loops, was improved by the increase of the thermal conductivity of the interphase region. Therefore, the ratios of loop and tail conformations are important to improving the thermal conductivity of semi-crystalline PEEK.

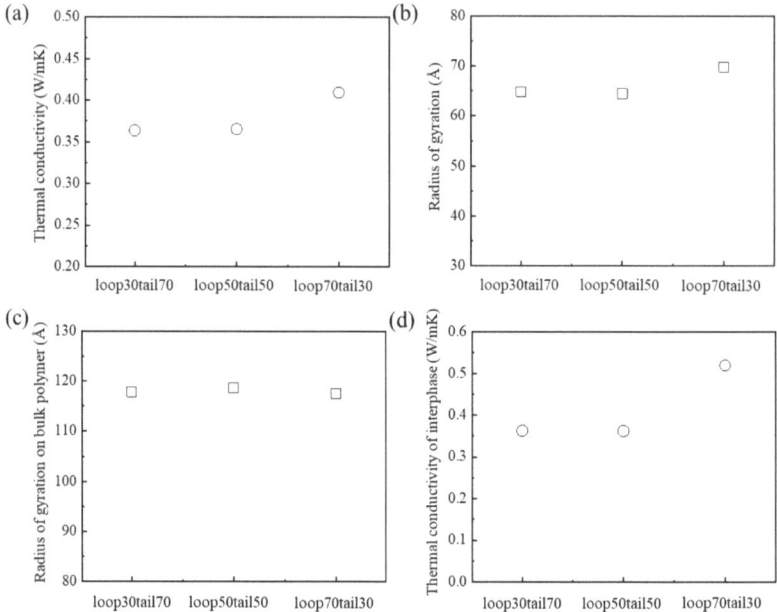

Figure 4. Thermal transport according to the ratio of loop and tail conformations: (**a**) thermal conductivities, (**b**) radius of gyration, (**c**) radius of gyration on the bulk polymer, and (**d**) thermal conductivities of interphase.

3.3. Effect of Bridge Conformation, Connecting Two Sides of the Crystalline Domain on Thermal Conductivity

The thermal conductivities for various ratios of bridge conformations were investigated to determine the optimal content of bridge conformation for improving the thermal conductivity. The bridge conformation content was set to 10, 30, 50, and 70%, relative to the total number of amorphous chains. The residual content of amorphous chains, except for the bridge conformation, consists of loop and tail conformations. The ratio of loop and tail conformations in loop70tail30 was used to determine the optimal content of the bridge conformation, which needed to achieve the highest thermal conductivity of semi-crystalline PEEK. The thermal conductivity of the model with a bridge content of 30% was about 0.48 W/mK. This was higher than those of the other models (Figure 5a). The thermal conductivities of models with bridge contents in the range of 10% to 30% increased as the $R_{a,g}$ values increased. Among the loop, tail, and bridge conformations, the bridge chain is the longest chain along the x-axis. Therefore, the $R_{a,g}$ of the amorphous chains increased in a linear form as the bridge content increased (Figure 5b). In addition, the PEEK model with a bridge conformation of 30% had the highest thermal conductivity due to the long thermal transport path induced by the wide distribution of polymer chains with low entanglement (Figure 5c). However, thermal conductivity decreased when the bridge chain content was over 30%. This result is related to the entanglement of the polymer chains. The R_{bulk}

values of models with bridge chain contents more than 30% gradually decreased due to the increase of the entanglement of polymer chains. The thermal conductivity decreased due to excess phonon scattering, induced by the increased entanglement of polymer chains [32]. These results indicate that a bridge conformation of 30% of the PEEK model with low phonon scattering and low entanglement of polymer chains was the optimal content for improving thermal conductivity.

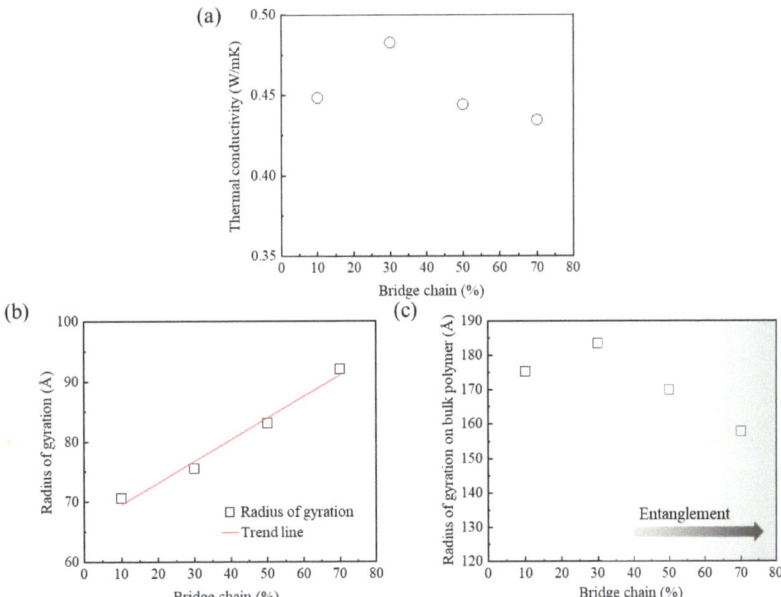

Figure 5. Thermal transport according to the content of bridge conformation: (**a**) thermal conductivities, (**b**) radius of gyration, and (**c**) radius of gyration on bulk polymer.

3.4. Effect of the Addition of BN Sheets on the Thermal Conductivity of the PEEK/BN Composite

The thermal conductivities of the PEEK/BN composites with the addition of BN sheets were investigated to verify how the filler affected high thermal conductivity. The semi-crystalline PEEK model with a bridge conformation of 30% was used to construct the PEEK/BN composite model using square BN sheets with a lateral length of 45 Å. The thermal conductivities of the PEEK/BN composites gradually increased when the contents of the BN sheets increased from 5 wt% to 10 wt%. It converged at about 0.58 W/mK when the BN sheet contents exceeded 10 wt% (Figure 6a). The thermal conductivities of composites with BN sheets contents of more than 10 wt% were improved by about 14% compared to the composite with 5 wt% BN sheets. The increasing content of the BN sheets with high thermal conductivity improved the thermal conductivity of the PEEK/BN composite. Figure 6b shows the phonon behaviors of the PEEK/BN composites according to the BN sheet content. When the BN sheet content was over 7 wt%, the phonon intensities increased in the frequency region of 79 THz. This result indicates that heat transfer was actively promoted by the increase of phonon vibration within the BN sheets. As a result, the thermal conductivity increased as the phonon intensities in the BN sheets increased. Additionally, the thermal conductivity of the composite increased as phonon scattering was reduced by the many heat transfer pathways created between the polymer chains and fillers [33]. Figure 6c shows that the $R_{a,g}$ increased with the addition of BN sheets. The π–π interaction between the aromatic rings of the polymer chains and the BN sheets facilitated the distribution of polymer chains on the surface of the BN sheets. As a result, the polymer chains became widely distributed along the surface of the BN sheets. Therefore, the long heat transfer path induced by the wide distribution of the polymer chains leads to an increase of thermal conductivity. In addition, the R_{bulk} increased with the increasing BN sheet content (Figure 6d). This result indicates that the wide distribution of polymer chains induced by the increase of BN sheet content reduced the entanglement of the polymer chains. The reduction of entanglement led to a reduction of the phonon scattering in the PEEK/BN composite. These results indicate that the increase of phonon vibration, the reduction of phonon scattering, and the wide distribution of polymer chains with the addition of BN sheets led to the improvement of thermal conductivity on the PEEK/BN composite.

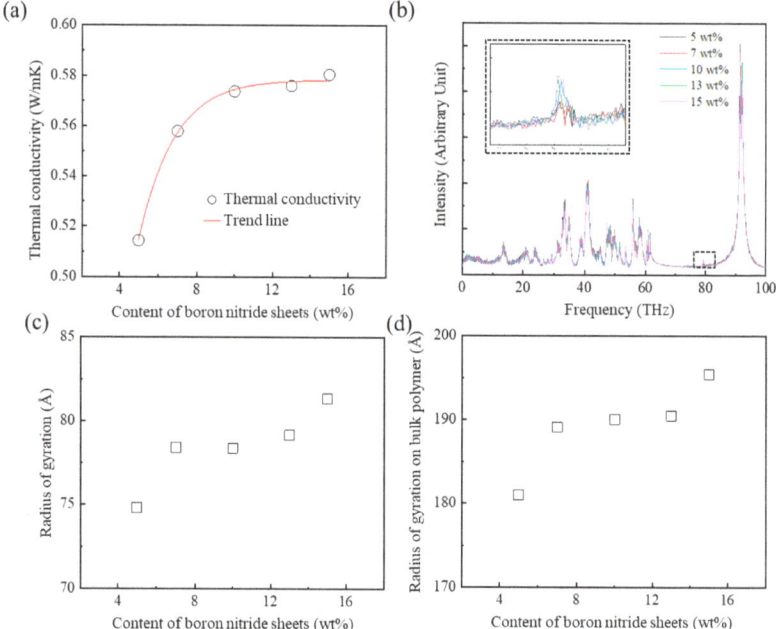

Figure 6. Thermal transport according to the content of boron nitride sheets in PEEK/BN composites: (**a**) thermal conductivities, (**b**) PDOS of composites, (**c**) radius of gyration, and (**d**) radius of gyration on the bulk polymer.

3.5. Effect of the Lateral Size of the Square BN Sheets on the Thermal Conductivity of the PEEK/BN Composite

Thermal transport simulations of the composites were performed to investigate the thermal conductivities for various lateral sizes of square BN sheets. The total BN sheet content was selected to be the 10 wt%. This was the point of convergence of the thermal conductivities. The lateral lengths of a square BN sheet ranged from 20 Å to 50 Å. The thermal conductivities of PEEK/BN composites with BN sheets with lateral lengths of more than 30 Å were improved by about 6% compared to the model with BN sheets with lateral lengths of 20 Å (Figure 7a). The $R_{a,g}$ of the polymers with increased BN sheet lateral size were almost identical (Figure 7b). This means that changing the lateral size of the BN sheet did not affect the distribution of the polymer chains. In addition, the R_{bulk} of polymers with various sizes of BN sheets were almost the same (Figure 7c). The degree of entanglement of the polymer chains was also almost the same because there was no change of the distribution of the polymer chains. Figure 7d shows the phonon behaviors of the PEEK/BN composites. The phonon intensity in the low frequency range indicates the intensity of phonon vibration within a BN sheet [27]. The phonon vibration within BN sheets increased as the lateral length of BN sheets increased in the low frequency range from 30 THz to 60 THz. The increase of phonon vibration resulted in an improvement of the thermal conductivity of the PEEK/BN composite. In addition, the phonon intensity in the high frequency region (about 92 THz) at the interface between the polymer chains and fillers increased as the lateral length of the BN sheets increased [34]. The increase of phonon intensity at the interface between the polymer chains and BN sheets led to the improvement of the thermal conductivities of the PEEK/BN composites.

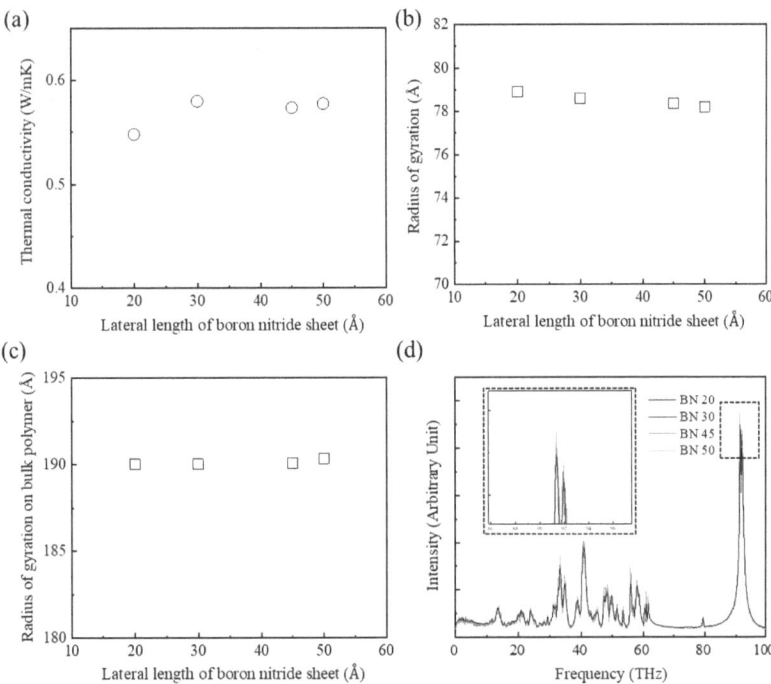

Figure 7. Thermal transport according to the lateral length of square BN sheets in the PEEK/BN composites: (**a**) thermal conductivities, (**b**) radius of gyration, (**c**) radius of gyration on the bulk polymer, and (**d**) PDOS of composites.

In addition, the dispersion of fillers affected the thermal conductivity of the composite [23]. The BN sheets in the composite models reinforced with BN sheets with lateral lengths of 20 Å and 30 Å were evenly dispersed, as shown in Figure 8a,b, respectively. The well-dispersed fillers facilitated more active heat transfer at the interface between the materials [23]. However, the thermal conductivities of the composites reinforced with BN sheets with lateral lengths of more than 30 Å gradually converged about 0.58 W/mK. The disconnection of the heat transfer path, due to aggregation of the BN sheets, leads to the convergence of the thermal conductivities (Figure 8c,d). These results mean that the improvement of thermal conductivity is limited by the aggregation of BN sheets, although the phonon transfer increased with the lateral size of the BN sheets. Therefore, the addition of BN sheets that are too large results in aggregation that obstructs the improvement of the thermal conductivity of the composite.

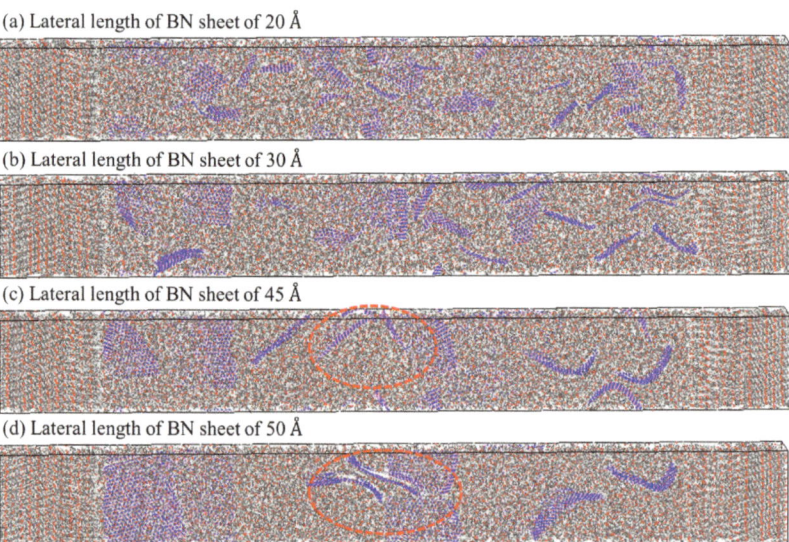

Figure 8. Equilibrium structure according to the lateral length of square BN sheets in PEEK/BN composites: (**a**) 20 Å (well-dispersed BN sheets), (**b**) 30 Å, (**c**) 45 Å, and (**d**) 50 Å (aggregated BN sheets as shown in the dotted red circle).

4. Conclusions

The important factors affecting the thermal conductivity of the semi-crystalline PEEK were revealed by the thermal transport simulations. The thermal conductivities of PEEK models with an amorphous chain with more than 70 repeat units were higher than the model with an amorphous chain of 10 repeat units. The long amorphous chains led to improved phonon transport by increasing the phonon vibration and collision probability. In addition, it was found that the wide distribution of long polymer chains with low entanglement induced long heat transfer paths. The thermal conductivity of the semi-crystalline polymer was related to the loop, tail, and bridge linkage conformations between the crystalline domain and the amorphous domain. The thermal conductivity of the model with high loop conformation content was enhanced by the increase of heat transfer paths, induced when both ends of the loop were connected with the crystalline domain. In addition, thermal conductivity increased as the bridge chain increased because both ends of the bridge chains were connected with the crystalline domains. However, an excessive content of bridge chains reduced thermal conductivity due to the entanglement of polymer chains. The thermal conductivities of the PEEK/BN composites were improved by better phonon transport after the addition of the BN sheets. In addition, the heat transfer path increased with the wide distribution of polymer chains along the surface of BN sheets. However, the thermal conductivity of the composite was related to the interfacial thermal transport between materials. Although phonon transfer improved as the lateral length of the BN sheets increased, the thermal conductivities of models containing BN sheets with lateral lengths of over 30 Å were converged due to the excessive aggregation of the BN sheets. These results mean that thermal conductivity was increased by the long heat transfer path induced by the long chain lengths, the high loop content, the optimal bridge content, and the addition of BN sheets. In addition, excessive bridge content and the aggregation of BN sheets caused polymer entanglement, resulting in low thermal transport. The simulation analyses of the structure of polymer are expected to assist the analysis of mechanisms for improving thermal properties of semi-crystalline polymer composites.

Supplementary Materials: The following supporting information can be downloaded at: https://www.mdpi.com/article/10.3390/polym15020450/s1, Figure S1: Energy of the bond stretching term using DFT and the N-DFF: (a) B–H bond and (b) N–H bond. Table S1: Parameters of the N-DFF used in the MD simulations. List of abbreviations. List of symbols.

Author Contributions: Conceptualization, Y.O. and J.Y.; methodology, Y.O.; software, Y.O. and K.J.B.; validation, Y.K. and J.Y.; formal analysis, Y.O. and J.Y.; investigation, Y.O.; data curation, Y.O.; writing—original draft preparation, Y.O.; writing—review and editing, Y.K. and J.Y.; supervision, J.Y.; project administration, J.Y.; funding acquisition, J.Y. All authors have read and agreed to the published version of the manuscript.

Funding: This study was supported by the Korea Institute of Science and Technology (KIST) Institutional Program. This study was also supported by the Technology Innovation Program 'Development of an ultra lightweight 19 carbon composite wheel using integrated braid preform manufacturing technology with tow prepreg material' (NO. 20021913) funded by the Ministry of Trade, Industry and Energy (MOTIE, Korea).

Data Availability Statement: The data presented in this study are available on request from the corresponding author.

Acknowledgments: This study was supported by the Korea Institute of Science and Technology (KIST) Institutional Program.

Conflicts of Interest: The authors declare no conflict of interest.

References

1. Ning, N.; Wang, M.; Zhou, G.; Qiu, Y.; Wei, Y. Effect of polymer nanoparticle morphology on fracture toughness enhancement of carbon fiber reinforced epoxy composites. *Compos. B Eng.* **2022**, *234*, 109749. [CrossRef]
2. Ramakrishna, S.; Mayer, J.; Wintermantel, E.; Leong, K.W. Biomedical applications of polymer-composite materials: A review. *Compos. Sci. Technol.* **2001**, *61*, 1189–1224. [CrossRef]
3. Wang, J.; Zhao, Z.; Weng, Q.; Wan, X. Insights on polymeric materials for the optimization of high-capacity anodes. *Compos. B Eng.* **2022**, *243*, 110131. [CrossRef]
4. Tabkhpaz, M.; Shajari, S.; Mahmoodi, M.; Park, D.Y.; Suresh, H.; Park, S.S. Thermal conductivity of carbon nanotube and hexagonal boron nitride polymer composites. *Compos. B Eng.* **2016**, *100*, 19–30. [CrossRef]
5. Dusunceli, N.; Colak, O.U. Modelling effects of degree of crystallinity on mechanical behavior of semicrystalline polymers. *Int. J. Plast.* **2008**, *24*, 1224–1242. [CrossRef]
6. O'Connor, T.C.; Elder, R.M.; Sliozberg, Y.R.; Sirk, T.W.; Andzelm, J.W.; Robbins, M.O. Molecular origins of anisotropic shock propagation in crystalline and amorphous polyethylene. *Phys. Rev. Mater.* **2018**, *2*, 035601. [CrossRef]
7. France-Lanord, A.; Merabia, S.; Albaret, T.; Lacroix, D.; Termentzidis, K. Thermal properties of amorphous/crystalline silicone superlattices. *J. Phys. Condens. Matter.* **2014**, *26*, 355801. [CrossRef]
8. Bao, Q.; Yang, Z.; Lu, Z. Molecular dynamics simulation of amorphous polyethylene (PE) under cyclic tensile-compressive loading below the glass transition temperature. *Polymer* **2020**, *186*, 121968. [CrossRef]
9. Zou, L.; Zhang, W. Molecular dynamics simulations of the effects of entanglement on polymer crystal nucleation. *Macromolecules* **2022**, *55*, 4899–4906. [CrossRef]
10. Wei, X.; Luo, T. Chain length effect on thermal transport in amorphous polymers and a structure-thermal conductivity relation. *Phys. Chem. Chem. Phys.* **2019**, *21*, 15523. [CrossRef]
11. Zhu, S.; Lempesis, N.; Veld, P.J.; Rutledge, G.C. Molecular simulation of thermoplastic polyurethanes under large tensile deformation. *Macromolecules* **2018**, *51*, 1850–1864. [CrossRef]
12. He, J.; Liu, J. Molecular dynamics simulation of thermal transport in semicrystalline polyethylene: Roles of strain and the crystalline-amorphous interphase region. *J. Appl. Phys.* **2021**, *130*, 225101. [CrossRef]
13. Guo, Y.; Ruan, K.; Shi, X.; Yang, X.; Gu, J. Factors affecting thermal conductivities of the polymers and polymer composites: A review. *Compos. Sci. Technol.* **2020**, *193*, 108374. [CrossRef]
14. Xu, Y.; Kraemer, D.; Song, B.; Jiang, Z.; Zhou, J.; Loomis, J.; Li, M.; Ghasemi, H.; Huang, X.; Li, X.; et al. Nanostructured polymer films with metal-like thermal conductivity. *Nat. Commun.* **2019**, *10*, 1771. [CrossRef] [PubMed]
15. Hwang, Y.; Kim, M.; Kim, J. Improvement of the mechanical properties and thermal conductivity of poly(ether-ether-ketone) with the addition of graphene oxide-carbon nanotube hybrid fillers. *Compos. Part A Appl. Sci. Manuf.* **2013**, *55*, 195–202. [CrossRef]
16. Liu, L.; Xiao, L.; Li, M.; Zhang, X.; Chang, Y.; Shang, L.; Ao, Y. Effect of hexagonal boron nitride on high-performance polyether ether ketone composites. *Colloid Polym. Sci.* **2016**, *294*, 127–133. [CrossRef]
17. Knobloch, T.; Illarionov, Y.Y.; Ducry, F.; Schleich, C.; Wachter, S.; Watanabe, K.; Taniguchi, T.; Mueller, T.; Waltl, M.; Lanza, M.; et al. The performance limits of hexagonal boron nitride as an insulator for scaled CMOS devices based on two-dimensional materials. *Nat. Electron.* **2021**, *4*, 98–108. [CrossRef]

18. Falin, A.; Cai, Q.; Santos, E.J.G.; Scullion, D.; Qian, D.; Zhang, R.; Yang, Z.; Huang, S.; Watanabe, K.; Taniguchi, T.; et al. Mechanical properties of atomically thin boron nitride and the role of interlayer interactions. *Nat. Commun.* **2017**, *8*, 15815. [CrossRef]
19. Ghosh, B.; Xu, F.; Hou, X. Thermally conductive poly(ether ether ketone)/boron nitride composites with low coefficient of thermal expansion. *J. Mater. Sci.* **2021**, *56*, 10326–10337. [CrossRef]
20. Mehra, N.; Mu, L.; Ji, T.; Yang, X.; Kong, J.; Gu, J.; Zhu, J. Thermal transport in polymeric materials and across composite interfaces. *Appl. Mater. Today* **2018**, *12*, 92–130. [CrossRef]
21. Li, Z.; Li, K.; Liu, J.; Hu, S.; Wen, S.; Liu, L.; Zhang, L. Tailoring the thermal conductivity of poly(dimethylsiloxane)/hexagonal boron nitride composite. *Polymer* **2019**, *177*, 262–273. [CrossRef]
22. Song, H.; Kim, B.G.; Kim, Y.S.; Bae, Y.S.; Kim, J.; Yoo, Y. Synergistic effects of various ceramic fillers on thermally conductive polyimide composite films and their model predictions. *Polymers* **2019**, *11*, 484. [CrossRef] [PubMed]
23. Fu, Y.X.; He, Z.X.; Mo, D.C.; Lu, S.S. Thermal conductivity enhancement with different fillers for epoxy resin adhesives. *Appl. Therm. Eng.* **2014**, *66*, 493–498. [CrossRef]
24. Doumeng, M.; Makhlouf, L.; Berthet, F.; Marsan, O.; Delbe, K.; Denape, J.; Chabert, F. A comparative study of the crystallinity of polyetheretherketone by using density, DSC, XRD, and Raman spectroscopy techniques. *Polym. Test.* **2021**, *93*, 106878. [CrossRef]
25. Blundell, D.J.; Osborn, B.N. The morphology of poly(aryl-ether-ether-ketone). *Polymer* **1983**, *24*, 953–958. [CrossRef]
26. Mayo, S.L.; Olafson, B.D.; Goddard, W.A. Dreiding: A generic force field for molecular simulations. *J. Phys. Chem.* **1990**, *94*, 8897–8909. [CrossRef]
27. Oh, Y.; Jung, H.; Bae, K.J.; Kim, Y.; Yu, J. Analysis of mechanical and thermal characterization of hexagonal boron nitride using a molecular dynamics simulation with the new Dreiding force field. *Mech. Adv. Mater. Struct.* **2022**, *29*, 6957–6965. [CrossRef]
28. Choi, H.K.; Jung, H.; Oh, Y.; Hong, H.; Yu, J.; Shin, E.S. Interfacial effects of nitrogen-doped carbon nanotubes on mechanical and thermal properties of nanocomposites: A molecular dynamics study. *Compos. B Eng.* **2019**, *167*, 615–620. [CrossRef]
29. Plimpton, S. Fast parallel algorithms for short-range molecular dynamics. *J. Comput. Phys.* **1995**, *117*, 1–19. [CrossRef]
30. Lin, S.; Cai, Z.; Wang, Y.; Zhao, L.; Zhai, C. Tailored morphology and highly enhanced phonon transport in polymer fibers: A multiscale computational framework. *NPJ Comput. Mater.* **2019**, *5*, 126. [CrossRef]
31. Zhao, J.; Jiang, J.W.; Wei, N.; Zhang, Y.; Rabczuk, T. Thermal conductivity dependence on chain length in amorphous polymers. *J. Appl. Phys.* **2013**, *113*, 184304. [CrossRef]
32. Xu, X.; Chen, J.; Zhou, J.; Li, B. Thermal conductivity of polymers and their nanocomposites. *Adv. Mater.* **2018**, *30*, 1705544. [CrossRef] [PubMed]
33. Jang, J.; Nam, H.E.; So, S.O.; Lee, H.; Kim, G.S.; Kim, S.Y.; Kim, S.H. Thermal percolation behavior in thermal conductivity of polymer nanocomposite with lateral size of graphene nanoplatelet. *Polymers* **2022**, *14*, 323. [CrossRef] [PubMed]
34. Chen, S.; Yang, M.; Liu, B.; Xu, M.; Zhang, T.; Zhuang, P.; Ding, D.; Huai, X.; Zhang, H. Enhanced thermal conductance at the graphene-water interface based on functionalized alkane chains. *RSC Adv.* **2019**, *9*, 4563. [CrossRef]

Disclaimer/Publisher's Note: The statements, opinions and data contained in all publications are solely those of the individual author(s) and contributor(s) and not of MDPI and/or the editor(s). MDPI and/or the editor(s) disclaim responsibility for any injury to people or property resulting from any ideas, methods, instructions or products referred to in the content.

MDPI AG
Grosspeteranlage 5
4052 Basel
Switzerland
Tel.: +41 61 683 77 34

Polymers Editorial Office
E-mail: polymers@mdpi.com
www.mdpi.com/journal/polymers

Disclaimer/Publisher's Note: The title and front matter of this reprint are at the discretion of the Guest Editors. The publisher is not responsible for their content or any associated concerns. The statements, opinions and data contained in all individual articles are solely those of the individual Editors and contributors and not of MDPI. MDPI disclaims responsibility for any injury to people or property resulting from any ideas, methods, instructions or products referred to in the content.

www.ingramcontent.com/pod-product-compliance
Lightning Source LLC
LaVergne TN
LVHW072321090526
838202LV00019B/2328